Visit classzone.com and get connected

Online resources provide instruction, practice, and learning support correlated to your text.

- **Misconceptions database** provides solutions for common student misconceptions about science and their world.
- **Professional development links,** including SciLinks, offer additional teaching resources.
- **Animations and visualizations** help improve comprehension.
- **Math Tutorial** helps strengthen students' math skills.
- **Flashcards** help students review vocabulary.
- **State test practice** prepares students for assessments.

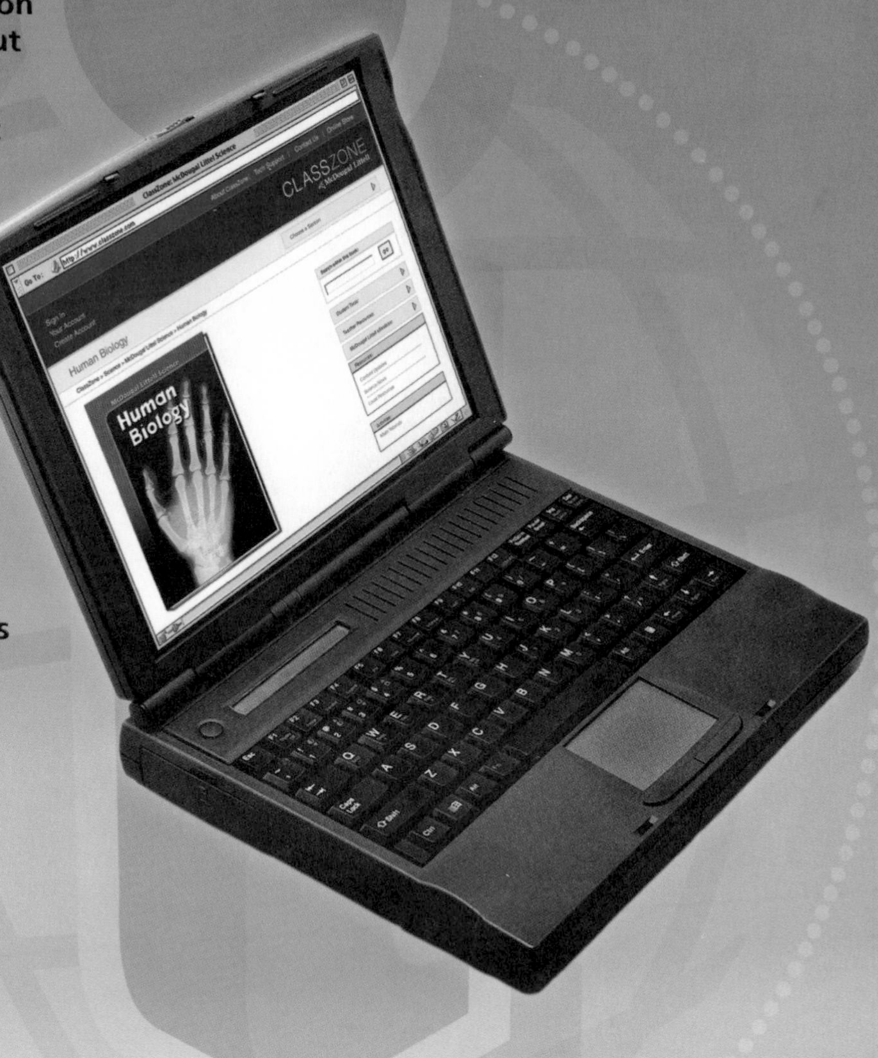

You have immediate access to *ClassZone's* teacher resources.

MCDTCOWDMSSZ

Use this code to create your own username and password.

Also visit *ClassZone* to learn more about these innovative and updated online resources.

- eEdition Plus Online
- eTest Plus Online
- EasyPlanner Plus Online
- Content Review Online

Now it all clicks!™

McDougal Littell

TEACHER'S EDITION

McDougal Littell Science

Human Biology

joint

tissue

HUMAN

(Homo sapiens)

skeletal system

Credits

5B *center* Illustration by Linda Nye; *right* Illustration by Bart Vallecoccia; **5C** Illustration by Debbie Maizels; **33B** *both* Illustrations by Bart Vallecoccia; © Martin Rotker/Phototake; *bottom left* © Michael Newman/PhotoEdit; **33C** Illustration by Bart Vallecoccia; **61B** Illustration by Linda Nye; **61C** Illustration by Debbie Maizels; **97B** Illustration by Steve Oh/KO Studios; **97C** Illustration by Linda Nye; **129B** © Tom Galliher/Corbis; **129C** © Dr. Kari Lounatmaa/Photo Researchers, Inc.

Acknowledgements

Excerpts and adaptations from National Science Education Standards by the National Academy of Sciences. Copyright © 1996 by the National Academy of Sciences. Reprinted with permission from the National Academies Press, Washington, D.C.

ISBN: 0-618-33432-7 1 2 3 4 5 6 7 8 VJM 08 07 06 05 04

Internet Web Site: http://www.mcdougallittell.com

McDougal Littell Science

Effective Science Instruction Tailored for Middle School Learners

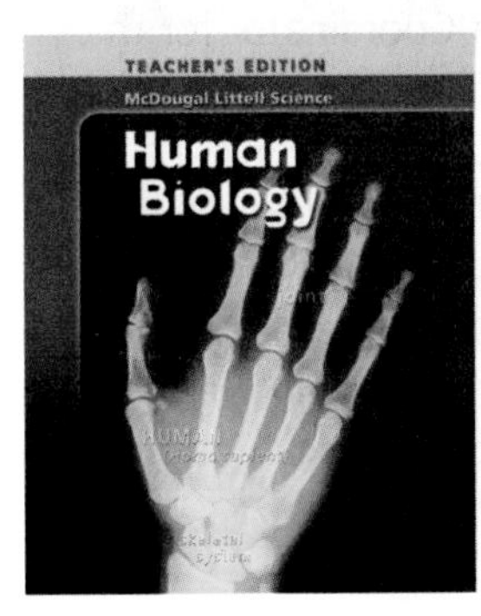

Human Biology
Teacher's Edition Contents

Consultants and Reviewers

Science Consultants

Chief Science Consultant

James Trefil, Ph.D. is the Clarence J. Robinson Professor of Physics at George Mason University. He is the author or co-author of more than 25 books, including *Science Matters* and *The Nature of Science.* Dr. Trefil is a member of the American Association for the Advancement of Science's Committee on the Public Understanding of Science and Technology. He is also a fellow of the World Economic Forum and a frequent contributor to *Smithsonian* magazine.

Rita Ann Calvo, Ph.D. is Senior Lecturer in Molecular Biology and Genetics at Cornell University, where for 12 years she also directed the Cornell Institute for Biology Teachers. Dr. Calvo is the 1999 recipient of the College and University Teaching Award from the National Association of Biology Teachers.

Kenneth Cutler, M.S. is the Education Coordinator for the Julius L. Chambers Biomedical Biotechnology Research Institute at North Carolina Central University. A former middle school and high school science teacher, he received a 1999 Presidential Award for Excellence in Science Teaching.

Instructional Design Consultants

Douglas Carnine, Ph.D. is Professor of Education and Director of the National Center for Improving the Tools of Educators at the University of Oregon. He is the author of seven books and over 100 other scholarly publications, primarily in the areas of instructional design and effective instructional strategies and tools for diverse learners. Dr. Carnine also serves as a member of the National Institute for Literacy Advisory Board.

Linda Carnine, Ph.D. consults with school districts on curriculum development and effective instruction for students struggling academically. A former teacher and school administrator, Dr. Carnine also co-authored a popular remedial reading program.

Sam Miller, Ph.D. is a middle school science teacher and the Teacher Development Liaison for the Eugene, Oregon, Public Schools. He is the author of curricula for teaching science, mathematics, computer skills, and language arts.

Donald Steely, Ph.D. serves as principal investigator at the Oregon Center for Applied Science (ORCAS) on federal grants for science and language arts programs. His background also includes teaching and authoring of print and multimedia programs in science, mathematics, history, and spelling.

Vicky Vachon, Ph.D. consults with school districts throughout the United States and Canada on improving overall academic achievement with a focus on literacy. She is also co-author of a widely used program for remedial readers.

Content Reviewers

John Beaver, Ph.D.
Ecology
Professor, Director of Science Education Center
College of Education and Human Services
Western Illinois University
Macomb, IL

Donald J. DeCoste, Ph.D.
Matter and Energy, Chemical Interactions
Chemistry Instructor
University of Illinois
Urbana-Champaign, IL

Dorothy Ann Fallows, Ph.D., MSc
Diversity of Living Things, Microbiology
Partners in Health
Boston, MA

Michael Foote, Ph.D.
The Changing Earth, Life Over Time
Associate Professor
Department of the Geophysical Sciences
The University of Chicago
Chicago, IL

Lucy Fortson, Ph.D.
Space Science
Director of Astronomy
Adler Planetarium and Astronomy Museum
Chicago, IL

Elizabeth Godrick, Ph.D.
Human Biology
Professor, CAS Biology
Boston University
Boston, MA

Isabelle Sacramento Grilo, M.S.
The Changing Earth
Lecturer, Department of the Geological Sciences
Montana State University
Bozeman, MT

David Harbster, MSc
Diversity of Living Things
Professor of Biology
Paradise Valley Community College
Phoenix, AZ

Richard D. Norris, Ph.D.
Earth's Waters
Professor of Paleobiology
Scripps Institution of Oceanography
University of California, San Diego
La Jolla, CA

Donald B. Peck, M.S.
Motion and Forces; Waves, Sound, and Light; Electricity and Magnetism
Director of the Center for Science Education (retired)
Fairleigh Dickinson University
Madison, NJ

Javier Penalosa, Ph.D.
Diversity of Living Things, Plants
Associate Professor, Biology Department
Buffalo State College
Buffalo, NY

Raymond T. Pierrehumbert, Ph.D.
Earth's Atmosphere
Professor in Geophysical Sciences (Atmospheric Science)
The University of Chicago
Chicago, IL

Brian J. Skinner, Ph.D.
Earth's Surface
Eugene Higgins Professor of Geology and Geophysics
Yale University
New Haven, CT

Nancy E. Spaulding, M.S.
Earth's Surface, The Changing Earth, Earth's Waters
Earth Science Teacher (retired)
Elmira Free Academy
Elmira, NY

Steven S. Zumdahl, Ph.D.
Matter and Energy, Chemical Interactions
Professor Emeritus of Chemistry
University of Illinois
Urbana-Champaign, IL

Susan L. Zumdahl, M.S.
Matter and Energy, Chemical Interactions
Chemistry Education Specialist
University of Illinois
Urbana-Champaign, IL

Safety Consultant

Juliana Texley, Ph.D.
Former K–12 Science Teacher and School Superintendent
Boca Raton, FL

English Language Advisor

Judy Lewis, M.A.
Director, State and Federal Programs for reading proficiency and high risk populations
Rancho Cordova, CA

Research-Based Solutions for Your Classroom

The distinguished program consultant team and a thorough, research-based planning and development process assure that *McDougal Littell Science* supports all students in learning science concepts, acquiring inquiry skills, and thinking scientifically.

Standards-Based Instruction

Concepts and skills were selected based on careful analysis of national and state standards.

- National Science Education Standards
- Project 2061 Benchmarks for Science Literacy
- Comprehensive database of state science standards

Standards and Benchmarks

Each chapter in **Human Biology** covers some of the learning goals that are described in the *National Science Education Standards* (NSES) and the Project 2061 *Benchmarks for Science Literacy*. Selected content and skill standards are show below in shortened form.The following National Science Education Standards are covered on pp. xii–xxvii, in Frontiers in Science, and in Timelines in Science, as well as in Ichapter features and laboratory investigations: Understandings About Scientific Inquiry (A.9), Understandings About Science and Technology (E.6), Science and Technology in Society (F.5), Nature of Science (G.2), and History of Science (G.3).

Content Standards

1 Systems, Support, and Movement

National Science Education Standards

C.1.a Levels of organization for living systems include: cells, tissues, organs, organ systems, whole organisms, and ecosystems.

C.1.d Specialized cells perform specialized functions in multicellular organisms.

C.1.e The human organism has systems: digestion, respiration, reproduction, circulation, excretion, movement, control and coordination, and protection.

Project 2061 Benchmarks

6.A.1 Like other animals, human beings have body systems for:
- obtaining and providing energy
- defense
- reproduction
- coordination of body functions

6.C.1 Organs and organ systems
- are made of cells
- help to provide all cells with basic needs

2 Absorption, Digestion, and Exchange

National Science Education Standards

C.1.e The human organism has systems: digestion, respiration, reproduction, circulation, excretion, movement, control and coordination, and protection.

F.1.e Food provides energy and nutrients for growth and development.

Project 2061 Benchmarks

6.C.2 For the body to use food energy and building materials, the food must first be digested into molecules that are absorbed and transported to cells.

6.C.3 Respiratory, urinary, and digestive systems remove wastes from the body.

3 Transport and Protection

National Science Education Standards

C.1.d Specialized cells perform specialized functions in multicellular organisms.

C.1.e The human organism has systems: digestion, respiration, reproduction, circulation, excretion, movement, control and coordination, and protection.

C.1.f Disease is a breakdown in structures or functions of an organism.

Project 2061 Benchmarks

6.C.3 The circulatory system moves substances to or from cells where they are needed or produced. It responds to changing demands.

6.C.4 Specialized cells and molecules they produce identify and destroy microbes.

6.E.4 White blood cells engulf invaders or produce antibodies that fight them.

x Unit: **Human Biology**

CHAPTER 1

Systems, Support, and Movement

the BIG idea

The human body is made up of systems that work together to perform necessary functions.

Key Concepts

SECTION 1.1 **The human body is complex.** Learn about the parts and systems in the human body.

SECTION 1.2 **The skeletal system provides support and protection.** Learn how the skeletal system is organized and what it does.

SECTION 1.3 **The muscular system makes movement possible.** Learn about the different types of muscles and how they work.

Internet Preview

CLASSZONE.COM

Chapter 1 online resources: Content Review, two Simulations, two Resource Centers, Math Tutorial, Test Practice

What systems make it possible for this racer to move so fast?
303

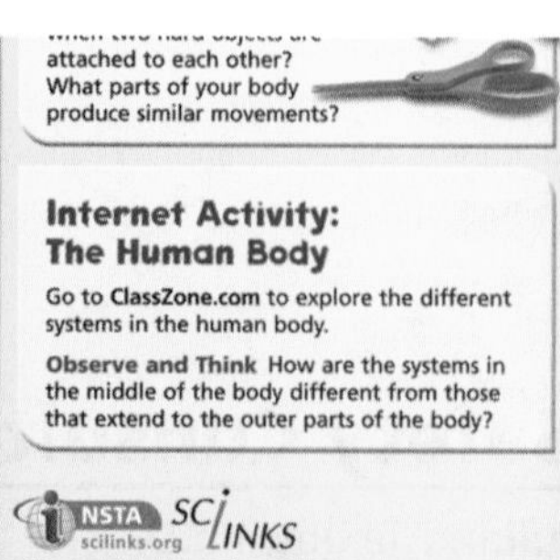

attached to each other? What parts of your body produce similar movements?
Internet Activity: The Human Body
Go to ClassZone.com to explore the different systems in the human body.
Observe and Think How are the systems in the middle of the body different from those that extend to the outer parts of the body?
NSTA scilinks.org SCILINKS
Tissues and Organs Code: MDL044

E 6 Unit: Human Biology

Chapter 1: Systems, Support, and Movement 7 E

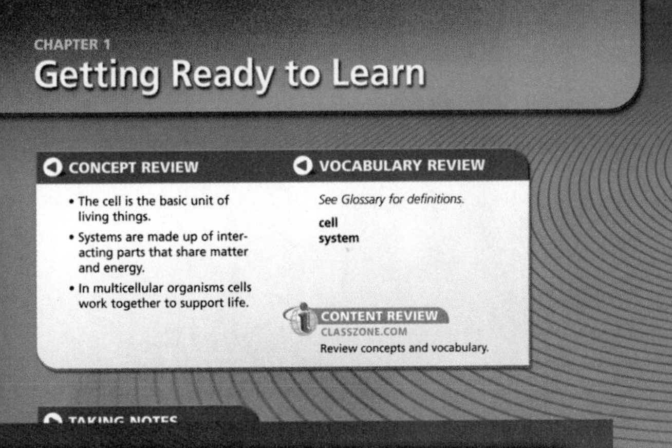
CHAPTER 1
Getting Ready to Learn
CONCEPT REVIEW
• The cell is the basic unit of living things.
• Systems are made up of interacting parts that share matter and energy.
• In multicellular organisms cells work together to support life.
VOCABULARY REVIEW
See Glossary for definitions.
cell
system
CONTENT REVIEW
CLASSZONE.COM
Review concepts and vocabulary.

Effective Instructional Strategies

McDougal Littell Science incorporates strategies that research shows are effective in improving student achievement. These strategies include

- Notetaking and nonlinguistic representations (Marzano, Pickering, and Pollock)
- A focus on big ideas (Kameenui and Carnine)
- Background knowledge and active involvement (Project CRISS)

Robert J. Marzano, Debra J. Pickering, and Jane E. Pollock, *Classroom Instruction that Works; Research-Based Strategies for Increasing Student Achievement* (ASCD, 2001)

Edward J. Kameenui and Douglas Carnine, *Effective Teaching Strategies that Accommodate Diverse Learners* (Pearson, 2002)

Project CRISS (Creating Independence through Student Owned Strategies)

4 Control and Reproduction

National Science Education Standards

C.1.d Specialized cells perform specialized functions in multicellular organisms.
C.1.e The human organism has systems: digestion, respiration, reproduction, circulation, excretion, movement, control and coordination, and protection.
C.2.a Reproduction is essential to the continuation of a species.
C.2.b Females produce eggs, and males sperm, which unite to begin a new individual.
C.2.c Every organism requires a set of instructions for specifying its traits.

Project 2061 Benchmarks

5.B.2 In sexual reproduction, a single specialized cell from a female merges with a specialized cell from a male.
6.B.1 Fertilization occurs when one of the sperm cells from the male enters the egg cell from the female.
6.B.3 After fertilization, cells divide and specialize as a fetus grows from the embryo following a set pattern of development.
6.C.5 Hormones are chemicals from glands that help the body respond to danger and regulate human growth, development, and reproduction.
6.C.6 Interactions among the senses, nerves, and brain make possible the learning that enables human beings to cope with changes in their environment.

5 Growth, Development, and Health

National Science Education Standards

C.1.e The human organism has systems: digestion, respiration, reproduction, circulation, excretion, movement, control and coordination, and protection.
C.1.f Disease is a breakdown in structures or functions of an organism.
C.3.a Organisms must obtain and use resources, grow, reproduce, and maintain stable internal conditions.
C.3.b Regulation of an organism's internal environment involves interactions with its external environment.
F.1.c The use of tobacco increases risk of illness.
F.1.d Alcohol and other drugs are often abused and can lead to addiction.
F.1.e Food provides energy and nutrients for growth and development.

Project 2061 Benchmarks

6.B.5 Changes occur as humans age. Many factors affect length and quality of life.
6.E.2 Toxic substances, diet, and behavior may harm one's health.
6.E.3 Viruses, bacteria, fungi, and parasites may interfere with the human body.
6.E.4 White blood cells engulf invaders or produce antibodies that fight them.

Skill Standards

National Science Education Standards

A.1 Identify questions that can be answered through scientific methods.
A.2 Design and conduct a scientific investigation.
A.3 Use appropriate tools and techniques to gather and interpret data.
A.4 Use evidence to describe, predict, explain, and model.
A.5 Think critically to find relationships between results and interpretations.
A.6 Give alternative explanations and predictions.
A.7 Communicate procedures, results, and conclusions.
A.8 Use mathematics in all aspects of scientific inquiry.
A.9.a Different kinds of questions suggest different kinds of scientific investigations.
A.9.c Mathematics is important in all aspects of scientific inquiry.
A.9.d Scientific explanations emphasize evidence, have logically consistent arguments, and use scientific principles, models, and theories.
E.6.b Many different people in different cultures have made and continue to make contributions to science and technology.
G.1.a Women and men of various social and ethnic backgrounds engage in the activities of science.
G.1.b Science requires different abilities. The work of science relies on basic human qualities, such as reasoning, insight, energy, skill, and creativity.

Project 2061 Benchmarks

1.C.1 Contributions to science have been made by different people, in different cultures, at different times.
12.A.1 Know why it is important in science to keep honest, clear, and accurate records.
12.A.2 Hypotheses are valuable, even if they turn out not to be true.
12.A.3 Different explanations can often be given for the same evidence.
12.C.3 Use appropriate units, use and read instruments that measure length, volume, weight, time, rate, and temperature.
12.D.1 Use tables and graphs to organize information and identify relationships.
12.D.2 Read, interpret, and describe tables and graphs.
12.D.3 Locate information in reference books and other resources.
12.D.4 Understand information that includes different types of charts and graphs, including circle charts, bar graphs, line graphs, data tables, diagrams, and symbols.

Standards and Benchmarks xi

the respiratory system to provide energy and materials. What other systems in your body can you compare to a system in the city?

The body has cells, tissues, and organs.

MAIN IDEA WEB
As you read this section, complete the main idea web begun on page 8.

Your body is made of many parts that work together as a system to help you grow and stay healthy. The simplest level of organization in your body is the cell. Next come tissues, then individual organs, and then systems that are made up of organs. The highest level of organization is the organism itself. You can think of the body as having five levels of organization: cells, tissues, organs, organ systems, and the organism. Although these levels seem separate from one another, they all work together.

CHECK YOUR
READING
What are five levels of organization in your body?

Chapter 1: Systems, Support, and Movement 9 E

Comprehensive Research, Review, and Field Testing

An ongoing program of research and review guided the development of *McDougal Littell Science.*

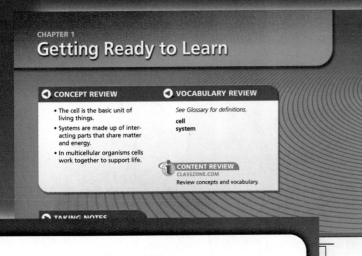

- Program plans based on extensive data from classroom visits, research surveys, teacher panels, and focus groups
- All pupil edition activities and labs classroom-tested by middle school teachers and students
- All chapters reviewed for clarity and scientific accuracy by the Content Reviewers listed on page T5
- Selected chapters field-tested in the classroom to assess student learning, ease of use, and student interest

Content Organized Around Big Ideas

Each chapter develops a big idea of science, helping students to place key concepts in context.

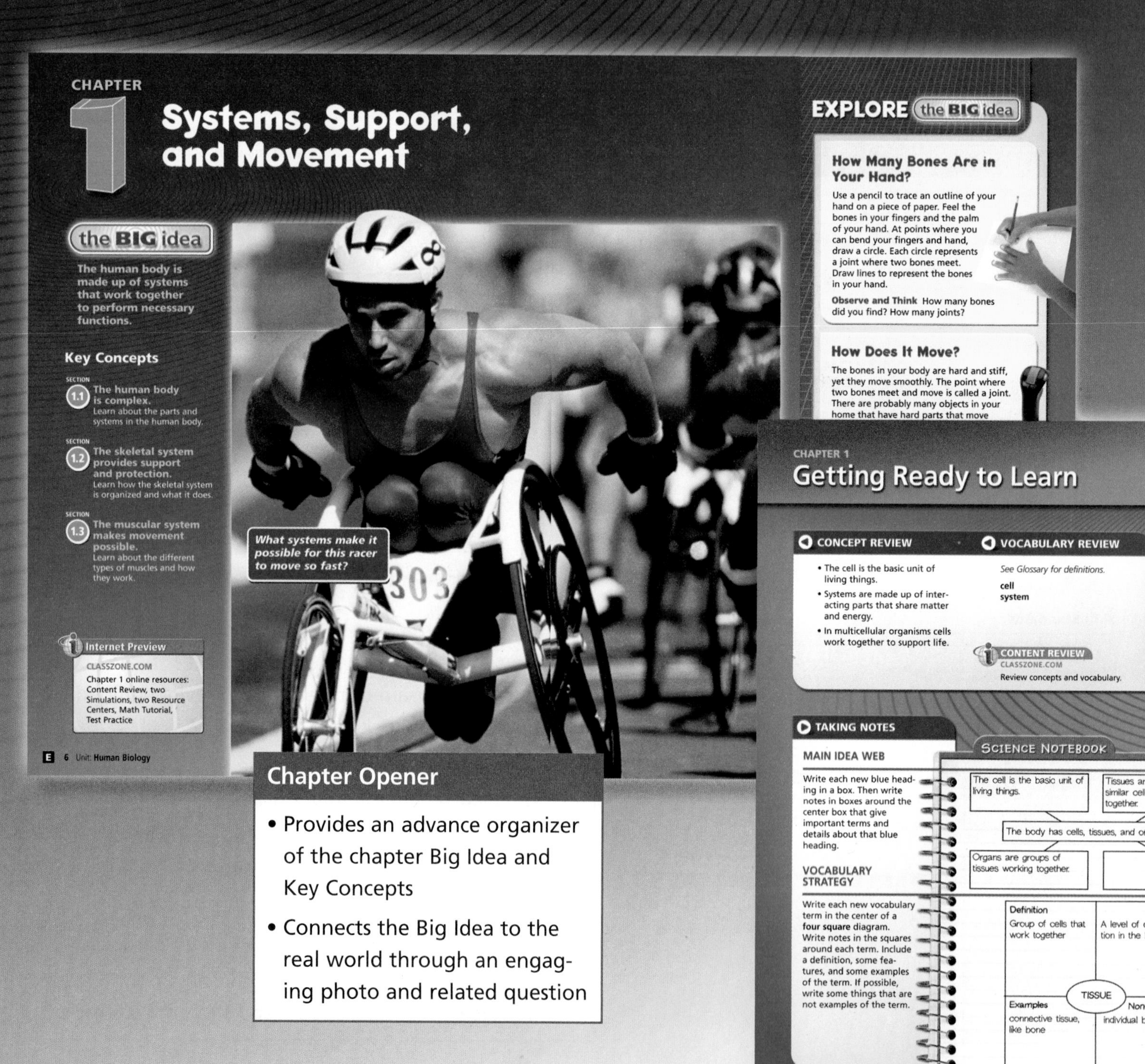

CHAPTER
1
Systems, Support, and Movement
the BIG idea
The human body is made up of systems that work together to perform necessary functions.
Key Concepts
SECTION 1.1 The human body is complex. Learn about the parts and systems in the human body.
SECTION 1.2 The skeletal system provides support and protection. Learn how the skeletal system is organized and what it does.
SECTION 1.3 The muscular system makes movement possible. Learn about the different types of muscles and how they work.
Internet Preview
CLASSZONE.COM
Chapter 1 online resources: Content Review, two Simulations, two Resource Centers, Math Tutorial, Test Practice
E 6 Unit: Human Biology
What systems make it possible for this racer to move so fast?
303
EXPLORE the BIG idea
How Many Bones Are in Your Hand?
Use a pencil to trace an outline of your hand on a piece of paper. Feel the bones in your fingers and the palm of your hand. At points where you can bend your fingers and hand, draw a circle. Each circle represents a joint where two bones meet. Draw lines to represent the bones in your hand.
Observe and Think How many bones did you find? How many joints?
How Does It Move?
The bones in your body are hard and stiff, yet they move smoothly. The point where two bones meet and move is called a joint. There are probably many objects in your home that have hard parts that move
CHAPTER 1
Getting Ready to Learn
CONCEPT REVIEW
• The cell is the basic unit of living things.
• Systems are made up of interacting parts that share matter and energy.
• In multicellular organisms cells work together to support life.
VOCABULARY REVIEW
See Glossary for definitions.
cell
system
CONTENT REVIEW
CLASSZONE.COM
Review concepts and vocabulary.
TAKING NOTES
MAIN IDEA WEB
Write each new blue heading in a box. Then write notes in boxes around the center box that give important terms and details about that blue heading.
VOCABULARY STRATEGY
Write each new vocabulary term in the center of a four square diagram. Write notes in the squares around each term. Include a definition, some features, and some examples of the term. If possible, write some things that are not examples of the term.
See the Note-Taking Handbook on pages R45–R51.
E 8 Unit: Human Biology
SCIENCE NOTEBOOK
The cell is the basic unit of living things.
The body has cells, tissues, and organs
Organs are groups of tissues working together.
Definition
Group of cells that work together
Examples
connective tissue, like bone
TISSUE

Chapter Opener

- Provides an advance organizer of the chapter Big Idea and Key Concepts
- Connects the Big Idea to the real world through an engaging photo and related question

Visual Summary

- Summarizes Key Concepts using both text and visuals
- Reinforces the connection of Key Concepts to the Big Idea

Chapter Review

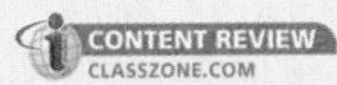

the BIG idea

The human body is made up of systems that work together to perform necessary functions.

CONTENT REVIEW CLASSZONE.COM

KEY CONCEPTS SUMMARY

1.1 The human body is complex.

You can think of the body as having five levels of organization: cells, tissues, organs, organ systems, and the whole organism itself. The different systems of the human body work together to maintain homeostasis.

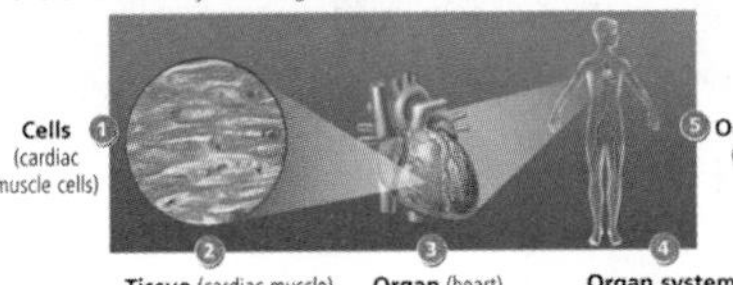

VOCABULARY
tissue p. 12
organ p. 12
organ system p. 12
homeostasis p. 12

1.2 The skeletal system provides support and protection.

Bones are living tissue. The skeleton is the body's framework and has two main divisions, the **axial skeleton** and the **appendicular skeleton**. Bones come together at joints.

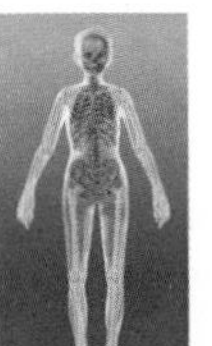

VOCABULARY
skeletal system p. 14
compact bone p. 14
spongy bone p. 15
axial skeleton p. 15
appendicular skeleton p. 16

1.3 The muscular system ma

Types of muscle
skeletal muscle, voluntary
smooth muscle, involuntary
cardiac muscle, involuntary

Reviewing Vocabulary

In one or two sentences describe how the vocabulary terms in each of the following pairs of words are related. Underline each vocabulary term in your answer.

1. cells, tissues
2. organs, organ systems
3. axial skeleton, appendicular skeleton
4. skeletal muscle, voluntary muscle
5. smooth muscle, involuntary muscle
6. compact bone, spongy bone

Reviewing Key Concepts

Multiple Choice *Choose the letter of the best answer.*

7. Which type of tissue carries electrical impulses from your brain?
   a. epithelial tissue
   b. muscle tissue
   c. nerve tissue
   d connective tissue
8. Connective tissue functions to provide
   a. support and strength
   b. messaging system
   c. movement
   d. heart muscle

Section Opener

- Highlights the Key Concept
- Connects new learning to prior knowledge
- Previews important vocabulary

KEY CONCEPT

1.1 The human body is complex.

BEFORE, you learned

- All living things are made of cells
- All living things need energy
- Living things meet their needs through interactions with the environment

NOW, you will learn

- About the organization of the human body
- About different types of tissues
- About the functions of organ systems

VOCABULARY
tissue p. 10
organ p. 11
organ system p. 12
homeostasis p. 12

THINK ABOUT

How is the human body like a city?

A city is made up of many parts that perform different functions. Buildings provide places to live and work. Transportation systems move people around. Electrical energy provides light and heat. Similarly, the human body is made of several systems. The skeletal system, like the framework of a building, provides support. The digestive system works with the respiratory system to provide energy and materials. What other systems in your body can you compare to a system in the city?

The body has cells, tissues, and organs.

MAIN IDEA WEB
As you read this section, complete the main idea web begun on page 8.

Your body is made of many parts that work together as a system to help you grow and stay healthy. The simplest level of organization in your body is the cell. Next come tissues, then individual organs, and then systems that are made up of organs. The highest level of organization is the organism itself. You can think of the body as having five levels of organization: cells, tissues, organs, organ systems, and the organism. Although these levels seem separate from one another, they all work together.

CHECK YOUR READING What are five levels of organization in your body?

Chapter 1: Systems, Support, and Movement 9 E

Thinking Critically

19. PROVIDE EXAMPLES What are the levels of organization of the human body from simple to most complex? Give an example of each.
20. CLASSIFY There are four types of tissue in the human body: epithelial, nerve, muscles, and connective. How would you classify blood? Explain your reasoning.
21. CONNECT A clam shell is made of a calcium compound. The material is hard, providing protection to the soft body of a clam. It is also lightweight. Describe three ways in which the human skeleton is similar to a seashell. What is one important way in which it is different?

Use the diagram below to answer the next two questions

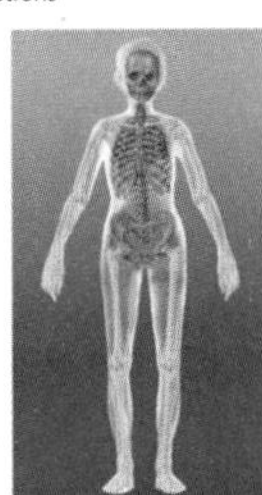

22. SYNTHESIZE Identify the type of joints that hold together the bones of the skull and sternum. How does this type of joint relate to the function of the skull and sternum?
23. SYNTHESIZE The human skeleton has two main divisions. Which skeleton do the arms and legs belong to? How do the joints that connect the arms to the shoulders and the legs to the hips relate to the function of this skeleton?
24. COMPARE AND CONTRAST How is the skeletal system of your body like the framework of a house or building? How is it different?
25. SUMMARIZE Describe three important functions of the skeleton.
26. APPLY The joints in the human body can be described as producing three types of movement. Relate these three types of movement to the action of brushing your teeth.
27. COMPARE AND CONTRAST When you stand, the muscles in you legs help to keep you balanced. Some of the muscles on both sides of your leg bones contract. How does this differ from how the muscles behave when you start to walk?
28. INFER Muscles are tissues that are made up of many muscle fibers. A muscle fiber can either be relaxed or contracted. Some movements you do require very little effort, like picking up a piece of paper. Others require a lot of effort, like picking up a book bag. How do you think a muscle produces the effort needed for a small task compared with a big task?

the BIG idea

29. INFER Look again at the picture on pages 6–7. Now that you have finished the chapter, how would you change or add details to your answer to the question on the photograph?
30. SUMMARIZE Write a paragraph explaining how skeletal muscles, bones, and joints work together to allow the body to move and be flexible. Underline the terms in your paragraph.

UNIT PROJECTS

If you are doing a unit project, make a folder for your project. Include in your folder a list of resources you will need, the date on which the project is due, and a schedule to track your progress. Begin gathering data.

E 32 Unit: Human Biology

The Big Idea Questions

- Help students connect their new learning back to the Big Idea
- Prompt students to synthesize and apply the Big Idea and Key Concepts

Many Ways to Learn

Because students learn in so many ways, *McDougal Littell Science* gives them a variety of experiences with important concepts and skills. Text, visuals, activities, and technology all focus on Big Ideas and Key Concepts.

Considerate Text

- Clear structure of meaningful headings
- Information clearly connected to main ideas
- Student-friendly writing style

Integrated Technology

- Interaction with Key Concepts through Simulations and Visualizations
- Easy access to relevant Web resources through Resource Centers and SciLinks
- Opportunities for review through Content Review and Math Tutorials

Structures in the respiratory system function together.

OUTLINE
Add *Structures in the respiratory system function together* to your outline. Be sure to include the six respiratory structures in your outline.

I. Main idea
A. Supporting idea
1. Detail
2. Detail
B. Supporting idea

The respiratory system is made up of many structures that allow you to move air in and out of your body, communicate, and keep out harmful materials.

Nose, Throat, and Trachea When you inhale, air enters your body through your nose or mouth. Inside your nose, tiny hairs called cilia filter dirt and other particles out of the air. Mucus, a sticky liquid in your nasal cavity, also helps filter air by trapping particles such as dirt and pollen as air passes by. The nasal cavity warms the air slightly before it moves down your throat toward a tubelike structure called the windpipe, or trachea (TRAY-kee-uh). A structure called the epiglottis (EHP-ih-GLAHT-ihs) keeps air from entering your stomach.

Lungs The lungs are two large organs located on either side of your heart. When you breathe, air enters the throat, passes through the trachea, and moves to the lungs through structures called bronchial tubes. Bronchial tubes branch throughout the lungs into smaller and smaller tubes. At the ends of the smallest tubes air enters tiny air sacs called alveoli. The walls of the alveoli are only one cell thick. In fact, one page in this book is much thicker than the walls of the alveoli. Oxygen passes from inside the alveoli through the thin walls and is dissolved into the blood. At the same time, carbon dioxide waste passes from the blood into the alveoli.

Through which structures does oxygen move into the lungs?

Ribs and Diaphragm If you put your hands on your ribs and take a deep breath, you can feel your ribs expand. The rib cage encloses a space inside your body called the thoracic (thu-RHAS-ihk) cavity. Some ribs are connected by cartilage to the breastbone or to each other, which makes the rib cage flexible. This flexibility allows the rib cage to expand when you breathe and make room for the lungs to expand and fill with air.

A large muscle called the diaphragm (DY-uh-FRAM) stretches across the floor of the thoracic cavity. When you inhale, your diaphragm contracts and pulls downward, which makes the thoracic cavity expand. This movement causes the lungs to push downward, filling the extra space. At the same time, other muscles draw the ribs outward and expand the lungs. Air rushes into the lungs, and inhalation is complete. When the diaphragm and other muscles relax, the process reverses and you exhale.

Explore the respiratory system.

Describe how the diaphragm and the rib cage move.

Visuals that Teach

- Information-rich visuals directly connected to the text
- Thoughtful pairing of diagrams and real-world photos
- Reading Visuals questions to support student learning

Respiratory System

The structures in the respiratory system allow this flutist to play music.

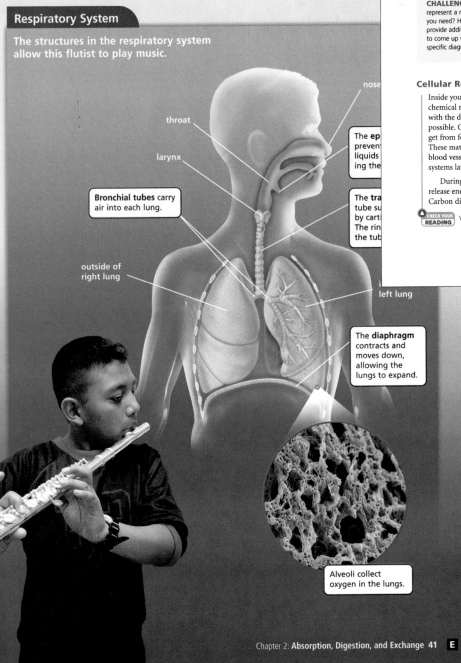

INVESTIGATE Lungs

How does air move in and out of lungs?

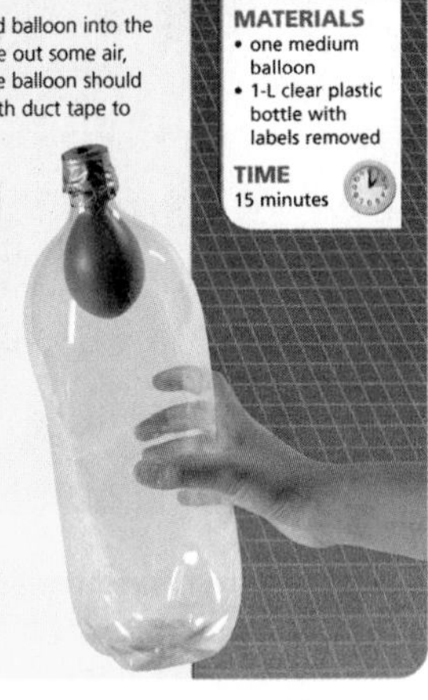

SKILL FOCUS
Modeling
MATERIALS
• one medium balloon
• 1-L clear plastic bottle with labels removed
TIME
15 minutes

PROCEDURE

1. Create a model of your lungs as shown. Insert an uninflated balloon into the top of the plastic bottle. While squeezing the bottle to force out some air, stretch the end of the balloon over the lip of the bottle. The balloon should still be open to the outside air. Tape the balloon in place with duct tape to make a tight seal
2. Release the bottle so that it expands back to its normal shape. Observe what happens to the balloon. Squeeze and release the bottle several times while observing the balloon. Record your observations.

WHAT DO YOU THINK?

- Describe, in words, what happens when you squeeze and release the bottle.
- How do you think your lungs move when you inhale? when you exhale?

CHALLENGE Design an addition to your model that could represent a muscle called the diaphragm. What materials do you need? How would this work? Your teacher may be able to provide additional materials so you can test your model. Be sure to come up with a comprehensive list of materials as well as a specific diagram.l as a specific diagram.

Cellular Respiration

Inside your cells, a process called **cellular respiration** uses oxygen in chemical reactions that release energy. The respiratory system works with the digestive and circulatory systems to make cellular respiration possible. Cellular respiration requires glucose, or sugars, which you get from food, in addition to oxygen, which you get from breathing. These materials are transported to every cell in your body through blood vessels. You will learn more about the digestive and circulatory systems later in this unit.

During cellular respiration, your cells use oxygen and glucose to release energy. Carbon dioxide is a waste product of the process. Carbon dioxide must be removed from cells.

VOCABULARY
Add a magnet diagram for *cellular respiration* to your notebook. Include the word *energy* in your diagram.

CHECK YOUR READING What three body systems are involved in cellular respiration?

Hands-on Learning

- Activities that reinforce Key Concepts
- Skill Focus for important inquiry and process skills
- Multiple activities in every chapter, from quick Explores to full-period Chapter Investigations

Differentiated Instruction

A full spectrum of resources for differentiating instruction supports you in reaching the wide range of learners in your classroom.

1.1 INSTRUCT

INVESTIGATE Systems

PURPOSE To understand how body systems work together

TIPS *20 min.*

- Allow students to use the chart from "Activate Prior Knowledge," p. 8.
- Remind students to be aware of their breathing and their hearts beating.
- Suggest that students bring in pictures of people performing everyday tasks, for follow-up discussion.

WHAT DO YOU THINK? *Sample answers: Talking on the phone involves the nervous system, the respiratory system, and the muscular system. When you sleep, your body breathes, excretes waste through the skin and the mouth, moves, and dreams.*

CHALLENGE *Answers will vary, but should group activities appropriately.*

 Datasheet, Systems, p. 20

Technology Resources

Customize this student lab as needed or look for an alternative. Print rubrics to assess student lab reports.

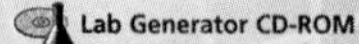

Lab Generator CD-ROM

Develop Critical Thinking

COMPARE Have students list ways in which people are like cells. *No two are exactly alike, both grow and can repair themselves, both have parts that serve to control all functions, both are living and have special jobs, both use food and oxygen to produce energy, both expel wastes, and both will eventually die.*

Ongoing Assessment

Describe the organization of the human body.

Ask: Why are the cells of complex organisms specialized? *Specialized cells perform different jobs necessary for survival.*

CHECK YOUR READING *Answer: Tissues are made of cells of the same type.*

INVESTIGATE Systems

How do the systems in your body interact?

PROCEDURE

1. Work with other classmates to make a list of everyday activities.
2. Discuss how your body responds to each task. Record your ideas.
3. Identify and count the systems in your body that you think are used to perform the task.
4. Have someone from your group make a chart of the different activities.

WHAT DO YOU THINK?

- Which systems did you name, and how did they work together to perform each activity?
- When you are asleep, what activities does your body perform?

CHALLENGE How could you make an experiment that would test your predictions?

SKILL FOCUS Predicting

MATERIALS large sheet of paper

TIME 20 minutes

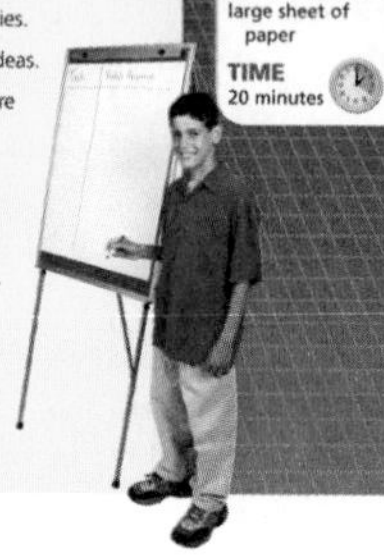

Cells

The cell is the basic unit of life. Cells make up all living things. Some organisms, such as bacteria, are made of only a single cell. In these organisms the single cell performs all of the tasks necessary for survival. That individual cell captures and releases energy, uses materials, and grows. In more complex organisms, such as humans and many other animals and plants, cells are specialized. Specialized cells perform special jobs. A red blood cell, for example, carries oxygen and other nutrients from the lungs throughout the body. A

Tissues

B A **tissue** is a group of similar cells that work together to perform a particular function. Think of a tissue as a brick wall and the cells within it as the individual bricks. Taken together, the bricks form something larger and more functional. But just as the bricks need to be placed in a certain way to form the wall, cells must be organized in a tissue.

 How are cells related to tissues?

The human body contains several types of tissues. These tissues are classified into four main groups according to their function: epithelial tissue, nerve tissue, muscle tissue, and connective tissue.

Epithelial (EHP-uh-THEE-lee-uhl) tissue functions as a boundary. It covers all of the inner and outer surfaces of your body. Each of your internal organs is covered with a layer of epithelial tissue. Another type of tissue, nerve tissue, functions as a messaging system. Cells in nerve tissue carry electrical impulses between your brain and the various parts of your body in response to changing conditions.

Muscle tissue functions in movement. Movement results when muscle cells contract, or shorten, and then relax. In some cases, such as throwing a ball, you control the movement. In other cases, such as the beating of your heart, the movement occurs without conscious control. Connective tissue functions to hold parts of the body together, providing support, protection, strength, padding, and insulation. Tendons and ligaments are connective tissues that hold bones and muscles together. Bone itself is another connective tissue. It supports and protects the soft parts of your body. C

Organs

D Groups of different tissues make up organs. An **organ** is a structure that is made up of two or more types of tissue that work together to carry out a function in the body. For example, the structures that carry blood around your body contain all four types of tissues. As in cells and tissues, the structure of an organ relates to its function. The stomach's bag-shaped structure and strong muscular walls make it suited for breaking down food. The walls of the heart are also muscular, allowing it to function as a pump.

Levels of Organization

The human body can be studied at different levels of organization.

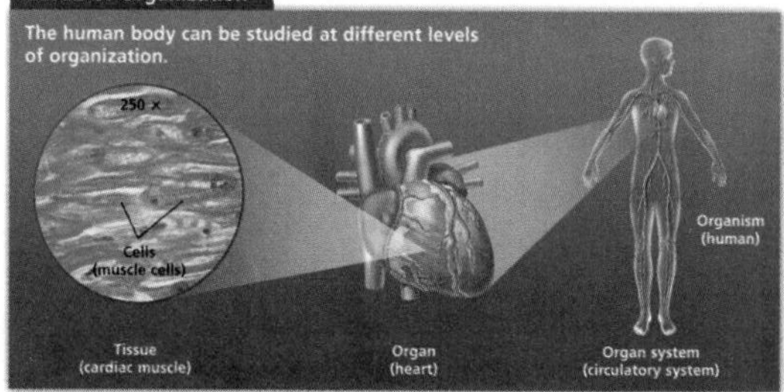

DIFFERENTIATE INSTRUCTION

More Reading Support

A Give an example of a specialized cell. *red blood cell*

B What do you call a group of similar cells that function together? *a tissue*

English Learners Help English learners with different word forms. For example, in the "Cells" paragraph, p. 10, *make* is used in a variety of forms: "Cells make up all living things;" "Some organisms, such as bacteria, are made of only a single cell." This verb can be found as *make, made, make up, made up, made up of,* or *made of.*

DIFFERENTIATE INSTRUCTION

More Reading Support

C What type of tissue provides support, protection, strength, and padding? *connective tissue*

D What do you call tissues that work together on one function? *organ*

Advanced Although this module presents the biology of a healthy human body, everyone has strengths and weaknesses. Explain that survival and high achievement are not dependent on full functioning of every organ. Discuss successful individuals with disabilities. Examples: Sir Steve Redgrave, Olympic gold medal for rowing: diabetes, dyslexia; Sir Stephen Hawking, theoretical physicist: amyotrophic lateral sclerosis (ALS); Andrea Bocelli, tenor: blindness; Tionne "T-Boz" Watkins, R&B singer: sickle-cell disease.

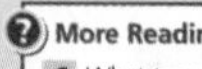 Challenge and Extension, p. 19

Teacher's Edition

- More Reading Support for below-level readers
- Strategies for below-level and advanced learners, English learners, and inclusion students

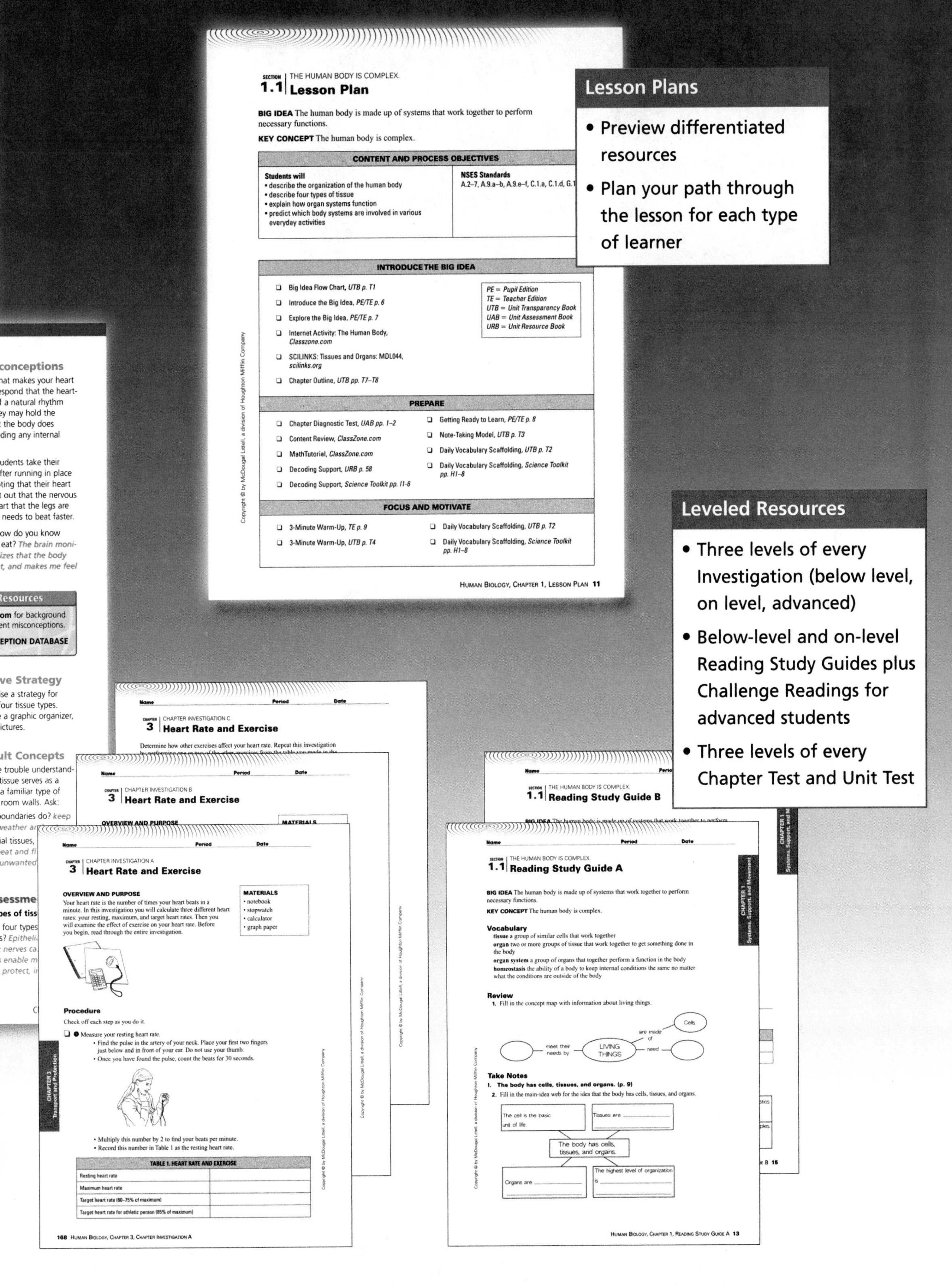

Lesson Plans

- Preview differentiated resources
- Plan your path through the lesson for each type of learner

SECTION 1.1 | THE HUMAN BODY IS COMPLEX.

Lesson Plan

BIG IDEA The human body is made up of systems that work together to perform necessary functions.

KEY CONCEPT The human body is complex.

CONTENT AND PROCESS OBJECTIVES	
Students will • describe the organization of the human body • describe four types of tissue • explain how organ systems function • predict which body systems are involved in various everyday activities	**NSES Standards** A.2–7, A.9.a–b, A.9.e–f, C.1.a, C.1.d, G.1

INTRODUCE THE BIG IDEA

- ❑ Big Idea Flow Chart, *UTB p. T1*
- ❑ Introduce the Big Idea, *PE/TE p. 6*
- ❑ Explore the Big Idea, *PE/TE p. 7*
- ❑ Internet Activity: The Human Body, *Classzone.com*
- ❑ SCILINKS: Tissues and Organs: MDL044, *scilinks.org*
- ❑ Chapter Outline, *UTB pp. T7–T8*

PE = Pupil Edition
TE = Teacher Edition
UTB = Unit Transparency Book
UAB = Unit Assessment Book
URB = Unit Resource Book

PREPARE

- ❑ Chapter Diagnostic Test, *UAB pp. 1–2*
- ❑ Content Review, *ClassZone.com*
- ❑ MathTutorial, *ClassZone.com*
- ❑ Decoding Support, *URB p. 58*
- ❑ Decoding Support, *Science Toolkit pp. I1-6*
- ❑ Getting Ready to Learn, *PE/TE p. 8*
- ❑ Note-Taking Model, *UTB p. T3*
- ❑ Daily Vocabulary Scaffolding, *UTB p. T2*
- ❑ Daily Vocabulary Scaffolding, *Science Toolkit pp. H1–8*

FOCUS AND MOTIVATE

- ❑ 3-Minute Warm-Up, *TE p. 9*
- ❑ 3-Minute Warm-Up, *UTB p. T4*
- ❑ Daily Vocabulary Scaffolding, *UTB p. T2*
- ❑ Daily Vocabulary Scaffolding, *Science Toolkit pp. H1–8*

Leveled Resources

- Three levels of every Investigation (below level, on level, advanced)
- Below-level and on-level Reading Study Guides plus Challenge Readings for advanced students
- Three levels of every Chapter Test and Unit Test

sconceptions
What makes your heart respond that the heart- of a natural rhythm hey may hold the at the body does eeding any internal

students take their after running in place noting that their heart nt out that the nervous eart that the legs are it needs to beat faster.

How do you know to eat? *The brain moni- alizes that the body ent, and makes me feel*

Resources
.com for background dent misconceptions.
CEPTION DATABASE

tive Strategy
vise a strategy for four tissue types. de a graphic organizer, pictures.

cult Concepts
ve trouble understand- l tissue serves as a s a familiar type of s room walls. Ask: boundaries do? *keep weather a*
elial tissues, *heat and fl r unwanted*

ssessme
ypes of tiss
e four types ns? *Epitheli s; nerves ca es enable m es protect, i*

Name Period Date

CHAPTER 3 | CHAPTER INVESTIGATION C

Heart Rate and Exercise

Determine how other exercises affect your heart rate. Repeat this investigation

Name Period Date

CHAPTER 3 | CHAPTER INVESTIGATION B

Heart Rate and Exercise

OVERVIEW AND PURPOSE
MATERIALS

Name Period Date

CHAPTER 3 | CHAPTER INVESTIGATION A

Heart Rate and Exercise

OVERVIEW AND PURPOSE
Your heart rate is the number of times your heart beats in a minute. In this investigation you will calculate three different heart rates: your resting, maximum, and target heart rates. Then you will examine the effect of exercise on your heart rate. Before you begin, read through the entire investigation.

MATERIALS
- notebook
- stopwatch
- calculator
- graph paper

Procedure
Check off each step as you do it.

❑ ❶ Measure your resting heart rate.
- Find the pulse in the artery of your neck. Place your first two fingers just below and in front of your ear. Do not use your thumb.
- Once you have found the pulse, count the beats for 30 seconds.

- Multiply this number by 2 to find your beats per minute.
- Record this number in Table 1 as the resting heart rate.

TABLE 1. HEART RATE AND EXERCISE	
Resting heart rate	
Maximum heart rate	
Target heart rate (60–75% of maximum)	
Target heart rate for athletic person (85% of maximum)	

CHAPTER 3 Transport and Protection

Name Period

SECTION 1.1 | THE HUMAN BODY IS COMPLEX.

Reading Study Guide B

CHAPTER 1 Systems, Support, and Movement

Name Period Date

SECTION 1.1 | THE HUMAN BODY IS COMPLEX.

Reading Study Guide A

BIG IDEA The human body is made up of systems that work together to perform necessary functions.

KEY CONCEPT The human body is complex.

Vocabulary
tissue a group of similar cells that work together
organ two or more groups of tissue that work together to get something done in the body
organ system a group of organs that together perform a function in the body
homeostasis the ability of a body to keep internal conditions the same no matter what the conditions are outside of the body

Review
1. Fill in the concept map with information about living things.

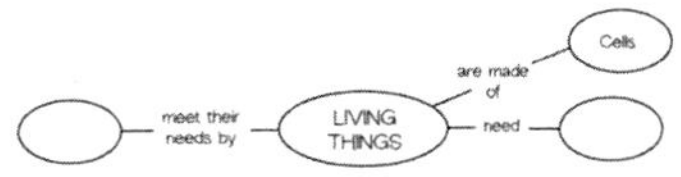

Take Notes
1. **The body has cells, tissues, and organs. (p. 9)**
2. Fill in the main-idea web for the idea that the body has cells, tissues, and organs.

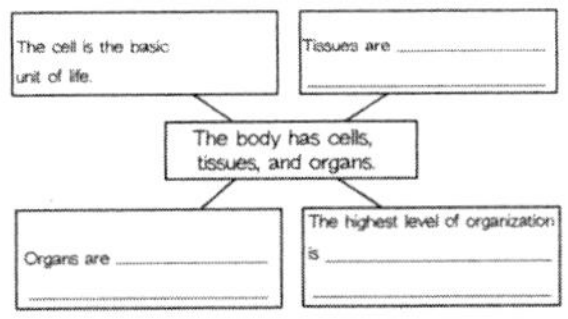

CHAPTER 1 Systems, Support, and Movement

Effective Assessment

McDougal Littell Science incorporates a comprehensive set of resources for assessing student knowledge and performance before, during, and after instruction.

Diagnostic Tests

- Assessment of students' prior knowledge
- Readiness check for concepts and skills in the upcoming chapter

Teach Difficult Concepts
Some students may have a hard time understanding homeostasis. Use a thermostat as an example. Point out that the heat or air conditioning is supplied just long enough to return an area to the set temperature.

EXPLORE (the BIG idea)
Revisit "Internet Activity: The Human Body" on p. 7. Have students discuss how body structures are related to their functions.

Ongoing Assessment
Explain how organ systems function.
Ask: How do organ systems meet the body's needs? *Each organ system performs a necessary function, and together the systems allow the human organism to grow, reproduce, and maintain life.*

Reinforce (the BIG idea)
Have students relate the section to the Big Idea.
Reinforcing Key Concepts, p. 21

1.1 ASSESS & RETEACH

Assess
Section 1.1 Quiz, p. 3

Reteach
Have students draw a flow chart of the progression from cells to tissues to organs to organ systems. In the final drawing, students should make a rough outline of the human body with the systems labeled. Have volunteers share their drawings in a class discussion.

Technology Resources
Have students visit **ClassZone.com** for reteaching of Key Concepts.
CONTENT REVIEW
CONTENT REVIEW CD-ROM

Organ Systems
An **organ system** is a group of organs that together perform a function that helps the body meet its needs for energy and materials. For example, your stomach, mouth, throat, large and small intestines, liver, and pancreas are all part of the organ system called the digestive system. The body is made up of many organ systems. In this unit, you will read about these systems. They include the skeletal, muscular, respiratory, digestive, urinary, circulatory, immune, nervous, and reproductive systems. Together, these systems allow the human organism to grow, reproduce, and maintain life.

The body's systems interact with one another.

READING TIP
VOCABULARY
The word *homeostasis* contains two word roots. *Homeo* comes from a root meaning "same." *Stasis* comes from a root meaning "stand still" or "stay."

The ability of your body to maintain internal conditions is called **homeostasis** (HOH-mee-oh-STAY-sihs). Your body is constantly regulating such things as your body temperature, the amount of sugar in your blood, even your posture. The processes that take place in your body occur within a particular set of conditions.

The body's many levels of organization, from cells to organ systems, work constantly to maintain the balance needed for the survival of the organism. For example, on a hot day, you may sweat. Sweating keeps the temperature inside your body constant, even though the temperature of your surroundings changes.

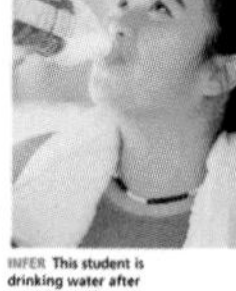

INFER This student is drinking water after exercising. Why is it important to drink fluids after you sweat?

1.1 Review

KEY CONCEPTS
1. Draw a diagram that shows the relationship among cells, tissues, organs, and organ systems.
2. Make a chart of the four basic tissue groups that includes names, functions, and examples.
3. Identify three functions performed by organ systems.

CRITICAL THINKING
4. **Apply** How does drinking water after you sweat help maintain homeostasis?
5. **Compare and Contrast** Compare and contrast the four basic tissue groups. How would all four types of tissue be involved in a simple activity, like raising your hand?

CHALLENGE
6. **Apply** Describe an object, such as a car, that can be used as a model of the human body. Explain how the parts of the model relate to the body.

B 12 Unit: Human Biology

ANSWERS

1. Diagrams should show organ systems as groups of organs, as groups of tissues, as groups of cells.
2. Charts should include epithelial, nerve, muscle, and connective tissues, and show understanding of each.
3. Sample answer: digestion, elimination, reproduction
4. When you sweat, your body loses water. Adding new water to the body helps maintain homeostasis.
5. Sample answer: When you raise your hand, nerve tissue carries a message from your brain, and muscle tissue contracts to move the bones, made of connective and epithelial tissues.
6. Answers should mention that the parts work together for the functioning of the whole.

Ongoing Assessment

- Check Your Reading questions for student self-check of comprehension
- Consistent Teacher Edition prompts for assessing understanding of Key Concepts

Reviewing Vocabulary

In one or two sentences describe how the vocabulary terms in each of the following pairs of words are related. Underline each vocabulary term in your answer.

1. cells, tissues
2. organs, organ systems
3. axial skeleton, appendicular skeleton
4. skeletal muscle, voluntary muscle
5. smooth muscle, involuntary muscle
6. compact bone, spongy bone

Reviewing Key Concepts

Multiple Choice *Choose the letter of the best answer.*

7. Which type of tissue carries electrical impulses from your brain?
   a. epithelial tissue
   b. muscle tissue
   c. nerve tissue
   d. connective tissue
8. Connective tissue functions to provide
   a. support and strength
   b. messaging system
   c. movement
   d. heart muscle
9. Inside bone cells is a network made of
   a. tendons
   b. calcium
   c. marrow
   d. joints
10. The marrow produces
   a. spongy bone
   b. red blood cells
   c. compact bone
   d. calcium
11. Which bones make up the axial skeleton?
   a. skull, shoulder blades, arm bones
   b. skull, spinal column, leg bones
   c. shoulder blades, spinal column, and hip bones
   d. skull, spinal column, ribs
12. Bones of the skeleton connect to each other at
   a. tendons
   b. ligaments
   c. joints
   d. muscles
13. How do muscles contribute to homeostasis?
   a. They keep parts of the body together.
   b. They control the amount of water in the body.
   c. They help you move.
   d. They produce heat when they contract.
14. Cardiac muscle is found in the
   a. heart
   b. stomach
   c. intestines
   d. arms and legs
15. The stomach is made up of
   a. cardiac muscle
   b. skeletal muscle
   c. smooth muscle
   d. voluntary muscle

Short Answer *Write a short answer to each question.*

16. What is the difference between spongy bone and compact bone?
17. The root word *homeo* means "same," and the root word *stasis* means "to stay." How do these root words relate to the definition of *homeostasis*?
18. Hold the upper part of one arm between your elbow and shoulder with your opposite hand. Feel the muscles there. What happens to those muscles as you bend your arm?

Section and Chapter Reviews

- Focus on Key Concepts and critical thinking skills
- A full range of question types and levels of thinking

Leveled Chapter and Unit Tests

- Three levels of test for every chapter and unit
- Same Big Ideas, Key Concepts, and essential skills assessed on all levels

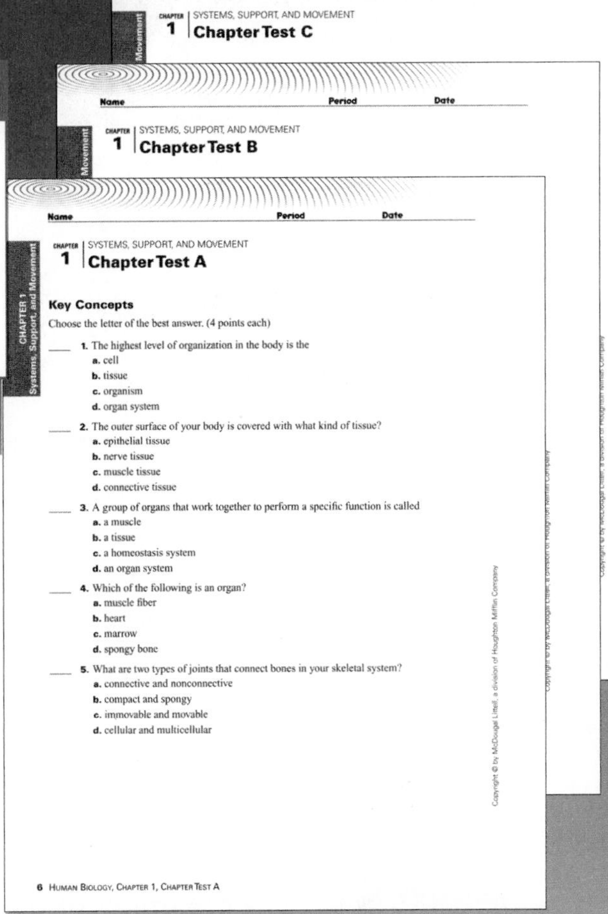

Name Period Date

CHAPTER 1 | SYSTEMS, SUPPORT, AND MOVEMENT
Chapter Test C

Name Period Date

CHAPTER 1 | SYSTEMS, SUPPORT, AND MOVEMENT
Chapter Test B

Name Period Date

CHAPTER 1 | SYSTEMS, SUPPORT, AND MOVEMENT
Chapter Test A

Key Concepts

Choose the letter of the best answer. (4 points each)

____ 1. The highest level of organization in the body is the
a. cell
b. tissue
c. organism
d. organ system

____ 2. The outer surface of your body is covered with what kind of tissue?
a. epithelial tissue
b. nerve tissue
c. muscle tissue
d. connective tissue

____ 3. A group of organs that work together to perform a specific function is called
a. a muscle
b. a tissue
c. a homeostasis system
d. an organ system

____ 4. Which of the following is an organ?
a. muscle fiber
b. heart
c. marrow
d. spongy bone

____ 5. What are two types of joints that connect bones in your skeletal system?
a. connective and nonconnective
b. compact and spongy
c. immovable and movable
d. cellular and multicellular

Copyright © by McDougal Littell, a division of Houghton Mifflin Company

6 Human Biology, Chapter 1, Chapter Test A

Thinking Critically

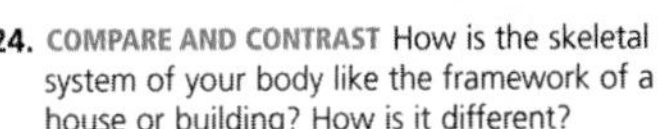

19. PROVIDE EXAMPLES What are the levels of organization of the human body from simple to most complex? Give an example of each.

20. CLASSIFY There are four types of tissue in the human body: epithelial, nerve, muscles, and connective. How would you classify blood? Explain your reasoning.

21. CONNECT A clam shell is made of a calcium compound. The material is hard, providing protection to the soft body of a clam. It is also lightweight. Describe three ways in which the human skeleton is similar to a seashell. What is one important way in which it is different?

Use the diagram below to answer the next two questions

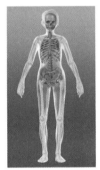

22. SYNTHESIZE Identify the type of joints that hold together the bones of the skull and sternum. How does this type of joint relate to the function of the skull and sternum?

23. SYNTHESIZE The human skeleton has two main divisions. Which skeleton do the arms and legs belong to? How do the joints that connect the arms to the shoulders and the legs to the hips relate to the function of this skeleton?

24. COMPARE AND CONTRAST How is the skeletal system of your body like the framework of a house or building? How is it different?

25. SUMMARIZE Describe three important functions of the skeleton.

26. APPLY The joints in the human body can be described as producing three types of movement. Relate these three types of movement to the action of brushing your teeth.

27. COMPARE AND CONTRAST When you stand, the muscles in you legs help to keep you balanced. Some of the muscles on both sides of your leg bones contract. How does this differ from how the muscles behave when you start to walk?

28. INFER Muscles are tissues that are made up of many muscle fibers. A muscle fiber can either be relaxed or contracted. Some movements you do require very little effort, like picking up a piece of paper. Others require a lot of effort, like picking up a book bag. How do you think a muscle produces the effort needed for a small task compared with a big task?

29. INFER Look again at the picture on pages 6–7. Now that you have finished the chapter, how would you change or add details to your answer to the question on the photograph?

30. SUMMARIZE Write a paragraph explaining how skeletal muscles, bones, and joints work together to allow the body to move and be flexible. Underline the terms in your paragraph.

If you are doing a unit project, make a folder for your project. Include in your folder a list of resources you will need, the date on which the

E 32 Unit: Human Biology

Rubrics

- Rubrics in Teacher Edition for all extended response questions
- Rubrics for all Unit Projects
- Alternative Assessment with rubric for each chapter
- A wide range of additional rubrics in the Science Toolkit

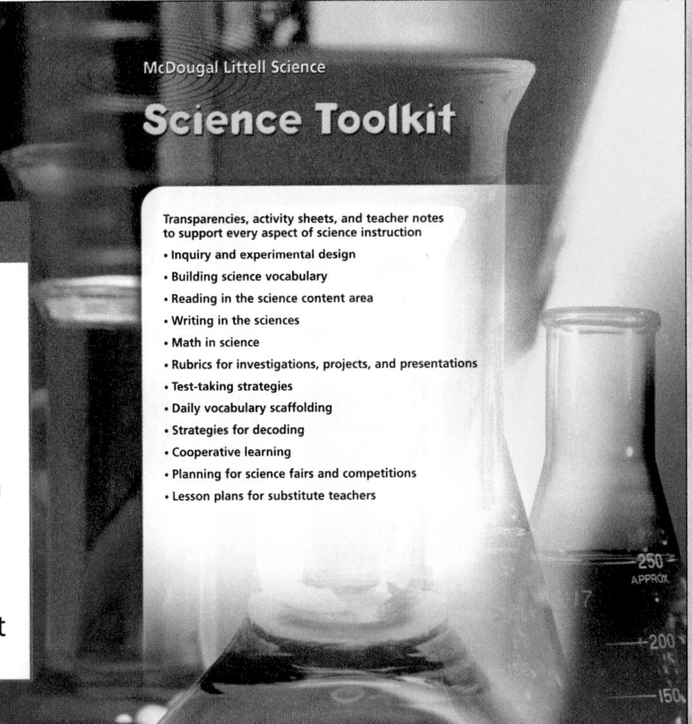

McDougal Littell Science
Science Toolkit
Transparencies, activity sheets, and teacher notes to support every aspect of science instruction
• Inquiry and experimental design
• Building science vocabulary
• Reading in the science content area
• Writing in the sciences
• Math in science
• Rubrics for investigations, projects, and presentations
• Test-taking strategies
• Daily vocabulary scaffolding
• Strategies for decoding
• Cooperative learning
• Planning for science fairs and competitions
• Lesson plans for substitute teachers
250
APPROX.
200
150

McDougal Littell Science Modular Series

McDougal Littell Science lets you choose the titles that match your curriculum. Each module in this flexible 15-book series takes an in-depth look at a specific area of life, earth, or physical science.

- Flexibility to match your curriculum
- Convenience of smaller books
- Complete Student Resource Handbooks in every module

Life Science Titles

A ▶ Cells and Heredity
1. The Cell
2. How Cells Function
3. Cell Division
4. Patterns of Heredity
5. DNA and Modern Genetics

B ▶ Life Over Time
1. The History of Life on Earth
2. Classification of Living Things
3. Population Dynamics

C ▶ Diversity of Living Things
1. Single-Celled Organisms and Viruses
2. Introduction to Multicellular Organisms
3. Plants
4. Invertebrate Animals
5. Vertebrate Animals

D ▶ Ecology
1. Ecosystems and Biomes
2. Interactions Within Ecosystems
3. Human Impact on Ecosystems

E ▶ Human Biology
1. Systems, Support, and Movement
2. Absorption, Digestion, and Exchange
3. Transport and Protection
4. Control and Reproduction
5. Growth, Development, and Health

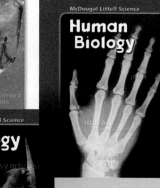

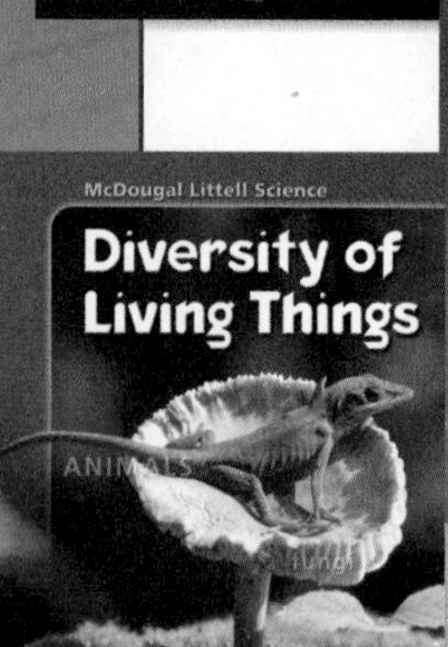

Earth Science Titles

A ▶ Earth's Surface
1. Views of Earth Today
2. Minerals
3. Rocks
4. Weathering and Soil Formation
5. Erosion and Deposition

B ▶ The Changing Earth
1. Plate Tectonics
2. Earthquakes
3. Mountains and Volcanoes
4. Views of Earth's Past
5. Natural Resources

C ▶ Earth's Waters
1. The Water Planet
2. Freshwater Resources
3. Ocean Systems
4. Ocean Environments

D ▶ Earth's Atmosphere
1. Earth's Changing Atmosphere
2. Weather Patterns
3. Weather Fronts and Storms
4. Climate and Climate Change

E ▶ Space Science
1. Exploring Space
2. Earth, Moon, and Sun
3. Our Solar System
4. Stars, Galaxies, and the Universe

Physical Science Titles

A ▶ Matter and Energy
1. Introduction to Matter
2. Properties of Matter
3. Energy
4. Temperature and Heat

B ▶ Chemical Interactions
1. Atomic Structure and the Periodic Table
2. Chemical Bonds and Compounds
3. Chemical Reactions
4. Solutions
5. Carbon in Life and Materials

C ▶ Motion and Forces
1. Motion
2. Forces
3. Gravity, Friction, and Pressure
4. Work and Energy
5. Machines

D ▶ Waves, Sound, and Light
1. Waves
2. Sound
3. Electromagnetic Waves
4. Light and Optics

E ▶ Electricity and Magnetism
1. Electricity
2. Circuits and Electronics
3. Magnetism

Teaching Resources

A wealth of print and technology resources help you adapt the program to your teaching style and to the specific needs of your students.

Book-Specific Print Resources

Unit Resource Book provides all of the teaching resources for the unit organized by chapter and section.

- Family Letters
- *Scientific American Frontiers* Video Guide
- Unit Projects
- Lesson Plans
- Reading Study Guides (Levels A and B)
- Spanish Reading Study Guides
- Challenge Readings
- Challenge and Extension Activities
- Reinforcing Key Concepts
- Vocabulary Practice
- Math Support and Practice
- Investigation Datasheets
- Chapter Investigations (Levels A, B, and C)
- Additional Investigations (Levels A, B, and C)
- Summarizing the Chapter

McDougal Littell Science

Unit Resource Book

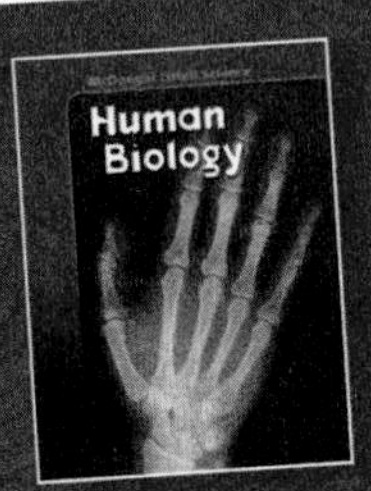

Human Biology

- Family Letters (English and Spanish)
- *Scientific American Frontiers* Video Guides
- Unit Projects (with Rubrics)
- Lesson Plans
- Reading Study Guides (Levels A and B and Spanish)
- Challenge Activities and Readings
- Reinforcing Key Concepts
- Vocabulary Practice and Decoding Support
- Math Support and Practice
- Investigation Datasheets
- Chapter Investigations (Levels A, B, and C)
- Additional Investigations (Levels A, B, and C)
- Summarizing the Chapter

Unit Assessment Book contains complete resources for assessing student knowledge and performance.

- Chapter Diagnostic Tests
- Section Quizzes
- Chapter Tests (Levels A, B, and C)
- Alternative Assessments
- Unit Tests (Levels A, B, and C)

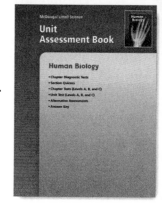

Unit Transparency Book includes instructional visuals for each chapter.

- Three-Minute Warm-Ups
- Note-Taking Models
- Daily Vocabulary Scaffolding
- Chapter Outlines
- Big Idea Flow Charts
- Chapter Teaching Visuals

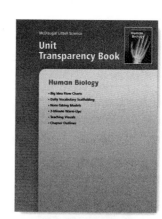

Unit Lab Manual

Unit Note-Taking/Reading Study Guide

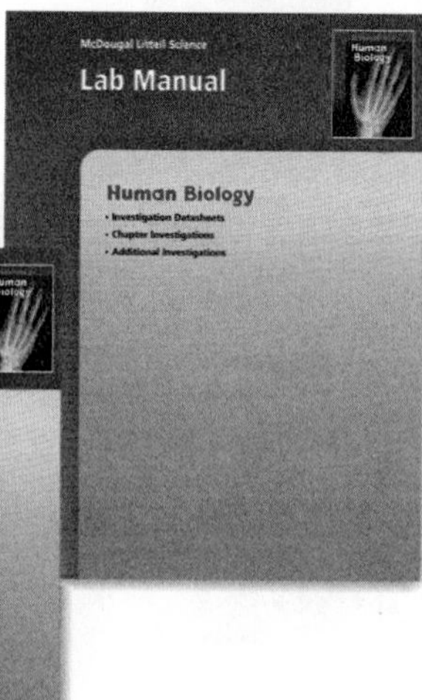

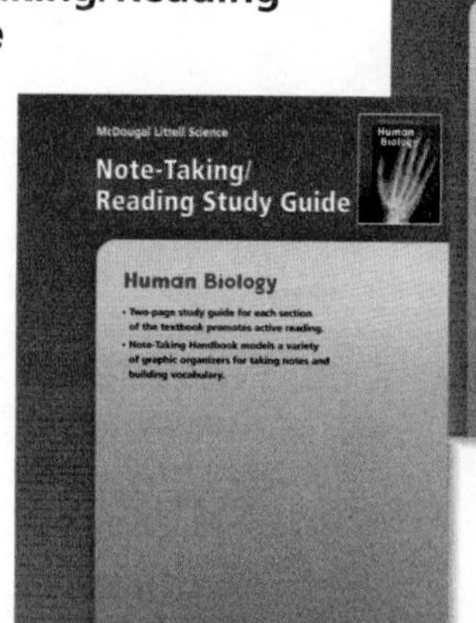

Program-Wide Print Resources

Process and Lab Skills

Problem Solving and Critical Thinking

Standardized Test Practice

Science Toolkit

City Science

Visual Glossary

Multi-Language Glossary

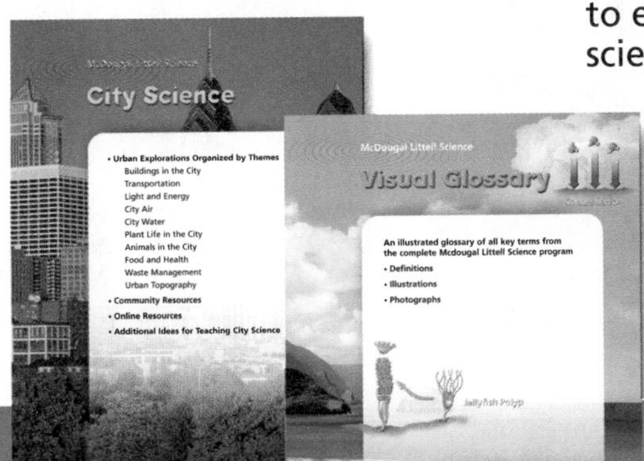

English Learners Package

***Scientific American Frontiers* Video Guide**

How Stuff Works Express
This quarterly magazine offers opportunities to explore current science topics.

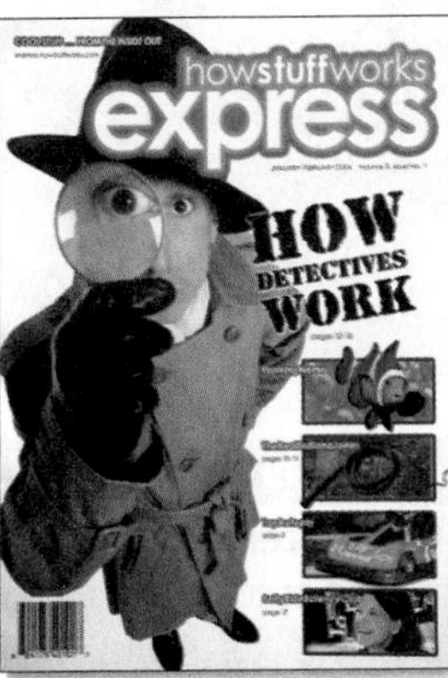

Technology Resources

***Scientific American Frontiers* Video Program**
Each specially-tailored segment from this award-winning PBS series correlates to a unit; available on VHS and DVD

Audio CDs Complete chapter texts read in both English and Spanish

Lab Generator CD-ROM
A searchable database of all activities from the program plus additional labs for each unit; edit and print your own version of labs

Test Generator CD-ROM

eEdition CD-ROM

EasyPlanner CD-ROM

Content Review CD-ROM

Power Presentations CD-ROM

Online Resources

- ClassZone.com
- Content Review Online
- eEdition Plus Online
- EasyPlanner Plus Online
- eTest Plus Online

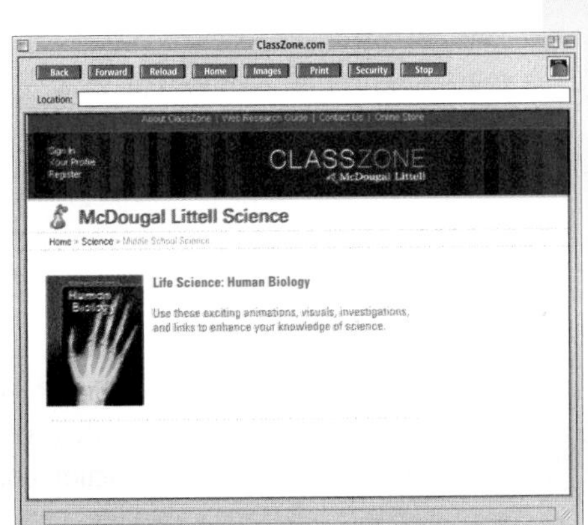

Correlation to National Science Education Standards

This chart provides an overview of how the five Life Science modules of *McDougal Littell Science* address the National Science Education Standards.

A Cells and Heredity
B Life Over Time
C Diversity of Living Things
D Ecology
E Human Biology

A. Science as Inquiry	Book, Chapter, and Section
A.1– A.8 **Abilities necessary to do scientific inquiry** Identify questions for investigation; design and conduct investigations; use evidence; think critically and logically; analyze alternative explanations; communicate; use mathematics.	All books (pp. R2–R44), All Chapter Investigations All Think Science features
A.9 **Understandings about scientific inquiry** Different kinds of investigations for different questions; investigations guided by current scientific knowledge; importance of mathematics and technology for data gathering and analysis; importance of evidence, logical argument, principles, models, and theories; role of legitimate skepticism; scientific investigations lead to new investigations.	All books (pp. xxii–xxv), A4.3, B1.3, C4.3, D2.2, E1.1

B. Physical Science	Book, Chapter, and Section
B.1 **Properties and changes of properties in matter** Physical properties; substances, elements, and compounds; chemical reactions.	A2.1, A3.1 (Connecting Sciences)
B.2 **Motions and forces** Position, speed, direction of motion; balanced and unbalanced forces.	C5.2 (Connecting Sciences)
B.3 **Transfer of energy** Energy transfer; forms of energy; heat and light; electrical circuits; sun as source of Earth's energy.	A2.2. D1.3, E4.2 (Connecting Sciences)

C. Life Science	Book, Chapter, and Section
C.1 **Structure and function in living systems** Systems; structure and function; levels of organization; cells and cell activities; specialization; human body systems; disease.	A1.1, A1.2, A1.3, A2.1, A2.2, A2.3, A3.1, B1, C1.1, C2, C4, D2.1, E1.1, E1.2, E1.3, E2.1, E2.2, E2.3, E3.1, E3.2, E.3.3, E4.1, E4.2, E4.3, E5.1, E5.2, E5.3
C.2 **Reproduction and heredity** Sexual and asexual reproduction; heredity and genes; traits determined by heredity and environment.	A3.2, A3.3, A4.1, A4.2, A4.3,, A5.1, A5.2, B2.1, C1.1, C1.2, C2.4, C3.3, C3.4, C4.2, C4.3, C4.4, C5.2, C5.3, C5.4, E4.3
C.3 **Regulation and behavior** Growth, reproduction, and maintenance of stable internal environment; regulation; behavior; evolution of behavior through adaptation to environment.	C1.1, C2.1, C4.1, C5.1, C5.2, C5.3, C5.4, E1.1, E4.1, E4.2, E5.1, E5.2
C.4 **Populations and ecosystems** Populations; ecosystems; producers, consumers, and decomposers; food webs; energy flow; population size and resource availability; population growth.	B3.1, B3.2, B3.3, C3.1, D1.1, D1.2, D1.3, D2.1, D2.2, D2.3, D3.1, D3.2, D3.3
C.5 **Diversity and adaptations of organisms** Unity and diversity; similarities in internal structures, chemical processes, and evidence of common ancestry; adaptation and biological evolution; extinction and fossil evidence.	B1.1, B1.1, B1.2, B1.3, B2.1, B2.2, B2.3, C4, C5

D. Earth and Space Science	Book, Chapter, and Section
D.1 **Structure of the earth system** Lithosphere, mantle, and core; plate movement and earthquakes, volcanoes, and mountain building; constructive and destructive forces on landforms; soil, weathering, and erosion; water and water cycle; atmosphere, weather, and climate; living organisms in earth system.	B3.3, D1.1
D.2 **Earth's history** Continuity of earth processes; impact of occasional catastrophes; fossil evidence.	B1.1, B (pp. R52–R57)
D.3 **Earth in the solar system** Sun, planets, asteroids, comets; regular and predictable motion and day, year, phases of the moon, and eclipses; gravity and orbits; sun as source of energy for earth; cause of seasons.	D1.3

E. Science and Technology	Book, Chapter, and Section
E.1– E.5 **Abilities of technological design** Identify problems; design a solution or product; implement a proposed design; evaluate completed designs or products; communicate the process of technological design.	All books (pp.xxvi–xxvii) B (p. 5)
E.6 **Understandings about science and technology** Similarities and differences between scientific inquiry and technological design; contributions of people in different cultures; reciprocal nature of science and technology; nonexistence of perfectly designed solutions; constraints, benefits, and unintended consequences of technological designs.	All books (pp. xxvi–xxvii) All books (Frontiers in Science, Timelines in Science) A5.3, D3.1, D3.2

F. Science in Personal and Social Perspectives	Book, Chapter, and Section
F.1 **Personal health** Exercise; fitness; hazards and safety; tobacco, alcohol, and other drugs; nutrition; STDs; environmental health.	C2.4, D3.2, E2.2, E5.2, E5.3
F.2 **Populations, resources, and environments** Overpopulation and resource depletion; environmental degradation.	B3.2, B3.3, D3.1, D3.2, D3.3
F.3 **Natural hazards** Earthquakes, landslides, wildfires, volcanic eruptions, floods, storms; hazards from human activity; personal and societal challenges.	D3.2
F.4 **Risks and benefits** Risk analysis; natural, chemical, biological, social, and personal hazards; decisions based on risks and benefits.	A5.3, E5.2, D3.3
F.5 **Science and technology in society** Science's influence on knowledge and world view; societal challenges and scientific research; technological influences on society; contributions from people of different cultures and times; work of scientists and engineers; ethical codes; limitations of science and technology.	A 5.3 All books (Timelines in Science)

G. History and Nature of Science	Book, Chapter, and Section
G.1 **Science as a human endeavor** Diversity of people working in science, technology, and related fields; abilities required by science .	All books (pp. xxii–xxv; Frontiers in Science)
G.2 **Nature of science** Observations, experiments, models; tentative nature of scientific ideas; differences in interpretation of evidence; evaluation of results of investigations, experiments, observations, theoretical models, and explanations; importance of questioning, response to criticism, and communication.	B1.2, B1.3, B2.3
G.3 **History of science** Historical examples of inquiry and relationships between science and society; scientists and engineers as valued contributors to culture; challenges of breaking through accepted ideas.	B1.2, E5.3 All books (Frontiers in Science; Timelines in Science)

Correlation to Benchmarks

This chart provides an overview of how the five Life Science modules of *McDougal Littell Science* address the Project 2061 Benchmarks for Science Literacy.

A Cells and Heredity
B Life Over Time
C Diversity of Living Things
D Ecology
E Human Biology

1. The Nature of Science	Book, Chapter, and Section
	The Nature of Science (pp. xxii–xxv); Scientific Thinking Handbook (pp. R2–R9); Lab Handbook (pp. R10–R35); Think Science Features: A1.3, A4.3, B1.3, C4.3, D2.2, E1.1

3. The Nature of Technology	Book, Chapter, and Section
	The Nature of Technology (pp. xxvi–xxvii); A5.3, B3.3, D3.3; Timelines in Science Features

4. The Physical Setting	Book, Chapter, and Section
4.B THE EARTH	B3.3, D1.1, D1.2, D3
4.C PROCESSES THAT SHAPE THE EARTH	B3.3, C3.4, D3.1, D3.2
4.D STRUCTURE OF MATTER	A2.1, D1.2
4.E ENERGY TRANSFORMATIONS	B1.3, C2.2

5. The Living Environment	Book, Chapter, and Section
5.A DIVERSITY OF LIFE	
5.A.1 Differences among plants, animals, and other organisms	C1.1, C2.2, C2.3, C2.4, D1.3
5.A.2 Body plans and internal structures for food and reproduction.	B2, C2.2, C2.3, C3, C4, C5
5.A.3 Similarities among organisms in internal and external structures used to determine relatedness	B2.1, B2.2, B2.3, C4.1, C4.2, C4.3, C4.4
5.A.4 For sexually reproducing organisms, a species comprises all organisms that can mate with one another to produce fertile offspring.	D2.1
5.A.5 Interconnected global food webs and cycles	C3.4, D1.3
5.B HEREDITY	
5.B.1 In some species, all the genes come from a single parent; whereas in organisms that have sexes, typically half of the genes come from each parent.	A3.3, A4
5.B.2 Sexual reproduction and the transmission of genetic information	A4.1, A4.3, A5.1, C2.1, E4.3
5.B.3 New varieties of plants and domestic animals from selective breeding	A5.2, A5.3, B1.2
5.C CELLS	
5.C.1 All living things are composed of cells; tissues and organs	A1, C1, C2.1
5.C.2 Cells divide to make more cells for growth and repair. Various organs and tissues function to serve the needs of cells for food, air, and waste removal.	A3.1, A3.2, A3.3, C2.1
5.C.3 Within cells, many basic functions of organisms are carried out; way in which cells function is similar in all living organisms.	A1.2, A2, C1, C2.1
5.C.4 About two thirds of the weight of cells is accounted for by water, which gives cells many of their properties.	A2.1

5.D2 INTERDEPENDENCE OF LIFE	
5.D.1 In all environments, organisms with similar needs may compete for resources; growth and survival of organisms depend on the physical conditions	B3.1, B3.2, D1.1, D2.2
5.D.2 Producer/consumer, predator/prey, or parasite/host relationships; scavengers and decomposers; competitive or mutually beneficial relationships.	C2.1, C2.2, C2.3, C2.4, D1.3, D2.2
5.E FLOW OF MATTER AND ENERGY	
5.E.1 Food molecules as fuel and building material for all organisms; photosynthesis and producers; consumers.	A2.1, A2.2, C2.2, C2.3, C3.1, D1.3
5.E.2 Flow of energy though living systems; amount of matter remains constant, even though its form and location change.	D1.2, D1.3
5.E.3 Energy can change from one form to another in living things; almost all food energy comes originally from sunlight.	A1.2, A2.2, C2.2, C2.3, D1.3
5.F EVOLUTION OF LIFE	
5.F.1 Small differences between parents and offspring can accumulate in successive generations so that descendants are very different from their ancestors.	A5.2, A5.3, B1.2, B1.3
5.F.2 Individual organisms with certain traits are more likely than others to survive and have offspring. Changes in environmental conditions can affect the survival of individual organisms and entire species.	B1.1, B1.2, C5
5.F.3 Sedimentary rock layers as evidence for the long history of the earth and of changing life forms whose remains are found in the rocks.	B1.1, B pp. R52–R57

6. The Human Organism	**Book, Chapter, and Section**
6.A HUMAN IDENTITY	E1, E2, E3, E4
6.B HUMAN DEVELOPMENT	C2.1, C5.4, E4, E5
6.C BASIC FUNCTIONS	C2.1, C5.4, E1, E2, E3, E4
6.E PHYSICAL HEALTH	C1.2, C1.3, C1.4, D3.2, E3, E5
8. The Designed World	**A5.3, B3.3, D1.3, D3.2, D3.39.**
9. The Mathematical World	**All Math in Science Features**
10. Historical Perspectives	**A1.1, A1.2**

12. Habits of Mind	**Book, Chapter, and Section**
12.A VALUES AND ATTITUDES	Think Science Features: A1.3, A4.3, B1.3, C4.3, D2.2, E1.1
12.B COMPUTATION AND ESTIMATION	All Math in Science Features, Lab Handbook (pp. R10–R35)
12.C MANIPULATION AND OBSERVATION	All Investigates and Chapter Investigations
12.D COMMUNICATION SKILLS	All Chapter Investigations, Lab Handbook (pp. R10–R35)
12.E CRITICAL-RESPONSE SKILLS	Think Science Features: A1.3, A4.3, B1.3, C4.3, D2.2, E1.1; Scientific Thinking Handbook (pp. R2–R9)

Planning the Unit

The Pacing Guide provides suggested pacing for all chapters in the unit as well as the two unit features shown below.

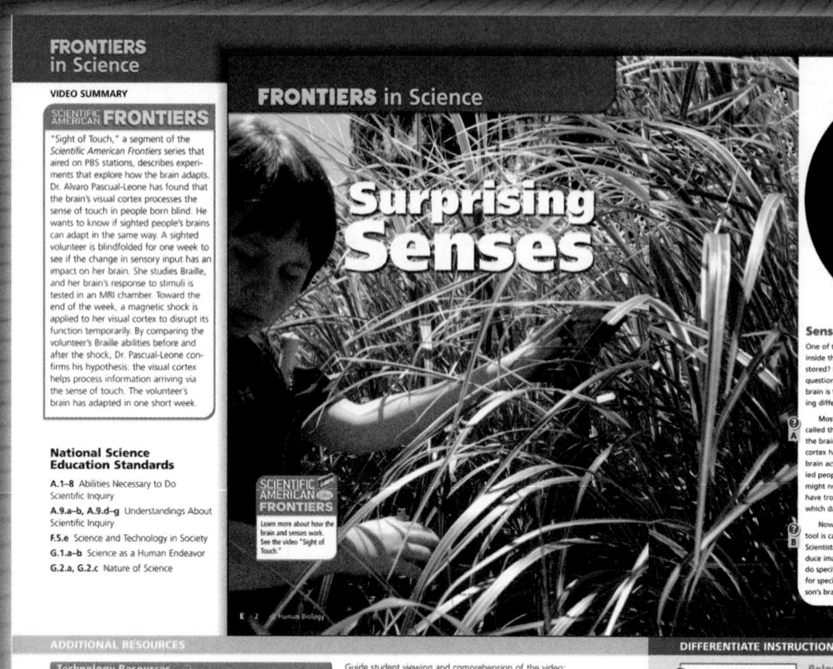

Frontiers in Science

- Features cutting-edge research as an engaging point of entry into the unit
- Connects to an accompanying *Scientific American Frontiers* video and viewing guide
- Introduces three options for unit projects.

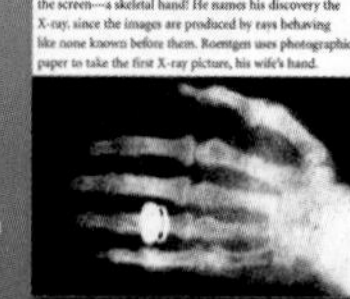

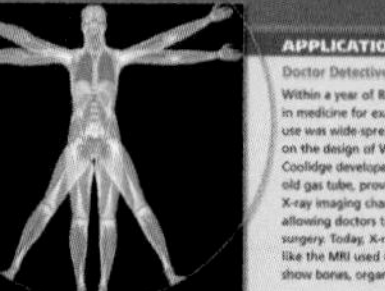

Timelines in Science

- Traces the history of key scientific discoveries
- Highlights interactions between science and technology.

Human Biology Pacing Guide

The following pacing guide shows how the chapters in ***Human Biology*** can be adapted to fit your specific course needs.

	TRADITIONAL SCHEDULE (DAYS)	BLOCK SCHEDULE (DAYS)
Frontiers in Science: Surprising Senses	1	0.5
Chapter 1 Systems, Support, and Movement		
1.1 The human body is complex.	2	1
1.2 The skeletal system provides support and protection.	2	1
1.3 The muscular system makes movement possible.	3	1.5
Chapter Investigation	1	0.5
Chapter 2 Absorption, Digestion, and Exchange		
2.1 The respiratory system gets oxygen and removes carbon dioxide.	2	1
2.2 The digestive system breaks down food.	2	1
2.3 The urinary system removes waste materials.	3	1.5
Chapter Investigation	1	0.5
Chapter 3 Transport and Protection		
3.1 The circulatory system transports materials.	2	1
3.2 The immune system defends the body.	2	1
3.3 The integumentary system shields the body.	3	1.5
Chapter Investigation	1	0.5
Timelines in Science: Seeing Inside the Body	1	0.5
Chapter 4 Control and Reproduction		
4.1 The nervous system responds and controls.	2	1
4.2 The endocrine system helps regulate body conditions.	2	1
4.3 The reproductive system allows the production of offspring.	3	1.5
Chapter Investigation	1	0.5
Chapter 5 Growth, Development, and Health		
5.1 The human body changes over time.	2	1
5.2 Systems in the body function to maintain health.	2	1
5.3 Science helps people prevent and treat disease.	3	1.5
Chapter Investigation	1	0.5
Total Days for Module	**42**	**21**

Planning the Chapter

Complete planning support precedes each chapter.

Previewing Content

- Section-by-section science background notes
- Common Misconceptions notes

CHAPTER

1 Systems, Support, and Movement

Life Science
UNIFYING PRINCIPLES

PRINCIPLE 1	PRINCIPLE 2	PRINCIPLE 3	PRINCIPLE 4
All living things share common characteristics.	All living things share common needs.	Living thing… their needs… interactions… environmen…	

Unit: Human Biology
BIG IDEAS

CHAPTER 1	CHAPTER 2	CHAPTER 3
Systems, Support, and Movement The human body is made up of systems that work together to perform necessary functions.	**Absorption, Digestion, and Exchange** Systems in the body obtain and process materials and remove waste.	**Transport and Protection** Systems function to transport materials and to defend and protect the body.

CHAPTER 1
KEY CONCEPTS

SECTION 1.1	SECTION 1.2
The human body is complex. 1. The body has cells, tissues, and organs. 2. The body's systems interact with one another.	**The skeletal system provides support and protection.** 1. Bones are living tissue. 2. The skeleton is the body's framewo… 3. The skeleton changes as the body develops and ages. 4. Joints connect parts of the skeletal system.

The Big Idea Flow Chart is available on p. T1 in the **UNIT TRANSPARENCY BOOK**

B 5A Unit: **Human Biology**

Previewing Content

SECTION

1.1 **The human body is complex.** pp. 9–13

1. **The body has cells, tissues, and organs.**
The cell is the basic unit of life, and the human body has many specialized cells that perform particular functions. A **tissue** is a group of cells that function together. There are four categories of tissue:
- epithelial, which forms boundaries
- muscle, which enables movement

SECTION 1.2

1. **Bones**
The **sk**… mainly
- **Com**…
- **Spor**…

Previewing Content

SECTION

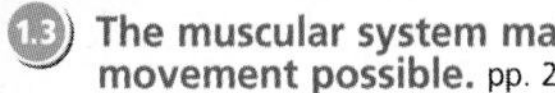

1.3 **The muscular system makes movement possible.** pp. 22–29

1. **Muscles perform important functions.**
Muscles are made of cells called muscle fibers. They perform three functions.
- Muscles move the body.
- Muscles help maintain the body's temperature by generating heat.
- Muscles maintain upright posture.

2. **Your body has different types of muscle.**
Skeletal muscles are also called **voluntary muscles.** They allow the body to move when you want it to.
- Slow-twitch muscles allow slow movement.
- Fast-twitch muscles allow rapid movement.

Smooth muscles, also called **involuntary muscles,** are found inside organs and enable the body to perform automatic movements.

Cardiac muscle makes up the heart. Its cells are arranged in webs that contract without conscious control to make the heart beat.

3. **Skeletal muscles and tendons allow bones to move.**
Tendons attach muscles to bones. Muscles work in pairs to cause bones to move. One muscle contracts, pulling a bone, while the other lengthens.

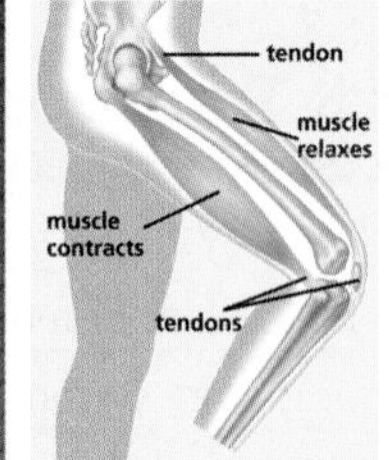

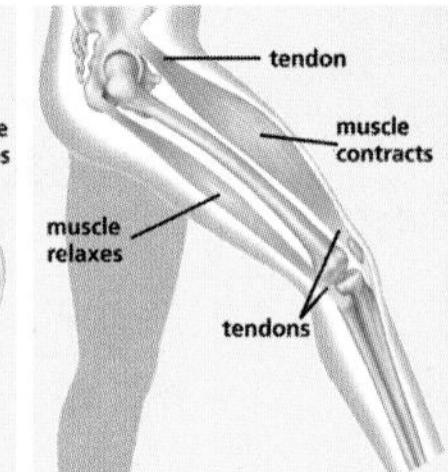

Muscles that cause movement always work in pairs or groups whose movements oppose each other.

4. **Muscles grow and heal.**
Muscles develop with age. Exercise increases muscle size.

Common Misconceptions

MUSCLES CONTRACT Most middle school students think that muscles can push and pull. In fact, muscles can only pull, which they do by contracting.

This misconception is addressed on p. 26.

MISCONCEPTION DATABASE
CLASSZONE.COM Background on student misconceptions

B 5C Unit: **Human Biology**

Previewing Chapter Resources

- Section-by-section listing of all print and technology resources
- Suggested pacing
- Correlations to National Science Education Standards

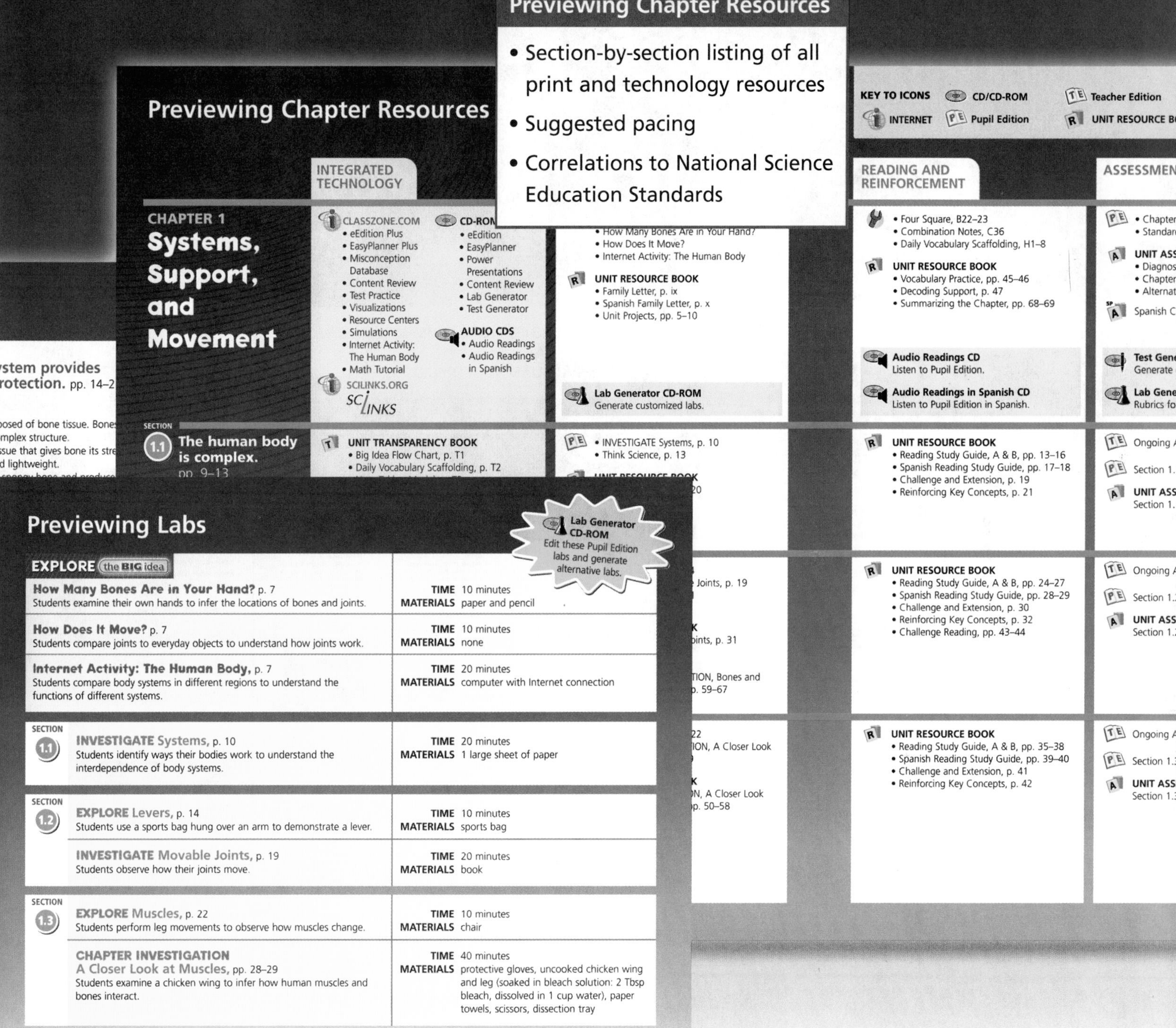

Previewing Chapter Resources

KEY TO ICONS: CD/CD-ROM; Teacher Edition; INTERNET; Pupil Edition; UNIT RESOURCE BOOK

CHAPTER 1
Systems, Support, and Movement

INTEGRATED TECHNOLOGY

CLASSZONE.COM
- eEdition Plus
- EasyPlanner Plus
- Misconception Database
- Content Review
- Test Practice
- Visualizations
- Resource Centers
- Simulations
- Internet Activity: The Human Body
- Math Tutorial

SCILINKS.ORG

CD-ROM
- eEdition
- EasyPlanner
- Power Presentations
- Content Review
- Lab Generator
- Test Generator

AUDIO CDS
- Audio Readings
- Audio Readings in Spanish

READING AND REINFORCEMENT
- Four Square, B22–23
- Combination Notes, C36
- Daily Vocabulary Scaffolding, H1–8

UNIT RESOURCE BOOK
- Vocabulary Practice, pp. 45–46
- Decoding Support, p. 47
- Summarizing the Chapter, pp. 68–69

Audio Readings CD
Listen to Pupil Edition.

Audio Readings in Spanish CD
Listen to Pupil Edition in Spanish.

- How Many Bones Are in Your Hand?
- How Does It Move?
- Internet Activity: The Human Body

UNIT RESOURCE BOOK
- Family Letter, p. ix
- Spanish Family Letter, p. x
- Unit Projects, pp. 5–10

Lab Generator CD-ROM
Generate customized labs.

SECTION 1.1 The human body is complex. pp. 9–13

UNIT TRANSPARENCY BOOK
- Big Idea Flow Chart, p. T1
- Daily Vocabulary Scaffolding, p. T2

- INVESTIGATE Systems, p. 10
- Think Science, p. 13

UNIT RESOURCE BOOK
- Reading Study Guide, A & B, pp. 13–16
- Spanish Reading Study Guide, pp. 17–18
- Challenge and Extension, p. 19
- Reinforcing Key Concepts, p. 21

UNIT RESOURCE BOOK
- Reading Study Guide, A & B, pp. 24–27
- Spanish Reading Study Guide, pp. 28–29
- Challenge and Extension, p. 30
- Reinforcing Key Concepts, p. 32
- Challenge Reading, pp. 43–44

UNIT RESOURCE BOOK
- Reading Study Guide, A & B, pp. 35–38
- Spanish Reading Study Guide, pp. 39–40
- Challenge and Extension, p. 41
- Reinforcing Key Concepts, p. 42

Previewing Labs

Lab Generator CD-ROM
Edit these Pupil Edition labs and generate alternative labs.

EXPLORE the BIG idea

How Many Bones Are in Your Hand? p. 7 Students examine their own hands to infer the locations of bones and joints.	TIME 10 minutes MATERIALS paper and pencil
How Does It Move? p. 7 Students compare joints to everyday objects to understand how joints work.	TIME 10 minutes MATERIALS none
Internet Activity: The Human Body, p. 7 Students compare body systems in different regions to understand the functions of different systems.	TIME 20 minutes MATERIALS computer with Internet connection
SECTION 1.1 **INVESTIGATE Systems,** p. 10 Students identify ways their bodies work to understand the interdependence of body systems.	TIME 20 minutes MATERIALS 1 large sheet of paper
SECTION 1.2 **EXPLORE Levers,** p. 14 Students use a sports bag hung over an arm to demonstrate a lever.	TIME 10 minutes MATERIALS sports bag
INVESTIGATE Movable Joints, p. 19 Students observe how their joints move.	TIME 20 minutes MATERIALS book
SECTION 1.3 **EXPLORE Muscles,** p. 22 Students perform leg movements to observe how muscles change.	TIME 10 minutes MATERIALS chair
CHAPTER INVESTIGATION A Closer Look at Muscles, pp. 28–29 Students examine a chicken wing to infer how human muscles and bones interact.	TIME 40 minutes MATERIALS protective gloves, uncooked chicken wing and leg (soaked in bleach solution: 2 Tbsp bleach, dissolved in 1 cup water), paper towels, scissors, dissection tray

Additional INVESTIGATION, Bones and Calcium, A, B, & C, pp. 59–67; Teacher Instructions, pp. 305–306

Previewing Labs

- Brief descriptions of all chapter labs and activities
- Time and materials required for each activity

Chapter 1: **Systems, Support, and Movement 5D B**

Planning the Lesson

Point-of-use support for each lesson provides a wealth of teaching options.

1. Prepare

- Concept and vocabulary review
- Note-taking and vocabulary strategies

2. Focus

- Set Learning Goals
- 3-Minute Warm-up

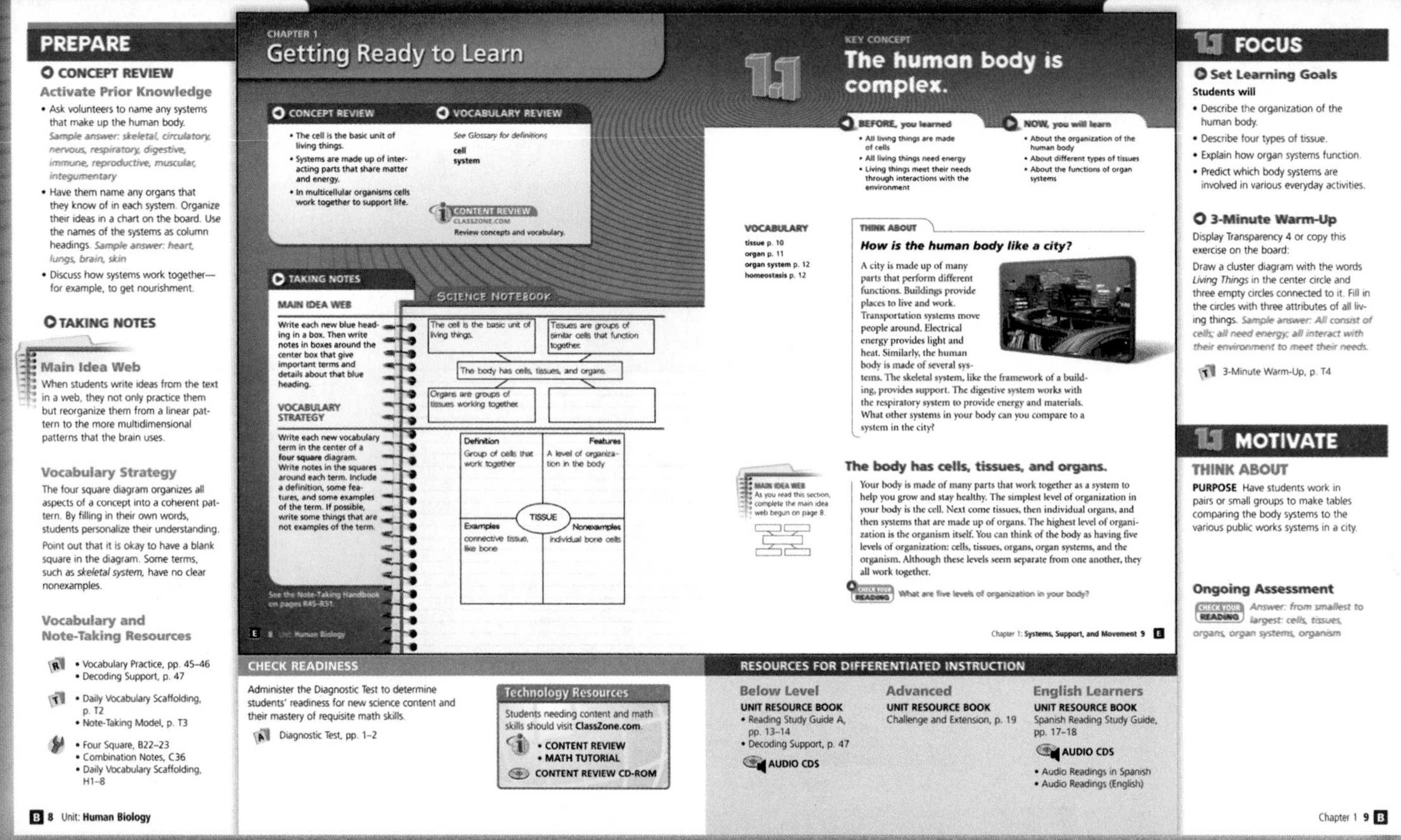

PREPARE

CONCEPT REVIEW

Activate Prior Knowledge

- Ask volunteers to name any systems that make up the human body. *Sample answer: skeletal, circulatory, nervous, respiratory, digestive, immune, reproductive, muscular, integumentary*
- Have them name any organs that they know of in each system. Organize their ideas in a chart on the board. Use the names of the systems as column headings. *Sample answer: heart, lungs, brain, skin*
- Discuss how systems work together—for example, to get nourishment.

TAKING NOTES

Main Idea Web

When students write ideas from the text in a web, they not only practice them but reorganize them from a linear pattern to the more multidimensional patterns that the brain uses.

Vocabulary Strategy

The four square diagram organizes all aspects of a concept into a coherent pattern. By filling in their own words, students personalize their understanding.

Point out that it is okay to have a blank square in the diagram. Some terms, such as *skeletal system,* have no clear nonexamples.

Vocabulary and Note-Taking Resources

- Vocabulary Practice, pp. 45–46
- Decoding Support, p. 47

- Daily Vocabulary Scaffolding, p. T2
- Note-Taking Model, p. T3

- Four Square, B22–23
- Combination Notes, C36
- Daily Vocabulary Scaffolding, H1–8

B 8 Unit: **Human Biology**

CHAPTER 1

Getting Ready to Learn

CONCEPT REVIEW

- The cell is the basic unit of living things.
- Systems are made up of interacting parts that share matter and energy.
- In multicellular organisms cells work together to support life.

VOCABULARY REVIEW

See Glossary for definitions

cell
system

CONTENT REVIEW CLASSZONE.COM
Review concepts and vocabulary.

TAKING NOTES

MAIN IDEA WEB

Write each new blue heading in a box. Then write notes in boxes around the center box that give important terms and details about that blue heading.

VOCABULARY STRATEGY

Write each new vocabulary term in the center of a **four square** diagram. Write notes in the squares around each term. Include a definition, some features, and some examples of the term. If possible, write some things that are not examples of the term.

See the Note-Taking Handbook on pages R45–R51.

SCIENCE NOTEBOOK

The cell is the basic unit of living things.
Tissues are groups of similar cells that function together.
The body has cells, tissues, and organs.
Organs are groups of tissues working together.
Definition
Group of cells that work together
Features
A level of organization in the body
TISSUE
Examples
connective tissue, like bone
Nonexamples
individual bone cells

E 8 Unit: Human Biology

KEY CONCEPT

1.1 The human body is complex.

BEFORE, you learned

- All living things are made of cells
- All living things need energy
- Living things meet their needs through interactions with the environment

NOW, you will learn

- About the organization of the human body
- About different types of tissues
- About the functions of organ systems

VOCABULARY

tissue p. 10
organ p. 11
organ system p. 12
homeostasis p. 12

THINK ABOUT

How is the human body like a city?

A city is made up of many parts that perform different functions. Buildings provide places to live and work. Transportation systems move people around. Electrical energy provides light and heat. Similarly, the human body is made of several systems. The skeletal system, like the framework of a building, provides support. The digestive system works with the respiratory system to provide energy and materials. What other systems in your body can you compare to a system in the city?

MAIN IDEA WEB
As you read this section, complete the main idea web begun on page 8.

The body has cells, tissues, and organs.

Your body is made of many parts that work together as a system to help you grow and stay healthy. The simplest level of organization in your body is the cell. Next come tissues, then individual organs, and then systems that are made up of organs. The highest level of organization is the organism itself. You can think of the body as having five levels of organization: cells, tissues, organs, organ systems, and the organism. Although these levels seem separate from one another, they all work together.

CHECK YOUR READING What are five levels of organization in your body?

Chapter 1: **Systems, Support, and Movement** 9 E

1.1 FOCUS

Set Learning Goals

Students will

- Describe the organization of the human body.
- Describe four types of tissue.
- Explain how organ systems function.
- Predict which body systems are involved in various everyday activities.

3-Minute Warm-Up

Display Transparency 4 or copy this exercise on the board:

Draw a cluster diagram with the words *Living Things* in the center circle and three empty circles connected to it. Fill in the circles with three attributes of all living things. *Sample answer: All consist of cells; all need energy; all interact with their environment to meet their needs.*

3-Minute Warm-Up, p. T4

1.1 MOTIVATE

THINK ABOUT

PURPOSE Have students work in pairs or small groups to make tables comparing the body systems to the various public works systems in a city.

Ongoing Assessment

CHECK YOUR READING *Answer: from smallest to largest: cells, tissues, organs, organ systems, organism*

Chapter 1 9 B

CHECK READINESS

Administer the Diagnostic Test to determine students' readiness for new science content and their mastery of requisite math skills.

Diagnostic Test, pp. 1–2

Technology Resources

Students needing content and math skills should visit **ClassZone.com**.

- CONTENT REVIEW
- MATH TUTORIAL

CONTENT REVIEW CD-ROM

RESOURCES FOR DIFFERENTIATED INSTRUCTION

Below Level
UNIT RESOURCE BOOK
- Reading Study Guide A, pp. 13–14
- Decoding Support, p. 47

AUDIO CDS

Advanced
UNIT RESOURCE BOOK
Challenge and Extension, p. 19

English Learners
UNIT RESOURCE BOOK
Spanish Reading Study Guide, pp. 17–18

AUDIO CDS
- Audio Readings in Spanish
- Audio Readings (English)

3. Motivate

- Engaging entry into the section
- Explore activity or Think About question

1.1 INSTRUCT

INVESTIGATE Systems

PURPOSE To understand how body systems work together

TIPS *20 min.*

- Allow students to use the chart from "Activate Prior Knowledge," p. 8.
- Remind students to be aware of their breathing and their hearts beating.
- Suggest that students bring in pictures of people performing everyday tasks, for follow-up discussion.

WHAT DO YOU THINK? *Sample answers: Talking on the phone involves the nervous system, the respiratory system, and the muscular system. When you sleep, your body breathes, excretes waste through the skin and the mouth, moves, and dreams.*

CHALLENGE *Answers will vary, but should group activities appropriately.*

Datasheet, Systems, p. 20

Technology Resources

Customize this student lab as needed or look for an alternative. Print rubrics to assess student lab reports.

Lab Generator CD-ROM

Develop Critical Thinking

COMPARE Have students list ways in which people are like cells. *No two are exactly alike, both grow and can repair themselves, both have parts that serve to control all functions, both are living and have special jobs, both use food and oxygen to produce energy, both expel wastes, and both will eventually die.*

Ongoing Assessment

Describe the organization of the human body.

Ask: Why are the cells of complex organisms specialized? *Specialized cells*

INVESTIGATE Systems

How do the systems in your body interact?

SKILL FOCUS Predicting

MATERIALS large sheet of paper

TIME 20 minutes

PROCEDURE

1. Work with other classmates to make a list of everyday activities.
2. Discuss how your body responds to each task. Record your ideas.
3. Identify and count the systems in your body that you think are used to perform the task.
4. Have someone from your group make a chart of the different activities.

WHAT DO YOU THINK?

- Which systems did you name, and how did they work together to perform each activity?
- When you are asleep, what activities does your body perform?

CHALLENGE How could you make an experiment that would test your predictions?

Cells

A The cell is the basic unit of life. Cells make up all living things. Some organisms, such as bacteria, are made of only a single cell. In these organisms the single cell performs all of the tasks necessary for survival. That individual cell captures and releases energy, uses materials, and grows. In more complex organisms, such as humans and many other animals and plants, cells are specialized. Specialized cells perform special jobs. A red blood cell, for example, carries oxygen and other nutrients from the lungs throughout the body.

Tissues

B A **tissue** is a group of similar cells that work together to perform a particular function. Think of a tissue as a brick wall and the cells within it as the individual bricks. Taken together, the bricks form something larger and more functional. But just as the bricks need to be placed in a certain way to form the wall, cells must be organized in a tissue.

CHECK YOUR READING How are cells related to tissues?

The human body contains several types of tissues. These tissues are classified into four main groups according to their function: epithelial tissue, nerve tissue, muscle tissue, and connective tissue.

B 10 Unit: Human Biology

Epithelial (EHP-uh-THEE-lee-uhl) tissue functions as a boundary. It covers all of the inner and outer surfaces of your body. Each of your internal organs is covered with a layer of epithelial tissue. Another type of tissue, nerve tissue, functions as a messaging system. Cells in nerve tissue carry electrical impulses between your brain and the various parts of your body in response to changing conditions.

C Muscle tissue functions in movement. Movement results when muscle cells contract, or shorten, and then relax. In some cases, such as throwing a ball, you control the movement. In other cases, such as the beating of your heart, the movement occurs without conscious control. Connective tissue functions to hold parts of the body together, providing support, protection, strength, padding, and insulation. Tendons and ligaments are connective tissues that hold bones and muscles together. Bone itself is another connective tissue. It supports and protects the soft parts of your body.

Organs

D Groups of different tissues make up organs. An **organ** is a structure that is made up of two or more types of tissue that work together to carry out a function in the body. For example, the structures that carry blood around your body contain all four types of tissues. As in cells and tissues, the structure of an organ relates to its function. The stomach's bag-shaped structure and strong muscular walls make it suited for breaking down food. The walls of the heart are also muscular, allowing it to function as a pump.

Levels of Organization

The human body can be studied at different levels of organization.

250 x; Cells (muscle cells); Tissue (cardiac muscle); Organ (heart); Organ system (circulatory system); Organism (human)

Chapter 1: Systems, Support, and Movement 11 B

Address Misconceptions

IDENTIFY Ask: What makes your heart beat? If students respond that the heartbeat is the result of a natural rhythm started at birth, they may hold the misconception that the body does things without needing any internal communications.

CORRECT Have students take their pulse before and after running in place for one minute, noting that their heart rate changes. Point out that the nervous system tells the heart that the legs are moving and that it needs to beat faster.

REASSESS Ask: How do you know when you need to eat? *The brain monitors the body, realizes that the body needs nourishment, and makes me feel hungry.*

Technology Resources

Visit **ClassZone.com** for background on common student misconceptions.

MISCONCEPTION DATABASE

Metacognitive Strategy

Have students devise a strategy for remembering the four tissue types. Possibilities include a graphic organizer, mnemonics, and pictures.

Teach Difficult Concepts

Students may have trouble understanding that epithelial tissue serves as a boundary. Discuss a familiar type of boundary, such as room walls. Ask:

- What do room boundaries do? *keep heat in or out, weather and noise out*
- What do epithelial tissues, such as skin, do? *keep heat and fluid in, bacteria and other unwanted things out*

Ongoing Assessment

Describe four types of tissue.

Ask: What are the four types of tissue and their functions? *Epithelial tissues act as boundaries; nerves carry messages; muscles enable movement; connective tissues protect, insulate, and*

Chapter 1 11 B

DIFFERENTIATE INSTRUCTION

More Reading Support

A Give an example of a specialized cell. *red blood cell*

English Learners Help English learners with different word forms. For example, in the "Cells" paragraph, p. 10, *make* is used in a variety of forms: "Cells make up all living things;" "Some organisms, such as bacteria, are made of only a single ... be found as *make, made, make up, made* ... *made of.*

DIFFERENTIATE INSTRUCTION

More Reading Support

C What type of tissue provides support, protection, strength, and padding? *connective tissue*

D What do you call tis-...

Advanced Although this module presents the biology of a healthy human body, everyone has strengths and weaknesses. Explain that survival and high achievement are not dependent on full functioning of every organ. Discuss successful individuals with disabilities. Examples: Sir Steve Redgrave, Olympic gold medal for rowing: diabetes, dyslexia; Sir Stephen Hawking, theoretical physicist: amyotrophic lateral sclerosis (ALS); Andrea ...

Teach Difficult Concepts

Some students may have a hard time understanding homeostasis. Use a thermostat as an example. Point out that the heat or air conditioning is supplied just long enough to return an area to the set temperature.

EXPLORE (the BIG idea)

Revisit "Internet Activity: The Human Body" on p. 7. Have students discuss how body structures are related to their functions.

Ongoing Assessment

Explain how organ systems function.

Ask: How do organ systems meet the body's needs? *Each organ system performs a necessary function, and together the systems allow the human organism to grow, reproduce, and maintain life.*

Reinforce (the BIG idea)

Have students relate the section to the Big Idea.

Reinforcing Key Concepts, p. 21

1.1 ASSESS & RETEACH

Assess

Section 1.1 Quiz, p. 3

Reteach

Have students draw a flow chart of the progression from cells to tissues to organs to organ systems. In the final drawing, students should make a rough outline of the human body with the systems labeled. Have volunteers share their drawings in a class discussion.

Technology Resources

Have students visit **ClassZone.com** for reteaching of Key Concepts.

CONTENT REVIEW

CONTENT REVIEW CD-ROM

B 12 Unit: Human Biology

Organ Systems

An **organ system** is a group of organs that together perform a function that helps the body meet its needs for energy and materials. For example, your stomach, mouth, throat, large and small intestines, liver, and pancreas are all part of the organ system called the digestive system. The body is made up of many organ systems. In this unit, you will read about these systems. They include the skeletal, muscular, respiratory, digestive, urinary, circulatory, immune, nervous, and reproductive systems. Together, these systems allow the human organism to grow, reproduce, and maintain life.

The body's systems interact with one another.

READING TIP VOCABULARY The word *homeostasis* contains two word roots. *Homeo* comes from a root meaning "same." *Stasis* comes from a root meaning "stand still" or "stay."

The ability of your body to maintain internal conditions is called **homeostasis** (HOH-mee-oh-STAY-sihs). Your body is constantly regulating such things as your body temperature, the amount of sugar in your blood, even your posture. The processes that take place in your body occur within a particular set of conditions.

The body's many levels of organization, from cells to organ systems, work constantly to maintain the balance needed for the survival of the organism. For example, on a hot day, you may sweat. Sweating keeps the temperature inside your body constant, even though the temperature of your surroundings changes.

INFER This student is drinking water after exercising. Why is it important to drink fluids after you sweat?

1.1 Review

KEY CONCEPTS

1. Draw a diagram that shows the relationship among cells, tissues, organs, and organ systems.
2. Make a chart of the four basic tissue groups that includes names, functions, and examples.
3. Identify three functions performed by organ systems.

CRITICAL THINKING

4. **Apply** How does drinking water after you sweat help maintain homeostasis?
5. **Compare and Contrast** Compare and contrast the four basic tissue groups. How would all four types of tissue be involved in a simple activity, like raising your hand?

CHALLENGE

6. **Apply** Describe an object, such as a car, that can be used as a model of the human body. Explain how the parts of the model relate to the body.

B 12 Unit: Human Biology

ANSWERS

1. Diagrams should show organ systems as groups of organs, as groups of tissues, as groups of cells.

2. Charts should include epithelial, nerve, muscle, and connective tissues, and show understanding of each.

3. Sample answer: digestion, elimination, reproduction

4. When you sweat, your body loses water. Adding new water to the body helps maintain homeostasis.

5. Sample answer: When you raise your hand, nerve tissue carries a message from your brain, and muscle tissue contracts to move the bones, made of connective and epithelial tissues.

6. Answers should mention that the parts work together for the functioning of the whole.

4. Instruct

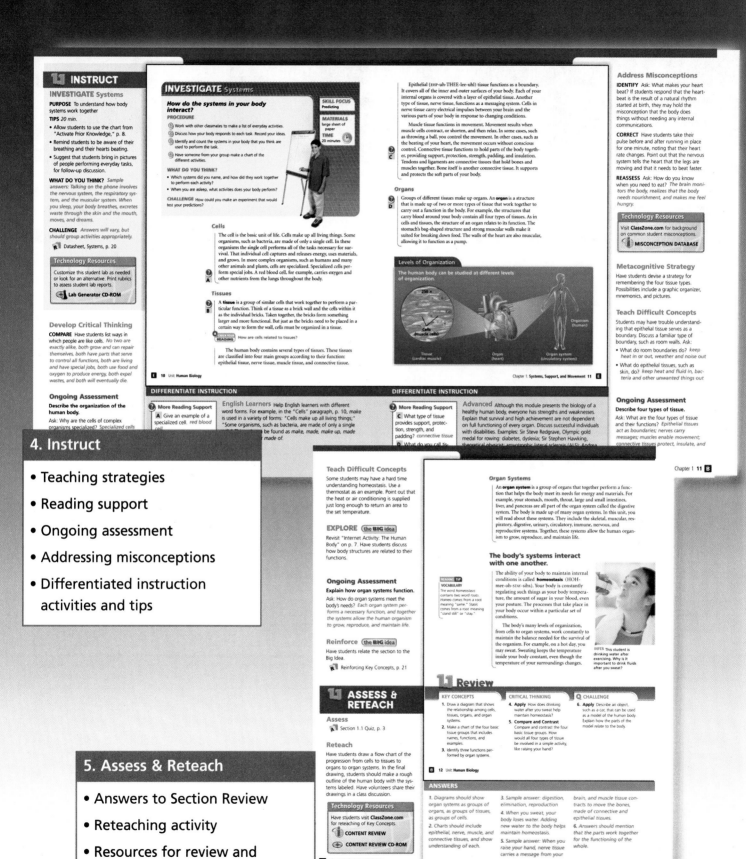

- Teaching strategies
- Reading support
- Ongoing assessment
- Addressing misconceptions
- Differentiated instruction activities and tips

5. Assess & Reteach

- Answers to Section Review
- Reteaching activity
- Resources for review and assessment

Lab Materials List

The following charts list the consumables, nonconsumables, and equipment needed for all activities. Quantities are per group of four students. Lab aprons, goggles, water, books, paper, pens, pencils, and calculators are assumed to be available for all activities.

Materials kits are available. For more information, please call McDougal Littell at 1-800-323-5435.

Consumables

Description	Quantity per Group	Explore *page*	Investigate *page*	Chapter Investigation *page*
albumen solution, 1%	20 mL			56
apple	1	83		
bacterial amylase	20 mL		47	
bag, clear plastic	4–6	52		
bag, paper lunch	3	101		
bag, zip-top sandwich	1	74		
balloon, medium	1		39	
bleach	2–4 cups			28
bottle, 1 liter, clear plastic	1		39	
chicken leg, uncooked	1			28
chicken wing, uncooked	1			28
cinnamon	1–3 sticks	101		
cloth, white cotton, 12" x 12"	1	74		
cotton ball	4–6		85	
cotton swab, sterile	2			108
cup, paper	1			108
filter paper, fine, 4"	1			56
food coloring, blue	1 bottle	74		108
glitter	4–6 tbs	148		
glucose solution, 1%	20 mL			56
hand soap	20 mL			154
Isopropyl alcohol	50 mL		85, 114	
knife, plastic	1		47	
liquid dish detergent	5 mL	45		
marker, permanent black	1		79	154
nutrient agar, sterile	100 mL			154
nutrition label, bag of fresh carrots	1		143	
nutrition label, canned spaghetti in sauce	1		143	
nutrition label, carbonated soft drink	1		143	
nutrition label, fruit juice	1		143	

Description	Quantity per Group	Explore *page*	Investigate *page*	Chapter Investigation *page*
nutrition label, plain popcorn kernel	1		143	
nutrition label, potato chips	1		143	
nutrition label, unsweetened apple sauce	1		143	
orange peel	1	101		
paper, graph, 8.5" x 11"	4	133	114, 137	72
paper, newsprint, 24" x 36"	1		10	
paper, white, 8.5" x 11"	1			108
paper towel	5			28, 108
pin	1	74		
pine branch, small	1	101		
potato, raw	1		47	
reinforcement ring, binder paper	1			108
salt solution, 1%	20 mL			56
slide, egg cell	1	118		
slide, sperm cell	1	118		
tape, duct	1 roll		39	
tape, masking	1 roll	52	79	154
test strip, glucose	2			56
test strip, protein	2			56
test strip, salinity	2			56
vegetable oil	5 mL	45		

Non-Consumables

Description	Quantity per Group	Explore *page*	Investigate *page*	Chapter Investigation *page*
beaker, 100 mL	1			56
bowl, large plastic	1	74		
computer graphing program	1		137	
container, small plastic with lid	4–6		79	
eyedropper	2	45	47, 85	
funnel, small plastic	1			56
graduated cylinder, 50 mL	1			56
hand lens	1			154
microscope	1	118		
petri dish with lid	3			154
scissors	1			28
stopwatch	1	52, 65	114	72
tape measure, flexible	1	133		
test tube rack	1	45		
test tube with stopper	1	45		
thermometer, mercury-free	1		114	
tray, dissection	1			28
vegetable peeler	1	83		

Safety Equipment

Description		Explore *page*	Investigate *page*	Chapter Investigation *page*
gloves				28

McDougal Littell Science

Human Biology

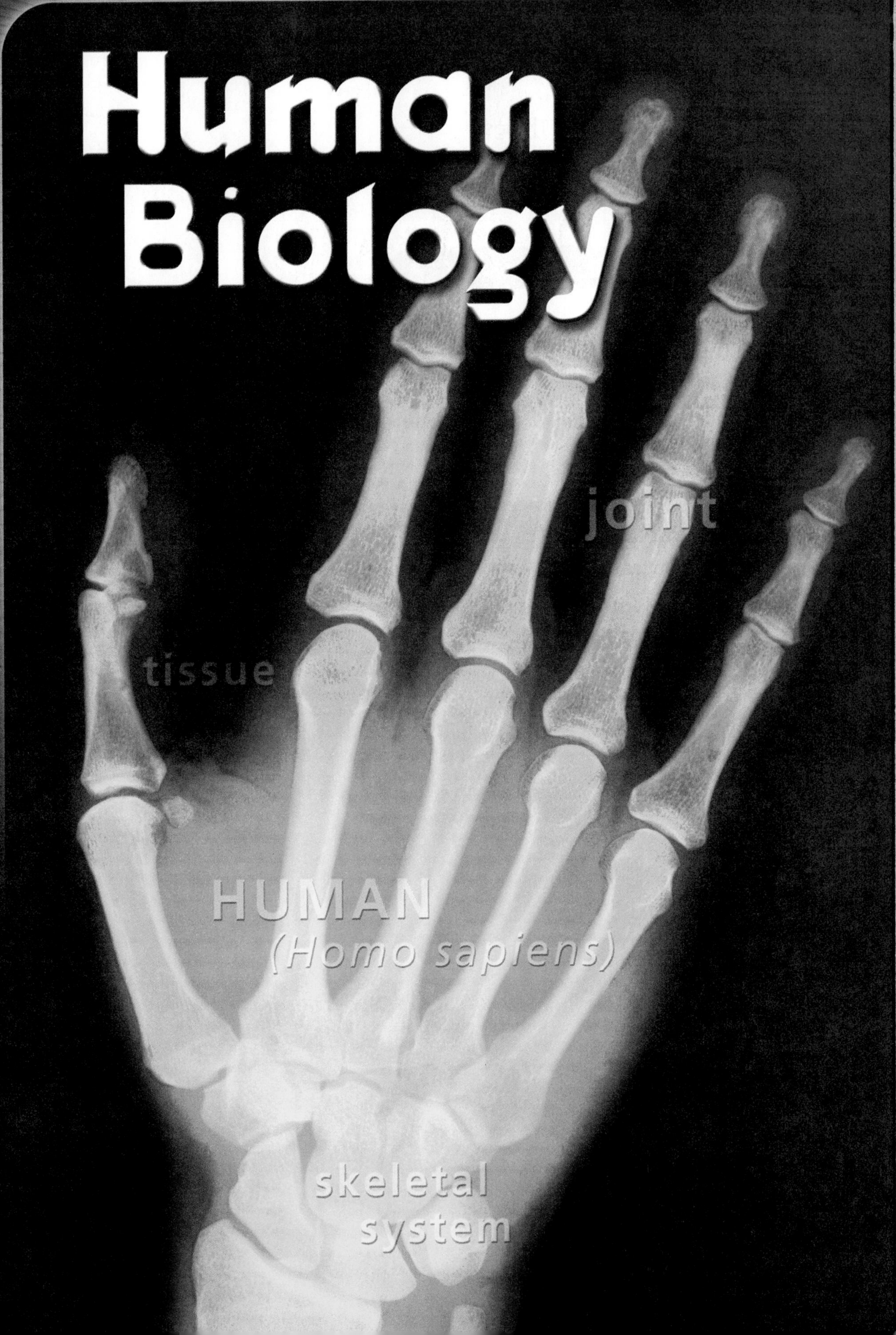

joint

tissue

HUMAN
(*Homo sapiens*)

skeletal
system

LIFE SCIENCE

A ▶ Cells and Heredity
B ▶ Life Over Time
C ▶ Diversity of Living Things
D ▶ Ecology
E ▶ Human Biology

EARTH SCIENCE

A ▶ Earth's Surface
B ▶ The Changing Earth
C ▶ Earth's Waters
D ▶ Earth's Atmosphere
E ▶ Space Science

PHYSICAL SCIENCE

A ▶ Matter and Energy
B ▶ Chemical Interactions
C ▶ Motion and Forces
D ▶ Waves, Sound, and Light
E ▶ Electricity and Magnetism

ISBN: 0-618-33431-9 1 2 3 4 5 6 7 8 VJM 08 07 06 05 04

Internet Web Site: http://www.mcdougallittell.com

ii

Science Consultants

Chief Science Consultant

James Trefil, Ph.D. is the Clarence J. Robinson Professor of Physics at George Mason University. He is the author or co-author of more than 25 books, including *Science Matters* and *The Nature of Science.* Dr. Trefil is a member of the American Association for the Advancement of Science's Committee on the Public Understanding of Science and Technology. He is also a fellow of the World Economic Forum and a frequent contributor to *Smithsonian* magazine.

Rita Ann Calvo, Ph.D. is Senior Lecturer in Molecular Biology and Genetics at Cornell University, where for 12 years she also directed the Cornell Institute for Biology Teachers. Dr. Calvo is the 1999 recipient of the College and University Teaching Award from the National Association of Biology Teachers.

Kenneth Cutler, M.S. is the Education Coordinator for the Julius L. Chambers Biomedical Biotechnology Research Institute at North Carolina Central University. A former middle school and high school science teacher, he received a 1999 Presidential Award for Excellence in Science Teaching.

Instructional Design Consultants

Douglas Carnine, Ph.D. is Professor of Education and Director of the National Center for Improving the Tools of Educators at the University of Oregon. He is the author of seven books and over 100 other scholarly publications, primarily in the areas of instructional design and effective instructional strategies and tools for diverse learners. Dr. Carnine also serves as a member of the National Institute for Literacy Advisory Board.

Linda Carnine, Ph.D. consults with school districts on curriculum development and effective instruction for students struggling academically. A former teacher and school administrator, Dr. Carnine also co-authored a popular remedial reading program.

Sam Miller, Ph.D. is a middle school science teacher and the Teacher Development Liaison for the Eugene, Oregon, Public Schools. He is the author of curricula for teaching science, mathematics, computer skills, and language arts.

Donald Steely, Ph.D. serves as principal investigator at the Oregon Center for Applied Science (ORCAS) on federal grants for science and language arts programs. His background also includes teaching and authoring of print and multimedia programs in science, mathematics, history, and spelling.

Vicky Vachon, Ph.D. consults with school districts throughout the United States and Canada on improving overall academic achievement with a focus on literacy. She is also co-author of a widely used program for remedial readers.

iii

Content Reviewers

John Beaver, Ph.D.
Ecology
Professor, Director of Science Education Center
College of Education and Human Services
Western Illinois University
Macomb, IL

Donald J. DeCoste, Ph.D.
Matter and Energy, Chemical Interactions
Chemistry Instructor
University of Illinois
Urbana-Champaign, IL

Dorothy Ann Fallows, Ph.D., MSc
Diversity of Living Things, Microbiology
Partners in Health
Boston, MA

Michael Foote, Ph.D.
The Changing Earth, Life Over Time
Associate Professor
Department of the Geophysical Sciences
The University of Chicago
Chicago, IL

Lucy Fortson, Ph.D.
Space Science
Director of Astronomy
Adler Planetarium and Astronomy Museum
Chicago, IL

Elizabeth Godrick, Ph.D.
Human Biology
Professor, CAS Biology
Boston University
Boston, MA

Isabelle Sacramento Grilo, M.S.
The Changing Earth
Lecturer, Department of the Geological Sciences
Montana State University
Bozeman, MT

David Harbster, MSc
Diversity of Living Things
Professor of Biology
Paradise Valley Community College
Phoenix, AZ

Richard D. Norris, Ph.D.
Earth's Waters
Professor of Paleobiology
Scripps Institution of Oceanography
University of California, San Diego
La Jolla, CA

Donald B. Peck, M.S.
Motion and Forces; Waves, Sound, and Light; Electricity and Magnetism
Director of the Center for Science Education (retired)
Fairleigh Dickinson University
Madison, NJ

Javier Penalosa, Ph.D.
Diversity of Living Things, Plants
Associate Professor, Biology Department
Buffalo State College
Buffalo, NY

Raymond T. Pierrehumbert, Ph.D.
Earth's Atmosphere
Professor in Geophysical Sciences (Atmospheric Science)
The University of Chicago
Chicago, IL

Brian J. Skinner, Ph.D.
Earth's Surface
Eugene Higgins Professor of Geology and Geophysics
Yale University
New Haven, CT

Nancy E. Spaulding, M.S.
Earth's Surface, The Changing Earth, Earth's Waters
Earth Science Teacher (retired)
Elmira Free Academy
Elmira, NY

Steven S. Zumdahl, Ph.D.
Matter and Energy, Chemical Interactions
Professor Emeritus of Chemistry
University of Illinois
Urbana-Champaign, IL

Susan L. Zumdahl, M.S.
Matter and Energy, Chemical Interactions
Chemistry Education Specialist
University of Illinois
Urbana-Champaign, IL

Safety Consultant

Juliana Texley, Ph.D.
Former K–12 Science Teacher and School Superintendent
Boca Raton, FL

English Language Advisor

Judy Lewis, M.A.
Director, State and Federal Programs for reading proficiency and high risk populations
Rancho Cordova, CA

iv

Teacher Panel Members

Carol Arbour
Tallmadge Middle School,
Tallmadge, OH

Patty Belcher
Goodrich Middle School,
Akron, OH

Gwen Broestl
Luis Munoz Marin Middle School,
Cleveland, OH

Al Brofman
Tehipite Middle School,
Fresno, CA

John Cockrell
Clinton Middle School,
Columbus, OH

Jenifer Cox
Sylvan Middle School,
Citrus Heights, CA

Linda Culpepper
Martin Middle School,
Charlotte, NC

Kathleen Ann DeMatteo
Margate Middle School,
Margate, FL

Melvin Figueroa
New River Middle School,
Ft. Lauderdale, FL

Doretha Grier
Kannapolis Middle School,
Kannapolis, NC

Robert Hood
Alexander Hamilton Middle School,
Cleveland, OH

Scott Hudson
Coverdale Elementary School,
Cincinnati, OH

Loretta Langdon
Princeton Middle School,
Princeton, NC

Carlyn Little
Glades Middle School,
Miami, FL

Ann Marie Lynn
Amelia Earhart Middle School,
Riverside, CA

James Minogue
Lowe's Grove Middle School,
Durham, NC

Joann Myers
Buchanan Middle School,
Tampa, FL

Barbara Newell
Charles Evans Hughes Middle School,
Long Beach, CA

Anita Parker
Kannapolis Middle School,
Kannapolis, NC

Greg Pirolo
Golden Valley Middle School,
San Bernardino, CA

Laura Pottmyer
Apex Middle School,
Apex, NC

Lynn Prichard
Booker T. Washington Middle Magnet
School, Tampa, FL

Jacque Quick
Walter Williams High School,
Burlington, NC

Robert Glenn Reynolds
Hillman Middle School,
Youngstown, OH

Theresa Short
Abbott Middle School,
Fayetteville, NC

Rita Slivka
Alexander Hamilton Middle School,
Cleveland, OH

Marie Sofsak
B F Stanton Middle School,
Alliance, OH

Nancy Stubbs
Sweetwater Union Unified School District,
Chula Vista, CA

Sharon Stull
Quail Hollow Middle School,
Charlotte, NC

Donna Taylor
Okeeheelee Middle School,
West Palm Beach, FL

Sandi Thompson
Harding Middle School,
Lakewood, OH

Lori Walker
Audubon Middle School & Magnet Center,
Los Angeles, CA

Teacher Lab Evaluators

Jill Brimm-Byrne
Albany Park Academy,
Chicago, IL

Gwen Broestl
Luis Munoz Marin Middle School,
Cleveland, OH

Al Brofman
Tehipite Middle School,
Fresno, CA

Michael A. Burstein
The Rashi School,
Newton, MA

Trudi Coutts
Madison Middle School,
Naperville, IL

Stacy Covert
Lufkin Road Middle School,
Apex, NC

Jenifer Cox
Sylvan Middle School,
Citrus Heights, CA

Larry Cwik
Madison Middle School,
Naperville, IL

Jennifer Donatelli
Kennedy Junior High School,
Lisle, IL

Paige Fullhart
Highland Middle School,
Libertyville, IL

Sue Hood
Glen Crest Middle School,
Glen Ellyn, IL

Ann Min
Beardsley Middle School,
Crystal Lake, IL

Aileen Mueller
Kennedy Junior High School,
Lisle, IL

Nancy Nega
Churchville Middle School,
Elmhurst, IL

Oscar Newman
Sumner Math and Science Academy,
Chicago, IL

Marina Penalver
Moore Middle School,
Portland, ME

Lynn Prichard
Booker T. Washington Middle Magnet
School, Tampa, FL

Jacque Quick
Walter Williams High School,
Burlington, NC

Seth Robey
Gwendolyn Brooks Middle School,
Oak Park, IL

Kevin Steele
Grissom Middle School,
Tinley Park, IL

v

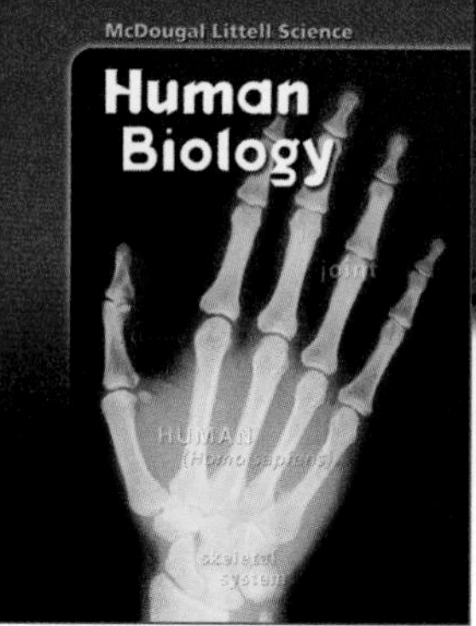

eEdition

Human Biology

Unit Features

What materials does your body need to function properly? page 34

vi

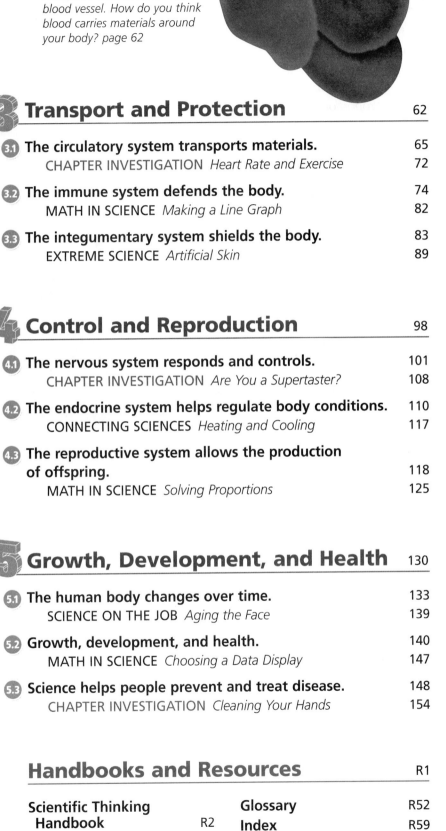

Red blood cells travel through a blood vessel. How do you think blood carries materials around your body? page 62

Features

Visual Highlights

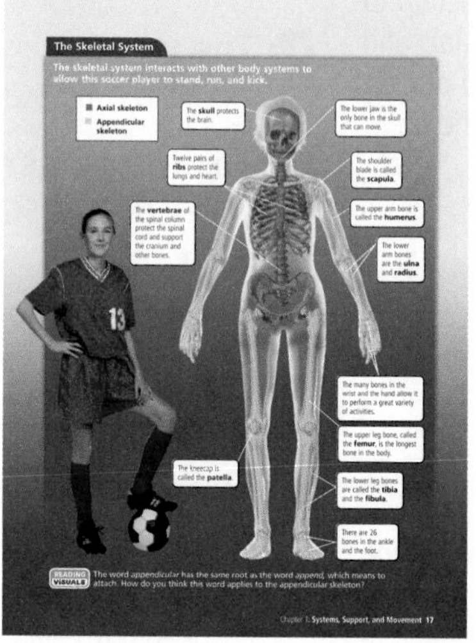

The Skeletal System
The skeletal system interacts with other body systems to allow this soccer player to stand, run, and kick.
Axial skeleton
Appendicular skeleton
The skull protects the brain.
The lower jaw is the only bone in the skull that can move.
Twelve pairs of ribs protect the lungs and heart.
The shoulder blade is called the scapula.
The vertebrae of the spinal column protect the spinal cord and support the cranium and other bones.
The upper arm bone is called the humerus.
The lower arm bones are the ulna and radius.
The many bones in the wrist and the hand allow it to perform a great variety of activities.
The upper leg bone, called the femur, is the longest bone in the body.
The kneecap is called the patella.
The lower leg bones are called the tibia and the fibula.
There are 26 bones in the ankle and the foot.
READING VISUALS The word appendicular has the same root as the word append, which means to attach. How do you think this word applies to the appendicular skeleton?
Chapter 1: Systems, Support, and Movement 17

Internet Resources @ ClassZone.com

INVESTIGATIONS AND ACTIVITIES

EXPLORE THE BIG IDEA

Chapter Opening Inquiries

CHAPTER INVESTIGATION

Full-Period Labs

EXPLORE

Introductory Inquiry Activities

INVESTIGATE

Skill Labs

ix

Standards and Benchmarks

Each chapter in **Human Biology** covers some of the learning goals that are described in the *National Science Education Standards* (NSES) and the Project 2061 *Benchmarks for Science Literacy*. Selected content and skill standards are show below in shortened form.The following National Science Education Standards are covered on pp. xii–xxvii, in Frontiers in Science, and in Timelines in Science, as well as in Ichapter features and laboratory investigations: Understandings About Scientific Inquiry (A.9), Understandings About Science and Technology (E.6), Science and Technology in Society (F.5), Nature of Science (G.2), and History of Science (G.3).

Content Standards

1 Systems, Support, and Movement

National Science Education Standards

C.1.a Levels of organization for living systems include: cells, tissues, organs, organ systems, whole organisms, and ecosystems.

C.1.d Specialized cells perform specialized functions in multicellular organisms.

C.1.e The human organism has systems: digestion, respiration, reproduction, circulation, excretion, movement, control and coordination, and protection.

Project 2061 Benchmarks

6.A.1 Like other animals, human beings have body systems for:
- obtaining and providing energy
- defense
- reproduction
- coordination of body functions

6.C.1 Organs and organ systems
- are made of cells
- help to provide all cells with basic needs

2 Absorption, Digestion, and Exchange

National Science Education Standards

C.1.e The human organism has systems: digestion, respiration, reproduction, circulation, excretion, movement, control and coordination, and protection.

F.1.e Food provides energy and nutrients for growth and development.

Project 2061 Benchmarks

6.C.2 For the body to use food energy and building materials, the food must first be digested into molecules that are absorbed and transported to cells.

6.C.3 Respiratory, urinary, and digestive systems remove wastes from the body.

3 Transport and Protection

National Science Education Standards

C.1.d Specialized cells perform specialized functions in multicellular organisms.

C.1.e The human organism has systems: digestion, respiration, reproduction, circulation, excretion, movement, control and coordination, and protection.

C.1.f Disease is a breakdown in structures or functions of an organism.

Project 2061 Benchmarks

6.C.3 The circulatory system moves substances to or from cells where they are needed or produced. It responds to changing demands.

6.C.4 Specialized cells and molecules they produce identify and destroy microbes.

6.E.4 White blood cells engulf invaders or produce antibodies that fight them.

4 Control and Reproduction

National Science Education Standards

C.1.d	Specialized cells perform specialized functions in multicellular organisms.
C.1.e	The human organism has systems: digestion, respiration, reproduction, circulation, excretion, movement, control and coordination, and protection.
C.2.a	Reproduction is essential to the continuation of a species.
C.2.b	Females produce eggs, and males sperm, which unite to begin a new individual.
C.2.c	Every organism requires a set of instructions for specifying its traits.

Project 2061 Benchmarks

5.B.2	In sexual reproduction, a single specialized cell from a female merges with a specialized cell from a male.
6.B.1	Fertilization occurs when one of the sperm cells from the male enters the egg cell from the female.
6.B.3	After fertilization, cells divide and specialize as a fetus grows from the embryo following a set pattern of development.
6.C.5	Hormones are chemicals from glands that help the body respond to danger and regulate human growth, development, and reproduction.
6.C.6	Interactions among the senses, nerves, and brain make possible the learning that enables human beings to cope with changes in their environment.

5 Growth, Development, and Health

National Science Education Standards

C.1.e	The human organism has systems: digestion, respiration, reproduction, circulation, excretion, movement, control and coordination, and protection.
C.1.f	Disease is a breakdown in structures or functions of an organism.
C.3.a	Organisms must obtain and use resources, grow, reproduce, and maintain stable internal conditions.
C.3.b	Regulation of an organism's internal environment involves interactions with its external environment.
F.1.c	The use of tobacco increases risk of illness.
F.1.d	Alcohol and other drugs are often abused and can lead to addiction.
F.1.e	Food provides energy and nutrients for growth and development.

Project 2061 Benchmarks

6.B.5	Changes occur as humans age. Many factors affect length and quality of life.
6.E.2	Toxic substances, diet, and behavior may harm one's health.
6.E.3	Viruses, bacteria, fungi, and parasites may interfere with the human body.
6.E.4	White blood cells engulf invaders or produce antibodies that fight them.

Skill Standards

National Science Education Standards

A.1	Identify questions that can be answered through scientific methods.
A.2	Design and conduct a scientific investigation.
A.3	Use appropriate tools and techniques to gather and interpret data.
A.4	Use evidence to describe, predict, explain, and model.
A.5	Think critically to find relationships between results and interpretations.
A.6	Give alternative explanations and predictions.
A.7	Communicate procedures, results, and conclusions.
A.8	Use mathematics in all aspects of scientific inquiry.
A.9.a	Different kinds of questions suggest different kinds of scientific investigations.
A.9.c	Mathematics is important in all aspects of scientific inquiry.
A.9.d	Scientific explanations emphasize evidence, have logically consistent arguments, and use scientific principles, models, and theories.
E.6.b	Many different people in different cultures have made and continue to make contributions to science and technology.
G.1.a	Women and men of various social and ethnic backgrounds engage in the activities of science.
G.1.b	Science requires different abilities. The work of science relies on basic human qualities, such as reasoning, insight, energy, skill, and creativity.

Project 2061 Benchmarks

1.C.1	Contributions to science have been made by different people, in different cultures, at different times.
12.A.1	Know why it is important in science to keep honest, clear, and accurate records.
12.A.2	Hypotheses are valuable, even if they turn out not to be true.
12.A.3	Different explanations can often be given for the same evidence.
12.C.3	Use appropriate units, use and read instruments that measure length, volume, weight, time, rate, and temperature.
12.D.1	Use tables and graphs to organize information and identify relationships.
12.D.2	Read, interpret, and describe tables and graphs.
12.D.3	Locate information in reference books and other resources.
12.D.4	Understand information that includes different types of charts and graphs, including circle charts, bar graphs, line graphs, data tables, diagrams, and symbols.

Introducing Life Science

Scientists are curious. Since ancient times, they have been asking and answering questions about the world around them. Scientists are also very suspicious of the answers they get. They carefully collect evidence and test their answers many times before accepting an idea as correct.

In this book you will see how scientific knowledge keeps growing and changing as scientists ask new questions and rethink what was known before. The following sections will help get you started.

UNIFYING PRINCIPLES of Life Science

What Is Life Science?

Life science is the study of living things. As you study life science, you will observe and read about a variety of organisms, from huge redwood trees to the tiny bacteria that cause sore throats. Because Earth is home to such a great variety of living things, the study of life science is rich and exciting.

But life science doesn't simply include learning the names of millions of organisms. It includes big ideas that help us to understand how all these livings things interact with their environment. Life science is the study of characteristics and needs that all living things have in common. It's also a study of changes—both daily changes as well as changes that take place over millions of years. Probably most important, in studying life science, you will explore the many ways that all living things—including you—depend upon Earth and its resources.

The text and visuals in this book will invite you into the world of living things and provide you with the key concepts you'll need in your study. Activities offer a chance for you to investigate some aspects of life science on your own. The four unifying principles listed below provide a way for you to connect the information and ideas in this program.

- **All living things share common characteristics.**
- **All living things share common needs.**
- **Living things meet their needs through interactions with the environment.**
- **The types and numbers of living things change over time.**

the BIG idea

Each chapter begins with a big idea. Keep in mind that each big idea relates to one or more of the unifying principles.

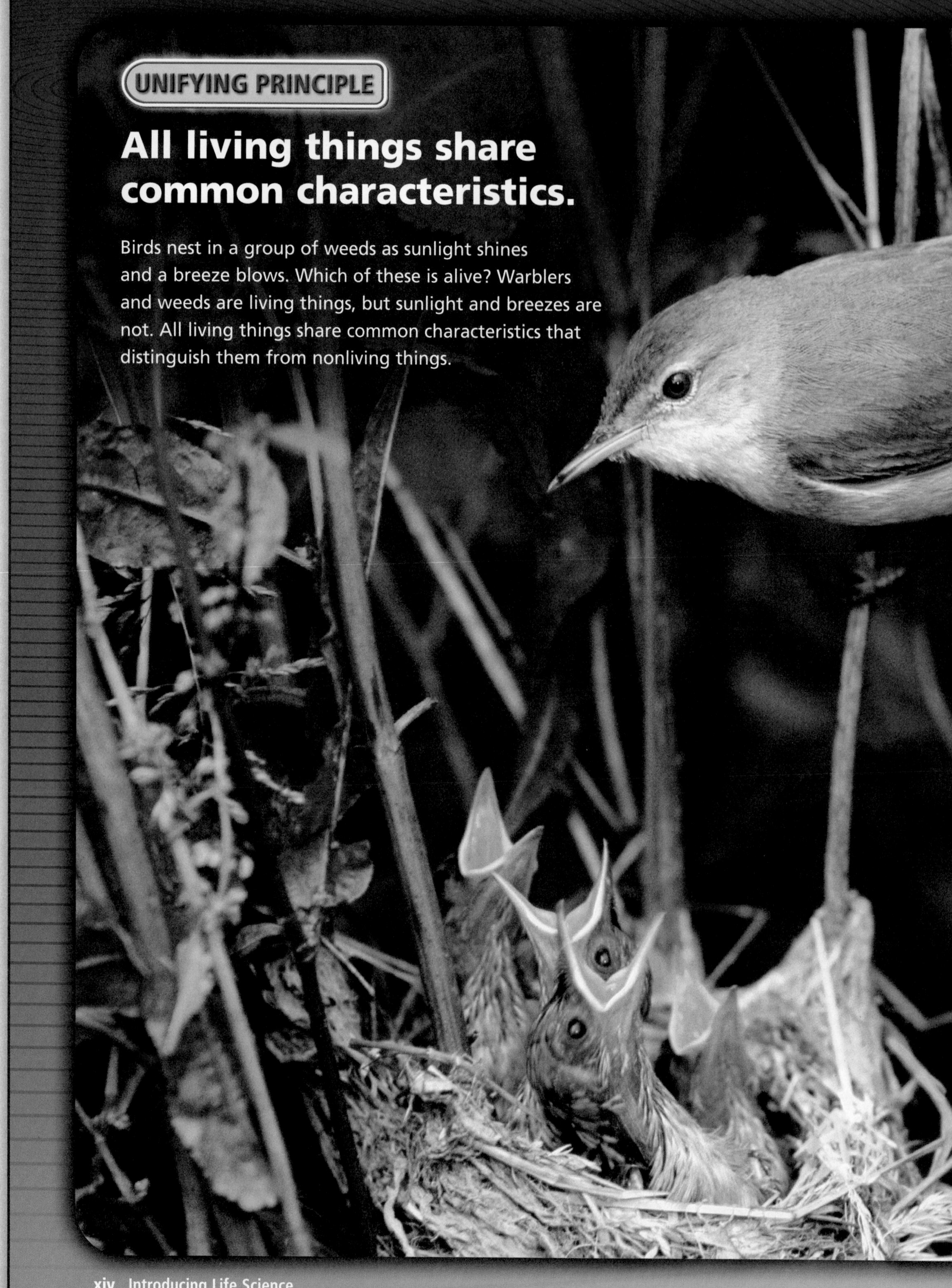

UNIFYING PRINCIPLE

All living things share common characteristics.

Birds nest in a group of weeds as sunlight shines and a breeze blows. Which of these is alive? Warblers and weeds are living things, but sunlight and breezes are not. All living things share common characteristics that distinguish them from nonliving things.

What It Means

This unifying principle helps you explore one of the biggest questions in science, "What is life?" Let's take a look at four characteristics that distinguish living things from nonliving things : organization, growth, reproduction, and response.

Organization

If you stand a short distance from a reed warbler's nest, you can observe the largest level of organization in a living thing—the **organism** itself. Each bird is an organism. If you look at a leaf under a microscope, you can observe the smallest level of organization capable of performing all the activities of life, a **cell.** All living things are made of cells.

Growth

Most living things grow and develop. Growth often involves not only an increase in size, but also an increase in complexity, such as a tadpole growing into a frog. If all goes well, the small warblers in the picture will grow to the size of their parent.

Reproduction

Most living things produce offspring like themselves. Those offspring are also able to reproduce. That means that reed warblers produce reed warblers which in turn reproduce more reed warblers.

Response

You've probably noticed that your body adjusts to changes in your surroundings. If you are exploring outside on a hot day, you may notice that you sweat. On a cold day, you may shiver. Sweating and shivering are examples of response.

Why It's Important

People of all ages experience the urge to explore and understand the living world. Understanding the characteristics of living things is a good way to start this exploration of life. In addition, knowing about the characteristics of living things helps you identify:

- similarities and differences among various organisms
- key questions to ask about any organism you study

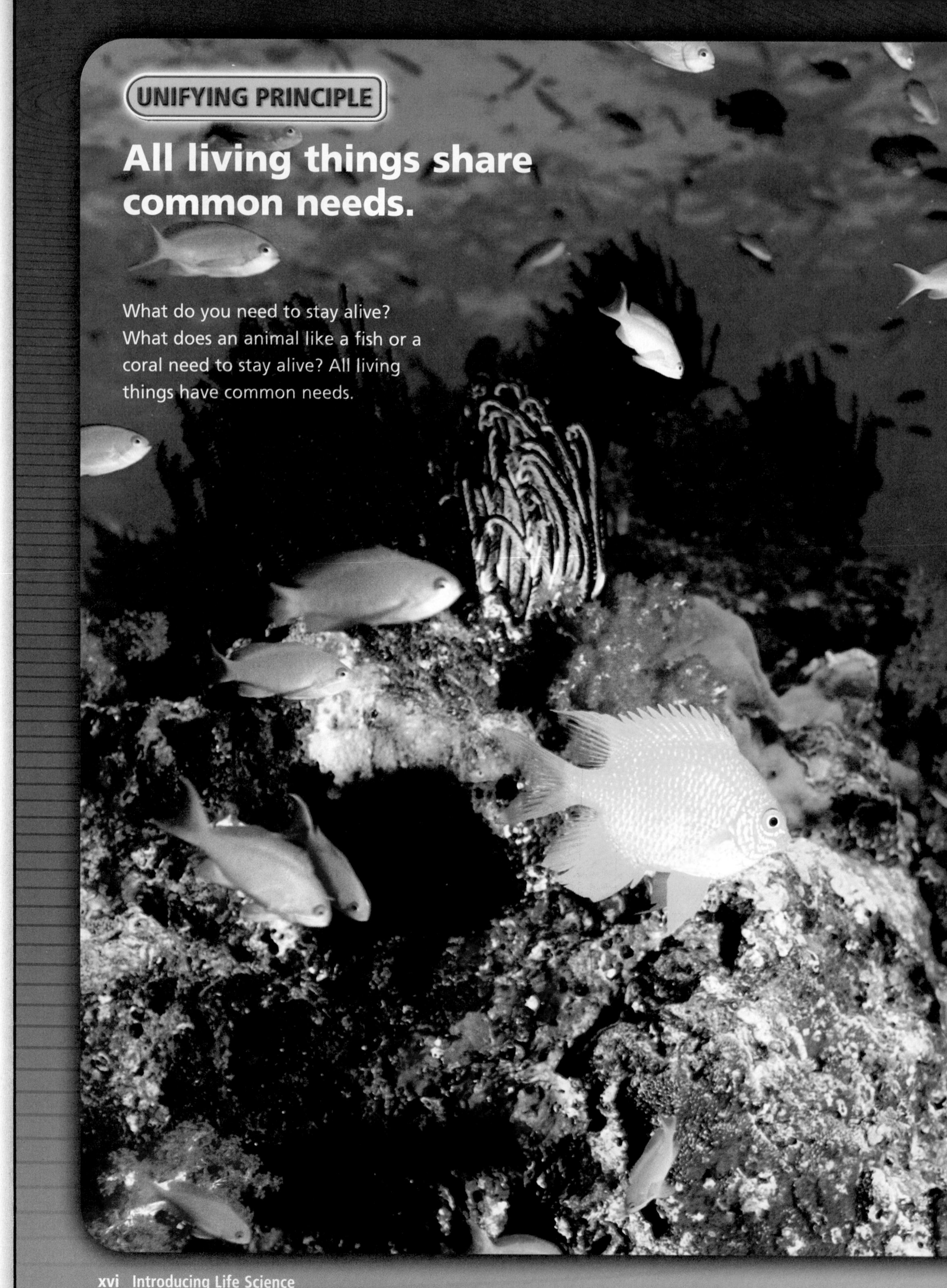

UNIFYING PRINCIPLE

All living things share common needs.

What do you need to stay alive? What does an animal like a fish or a coral need to stay alive? All living things have common needs.

What It Means

Inside every living thing, chemical reactions constantly change materials into new materials. For these reactions to occur, an organism needs energy, water and other materials, and living space.

Energy

You use energy all the time. Movement, growth, and sleep all require energy, which you get from food. Plants use the energy of sunlight to make their own food. All animals get their energy by eating either plants or other animals that eat plants.

Water and Other Materials

Water is the main ingredient in the cells of all living things. The chemical reactions inside cells take place in water, and water plays a part in moving materials around within organisms.

Other materials are also essential for life. For example, plants must have carbon dioxide from the air to make their own food. Plants and animals both use oxygen to release the energy stored in their food. You and other animals that live on land get oxygen when you breathe in air. The fish swimming around the coral reef in the picture have gills, which allow them to get oxygen that is dissolved in the water.

Living Space

You can think of living space as a home—a space that protects you from external conditions and a place where you can get materials such as water and air. The ocean provides living space for the coral that makes up this coral reef. The coral itself provides living space for many other organisms.

Why It's Important

Understanding the needs of living things helps people make wise decisions about resources. This knowledge can also help you think carefully about

- the different ways in which various organisms meet their needs for energy and materials
- the effects of adding chemicals to the water and air around us
- the reasons why some types of plants or animals may disappear from an area

xvii

UNIFYING PRINCIPLE

Living things meet their needs through interactions with the environment.

A moose chomps on the leaves of a plant. This ordinary event involves many interactions among living and nonliving things within the forest.

What It Means

To understand this unifying principle, take a closer look at the words *environment* and *interactions.*

Environment

The **environment** is everything that surrounds a living thing. An environment is made up of both living and nonliving factors. For example, the environment in this forest includes rainfall, rocks, and soil as well as the moose, the evergreen trees, and the birch trees. In fact, the soil in these forests is called "moose and spruce" soil because it contains materials provided by the animals and evergreens in the area.

Interaction

All living things in an environment meet their needs through interactions. An **interaction** occurs when two or more things act in ways that affect one another. For example, trees and other forest plants can meet their need for energy and materials through interactions with materials in soil and air, and light from the Sun. New plants get living space as birds, wind, and other factors carry seeds from one location to another.

Animals like this moose meet their need for food through interactions with other living things. The moose gets food by eating leaves off trees and other plants. In turn, the moose becomes food for wolves.

Why It's Important

Learning about living things and their environment helps scientists and decision makers address issues such as:

- developing land for human use without damaging the environment
- predicting how a change in the moose population would affect the soil in the forest
- determining the ways in which animals harm or benefit the trees in a forest

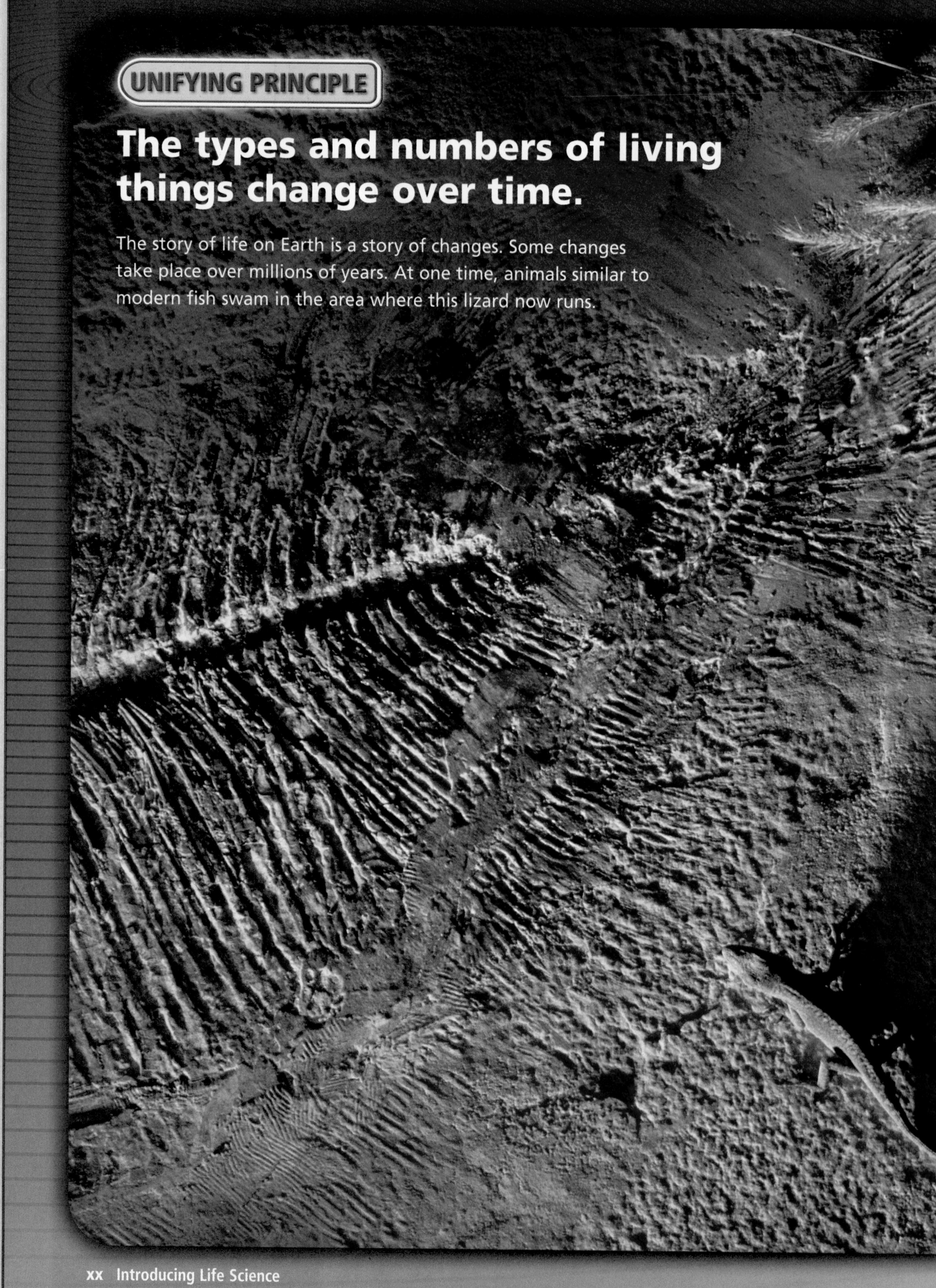

UNIFYING PRINCIPLE

The types and numbers of living things change over time.

The story of life on Earth is a story of changes. Some changes take place over millions of years. At one time, animals similar to modern fish swam in the area where this lizard now runs.

What It Means

To understand how living things change over time, let's look closely at the terms *diversity* and *adaptation.*

Diversity

You are surrounded by an astonishing variety of living things. This variety is called **biodiversity.** Today, scientists have described and named 1.4 million species. There are even more species that haven't been named. Scientists use the term *species* to describe a group of closely related living things. Members of a **species** are so similar that they can reproduce offspring that are able to reproduce. Lizards, such as the one you see in the photograph, are so diverse that they make up many different species.

Over the millions of years that life has existed on Earth, new species have originated and others have disappeared. The disappearance of a species is called **extinction.** Fossils, like the one in the photograph, provide evidence that some types of organisms that lived millions of years ago became extinct.

Adaptation

Scientists use the term **adaptation** to mean a characteristic of a species that allows members of that species to survive in a particular environment. Adaptations are related to needs. A salamander's legs are an adaptation that allows it to move on land.

Over time, species either develop adaptations to changing environments or they become extinct. The history of living things on Earth is related to the history of the changing Earth. The presence of a fish-like fossil indicates that the area shown in this photograph was once covered by water.

Why It's Important

By learning how living things change over time, you will gain a better understanding of the life that surrounds you and how it survives. Discovering more about the history of life helps scientists to

- identify patterns of relationships among various species
- predict how changes in the environment may affect species in the future

The Nature of Science

You may think of science as a body of knowledge or a collection of facts. More important, however, science is an active process that involves certain ways of looking at the world.

Scientific Habits of Mind

Scientists are curious. They are always asking questions. A scientist who observes that the number of plants in a forest preserve has decreased might ask questions such as, "Are more animals eating the plants?" or "Has the way the land is used affected the numbers of plants?" Scientists around the world investigate these and other important questions.

Scientists are observant. They are always looking closely at the world around them. A scientist who studies plants often sees details such as the height of a plant, its flowers, and how many plants live in a particular area.

Scientists are creative. They draw on what they know to form a possible explanation for a pattern, event, or behavior that they have observed. Then scientists create a plan for testing their ideas.

Scientists are skeptical. Scientists don't accept an explanation or answer unless it is based on evidence and logical reasoning. They continually question their own conclusions as well as conclusions suggested by other scientists. Scientists trust only evidence that is confirmed by other people or methods.

A white-tailed deer feeds on many plants including the trillium shown here.

By measuring the growth of this tree, a scientist can study interactions in the ecosystem.

Science Processes at Work

You can think of science as a continuous cycle of asking and seeking answers to questions about the world. Although there are many processes that scientists use, scientists typically do each of the following:

- Ask a question
- Determine what is known
- Investigate
- Interpret results
- Share results

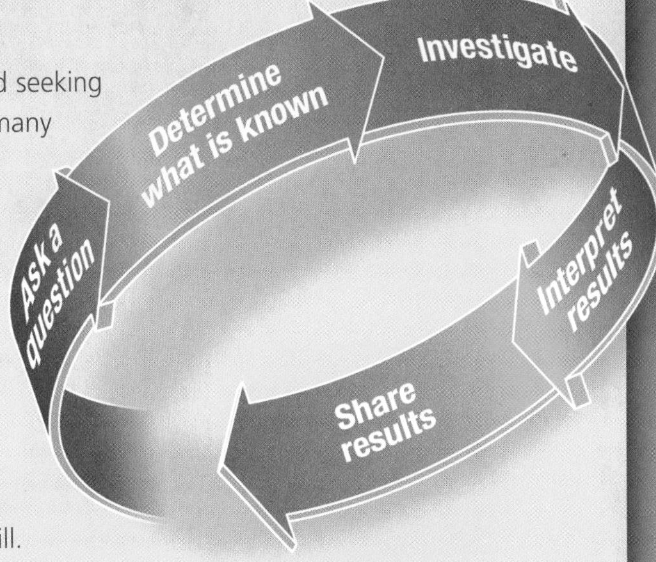

Ask a Question

It may surprise you that asking questions is an important skill. A scientific investigation may start when a scientist asks a question. Perhaps scientists observe an event or process that they don't understand, or perhaps answering one question leads to another.

Determine What Is Known

When beginning an inquiry, scientists find out what is already known about a question. They study results from other scientific investigations, read journals, and talk with other scientists. A biologist who is trying to understand how the change in the number of deer in an area affects plants will study reports of censuses taken for both plants and animals.

Investigate

Investigating is the process of collecting evidence. Two important ways of collecting evidence are observing and experimenting.

Observing is the act of noting and recording an event, characteristic, behavior, or anything else detected with an instrument or with the senses. For example, a scientist notices that plants in one part of the forest are not thriving. She sees broken plants and compares the height of the plants in one area with those in another.

An **experiment** is an organized procedure during which all factors but the one being studied are controlled. For example, the scientist thinks the reason that some plants in the forest are not thriving may be because deer are eating the flowers off the plants. An experiment she might try is to mark two similar parts of an area where the plants grow and then to build a fence around one part so the deer can't get to the plants there. The fence must be constructed so the same amount of light, air, and water reach the plants. The only factor that changes is contact between plants and the deer.

Close observation of the Colorado Potatobeetle led scientists to answers that help farmers control this insect pest.

Forming hypotheses and making predictions are two other skills involved in scientific investigations. A **hypothesis** is a tentative explanation for an observation or a scientific problem that can be tested by further investigation. For example, since at least 1900, Colorado Potatobeetles were able to resist chemical insecticides. It was hypothesized that bacteria living in the beetles' environment were killing many beetles, because otherwise the beetles would be found in larger numbers. A **predicton** is an expectation of what will be observed or what will happen, and can be used to test a hypothesis. It was predicted that certain bacteria would kill Colorado Potatobeetles. This prediction was confirmed when bacteria called *Bt* was discovered to kill Colorado Potatobeetles and other insect pests.

Interpret Results

As scientists investigate, they analyze their evidence, or data, and begin to draw conclusions. **Analyzing data** involves looking at the evidence gathered through observations or experiments and trying to identify any patterns that might exist in the data. Often scientists need to make additional observations or perform more experiments before they are sure of their conclusions. Many times scientists make new predictions or revise their hypotheses.

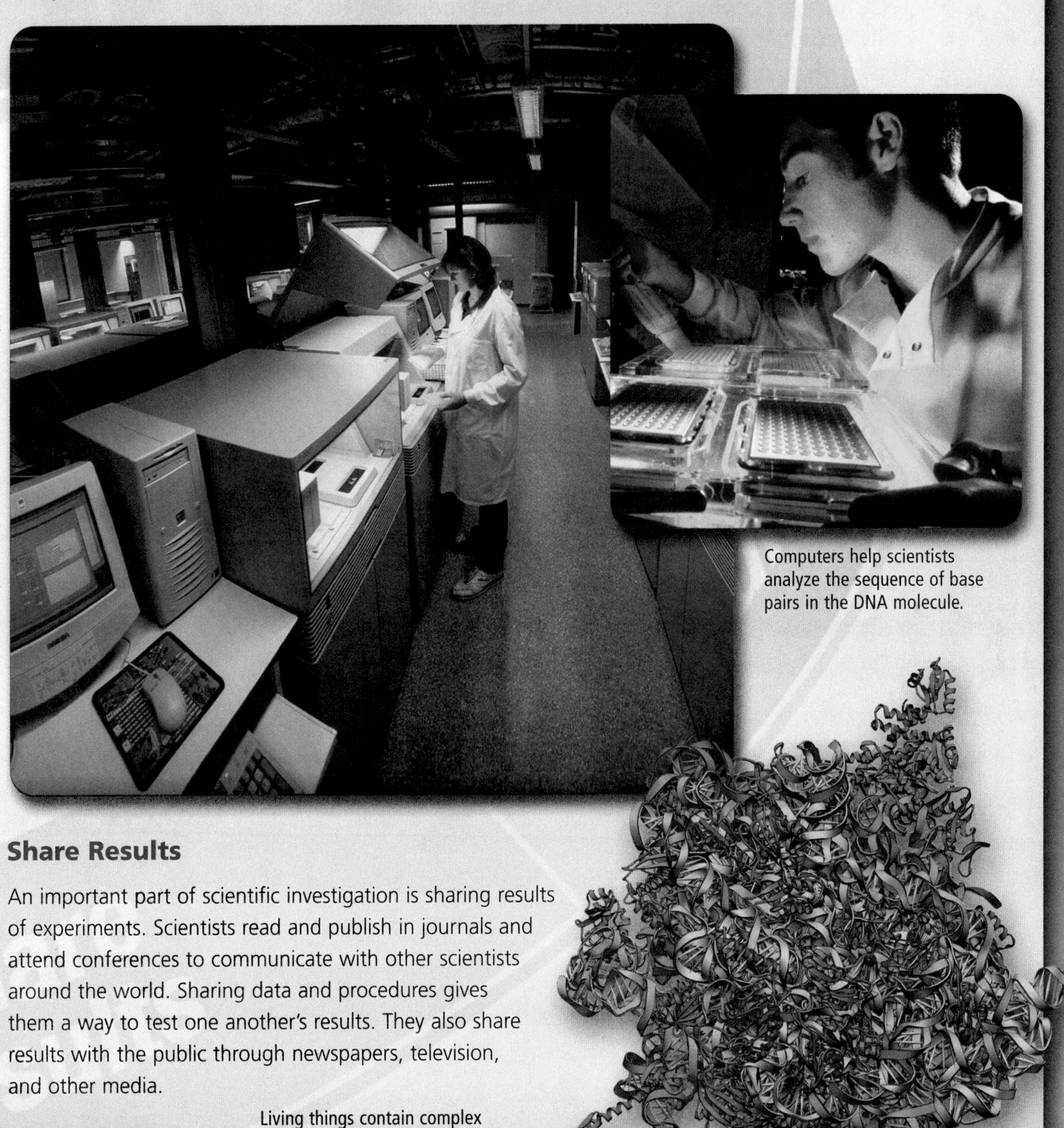

Computers help scientists analyze the sequence of base pairs in the DNA molecule.

Share Results

An important part of scientific investigation is sharing results of experiments. Scientists read and publish in journals and attend conferences to communicate with other scientists around the world. Sharing data and procedures gives them a way to test one another's results. They also share results with the public through newspapers, television, and other media.

Living things contain complex molecules such as RNA and DNA. To study them scientists often use models like the one shown here.

The Nature of Technology

Imagine what life would be like without cars, computers, and cell phones. Imagine having no refrigerator or radio. It's difficult to think of a world without these items we call technology. Technology, however, is more than just machines that make our daily activities easier. Like science, technology is also a process. The process of technology uses scientific knowledge to design solutions to real-world problems.

Science and Technology

Science and technology go hand in hand. Each depends upon the other. Even designing a device as simple as a toaster requires knowledge of how heat flows and which materials are the best conductors of heat. Scientists also use a number of devices to help them collect data. Microscopes, telescopes, spectrographs, and computers are just a few of the tools that help scientists learn more about the world. The more information these tools provide, the more devices can be developed to aid scientific research and to improve modern lives.

The Process of Technological Design

Heart disease is among the leading causes of death today. Doctors have successfully replaced damaged hearts with hearts from donors. Medical engineers have developed pacemakers that improve the ability of a damaged heart to pump blood. But none of these solutions is perfect. Although it is very complex, the heart is really a pump for blood, thus, using technology to build a better replacement pump should be possible. The process of technological design involves many choices. In the case of an artificial heart, choices about how and what to develop involve cost, safety, and patient preference. What kind of technology will result in the best quality of life for the patient?

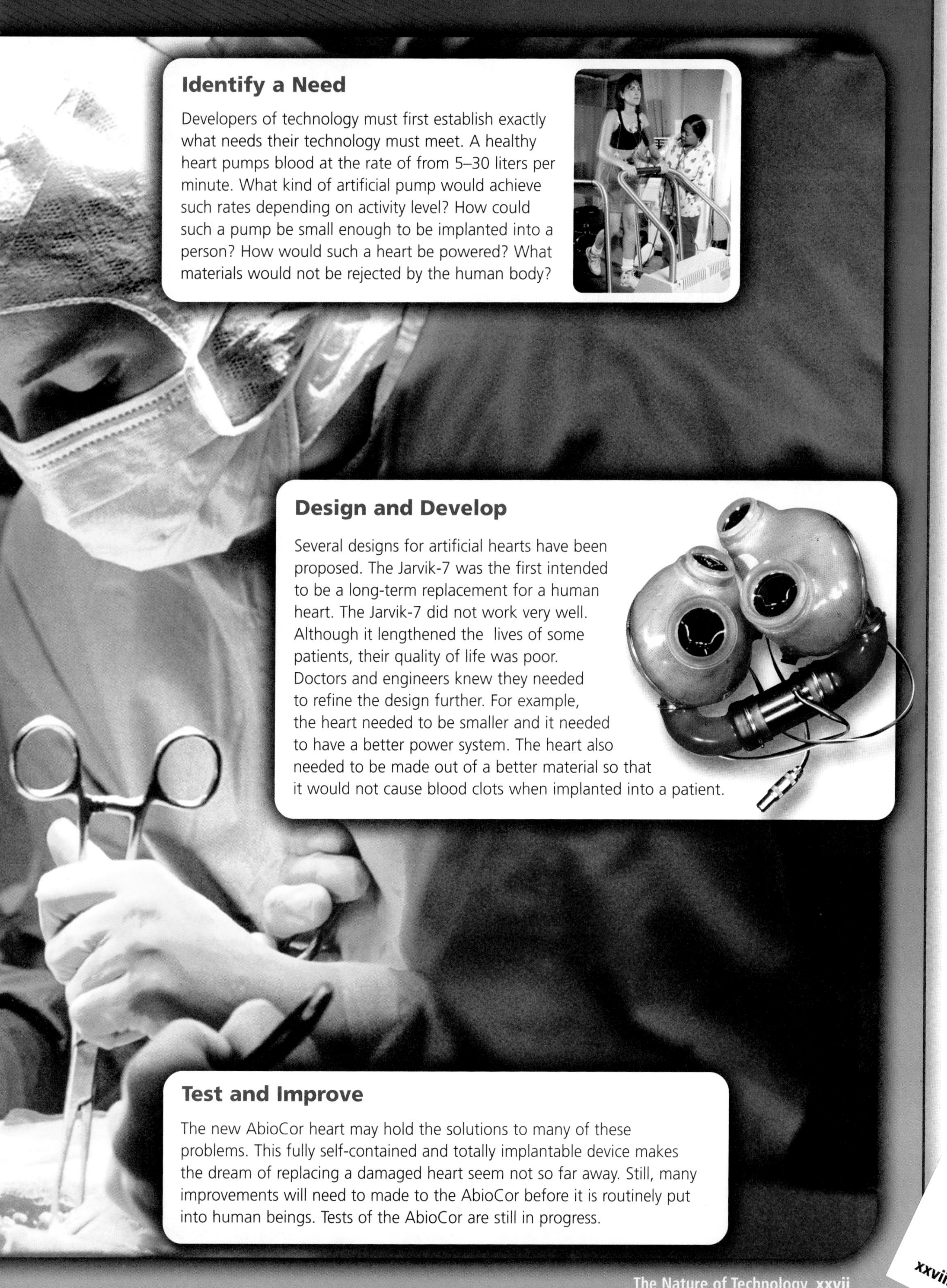

Identify a Need

Developers of technology must first establish exactly what needs their technology must meet. A healthy heart pumps blood at the rate of from 5–30 liters per minute. What kind of artificial pump would achieve such rates depending on activity level? How could such a pump be small enough to be implanted into a person? How would such a heart be powered? What materials would not be rejected by the human body?

Design and Develop

Several designs for artificial hearts have been proposed. The Jarvik-7 was the first intended to be a long-term replacement for a human heart. The Jarvik-7 did not work very well. Although it lengthened the lives of some patients, their quality of life was poor. Doctors and engineers knew they needed to refine the design further. For example, the heart needed to be smaller and it needed to have a better power system. The heart also needed to be made out of a better material so that it would not cause blood clots when implanted into a patient.

Test and Improve

The new AbioCor heart may hold the solutions to many of these problems. This fully self-contained and totally implantable device makes the dream of replacing a damaged heart seem not so far away. Still, many improvements will need to made to the AbioCor before it is routinely put into human beings. Tests of the AbioCor are still in progress.

Using McDougal Littell Science

Reading Text and Visuals

This book is organized to help you learn. Use these boxed pointers as a path to help you learn and remember the **Big Ideas** and **Key Concepts.**

Take notes.

Use the strategies on the **Getting Ready to Learn** page.

Read the Big Idea.

As you read **Key Concepts** for the chapter, relate them to **the Big Idea.**

CHAPTER 3
Transp
Protec

the BIG idea

Systems function to transport materials and to defend and protect the body.

Key Concepts

SECTION 3.1 The circulatory system transports materials. Learn how materials move through blood vessels.

SECTION 3.2 The immune system defends the body. Learn about the body's defenses and responses to foreign materials.

SECTION 3.3 The integumentary system shields the body. Learn about the structure of skin and how it protects the body.

Internet Preview
CLASSZONE.COM
Chapter 3 online resources: Content Review, two Visualizations, four Resource Centers, Math Tutorial, Test Practice

E 62 Unit: Human Biology

CHAPTER 3
Getting Ready to Learn

CONCEPT REVIEW
- The body's systems interact.
- The body's systems work to maintain internal conditions.
- The digestive system breaks down food.
- The respiratory system gets oxygen and removes carbon dioxide.

VOCABULARY REVIEW
organ p. 11
organ system p. 12
homeostasis p. 12
nutrient p. 45

Review concepts and vocabulary.

TAKING NOTES

MAIN IDEA AND DETAIL NOTES

Make a two-column chart. Write the main ideas, such as those in the blue headings, in the column on the left. Write details about each of those main heads in the column on the right.

VOCABULARY STRATEGY

Write each new vocabulary term in the center of a **frame game** diagram. Decide what information to frame it with. Use examples, descriptions, parts, sentences that use the term in context, or pictures. You can change the frame to fit each term.

See the Note-Taking Handbook on pages R45–R51.

SCIENCE NOTEBOOK

MAIN IDEAS	DETAIL NOTES
1. The circulatory system works with other body systems.	1. Transports materials f digestive and respirat systems to cells 2. Blood is fluid that car materials and wastes 3. Blood is always movi through the body 4. Blood delivers oxyger takes away carbon d

carries material to cells
moves continuously through body
BLOOD
carries waste away fro cells
circulatory system

E 64 Unit: Human Biology

Read each heading.

See how it fits in the outline of the chapter.

KEY CONCEPT

3.1 The circulatory system transports materials.

BEFORE, you learned
- The urinary system removes waste
- The kidneys play a role in homeostasis

NOW, you will learn
- How different structures of the circulatory system work together
- About the structure and function of blood
- What blood pressure is and why it is important

VOCABULARY

circulatory system p. 65
blood p. 65
red blood cell p. 67
artery p. 69
vein p. 69
capillary p. 69

EXPLORE The Circulatory System

How fast does your heart beat?

PROCEDURE

1. Hold out your left hand with your palm facing up.
2. Place the first two fingers of your right hand on your left wrist below your thumb. Move your fingertips slightly until you can feel your pulse.
3. Use the stopwatch to determine how many times your heart beats in one minute.

WHAT DO YOU THINK?
- How many times did your heart beat?
- What do you think you would find if you took your pulse after exercising?

MATERIALS
stopwatch

The circulatory system works with other body systems.

VOCABULARY
Add a frame game diagram for the term *circulatory system* to your notebook.

You have read that the systems in your body provide materials and energy. The digestive system breaks down food and nutrients, and the respiratory system provides the oxygen that cells need to release energy. Another system, called the **circulatory system,** transports products from the digestive and the respiratory systems to the cells.

Materials and wastes are carried in a fluid called **blood**. Blood moves continuously through the body, delivering oxygen and other materials to cells and removing carbon dioxide and other wastes from cells.

Chapter 3: **Transport and Protection** 65 **E**

Remember what you know.

Think about concepts you learned earlier and preview what you'll learn now.

Try the activities.

They will introduce you to science concepts.

Learn the vocabulary.

Take notes on each term.

Reading Text and Visuals

Read one paragraph at a time.

Look for a topic sentence that explains the **main idea** of the paragraph. Figure out how **the details** relate to that idea. One paragraph might have several important ideas; you may have to reread to understand.

Answer the questions.

Check Your Reading questions will help you remember what you read.

Study the visuals.

- Read the title.
- Read all labels and captions.
- Figure out what the picture is showing. Notice the information in the captions.

Exchanging Oxygen and Carbon Dioxide

Like almost all living things, the human body needs oxygen to survive. Without oxygen, cells in the body die quickly. How does the oxygen you need get to your cells? Oxygen, along with other gases, enters the body when you inhale. Oxygen is then transported to cells throughout the body.

The air that you breathe contains only about 20 percent oxygen and less than 1 percent carbon dioxide. Almost 80 percent of air is nitrogen gas. The air that you exhale contains more carbon dioxide and less oxygen than the air that you inhale. It's important that you exhale carbon dioxide because high levels of it will damage, even destroy, cells.

In cells and tissues, proper levels of both oxygen and carbon dioxide are essential. Recall that systems in the body work together to maintain homeostasis. If levels of oxygen or carbon dioxide change, your brain or blood vessels signal the body to breathe faster or slower.

The photograph shows how someone underwater maintains proper levels of carbon dioxide and oxygen. The scuba diver needs to inhale oxygen from a tank. She removes carbon dioxide wastes with other gases when she exhales into the water. The bubbles you see in the water are formed when she exhales.

CHECK YOUR READING What gases are in the air that you breathe?

Gas Exchange

This scuba diver breathes the same mixture of gases present in air.

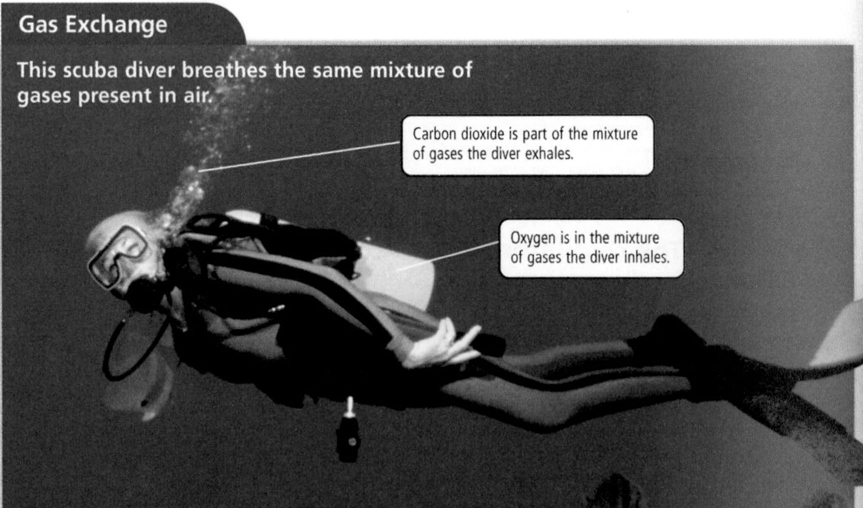

E 38 Unit: **Human Biology**

Doing Labs

To understand science, you have to see it in action. Doing labs helps you understand how things really work.

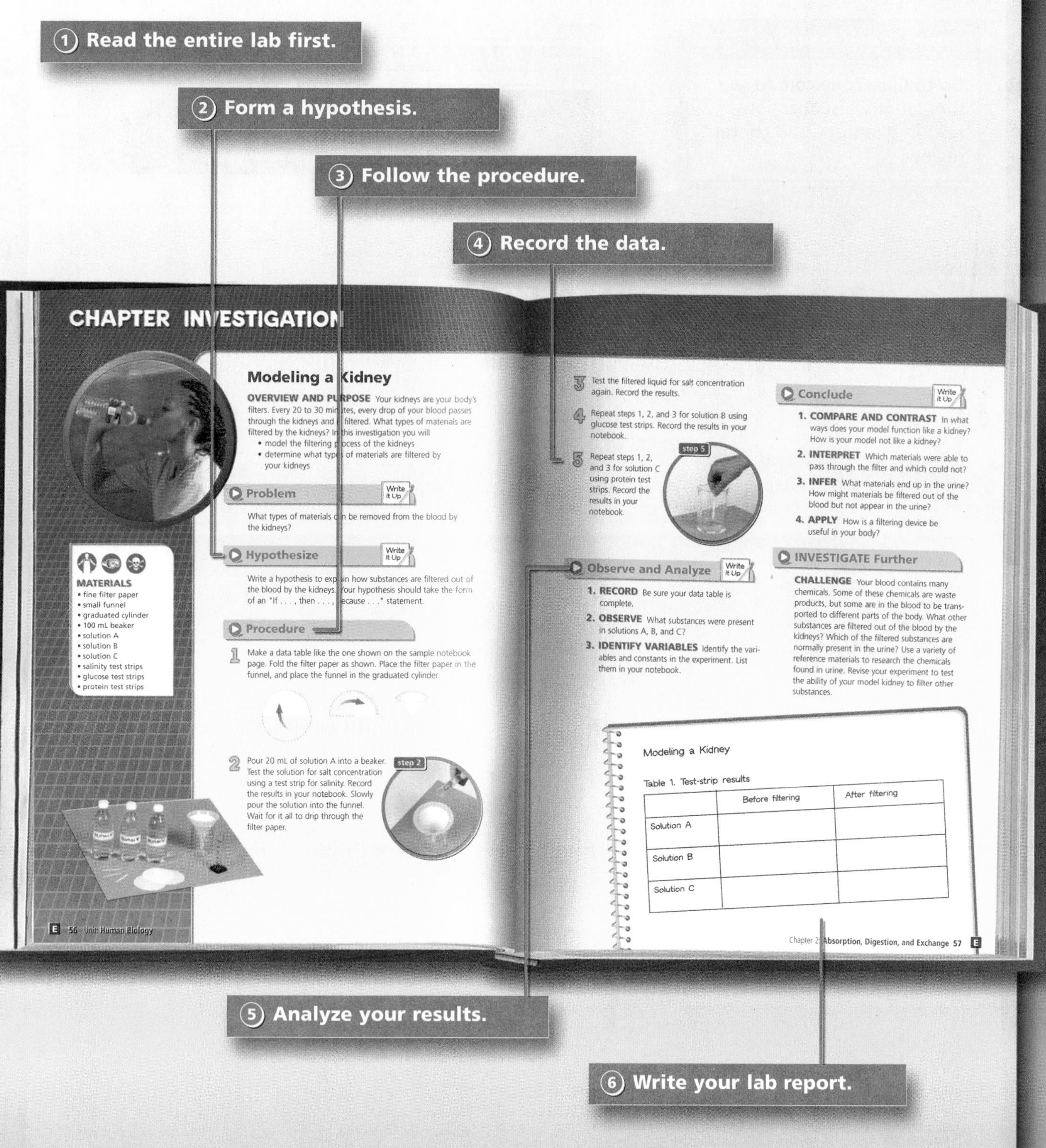

1 Read the entire lab first.
2 Form a hypothesis.
3 Follow the procedure.
4 Record the data.
5 Analyze your results.
6 Write your lab report.

CHAPTER INVESTIGATION

Modeling a Kidney

OVERVIEW AND PURPOSE Your kidneys are your body's filters. Every 20 to 30 minutes, every drop of your blood passes through the kidneys and is filtered. What types of materials are filtered by the kidneys? In this investigation you will

- model the filtering process of the kidneys
- determine what types of materials are filtered by your kidneys

Problem Write It Up

What types of materials can be removed from the blood by the kidneys?

Hypothesize Write It Up

Write a hypothesis to explain how substances are filtered out of the blood by the kidneys. Your hypothesis should take the form of an "If . . . , then . . . , because . . ." statement.

MATERIALS
- fine filter paper
- small funnel
- graduated cylinder
- 100 mL beaker
- solution A
- solution B
- solution C
- salinity test strips
- glucose test strips
- protein test strips

Procedure

1. Make a data table like the one shown on the sample notebook page. Fold the filter paper as shown. Place the filter paper in the funnel, and place the funnel in the graduated cylinder.
2. Pour 20 mL of solution A into a beaker. Test the solution for salt concentration using a test strip for salinity. Record the results in your notebook. Slowly pour the solution into the funnel. Wait for it all to drip through the filter paper. (step 2)
3. Test the filtered liquid for salt concentration again. Record the results.
4. Repeat steps 1, 2, and 3 for solution B using glucose test strips. Record the results in your notebook.
5. Repeat steps 1, 2, and 3 for solution C using protein test strips. Record the results in your notebook. (step 5)

Observe and Analyze Write It Up

1. **RECORD** Be sure your data table is complete.
2. **OBSERVE** What substances were present in solutions A, B, and C?
3. **IDENTIFY VARIABLES** Identify the variables and constants in the experiment. List them in your notebook.

Conclude Write It Up

1. **COMPARE AND CONTRAST** In what ways does your model function like a kidney? How is your model not like a kidney?
2. **INTERPRET** Which materials were able to pass through the filter and which could not?
3. **INFER** What materials end up in the urine? How might materials be filtered out of the blood but not appear in the urine?
4. **APPLY** How is a filtering device be useful in your body?

INVESTIGATE Further

CHALLENGE Your blood contains many chemicals. Some of these chemicals are waste products, but some are in the blood to be transported to different parts of the body. What other substances are filtered out of the blood by the kidneys? Which of the filtered substances are normally present in the urine? Use a variety of reference materials to research the chemicals found in urine. Revise your experiment to test the ability of your model kidney to filter other substances.

Modeling a Kidney

Table 1. Test-strip results

	Before filtering	After filtering
Solution A		
Solution B		
Solution C		

E 56 Unit: Human Biology

Chapter 2: Absorption, Digestion, and Exchange 57 E

Using Technology

The Internet is a great source of information about up-to-date science. The ClassZone Web site and NSTA SciLinks have exciting sites for you to explore. Video clips and simulations can make science come alive.

Look for red banners.

Go to **ClassZone.com** to see simulations, visualizations, resource centers, and content review.

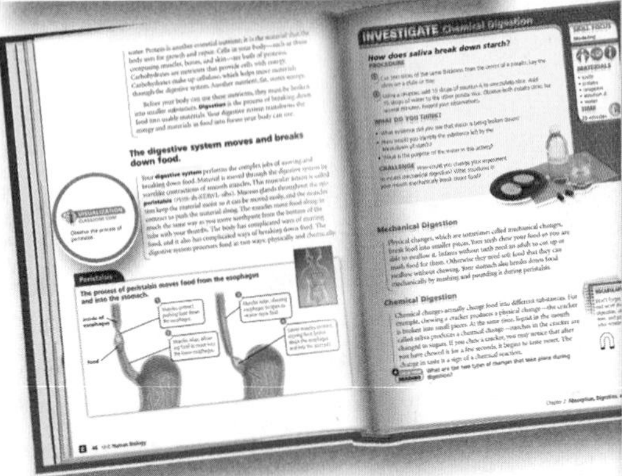

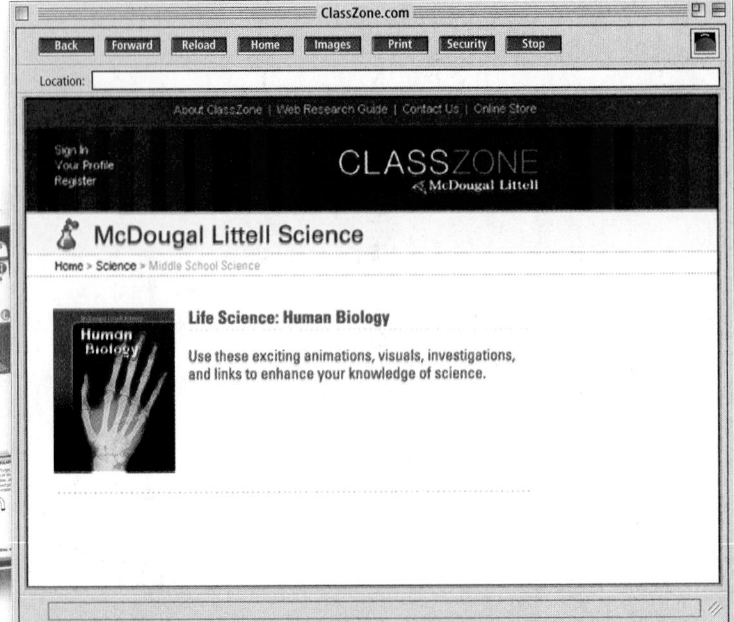

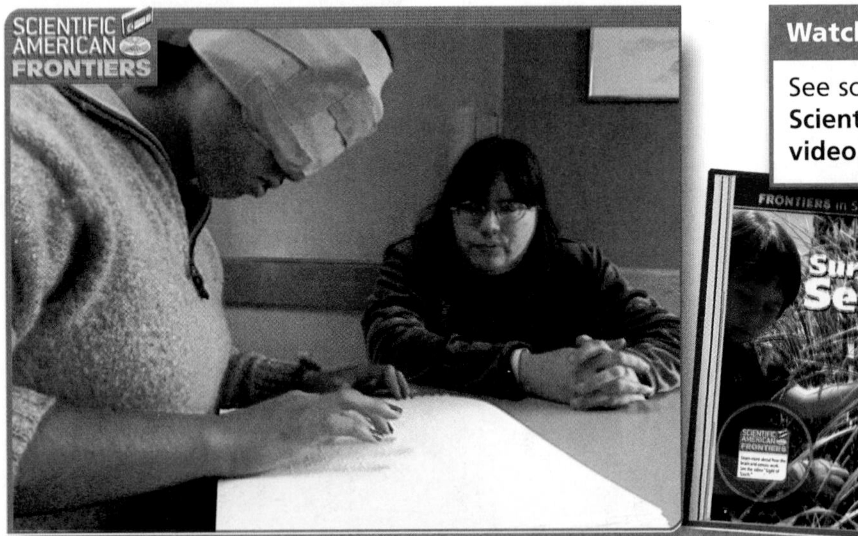

Watch the video.

See science at work in the **Scientific American Frontiers video.**

Look up SciLinks.

Go to **scilinks.org** to explore the topic.

Tissues and Organs **Code: MDL044**

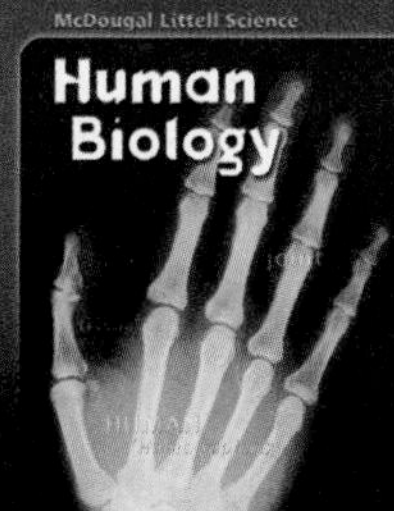

Human Biology Contents Overview

Unit Features

the BIG idea

The human body is made up of systems that work together to perform necessary functions.

the BIG idea

Systems in the body obtain and process materials and remove waste.

the BIG idea

Systems function to transport materials and to defend and protect the body.

the BIG idea

The nervous and endocrine systems allow the body to respond to internal and external conditions.

the BIG idea

The body develops and maintains itself over time.

VIDEO SUMMARY

SCIENTIFIC AMERICAN FRONTIERS

"Sight of Touch," a segment of the *Scientific American Frontiers* series that aired on PBS stations, describes experiments that explore how the brain adapts. Dr. Alvaro Pascual-Leone has found that the brain's visual cortex processes the sense of touch in people born blind. He wants to know if sighted people's brains can adapt in the same way. A sighted volunteer is blindfolded for one week to see if the change in sensory input has an impact on her brain. She studies Braille, and her brain's response to stimuli is tested in an MRI chamber. Toward the end of the week, a magnetic shock is applied to her visual cortex to disrupt its function temporarily. By comparing the volunteer's Braille abilities before and after the shock, Dr. Pascual-Leone confirms his hypothesis: the visual cortex helps process information arriving via the sense of touch. The volunteer's brain has adapted in one short week.

National Science Education Standards

A.1–8 Abilities Necessary to Do Scientific Inquiry

A.9.a–b, A.9.d–g Understandings About Scientific Inquiry

F.5.e Science and Technology in Society

G.1.a–b Science as a Human Endeavor

G.2.a, G.2.c Nature of Science

FRONTIERS in Science

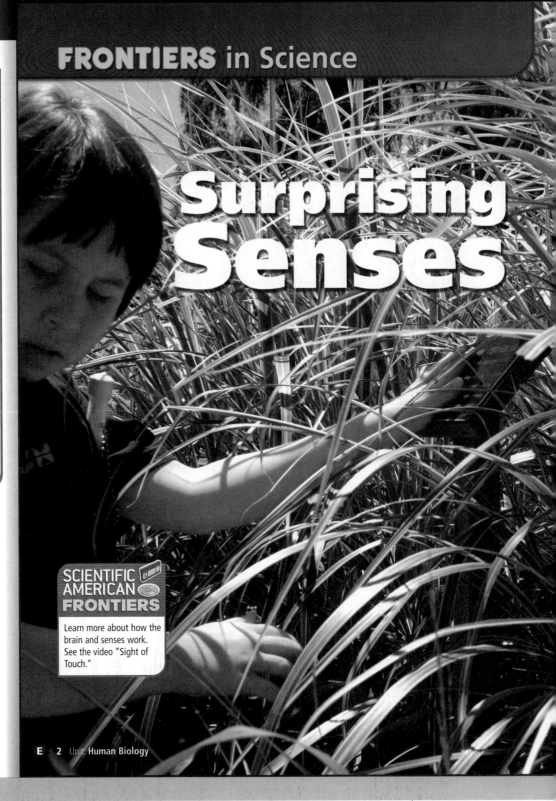

Surprising Senses

SCIENTIFIC AMERICAN FRONTIERS

Learn more about how the brain and senses work. See the video "Sight of Touch."

ADDITIONAL RESOURCES

Technology Resources

Scientific American Frontiers Video: *Sight of Touch:* 11-minute video segment that introduces the unit.

ClassZone.com
CAREER LINK, Neurobiologist

Guide student viewing and comprehension of the video:

Video Teaching Guide, pp. 1–2; Video Viewing Guide, p. 3; Video Wrap-Up, p. 4

Scientific American Frontiers Video Guide, pp. 15–18

Unit projects procedures and rubrics:

Unit Projects, pp. 5–10

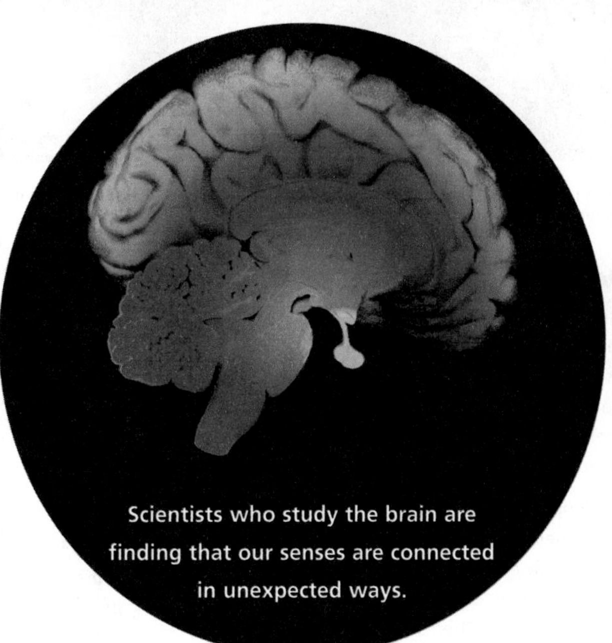

Scientists who study the brain are finding that our senses are connected in unexpected ways.

Senses and the Brain

One of the great mysteries still unsolved in science is what happens inside the brain. What is a thought? How is it formed? Where is it stored? How do our senses shape our thoughts? There are far more questions than answers. One way to approach questions about the brain is to study brain activity at times when the body is performing different functions.

A Most advanced brain functions happen in the part of the brain called the cerebral cortex (suh-REE-bruhl KOR-tehks). That's where the brain interprets information from the senses. The cerebral cortex has many specialized areas. Each area controls one type of brain activity. Scientists are mapping these areas. At first, they studied people with brain injuries. A person with an injury to one area might not be able to speak. Someone with a different injury might have trouble seeing or hearing. Scientists mapped the areas in which damage seemed to cause each kind of problem.

B Now scientists have even more tools to study the brain. One tool is called functional magnetic resonance imaging, or FMRI. Scientists put a person into a machine that uses radio waves to produce images of the person's brain. Scientists then ask the person to do specific activities, such as looking at pictures of faces or listening for specific sounds. The FMRI images show what parts of the person's brain are most active during each activity.

FOCUS

Set Learning Goals

Students will

- Observe how the brain processes information from the senses.
- Analyze how imaging technology is used to map the brain.
- Analyze and model body systems.

The "Sight of Touch" video shows scientists collecting data to use in mapping the functions of areas of the brain. Point out that maps are often a valuable first step in exploring a frontier.

INSTRUCT

Scientific Process

Ask students to identify the skills scientists used when they studied injured people to map the brain. *observation, collecting and analyzing data* Ask: Once scientists realized that they could observe injured people for clues to brain function, what skill did they use to connect an area of the brain with its function? *They made inferences from observations.*

Technology Design

Studying animal brains is not enough to completely understand the human brain. Scientists in the fields of neurology, psychiatry, physiology, and pharmacology all need the information that FMRI studies provide. Aside from information about the senses, researchers wish to understand—among other things—pain, depression, epilepsy, Alzheimer's disease, and how people recover from a stroke. Ask students to discuss how FMRIs could be used to research other areas of study.

DIFFERENTIATE INSTRUCTION

More Reading Support

A Where in the brain do most advanced functions happen? *the cerebral cortex*

B How do scientists use FMRI? *as a tool to study brain activity*

Below Level Have students list three or four common activities and then analyze what senses they use in accomplishing them. You might also suggest that they try tasting their lunches while holding their noses.

Teach from Visuals

To help students interpret the four images of brain activity, focus attention on the image that shows Braille reading (third from left). Explain that when observing people read Braille, it might appear that the sense of touch is stimulated, but in reality many areas of the brain are active. The PET scans show that part of the vision area, part of the hearing area, and part of the thought area all combine in the act of processing Braille.

Scientific Process

Point out that it is not safe to assume that every person has the same brain functioning as another. But you can use information gained from one study subject to form a hypothesis about others. Ask: What hypothesis did Alvaro Pascual-Leone form after studying Gil's brain activity? *The visual center helps the touch center read Braille.* Ask: What skill was Pascual-Leone using when he asked a volunteer to wear a blindfold for a week? *testing a hypothesis*

Asking a Question

Suggest that other situations exist where using a sense to replace one that is impaired might help. Have students form a question and a testable hypothesis that might be investigated to help stroke victims who have trouble speaking.

Technology Design

Translating documents and books into Braille used to be expensive and time-consuming. Software developers have used computers to accomplish what it used to take trained transcriptionists long hours to do. The process now is highly automated, and therefore much faster and cheaper. Ask: Why is this task a good choice for automation? *It is relatively simple, with one-for-one transcription of characters.*

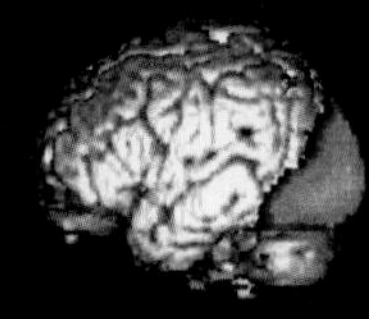
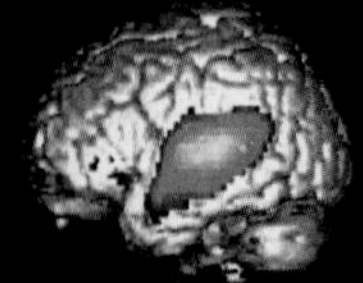
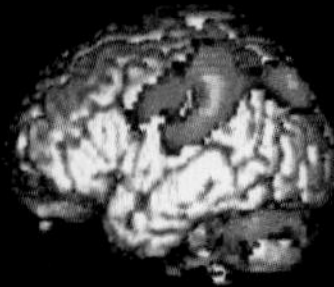
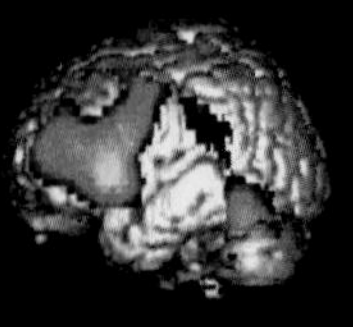

From left to right, these PET scans show brain activity during sight, hearing, Braille reading, and thought. Braille is a textured alphabet which is read by the fingers.

Double Duty

Using FMRI and other tools, scientists have identified the parts of the cerebral cortex that are responsible for each of the senses. The vision area is located at the back of the brain. The smell, taste, touch, and hearing areas are all close together in the middle part of the brain. C

People don't usually use just one sense at a time. Scientists have found some unexpected connections. In one study, Marisa Taylor-Clarke poked the arms of some volunteers with either one or two pins. Then she asked them how many pins they felt. Taylor-Clarke found that people who looked at their arms before the test did better than those who didn't. FMRI showed that the part of their brains responsible for touch was also more active when they used their sense of sight.

These connections in the brain show up even when one sense doesn't work. Many people who have hearing impairments read lips to understand what other people are saying. Scientists using FMRI discovered that these people use the part of the brain normally used for hearing to help them understand what they see. This is even true for people who have never been able to hear.

Scrambled Senses

Some people have more connections between their senses than most people have. They may look at numbers and see colors, or associate smells with shapes. Some even get a taste in their mouths when they touch something. All these are examples of synesthesia (sin-uhs-THEE-zhuh). About 1 in 200 people have some kind of synesthesia.

View the "Sight of Touch" segment of your Scientific American Frontiers video to learn about another example of connections between the senses.

IN THIS SCENE FROM THE VIDEO Michelle, a research subject, reads Braille with her fingers after wearing a blindfold for three days.

SEEING BY TOUCHING Many blind people read using Braille, a system of raised dots used to represent letters. Some, such as Braille proofreader Gil Busch, can read Braille at astonishing speeds. Scientist Alvaro Pascual-Leone used MRI to study Gil's brain. The visual area of Gil's brain was active while he read Braille.

Gil has been blind since birth, so his brain has had a long time to adjust. Pascual-Leone wanted to know whether the brain could rewire itself in a shorter time. He asked volunteer Michelle Geronimo to wear a blindfold for a week. During that time, she learned to read Braille and experienced the world as a blind person does. At the end of the week, Pascual-Leone did an MRI of Michelle. The activity of her brain had changed too. Her visual center was active while she read Braille.

DIFFERENTIATE INSTRUCTION

More Reading Support

C Which sense areas are close together in the brain? *smell, taste, touch, and hearing*

D What is synesthesia? *connecting senses in unusual ways*

Below Level Have students make a poster or bulletin board with a diagram and short summary of the purposes of the different areas of the brain.

E FMRI has made it possible for scientists to learn more about synesthesia. One group of scientists studied people who saw colors when they heard words. FMRI showed that the visual areas of their brains were active along with the hearing areas. (For most people, only the hearing area would be active.)

But why does synesthesia happen? Some scientists think that people with synesthesia have more connections between areas of their brains. Every person has extra connections when they're born, but most people lose many of them in childhood. Perhaps people with synesthesia keep theirs. Another theory suggests that their brains are "cross-wired," so information goes in unusual directions.

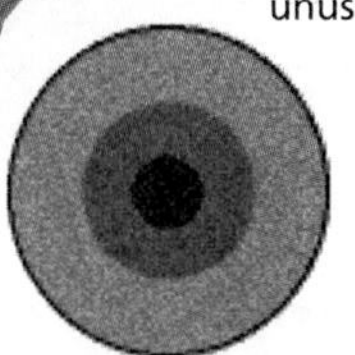

Some people with synesthesia see this colorful pattern when they hear a dog bark.

As scientists explore synesthesia and other connections between the senses, they learn more about how the parts of the brain work together. The human body is complex. And the brain along with the rest of the nervous system, has yet to be fully understood.

UNANSWERED Questions

Scientists have learned a lot about how senses are connected. Their research leads to new questions.

- How does information move between different areas of the brain?
- How and why does the brain rewire itself?
- How does cross-wired sensing (synesthesia) happen?

UNIT PROJECTS

As you study this unit, work alone or in a group on one of the projects below.

Your Body System

Create one or several models showing important body systems.

- Draw the outline of your own body on a large piece of craft paper.
- Use reference materials to help you place everything correctly. Label each part.

The Brain: "Then and Now"

Compare and contrast past and present understandings of the brain.

- One understanding is that each part of the brain is responsible for different body functions. This understanding has changed over time.
- Research the history of this idea.
- Prepare diagrams of then and now. Share your presentation.

Design an Experiment

Design an experiment that will test one of the five senses. You should first identify a problem question you want to explore.

- The experiment may include a written introduction, materials procedure, and a plan for recording and presenting outcomes.
- Prepare a blank written experiment datasheet for your classmates to use.

Learn more about careers in neurobiology.

UNANSWERED Questions

Have students read the questions and think of some of their own. Remind them that scientists always end up with more questions—that inquiry is the driving force of science.

- With the class, generate on the board a list of new questions.
- Students can add to the list after they watch the Scientific American Frontiers Video.
- Students can use the list as a springboard for choosing their Unit Projects.

UNIT PROJECTS

Encourage students to pick the project that most appeals to them. Point out that each is long-term and will take several weeks to complete. You might group or pair students to work on projects and in some cases guide student choice. Some of the projects have student choice built into them. Each project has two worksheet pages, including a rubric. Use the pages to guide students through criteria, process, and schedule.

 Unit Projects, pp. 5–10

REVISIT concepts introduced in this article:

Chapter 1
- The human body is made up of systems that work together to perform necessary functions, pp. 6–12

Timelines in Science
- Seeing inside the body, pp. 94–97

Chapter 4
- The nervous system responds and controls, pp. 101–109
- Sight, hearing, and touch, pp. 102–103
- The central nervous system controls functions, pp. 104–105

DIFFERENTIATE INSTRUCTION

More Reading Support

E What technology has been used to study synesthesia? *FMRI*

Differentiate Unit Projects Projects are appropriate for varying abilities. Allow students to choose the ones that interest them most and let them vary their product. Encourage below level students to give visual or oral presentations or to record audio presentations about their topic.

Below Level Encourage students to try "Your Body System."

Advanced Challenge students to complete "Design an Experiment."

CHAPTER

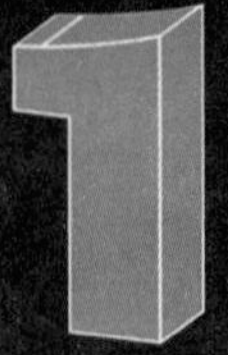

Systems, Support, and Movement

Life Science UNIFYING PRINCIPLES

PRINCIPLE 1
All living things share common characteristics.

PRINCIPLE 2
All living things share common needs.

PRINCIPLE 3
Living things meet their needs through interactions with the environment.

PRINCIPLE 4
The types and numbers of living things change over time.

Unit: Human Biology BIG IDEAS

CHAPTER 1
Systems, Support, and Movement
The human body is made up of systems that work together to perform necessary functions.

CHAPTER 2
Absorption, Digestion, and Exchange
Systems in the body obtain and process materials and remove waste.

CHAPTER 3
Transport and Protection
Systems function to transport materials and to defend and protect the body.

CHAPTER 4
Control and Reproduction
The nervous and endocrine systems allow the body to respond to internal and external conditions.

CHAPTER 5
Growth, Development, and Health
The body develops and maintains itself over time.

CHAPTER 1 KEY CONCEPTS

SECTION 1.1

The human body is complex.

1. The body has cells, tissues, and organs.
2. The body's systems interact with one another.

SECTION 1.2

The skeletal system provides support and protection.

1. Bones are living tissue.
2. The skeleton is the body's framework.
3. The skeleton changes as the body develops and ages.
4. Joints connect parts of the skeletal system.

SECTION 1.3

The muscular system makes movement possible.

1. Muscles perform important functions.
2. Your body has different types of muscle.
3. Skeletal muscles and tendons allow bones to move.
4. Muscles grow and heal.

The Big Idea Flow Chart is available on p. T1 in the **UNIT TRANSPARENCY BOOK.**

Previewing Content

SECTION

The human body is complex. pp. 9–13

1. **The body has cells, tissues, and organs.**
   The cell is the basic unit of life, and the human body has many specialized cells that perform particular functions. A **tissue** is a group of cells that function together. There are four categories of tissue:
   - epithelial, which forms boundaries
   - muscle, which enables movement
   - nerve, which carries messages
   - connective, which supports, protects, pads, insulates, and stores energy

   An **organ** is two or more types of tissue that work together to perform a function. For instance, a vein consists of epithelial, muscle, nerve, and connective tissues. Organs work together in **organ systems,** such as the circulatory system, shown below.

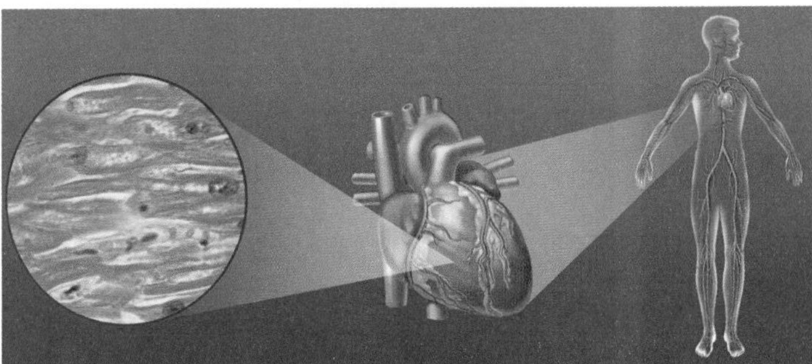

Tissue (cardiac muscle cells) — Organ (heart) — Body system (circulatory system)

2. **The body's systems interact with one another.**
   The body's levels of organization work together constantly to maintain an internal balance called **homeostasis.** An example of homeostasis is the maintenance of body temperature.

SECTION

The skeletal system provides support and protection. pp. 14–21

1. **Bones are living tissue.**
   The **skeletal system** is composed of bone tissue. Bones are mainly calcium but have a complex structure.
   - **Compact bone** is dense tissue that gives bone its strength.
   - **Spongy bone** is strong and lightweight.
   - Marrow occupies spaces in spongy bone and produces all the body's red blood cells.
   - Blood vessels supply oxygen and nutrients to bone and carry away new blood cells.

2. **The skeleton is the body's framework.**
   The skeleton has two main divisions:
   - The axial skeleton runs top to bottom and contains the skull (cranium), the vertebrae, and the ribs.
   - The appendicular skeleton runs side to side and contains the bones of the shoulders, arms, hips, and legs.

3. **The skeleton changes as the body develops and ages.**
   The skeleton changes as the body ages. In infancy, to allow for rapid growth of the brain, the skull has spaces between the bones, which close up as a baby ages. The bones of the arms and legs, called the long bones, have growth plates at their ends, which allow these bones to grow longer until puberty.

   Bone is constantly being broken down and replaced with new bone. As you get older, the replacement may not keep up with loss, and bone density may decrease.

4. **Joints connect parts of the skeletal system.**
   Bones connect to each other at joints.
   - Immovable joints lock bones together at places called sutures. The skull is made up of immovable joints.
   - Slightly movable joints can flex slightly.
   - Tissues called ligaments hold bones together at freely movable joints.
   - Freely movable joints are classified by the types of movement they produce.

Common Misconceptions

MISCONCEPTION DATABASE
CLASSZONE.COM Background on student misconceptions

FEEDBACK MECHANISMS Students sometimes think that the body performs its functions without needing any internal communication mechanism. The body is constantly monitoring itself and making adjustments, without our conscious help.

 This misconception is addressed on p. 11.

BONES ARE ALIVE Students sometimes conceive of bones as being solid and inert rather than living and needing nutrients and oxygen. A bone contains different types of tissue that have different densities. A bone is made up of living cells that need energy.

 This misconception is addressed on pp. 15 and 16.

Previewing Content

SECTION

The muscular system makes movement possible. pp. 22–29

1. Muscles perform important functions.
Muscles are made of cells called muscle fibers. They perform three functions.

- Muscles move the body.
- Muscles help maintain the body's temperature by generating heat.
- Muscles maintain upright posture.

2. Your body has different types of muscle.
Skeletal muscles are also called **voluntary muscles.** They allow the body to move when you want it to.

- Slow-twitch muscles allow slow movement.
- Fast-twitch muscles allow rapid movement.

Smooth muscles, also called **involuntary muscles,** are found inside organs and enable the body to perform automatic movements.

Cardiac muscle makes up the heart. Its cells are arranged in webs that contract without conscious control to make the heart beat.

3. Skeletal muscles and tendons allow bones to move.
Tendons attach muscles to bones. Muscles work in pairs to cause bones to move. One muscle contracts, pulling a bone, while the other lengthens.

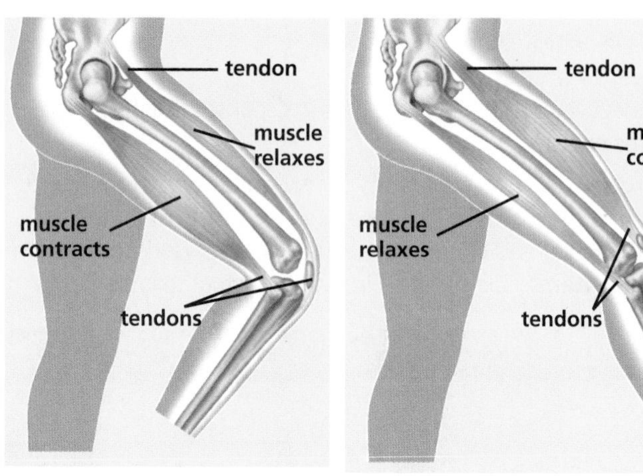

Muscles that cause movement always work in pairs or groups whose movements oppose each other.

4. Muscles grow and heal.
Muscles develop with age. Exercise increases muscle size.

Common Misconceptions

MISCONCEPTION DATABASE
CLASSZONE.COM Background on student misconceptions

MUSCLES CONTRACT Most middle school students think that muscles can push and pull. In fact, muscles can only pull, which they do by contracting.

 This misconception is addressed on p. 26.

Previewing Labs

Lab Generator CD-ROM
Edit these Pupil Edition labs and generate alternative labs.

EXPLORE the BIG idea

Lab	Time and Materials
How Many Bones Are in Your Hand? p. 7 Students examine their own hands to infer the locations of bones and joints.	**TIME** 10 minutes **MATERIALS** paper and pencil
How Does It Move? p. 7 Students compare joints to everyday objects to understand how joints work.	**TIME** 10 minutes **MATERIALS** none
Internet Activity: The Human Body, p. 7 Students compare body systems in different regions to understand the functions of different systems.	**TIME** 20 minutes **MATERIALS** computer with Internet connection

Section	Lab	Time and Materials
SECTION 1.1	**INVESTIGATE Systems,** p. 10 Students identify ways their bodies work to understand the interdependence of body systems.	**TIME** 20 minutes **MATERIALS** 1 large sheet of paper
SECTION 1.2	**EXPLORE Levers,** p. 14 Students use a sports bag hung over an arm to demonstrate a lever.	**TIME** 10 minutes **MATERIALS** sports bag
	INVESTIGATE Movable Joints, p. 19 Students observe how their joints move.	**TIME** 20 minutes **MATERIALS** book
SECTION 1.3	**EXPLORE Muscles,** p. 22 Students perform leg movements to observe how muscles change.	**TIME** 10 minutes **MATERIALS** chair
	CHAPTER INVESTIGATION **A Closer Look at Muscles,** pp. 28–29 Students examine a chicken wing to infer how human muscles and bones interact.	**TIME** 40 minutes **MATERIALS** protective gloves, uncooked chicken wing and leg (soaked in bleach solution: 2 Tbsp bleach, dissolved in 1 cup water), paper towels, scissors, dissection tray

Additional INVESTIGATION, Bones and Calcium, A, B, & C, pp. 59–67; Teacher Instructions, pp. 305–306

Previewing Chapter Resources

	INTEGRATED TECHNOLOGY	LABS AND ACTIVITIES
CHAPTER 1 **Systems, Support, and Movement**	 CLASSZONE.COM • eEdition Plus • EasyPlanner Plus • Misconception Database • Content Review • Test Practice • Visualizations • Resource Centers • Simulations • Internet Activity: The Human Body • Math Tutorial  SCILINKS.ORG SCILINKS  **CD-ROMS** • eEdition • EasyPlanner • Power Presentations • Content Review • Lab Generator • Test Generator  **AUDIO CDS** • Audio Readings • Audio Readings in Spanish	 EXPLORE the Big Idea, p. 7 • How Many Bones Are in Your Hand? • How Does It Move? • Internet Activity: The Human Body  **UNIT RESOURCE BOOK** • Family Letter, p. ix • Spanish Family Letter, p. x • Unit Projects, pp. 5–10 **Lab Generator CD-ROM** Generate customized labs.
SECTION  1.1 **The human body is complex.** pp. 9–13 Time: 2 periods (1 block)  Lesson Plan, pp. 11–12	 **UNIT TRANSPARENCY BOOK** • Big Idea Flow Chart, p. T1 • Daily Vocabulary Scaffolding, p. T2 • Note-Taking Model, p. T3 • 3-Minute Warm-Up, p. T4	 • INVESTIGATE Systems, p. 10 • Think Science, p. 13  **UNIT RESOURCE BOOK** Datasheet, Systems, p. 20
SECTION  1.2 **The skeletal system provides support and protection.** pp. 14–21 Time: 2 periods (1 block)  Lesson Plan, pp. 22–23	 • **RESOURCE CENTER,** Skeletal System • **SIMULATION,** Assemble a Skeleton • **MATH TUTORIAL**  **UNIT TRANSPARENCY BOOK** • Daily Vocabulary Scaffolding, p. T2 • 3-Minute Warm-Up, p. T4 • "A Close Look at Bone" Visual, p. T6	 • EXPLORE Levers, p. 14 • INVESTIGATE Movable Joints, p. 19 • Math in Science, p. 21  **UNIT RESOURCE BOOK** • Datasheet, Movable Joints, p. 31 • Math Support, p. 48 • Math Practice, p. 49 • Additional INVESTIGATION, Bones and Calcium, A, B, & C, pp. 59–67
SECTION   1.3 **The muscular system makes movement possible.** pp. 22–29 Time: 4 periods (2 blocks)  Lesson Plan, pp. 33–34	 • **RESOURCE CENTER,** Muscles • **SIMULATION,** Skeletal Muscles  **UNIT TRANSPARENCY BOOK** • Big Idea Flow Chart, p. T1 • Daily Vocabulary Scaffolding, p. T2 • 3-Minute Warm-Up, p. T5 • Chapter Outline, pp. T7–T8	 • EXPLORE Muscles, p. 22 • CHAPTER INVESTIGATION, A Closer Look at Muscles, pp. 28–29  **UNIT RESOURCE BOOK** CHAPTER INVESTIGATION, A Closer Look at Muscles, A, B, & C, pp. 50–58

KEY TO ICONS 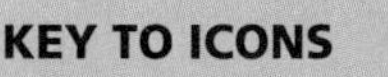 INTERNET  CD/CD-ROM **Pupil Edition**  **Teacher Edition** UNIT RESOURCE BOOK  UNIT TRANSPARENCY BOOK UNIT ASSESSMENT BOOK  SPANISH ASSESSMENT BOOK SCIENCE TOOLKIT

READING AND REINFORCEMENT

ASSESSMENT

STANDARDS

- Four Square, B22–23
- Combination Notes, C36
- Daily Vocabulary Scaffolding, H1–8

UNIT RESOURCE BOOK

- Vocabulary Practice, pp. 45–46
- Decoding Support, p. 47
- Summarizing the Chapter, pp. 68–69

Audio Readings CD
Listen to Pupil Edition.

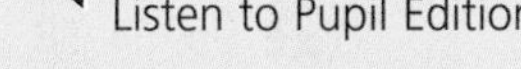

Audio Readings in Spanish CD
Listen to Pupil Edition in Spanish.

- Chapter Review, pp. 30–32
- Standardized Test Practice, p. 33

UNIT ASSESSMENT BOOK

- Diagnostic Test, pp. 1–2
- Chapter Test, A, B, & C, pp. 6–17
- Alternative Assessment, pp. 18–19

Spanish Chapter Test, pp. 65–68

Test Generator CD-ROM
Generate customized tests.

Lab Generator CD-ROM
Rubrics for Labs

National Standards
A.1–8, A.9.a–c, A.9.e–g, C.1.a, C.1.d–e, G.1.b

See p. 6 for the standards.

UNIT RESOURCE BOOK

- Reading Study Guide, A & B, pp. 13–16
- Spanish Reading Study Guide, pp. 17–18
- Challenge and Extension, p. 19
- Reinforcing Key Concepts, p. 21

Ongoing Assessment, pp. 9–12

Section 1.1 Review, p. 12

UNIT ASSESSMENT BOOK
Section 1.1 Quiz, p. 3

National Standards
A.2–7, A.9.a–b, A.9.e–f, C.1.a, C.1.d, G.1.b

UNIT RESOURCE BOOK

- Reading Study Guide, A & B, pp. 24–27
- Spanish Reading Study Guide, pp. 28–29
- Challenge and Extension, p. 30
- Reinforcing Key Concepts, p. 32
- Challenge Reading, pp. 43–44

Ongoing Assessment, pp. 15–17, 19–20

Section 1.2 Review, p. 20

UNIT ASSESSMENT BOOK
Section 1.2 Quiz, p. 4

National Standards
A.2–8, A.9.a–c, A.9.e–f, C.1.d–e, G.1.b

UNIT RESOURCE BOOK

- Reading Study Guide, A & B, pp. 35–38
- Spanish Reading Study Guide, pp. 39–40
- Challenge and Extension, p. 41
- Reinforcing Key Concepts, p. 42

Ongoing Assessment, pp. 22–25, 27

Section 1.3 Review, p. 27

UNIT ASSESSMENT BOOK
Section 1.3 Quiz, p. 5

National Standards
A.1–7, A.9.a–b, A.9.e–g, C.1.d–e, G.1.b

Previewing Resources for Differentiated Instruction

CHAPTER INVESTIGATION

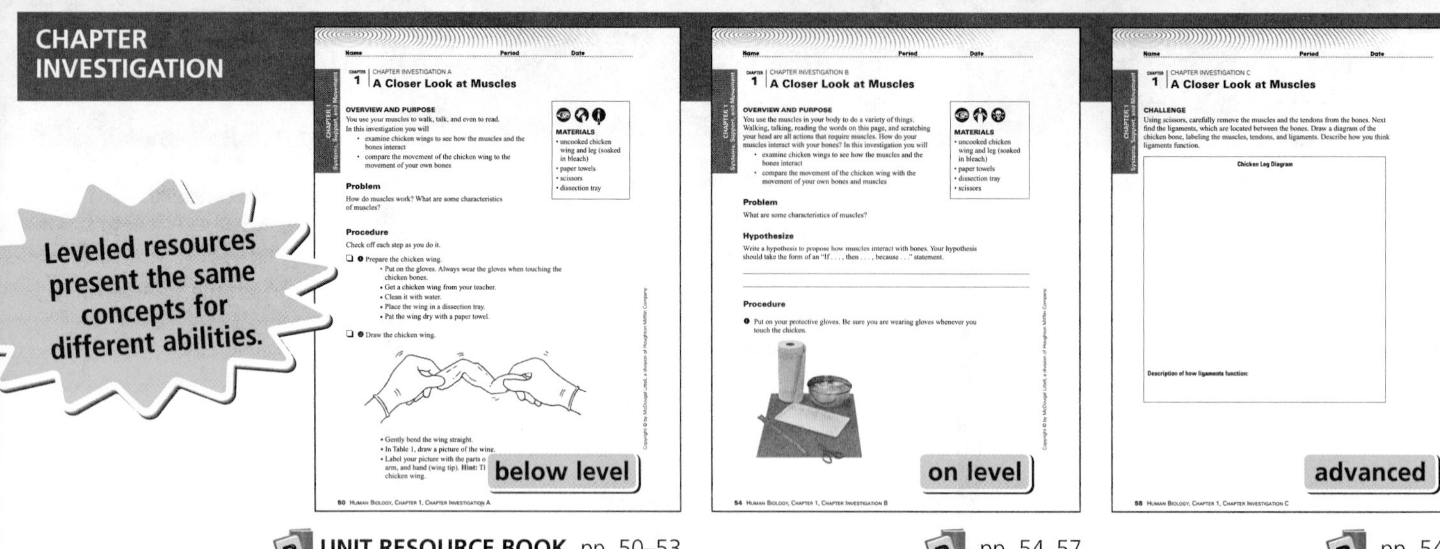

UNIT RESOURCE BOOK, pp. 50–53 | pp. 54–57 | pp. 54–58

READING STUDY GUIDE

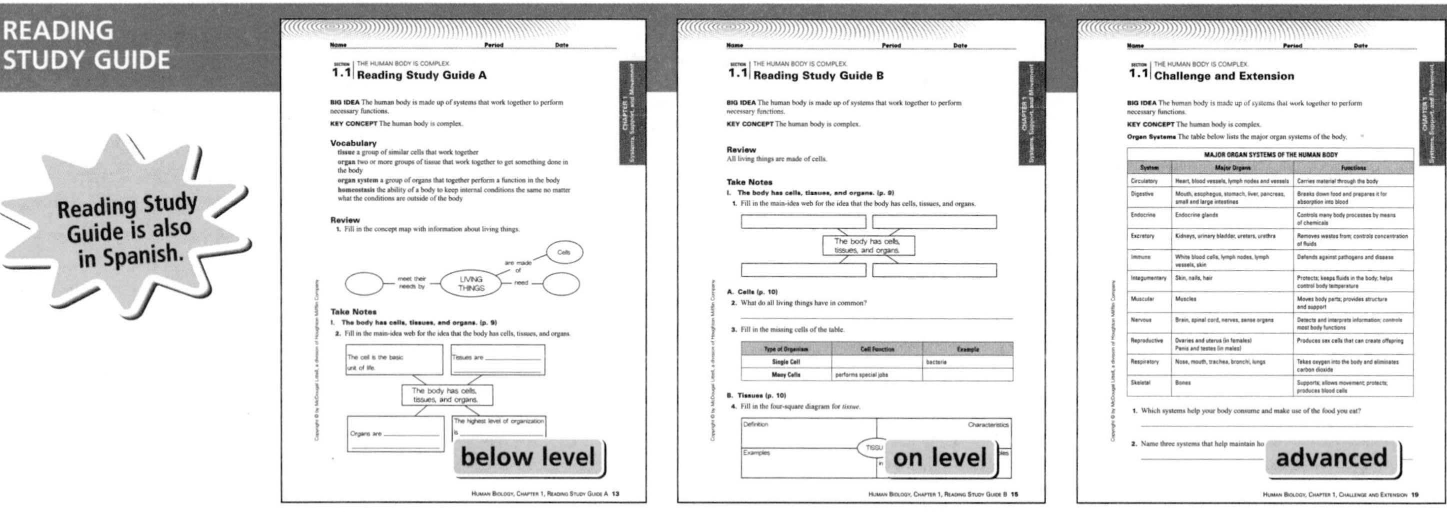

UNIT RESOURCE BOOK, pp. 13–14 | pp. 15–16 | p. 19

CHAPTER TEST

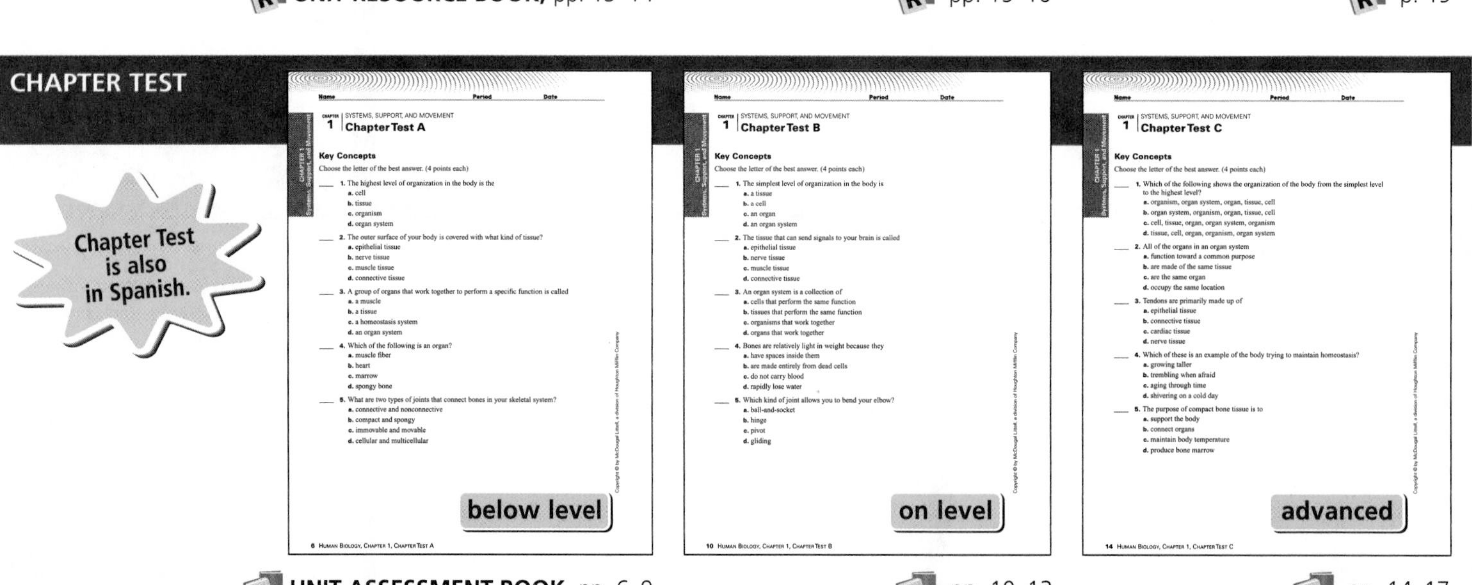

UNIT ASSESSMENT BOOK, pp. 6–9 | pp. 10–13 | pp. 14–17

TECHNOLOGY

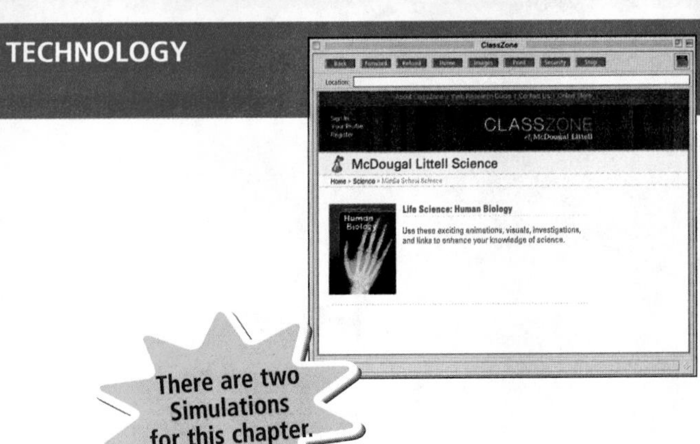

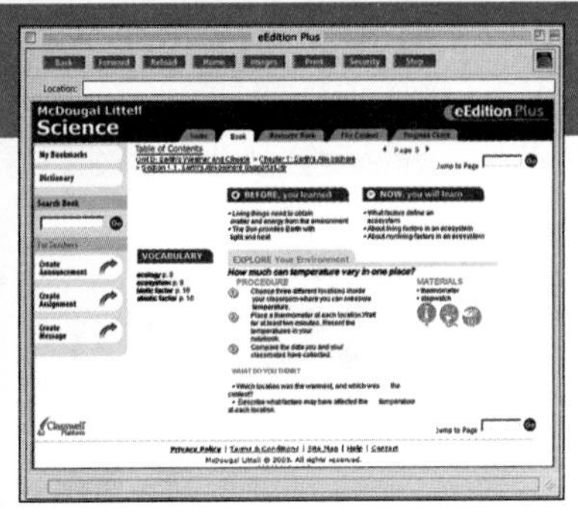

VISUAL CONTENT

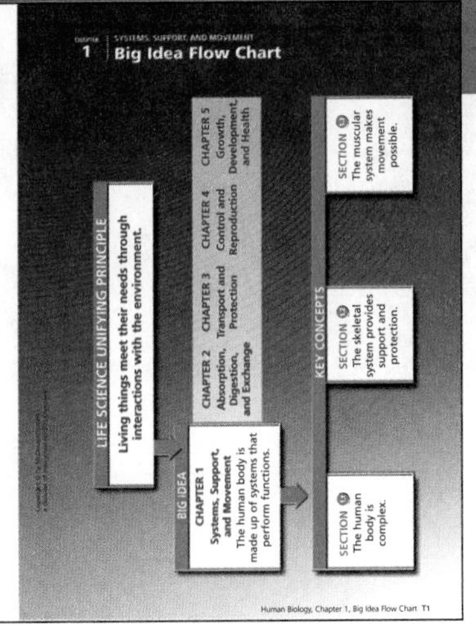

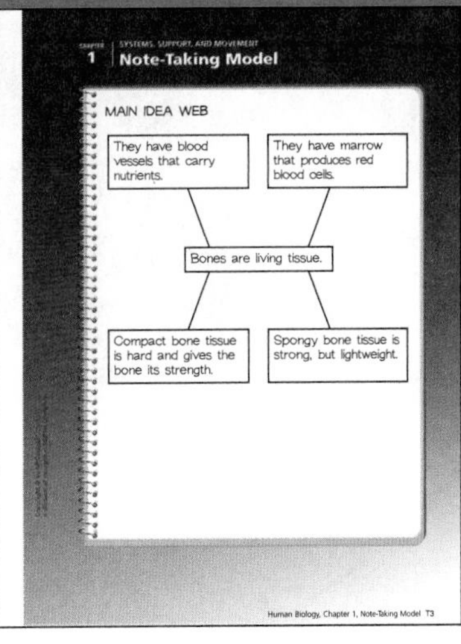

CHAPTER 1 SYSTEMS, SUPPORT, AND MOVEMENT
Note-Taking Model
MAIN IDEA WEB
They have blood vessels that carry nutrients.
They have marrow that produces red blood cells.
Bones are living tissue.
Compact bone tissue is hard and gives the bone its strength.
Spongy bone tissue is strong, but lightweight.
Human Biology, Chapter 1, Note-Taking Model T3

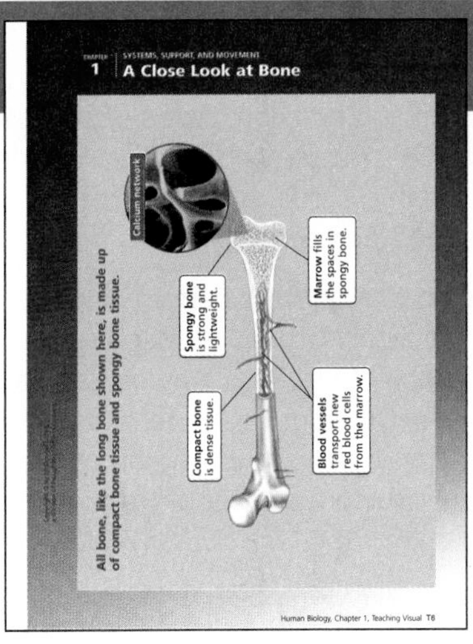

UNIT TRANSPARENCY BOOK, p. T1

p. T3

p. T6

MORE SUPPORT

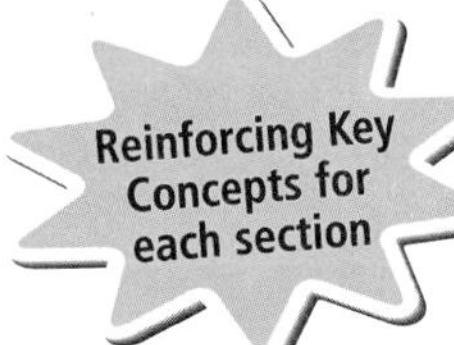

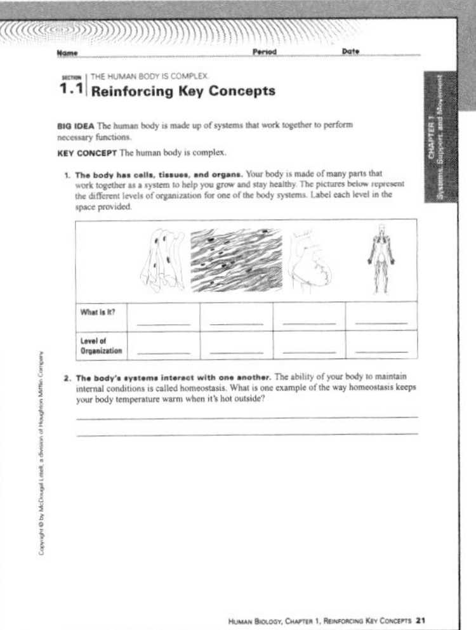

Name Period Date

SECTION 1.1 THE HUMAN BODY IS COMPLEX

Reinforcing Key Concepts

BIG IDEA The human body is made up of systems that work together to perform necessary functions.

KEY CONCEPT The human body is complex.

1. **The body has cells, tissues, and organs.** Your body is made of many parts that work together as a system to help you grow and stay healthy. The pictures below represent the different levels of organization for one of the body systems. Label each level in the space provided.

What is it?				
Level of Organization				

2. **The body's systems interact with one another.** The ability of your body to maintain internal conditions is called homeostasis. What is one example of the way homeostasis keeps your body temperature warm when it's hot outside?

Human Biology, Chapter 1, Reinforcing Key Concepts 21

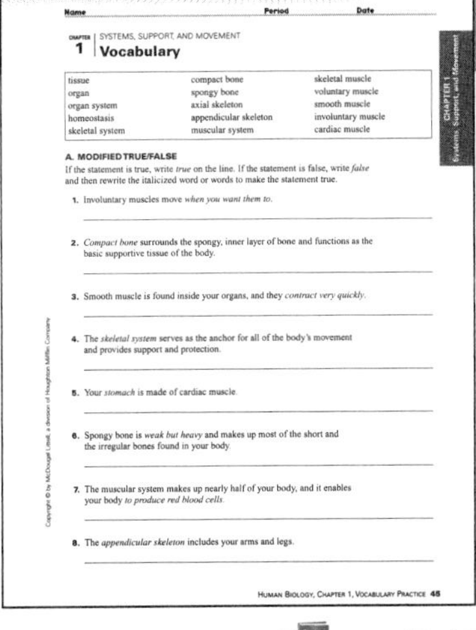

Name Period Date

CHAPTER 1 SYSTEMS, SUPPORT, AND MOVEMENT

Vocabulary

tissue	compact bone	skeletal muscle
organ	spongy bone	voluntary muscle
organ system	axial skeleton	smooth muscle
homeostasis	appendicular skeleton	involuntary muscle
skeletal system	muscular system	cardiac muscle

A. MODIFIED TRUE/FALSE

If the statement is true, write *true* on the line. If the statement is false, write *false* and then rewrite the italicized word or words to make the statement true.

1. Involuntary muscles move *when you want them to.*
2. *Compact bone* surrounds the spongy, inner layer of bone and functions as the basic supportive tissue of the body.
3. Smooth muscle is found inside your organs, and they *contract very quickly.*
4. The *skeletal system* serves as the anchor for all of the body's movement and provides support and protection.
5. Your *stomach* is made of cardiac muscle.
6. Spongy bone is *weak but heavy* and makes up most of the short and the irregular bones found in your body.
7. The muscular system makes up nearly half of your body, and it enables your body *to produce red blood cells.*
8. The *appendicular skeleton* includes your arms and legs.

Human Biology, Chapter 1, Vocabulary Practice 45

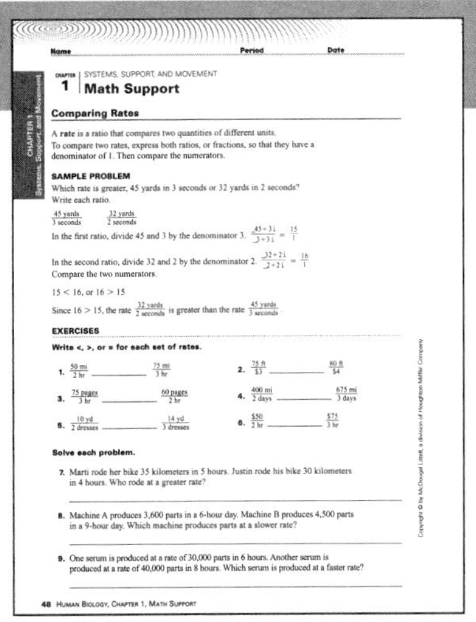

Name Period Date

CHAPTER 1 SYSTEMS, SUPPORT, AND MOVEMENT

Math Support

Comparing Rates

A **rate** is a ratio that compares two quantities of different units. To compare two rates, express both ratios, or fractions, so that they have a denominator of 1. Then compare the numerators.

SAMPLE PROBLEM

Which rate is greater, 45 yards in 3 seconds or 32 yards in 2 seconds? Write each ratio.

$\frac{45 \text{ yards}}{3 \text{ seconds}}$ $\frac{32 \text{ yards}}{2 \text{ seconds}}$

In the first ratio, divide 45 and 3 by the denominator 3. $\frac{45 \div 3}{3 \div 3} = \frac{15}{1}$

In the second ratio, divide 32 and 2 by the denominator 2. $\frac{32 \div 2}{2 \div 2} = \frac{16}{1}$

Compare the two numerators.

$15 < 16$, or $16 > 15$

Since $16 > 15$, the rate $\frac{32 \text{ yards}}{2 \text{ seconds}}$ is greater than the rate $\frac{45 \text{ yards}}{3 \text{ seconds}}$

EXERCISES

Write <, >, or = for each set of rates.

Solve each problem.

7. Marti rode her bike 35 kilometers in 5 hours. Justin rode his bike 30 kilometers in 4 hours. Who rode at a greater rate?
8. Machine A produces 3,600 parts in a 6-hour day. Machine B produces 4,500 parts in a 9-hour day. Which machine produces parts at a slower rate?
9. One serum is produced at a rate of 30,000 parts in 6 hours. Another serum is produced at a rate of 40,000 parts in 8 hours. Which serum is produced at a faster rate?

48 Human Biology, Chapter 1, Math Support

UNIT RESOURCE BOOK, p. 21

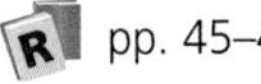

pp. 45–46

p. 48

CHAPTER 1

Systems, Support, and Movement

INTRODUCE

the BIG idea

Have students look at the photograph of the wheelchair racer and discuss how the question in the box links to the Big Idea:

- How are joints helping this racer?
- What is the function of the skeletal system? of the muscular system?

National Science Education Standards

Content

C.1.a All living systems demonstrate structure and function. Levels of organization for structure and function include cells, organs, tissues, organ systems, organisms, and ecosystems.

C.1.d Specialized cells perform specialized functions. Groups of specialized cells cooperate to form a tissue, such as a muscle. Different tissues form organs. Each type of cell, tissue, and organ has a distinct structure and set of functions.

C.1.e The human organism has systems for digestion, respiration, reproduction, circulation, excretion, movement, control and coordination, and protection.

Process

A.1–8 Identify questions that can be answered through scientific investigations; design and conduct an investigation; use tools to gather and interpret data; use evidence to describe, predict, explain, model; think critically to make relationships between evidence and explanation; recognize different explanations and predictions; communicate scientific procedures and explanations; use mathematics.

A.9.a–c, A.9.e–g Understand scientific inquiry by using different investigations, methods, mathematics, and explanations based on logic, evidence, and skepticism. Data often results in new investigations.

G.1.b Science requires different abilities.

CHAPTER

Systems, Support, and Movement

the BIG idea

The human body is made up of systems that work together to perform necessary functions.

Key Concepts

SECTION 1.1 **The human body is complex.**
Learn about the parts and systems in the human body.

SECTION 1.2 **The skeletal system provides support and protection.**
Learn how the skeletal system is organized and what it does.

SECTION 1.3 **The muscular system makes movement possible.**
Learn about the different types of muscles and how they work.

Internet Preview

CLASSZONE.COM

Chapter 1 online resources: Content Review, two Simulations, two Resource Centers, Math Tutorial, Test Practice

What systems make it possible for this racer to move so fast?

INTERNET PREVIEW

CLASSZONE.COM For student use with the following pages:

Review and Practice
- Content Review, pp. 8, 30
- Math Tutorial: Comparing Rates, p. 21
- Test Practice, p. 33

Activities and Resources
- Internet Activity: The Human Body, p. 7
- Resource Centers: Skeletal System, p. 15; Muscles, p. 23
- Simulations: Assemble a Skeleton, p. 16; Skeletal Muscles, p. 26

Tissues and Organs
Code: MDL044

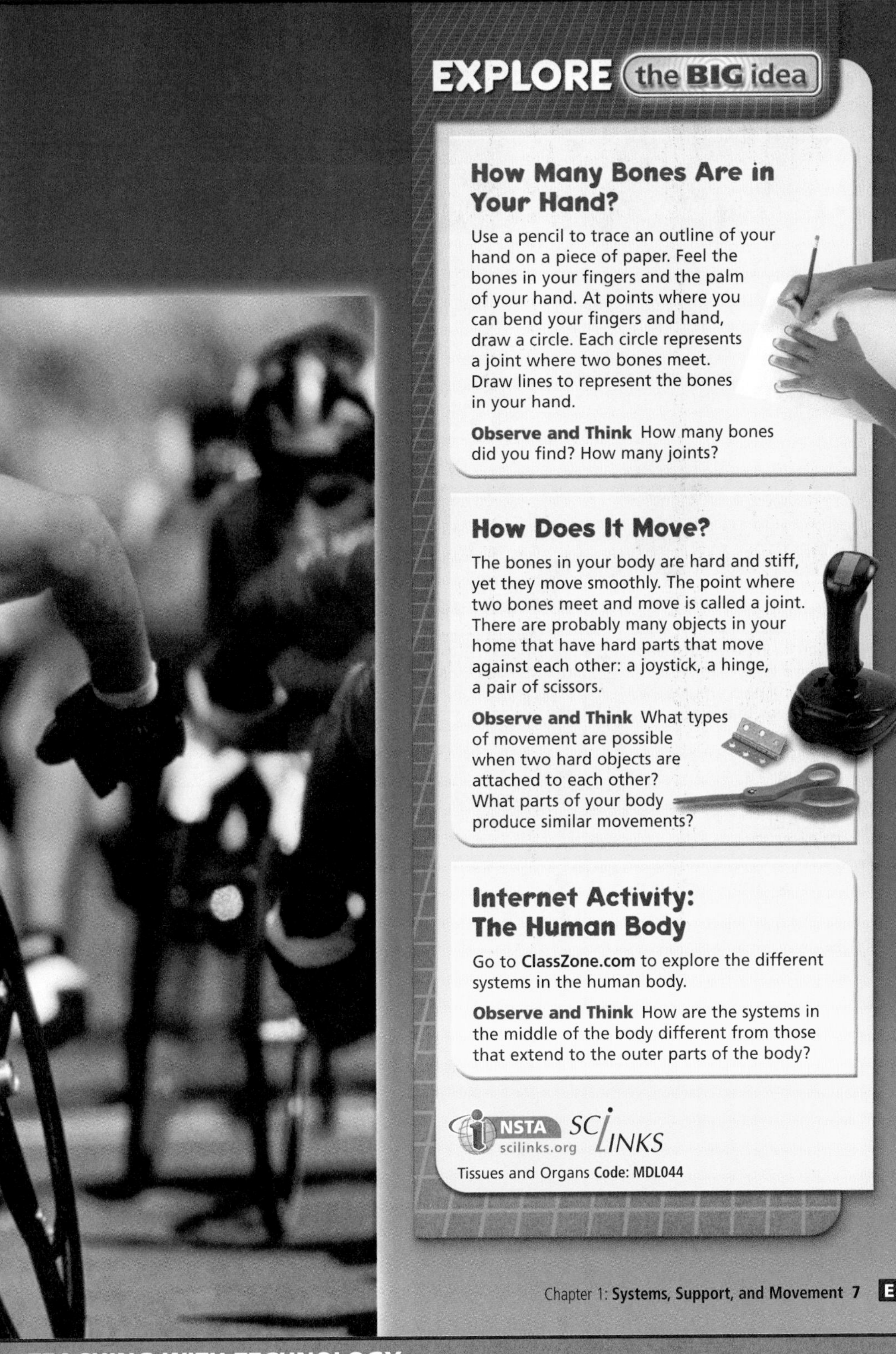
EXPLORE the BIG idea

How Many Bones Are in Your Hand?

Use a pencil to trace an outline of your hand on a piece of paper. Feel the bones in your fingers and the palm of your hand. At points where you can bend your fingers and hand, draw a circle. Each circle represents a joint where two bones meet. Draw lines to represent the bones in your hand.

Observe and Think How many bones did you find? How many joints?

How Does It Move?

The bones in your body are hard and stiff, yet they move smoothly. The point where two bones meet and move is called a joint. There are probably many objects in your home that have hard parts that move against each other: a joystick, a hinge, a pair of scissors.

Observe and Think What types of movement are possible when two hard objects are attached to each other? What parts of your body produce similar movements?

Internet Activity: The Human Body

Go to **ClassZone.com** to explore the different systems in the human body.

Observe and Think How are the systems in the middle of the body different from those that extend to the outer parts of the body?

NSTA scilinks.org SCILINKS
Tissues and Organs **Code: MDL044**

Chapter 1: **Systems, Support, and Movement 7** **E**

TEACHING WITH TECHNOLOGY

CBL and Probeware For the demonstration of leverage that students perform on p. 14, you may wish to have them take measurements with a force sensor to compare forces.

Digital Camera For the investigation on p. 19, you might want to have students use a digital camera to take pictures of joint movements. They can print the photographs and add labels for their activity write-ups.

EXPLORE the BIG idea

These inquiry-based activities are appropriate for use at home or as a supplement to classroom instruction.

How Many Bones Are in Your Hand?

PURPOSE To introduce students to the structure of bones and joints. By examining their own hands, students will see that movement occurs at joints between bones.

TIP *10 min.* Encourage students to find all the joints in their hands, then have them infer whether the same number might exist in their feet.

Answer: There are 19 bones in a hand and 14 joints. A joint allows movement between two bones.

REVISIT after p. 17.

How Does It Move?

PURPOSE To introduce students to how joints allow movement. Students observe the movements of common household objects and compare them to body movements.

TIP *10 min.* You might display a group of objects from your home or school (hinge, scissors, joystick, tracking ball) instead of eliciting recollection.

Answer: Movement is limited in direction by how the joint is put together. Elbow, knee, and fingers are like hinges; the neck is like the pivot on a pair of scissors; the shoulder and hip are like the joystick.

REVISIT after p. 19.

Internet Activity: The Human Body

PURPOSE To introduce students to the structures and functions of the different systems of the human body.

TIP *20 min.* You might wish to have each student take a different system and then compare notes with other students.

Sample answer: The systems in the middle of the body are more essential to body function than the systems toward the outer parts of the body.

REVISIT after p. 12.

PREPARE

CONCEPT REVIEW

Activate Prior Knowledge

- Ask volunteers to name any systems that make up the human body. *Sample answer: skeletal, circulatory, nervous, respiratory, digestive, immune, reproductive, muscular, integumentary*
- Have them name any organs that they know of in each system. Organize their ideas in a chart on the board. Use the names of the systems as column headings. *Sample answer: heart, lungs, brain, skin*
- Discuss how systems work together—for example, to get nourishment.

TAKING NOTES

Main Idea Web

When students write ideas from the text in a web, they not only practice them but reorganize them from a linear pattern to the more multidimensional patterns that the brain uses.

Vocabulary Strategy

The four square diagram organizes all aspects of a concept into a coherent pattern. By filling in their own words, students personalize their understanding.

Point out that it is okay to have a blank square in the diagram. Some terms, such as *skeletal system,* have no clear nonexamples.

Vocabulary and Note-Taking Resources

- Vocabulary Practice, pp. 45–46
- Decoding Support, p. 47

- Daily Vocabulary Scaffolding, p. T2
- Note-Taking Model, p. T3

- Four Square, B22–23
- Combination Notes, C36
- Daily Vocabulary Scaffolding, H1–8

CHAPTER 1

Getting Ready to Learn

CONCEPT REVIEW

- The cell is the basic unit of living things.
- Systems are made up of interacting parts that share matter and energy.
- In multicellular organisms cells work together to support life.

VOCABULARY REVIEW

See Glossary for definitions.

cell
system

CONTENT REVIEW
CLASSZONE.COM
Review concepts and vocabulary.

TAKING NOTES

MAIN IDEA WEB

Write each new blue heading in a box. Then write notes in boxes around the center box that give important terms and details about that blue heading.

VOCABULARY STRATEGY

Write each new vocabulary term in the center of a **four square** diagram. Write notes in the squares around each term. Include a definition, some features, and some examples of the term. If possible, write some things that are not examples of the term.

SCIENCE NOTEBOOK

The cell is the basic unit of living things.

Tissues are groups of similar cells that function together.

The body has cells, tissues, and organs.

Organs are groups of tissues working together.

Definition	Features
Group of cells that work together	A level of organization in the body
Examples	**Nonexamples**
connective tissue, like bone	individual bone cells

TISSUE

See the Note-Taking Handbook on pages R45–R51.

CHECK READINESS

Administer the Diagnostic Test to determine students' readiness for new science content and their mastery of requisite math skills.

Diagnostic Test, pp. 1–2

Technology Resources

Students needing content and math skills should visit **ClassZone.com**.

- **CONTENT REVIEW**
- **MATH TUTORIAL**

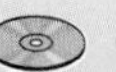

CONTENT REVIEW CD-ROM

KEY CONCEPT

The human body is complex.

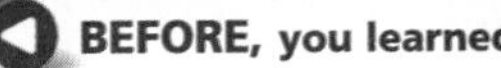

BEFORE, you learned

- All living things are made of cells
- All living things need energy
- Living things meet their needs through interactions with the environment

NOW, you will learn

- About the organization of the human body
- About different types of tissues
- About the functions of organ systems

VOCABULARY

tissue p. 10
organ p. 11
organ system p. 12
homeostasis p. 12

THINK ABOUT

How is the human body like a city?

A city is made up of many parts that perform different functions. Buildings provide places to live and work. Transportation systems move people around. Electrical energy provides light and heat. Similarly, the human body is made of several systems. The skeletal system, like the framework of a building, provides support. The digestive system works with the respiratory system to provide energy and materials. What other systems in your body can you compare to a system in the city?

The body has cells, tissues, and organs.

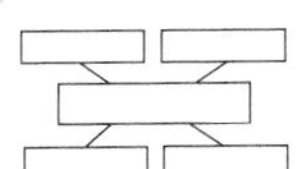

MAIN IDEA WEB
As you read this section, complete the main idea web begun on page 8.

Your body is made of many parts that work together as a system to help you grow and stay healthy. The simplest level of organization in your body is the cell. Next come tissues, then individual organs, and then systems that are made up of organs. The highest level of organization is the organism itself. You can think of the body as having five levels of organization: cells, tissues, organs, organ systems, and the organism. Although these levels seem separate from one another, they all work together.

CHECK YOUR READING What are five levels of organization in your body?

FOCUS

Set Learning Goals

Students will

- Describe the organization of the human body.
- Describe four types of tissue.
- Explain how organ systems function.
- Predict which body systems are involved in various everyday activities.

3-Minute Warm-Up

Display Transparency 4 or copy this exercise on the board:

Draw a cluster diagram with the words *Living Things* in the center circle and three empty circles connected to it. Fill in the circles with three attributes of all living things. *Sample answer: All consist of cells; all need energy; all interact with their environment to meet their needs.*

3-Minute Warm-Up, p. T4

MOTIVATE

THINK ABOUT

PURPOSE Have students work in pairs or small groups to make tables comparing the body systems to the various public works systems in a city.

Ongoing Assessment

CHECK YOUR READING *Answer: from smallest to largest: cells, tissues, organs, organ systems, organism*

RESOURCES FOR DIFFERENTIATED INSTRUCTION

Below Level

UNIT RESOURCE BOOK
- Reading Study Guide A, pp. 13–14
- Decoding Support, p. 47

Advanced

UNIT RESOURCE BOOK
Challenge and Extension, p. 19

English Learners

UNIT RESOURCE BOOK
Spanish Reading Study Guide, pp. 17–18

- Audio Readings in Spanish
- Audio Readings (English)

1.1 INSTRUCT

INVESTIGATE Systems

PURPOSE To understand how body systems work together

TIPS *20 min.*

- Allow students to use the chart from "Activate Prior Knowledge," p. 8.
- Remind students to be aware of their breathing and their hearts beating.
- Suggest that students bring in pictures of people performing everyday tasks, for follow-up discussion.

WHAT DO YOU THINK? *Sample answers: Talking on the phone involves the nervous system, the respiratory system, and the muscular system. When you sleep, your body breathes, excretes waste through the skin and the mouth, moves, and dreams.*

CHALLENGE *Answers will vary, but should group activities appropriately.*

Datasheet, Systems, p. 20

Technology Resources

Customize this student lab as needed or look for an alternative. Print rubrics to assess student lab reports.

Lab Generator CD-ROM

Develop Critical Thinking

COMPARE Have students list ways in which people are like cells. *No two are exactly alike, both grow and can repair themselves, both have parts that serve to control all functions, both are living and have special jobs, both use food and oxygen to produce energy, both expel wastes, and both will eventually die.*

Ongoing Assessment

Describe the organization of the human body.

Ask: Why are the cells of complex organisms specialized? *Specialized cells perform different jobs necessary for survival.*

Answer: Tissues are made of cells of the same type.

INVESTIGATE Systems

How do the systems in your body interact?

SKILL FOCUS
Predicting

MATERIALS
large sheet of paper

TIME
20 minutes

PROCEDURE

1. Work with other classmates to make a list of everyday activities.
2. Discuss how your body responds to each task. Record your ideas.
3. Identify and count the systems in your body that you think are used to perform the task.
4. Have someone from your group make a chart of the different activities.

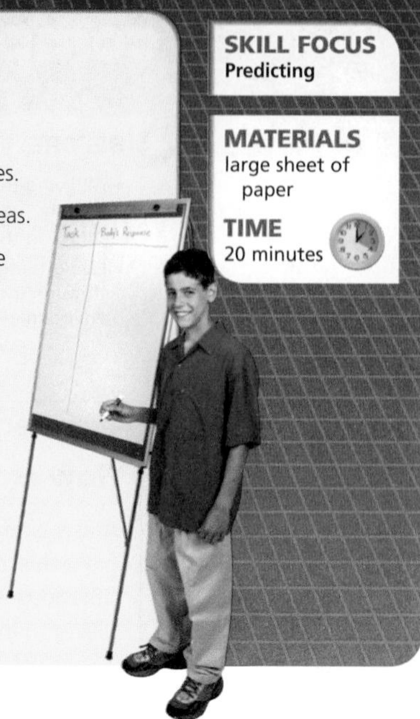

WHAT DO YOU THINK?

- Which systems did you name, and how did they work together to perform each activity?
- When you are asleep, what activities does your body perform?

CHALLENGE How could you make an experiment that would test your predictions?

Cells

The cell is the basic unit of life. Cells make up all living things. Some organisms, such as bacteria, are made of only a single cell. In these organisms the single cell performs all of the tasks necessary for survival. That individual cell captures and releases energy, uses materials, and grows. In more complex organisms, such as humans and many other animals and plants, cells are specialized. Specialized cells perform special jobs. A red blood cell, for example, carries oxygen and other nutrients from the lungs throughout the body.

Tissues

A **tissue** is a group of similar cells that work together to perform a particular function. Think of a tissue as a brick wall and the cells within it as the individual bricks. Taken together, the bricks form something larger and more functional. But just as the bricks need to be placed in a certain way to form the wall, cells must be organized in a tissue.

How are cells related to tissues?

The human body contains several types of tissues. These tissues are classified into four main groups according to their function: epithelial tissue, nerve tissue, muscle tissue, and connective tissue.

DIFFERENTIATE INSTRUCTION

More Reading Support

A Give an example of a specialized cell. *red blood cell*

B What do you call a group of similar cells that function together? *a tissue*

English Learners Help English learners with different word forms. For example, in the "Cells" paragraph, p. 10, *make* is used in a variety of forms: "Cells make up all living things;" "Some organisms, such as bacteria, are made of only a single cell." This verb can be found as *make, made, make up, made up, made up of,* or *made of.*

Epithelial (EHP-uh-THEE-lee-uhl) tissue functions as a boundary. It covers all of the inner and outer surfaces of your body. Each of your internal organs is covered with a layer of epithelial tissue. Another type of tissue, nerve tissue, functions as a messaging system. Cells in nerve tissue carry electrical impulses between your brain and the various parts of your body in response to changing conditions.

Muscle tissue functions in movement. Movement results when muscle cells contract, or shorten, and then relax. In some cases, such as throwing a ball, you control the movement. In other cases, such as the beating of your heart, the movement occurs without conscious control. Connective tissue functions to hold parts of the body together, providing support, protection, strength, padding, and insulation. Tendons and ligaments are connective tissues that hold bones and muscles together. Bone itself is another connective tissue. It supports and protects the soft parts of your body.

Organs

Groups of different tissues make up organs. An **organ** is a structure that is made up of two or more types of tissue that work together to carry out a function in the body. For example, the structures that carry blood around your body contain all four types of tissues. As in cells and tissues, the structure of an organ relates to its function. The stomach's bag-shaped structure and strong muscular walls make it suited for breaking down food. The walls of the heart are also muscular, allowing it to function as a pump.

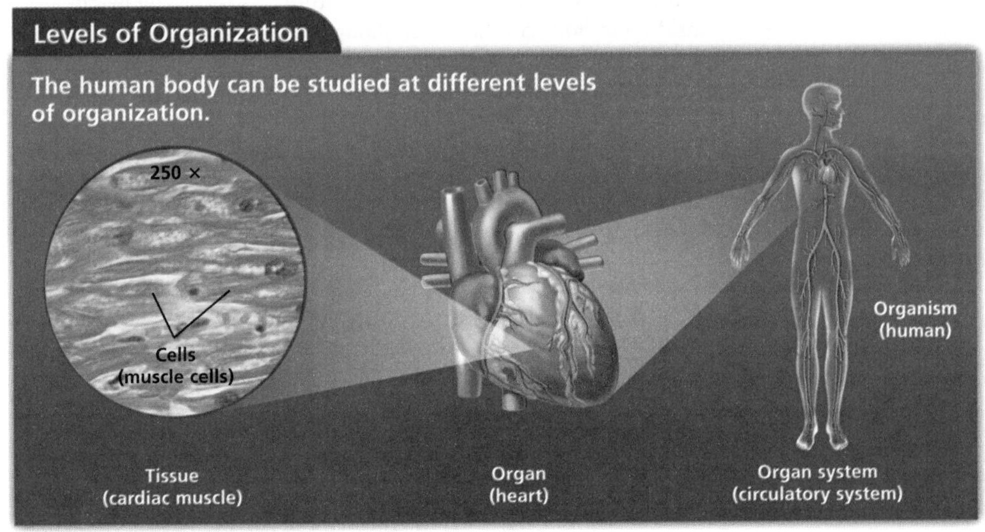

DIFFERENTIATE INSTRUCTION

More Reading Support

C What type of tissue provides support, protection, strength, and padding? *connective tissue*

D What do you call tissues that work together on one function? *organ*

Advanced Although this module presents the biology of a healthy human body, everyone has strengths and weaknesses. Explain that survival and high achievement are not dependent on full functioning of every organ. Discuss successful individuals with disabilities. Examples: Sir Steve Redgrave, Olympic gold medal for rowing: diabetes, dyslexia; Sir Stephen Hawking, theoretical physicist: amyotrophic lateral sclerosis (ALS); Andrea Bocelli, tenor: blindness; Tionne "T-Boz" Watkins, R&B singer: sickle-cell disease.

R Challenge and Extension, p. 19

Address Misconceptions

IDENTIFY Ask: What makes your heart beat? If students respond that the heartbeat is the result of a natural rhythm started at birth, they may hold the misconception that the body does things without needing any internal communications.

CORRECT Have students take their pulse before and after running in place for one minute, noting that their heart rate changes. Point out that the nervous system tells the heart that the legs are moving and that it needs to beat faster.

REASSESS Ask: How do you know when you need to eat? *The brain monitors the body, realizes that the body needs nourishment, and makes me feel hungry.*

Technology Resources

Visit **ClassZone.com** for background on common student misconceptions.

MISCONCEPTION DATABASE

Metacognitive Strategy

Have students devise a strategy for remembering the four tissue types. Possibilities include a graphic organizer, mnemonics, and pictures.

Teach Difficult Concepts

Students may have trouble understanding that epithelial tissue serves as a boundary. Discuss a familiar type of boundary, such as room walls. Ask:

- What do room boundaries do? *keep heat in or out, weather and noise out*
- What do epithelial tissues, such as skin, do? *keep heat and fluid in, bacteria and other unwanted things out*

Ongoing Assessment

Describe four types of tissue.

Ask: What are the four types of tissue and their functions? *Epithelial tissues act as boundaries; nerves carry messages; muscles enable movement; connective tissues protect, insulate, and support.*

Teach Difficult Concepts

Some students may have a hard time understanding homeostasis. Use a thermostat as an example. Point out that the heat or air conditioning is supplied just long enough to return an area to the set temperature.

EXPLORE the BIG idea

Revisit "Internet Activity: The Human Body" on p. 7. Have students discuss how body structures are related to their functions.

Ongoing Assessment

Explain how organ systems function.

Ask: How do organ systems meet the body's needs? *Each organ system performs a necessary function, and together the systems allow the human organism to grow, reproduce, and maintain life.*

Reinforce the BIG idea

Have students relate the section to the Big Idea.

 Reinforcing Key Concepts, p. 21

1.1 ASSESS & RETEACH

Assess

 Section 1.1 Quiz, p. 3

Reteach

Have students draw a flow chart of the progression from cells to tissues to organs to organ systems. In the final drawing, students should make a rough outline of the human body with the systems labeled. Have volunteers share their drawings in a class discussion.

Technology Resources

Have students visit **ClassZone.com** for reteaching of Key Concepts.

 CONTENT REVIEW

 CONTENT REVIEW CD-ROM

Organ Systems

An **organ system** is a group of organs that together perform a function that helps the body meet its needs for energy and materials. For example, your stomach, mouth, throat, large and small intestines, liver, and pancreas are all part of the organ system called the digestive system. The body is made up of many organ systems. In this unit, you will read about these systems. They include the skeletal, muscular, respiratory, digestive, urinary, circulatory, immune, nervous, and reproductive systems. Together, these systems allow the human organism to grow, reproduce, and maintain life.

The body's systems interact with one another.

READING TIP

VOCABULARY
The word *homeostasis* contains two word roots. *Homeo* comes from a root meaning "same." *Stasis* comes from a root meaning "stand still" or "stay."

The ability of your body to maintain internal conditions is called **homeostasis** (HOH-mee-oh-STAY-sihs). Your body is constantly regulating such things as your body temperature, the amount of sugar in your blood, even your posture. The processes that take place in your body occur within a particular set of conditions.

The body's many levels of organization, from cells to organ systems, work constantly to maintain the balance needed for the survival of the organism. For example, on a hot day, you may sweat. Sweating keeps the temperature inside your body constant, even though the temperature of your surroundings changes.

INFER This student is drinking water after exercising. Why is it important to drink fluids after you sweat?

1.1 Review

KEY CONCEPTS

1. Draw a diagram that shows the relationship among cells, tissues, organs, and organ systems.
2. Make a chart of the four basic tissue groups that includes names, functions, and examples.
3. Identify three functions performed by organ systems.

CRITICAL THINKING

4. **Apply** How does drinking water after you sweat help maintain homeostasis?
5. **Compare and Contrast** Compare and contrast the four basic tissue groups. How would all four types of tissue be involved in a simple activity, like raising your hand?

CHALLENGE

6. **Apply** Describe an object, such as a car, that can be used as a model of the human body. Explain how the parts of the model relate to the body.

ANSWERS

1. Diagrams should show organ systems as groups of organs, as groups of tissues, as groups of cells.

2. Charts should include epithelial, nerve, muscle, and connective tissues, and show understanding of each.

3. Sample answer: digestion, elimination, reproduction

4. When you sweat, your body loses water. Adding new water to the body helps maintain homeostasis.

5. Sample answer: When you raise your hand, nerve tissue carries a message from your brain, and muscle tissue contracts to move the bones, made of connective and epithelial tissues.

6. Answers should mention that the parts work together for the functioning of the whole.

SKILL: INFERRING

What Does the Body Need to Survive?

In 1914, Ernest Shackleton and 27 men set sail for Antarctica. Their goal was to cross the continent by foot and sled. The crew never set foot on Antarctica. Instead, the winter sea froze around their ship, crushing it until it sank. They were stranded on floating ice, over 100 miles from land. How long could they survive? How would their bodies respond? What would they need to stay alive?

You can make inferences in answer to any of these questions. First you need to recall what you know. Then you need new evidence. What was available to the explorers? Did they save supplies from their ship? What resources existed in the environment?

Prior Knowledge

- The human body needs air, water, and food.
- The human body needs to maintain its temperature. The body can be harmed if it loses too much heat.

Observations

Several of Shackleton's explorers kept diaries. From the diaries we know that the following:

- The crew hunted seals and penguins for fresh meat.
- The temperature was usually below freezing.
- Tents and overturned lifeboats sheltered the crew from the wind.
- Their clothes were made of thick fabric and animal skins and furs.
- They melted snow and ice in order to have fresh water.

Make Inferences

On Your Own Describe how the explorers met each of the needs of the human body.

As a Group How long do you think these 28 men could have survived these conditions? Use evidence and inferences in your answer.

CHALLENGE How might survival needs differ for sailors shipwrecked in the tropics compared to the Antarctic?

Learn more about Shackleton's expedition.

13 E

ANSWERS

ON YOUR OWN *For food, they hunted; for shelter, they used tents and overturned lifeboats; their clothes protected them from the cold; for water, they melted snow and ice.*

AS A GROUP *Sample answer: These men could have survived as long as seals and penguins were available for food. Diseases and injuries may have threatened their survival.*

CHALLENGE *Protection from sunburn and heat stroke and from harmful insects and other animals would be necessary. Obtaining fresh water might be difficult.*

THINK SCIENCE
Scientific Methods of Thinking

Set Learning Goal

To make inferences, using prior knowledge and new evidence

Present the Science

Shackleton and his men rescued themselves. When they had to abandon their ship, they took three small open boats with them and dragged them along as they made their way across the ice that had frozen over the Weddell Sea. When they reached open water, they sailed to Elephant Island. One boat, with Shackleton as part of its crew, set out for South Georgia, 800 miles away. It took them 14 days to reach the island. Then three of them traveled 17 miles across South Georgia's mountains and glaciers to reach a whaling station. It took Shackleton four tries to get back to Elephant Island, but in the end not one member of the expedition was lost.

Guide the Activity

- Suggest that students remember times when they were cold and think about what they needed then.
- Have students share any experiences of outdoor exposure they may have had, such as camping in the rain or extreme heat or cold or being caught outdoors in a storm.
- After they have answered the questions on the page, students may do a Web quest for Shackleton's diary entries.

Close

Ask: How might an expedition today be better equipped? *Sample answer: It might carry global positioning communication devices, special fabrics for warmth and dryness, and compact emergency food.*

Technology Resources

Have students visit **ClassZone.com** to learn more about Shackleton's expedition.

RESOURCE CENTER

1.2 FOCUS

Set Learning Goals

Students will

- Describe different types of bone tissue.
- Explain how the human skeleton is organized.
- Describe how joints allow movement.
- Observe how joints help them perform several activities.

3-Minute Warm-Up

Display Transparency 4 or copy this exercise on the board:

Draw a sequence diagram showing the levels of organization of the human body, from simplest to most complex.

Diagram should show cells, tissues, organs, organ systems, and the organism.

3-Minute Warm-Up, p. T4

1.2 MOTIVATE

EXPLORE Levers

PURPOSE To observe that bones can act as levers

TIP *10 min.* Students can use a backpack instead of a sports bag.

WHAT DO YOU THINK? *Nearest to the shoulder; in the hand; the action is slowed as the weight travels farther.*

Teaching with Technology

For the demonstration of levers, you may wish to have students use a force sensor if one is available.

KEY CONCEPT

1.2 The skeletal system provides support and protection.

BEFORE, you learned

- The body is made of cells, tissues, organs, and systems
- Cells, tissues, organs, and organ systems work together
- Systems in the body interact

NOW, you will learn

- About different types of bone tissue
- How the human skeleton is organized
- How joints allow movement

VOCABULARY

skeletal system p. 14
compact bone p. 15
spongy bone p. 15
axial skeleton p. 16
appendicular skeleton p. 16

EXPLORE Levers

How can a bone act as a lever?

PROCEDURE

1. A lever is a stiff rod that pivots about a fixed point. Hold the bag in your hand and keep your arm straight, like a lever. Move the bag up and down.
2. Move the handles of the bag over your elbow. Again hold your arm straight and move the bag up and down.
3. Now move the bag to the top of your arm and repeat the procedure.

MATERIALS

sports bag

WHAT DO YOU THINK?

- At which position is it easiest to move the bag?
- At which position does the bag move the farthest?
- How does the position of a load affect the action of a lever?

MAIN IDEA WEB

Make a web of the important terms and details about the main idea: *Bones are living tissue.*

Bones are living tissue.

Every movement of the human body is possible because of the interaction of muscles with the **skeletal system.** Made up of a strong connective tissue called bone, the skeletal system serves as the anchor for all of the body's movement, provides support, and protects soft organs inside the body. Bones can be classified as long bones, short bones, irregular bones, and flat bones. Long bones are found in the arms and legs. Short bones are found in the feet and hands. Irregular bones are found in the spine. Flat bones are found in the ribs and skull.

You might think that bones are completely solid and made up of dead tissue. They actually are made of both hard and soft materials.

RESOURCES FOR DIFFERENTIATED INSTRUCTION

Below Level

UNIT RESOURCE BOOK

- Reading Study Guide A, pp. 24–25
- Decoding Support, p. 47

AUDIO CDS

Additional INVESTIGATION, Bones and Calcium, A, B, & C, pp. 59–67; Teacher Instructions, pp. 308–309

Advanced

UNIT RESOURCE BOOK

- Challenge and Extension, p. 30
- Challenge Reading, pp. 43–44

English Learners

UNIT RESOURCE BOOK

Spanish Reading Study Guide, pp. 28–29

AUDIO CDS

- Audio Readings in Spanish
- Audio Readings (English)

Like your heart or skin, bones are living tissue. Bones are not completely solid, either; they have spaces inside. The spaces allow blood cells carrying nutrients to travel throughout the bones. Because bones have spaces, they weigh much less than they would if they were solid.

CLASSZONE.COM

Explore the skeletal system.

Two Types of Bone Tissue

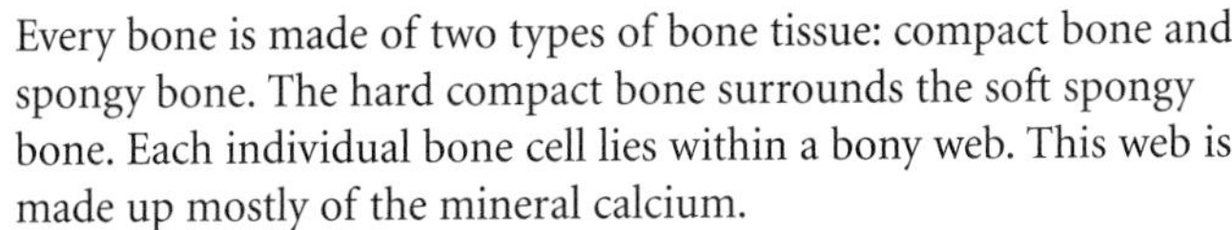

A Every bone is made of two types of bone tissue: compact bone and spongy bone. The hard compact bone surrounds the soft spongy bone. Each individual bone cell lies within a bony web. This web is made up mostly of the mineral calcium.

Compact Bone Surrounding the spongy, inner layer of the bone is a hard layer called **compact bone.** Compact bone functions as the basic supportive tissue of the body, the part of the body you call the skeleton. The outer layer of compact bone is very hard and tough. It covers the outside of most bones.

Spongy Bone Inside the bone, the calcium network is less dense. This tissue is called **spongy bone.** Spongy bone is strong but lightweight. It makes up most of the short and the irregular bones found in your body. It also makes up the ends of long bones.

Marrow and Blood Cells

B Within the spongy bone tissue is marrow, the part of the bone that produces red blood cells. The new red blood cells travel from the marrow into the blood vessels that run throughout the bone. The blood cells bring nutrients to the bone cells and carry waste materials away.

A Close Look at Bone

All bone, like the long bone shown here, is made up of compact bone tissue and spongy bone tissue.

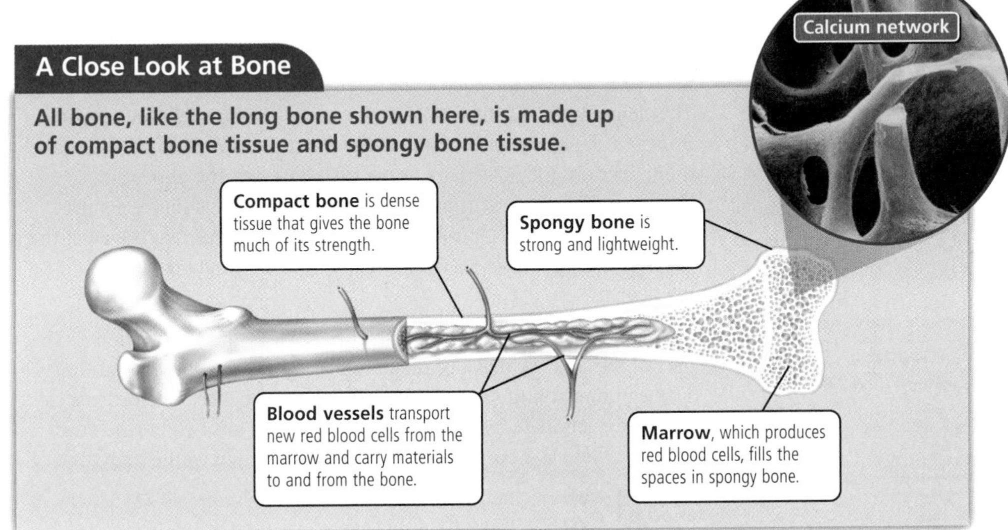

Calcium network
Compact bone is dense tissue that gives the bone much of its strength.
Spongy bone is strong and lightweight.
Blood vessels transport new red blood cells from the marrow and carry materials to and from the bone.
Marrow, which produces red blood cells, fills the spaces in spongy bone.

1.2 INSTRUCT

Teach from Visuals

To help students interpret the bone diagram, ask:

- Why is the compact bone able to give the bone strength? *It is dense.*
- How does bone receive nourishment? *Blood vessels bring in nutrients.*
- Where are red blood cells formed? *in the marrow*

Address Misconceptions

IDENTIFY Show students a bone and a solid white rock. Ask: What are some differences between a bone and a rock? If students do not mention the facts that bones have hollow spaces and soft marrow inside them and are living tissues made up of cells, they may hold the misconception that bones are solid and nonliving.

CORRECT Refer students to the bone diagram on p. 15. Have them identify specific differences between rock and bone.

REASSESS Ask: What are the parts of a bone, and what is the function of each? *Dense compact bone gives bones most of their strength; spongy bone is strong but less dense to reduce the weight of bones; marrow produces red blood cells; blood vessels bring nutrients to the bone and carry away waste materials.*

Technology Resources

Visit **ClassZone.com** for background on common student misconceptions.

MISCONCEPTION DATABASE

Ongoing Assessment

Describe different types of bone tissue.

Ask: What kinds of bone tissue make up bones? *Compact bone is dense and gives bones strength; spongy bone is less dense.*

DIFFERENTIATE INSTRUCTION

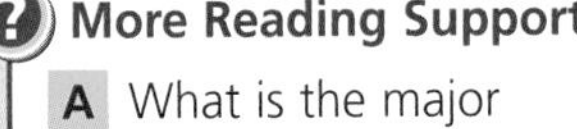

More Reading Support

A What is the major ingredient of bone? *calcium*

B Name the four parts of a bone. *compact bone, spongy bone, marrow, and blood vessels*

English Learners Demonstrate for English learners how some English sentences can be rearranged without changing the meaning. For example, the sentence "Surrounding the spongy, inner layer of the bone is a hard layer called compact bone," can be rearranged to read "A hard layer called compact bone surrounds the spongy, inner layer of the bone." Write the two sentences on the board and ask students to explain how they are different.

Real World Example

All blood cells begin as stem cells. A disease like leukemia interferes with stem-cell growth. In such cases, a bone marrow transplant can help. A donor provides healthy bone marrow, which is introduced into the patient intravenously. In a successful transplant, the marrow cells find their way to the bone spaces and begin to produce new stem cells after a few weeks.

Develop Critical Thinking

APPLY Have students discuss how the length of one's bones might affect one's athletic abilities. Ask: Would a short femur be more advantageous to a basketball player or to a jockey? *jockey*

Address Misconceptions

IDENTIFY Ask: Why do bones need food and oxygen? If students respond that bones convey food and oxygen to other organs or that they "store" food and oxygen, they may hold the misconception that bones are not alive and do not need nourishment.

CORRECT Ask students to discuss what happens to a bone after it breaks. Remind them that a bone must be alive in order to heal.

REASSESS Ask: How do bones use food and oxygen? *They use them in order to perform their special functions and to grow new cells.*

Technology Resources

Visit **ClassZone.com** for background on common student misconceptions.

MISCONCEPTION DATABASE

Ongoing Assessment

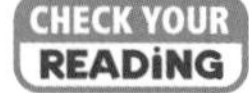
CHECK YOUR READING *Answer: Both divisions of the skeleton work with other systems to allow movement. The axial skeleton provides mainly protection and support, while the appendicular skeleton functions mainly to allow movement.*

The skeleton is the body's framework.

Like the frame of a building, the skeleton provides the body's shape. The skeleton also works with other systems to allow movement. Scientists have identified two main divisions in the skeleton. These are the axial (AK-see-uhl) skeleton, which is the central part of the skeleton, and the appendicular (AP-uhn-DIHK-yuh-luhr) skeleton. Bones in the appendicular skeleton are attached to the axial skeleton. The diagram on page 17 labels some of the important bones in your skeleton.

VOCABULARY
Remember to add four squares for *axial skeleton* and *appendicular skeleton* to your notebook.

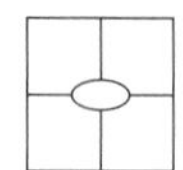

The Axial Skeleton

Imagine a line straight down your back. You can think of that line as an axis. Sitting, standing, and twisting are some of the motions that turn around the axis. The **axial skeleton** is the part of the skeleton that forms the axis. It provides support and protection. In the diagram, parts of the axial skeleton are colored in red.

The axial skeleton includes the skull, or the cranium (KRAY-nee-uhm). The major function of the cranium is protection of the brain. Most of the bones in the cranium do not move. The skull connects to the spinal column in a way that allows the head to move up and down as well as right to left.

Your spinal column makes up the main portion of the axial skeleton. The spinal column is made up of many bones called vertebrae. The many bones allow flexibility. If you run your finger along your back you will feel the vertebrae. Another set of bones belonging to the axial skeleton are the rib bones. The ribs function to protect the soft internal organs, such as the heart and lungs.

The Appendicular Skeleton

The diagram shows the bones in the appendicular skeleton in yellow. Bones in the **appendicular skeleton** function mainly to allow movement. The shoulder belongs to the upper part of the appendicular skeleton. The upper arm bone that connects to the shoulder is the longest bone in the upper body. It connects with the two bones of the lower arm. The wristbone is the end of one of these bones in the lower arm.

SIMULATION
CLASSZONE.COM
Assemble a skeleton.

The lower part of the body includes the legs and the hip bones. This part of the body bears all of the body's weight when you are standing. The leg bones are the strongest of all the bones in the skeleton. Just as the lower arm includes two bones, the lower leg has two bones. The larger of these two bones carries most of the weight of the body.

CHECK YOUR READING How are the axial and appendicular skeletons alike? How are they different?

DIFFERENTIATE INSTRUCTION

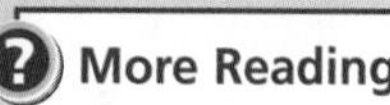
More Reading Support

C What is an advantage of having many bones in the spinal column? *flexibility*

Below Level Have students prepare a Venn diagram to compare and contrast the axial skeleton and the appendicular skeleton.

Axial Skeleton	Both	Appendicular Skeleton
• central part of skeleton • includes skull, spinal column, and ribs • main function is to provide support and protection	• made up of bones • provides body's shape	• attached to axial skeleton • includes bones of the shoulders, hips, arms and legs • main function is to allow movement

The Skeletal System

The skeletal system interacts with other body systems to allow this soccer player to stand, run, and kick.

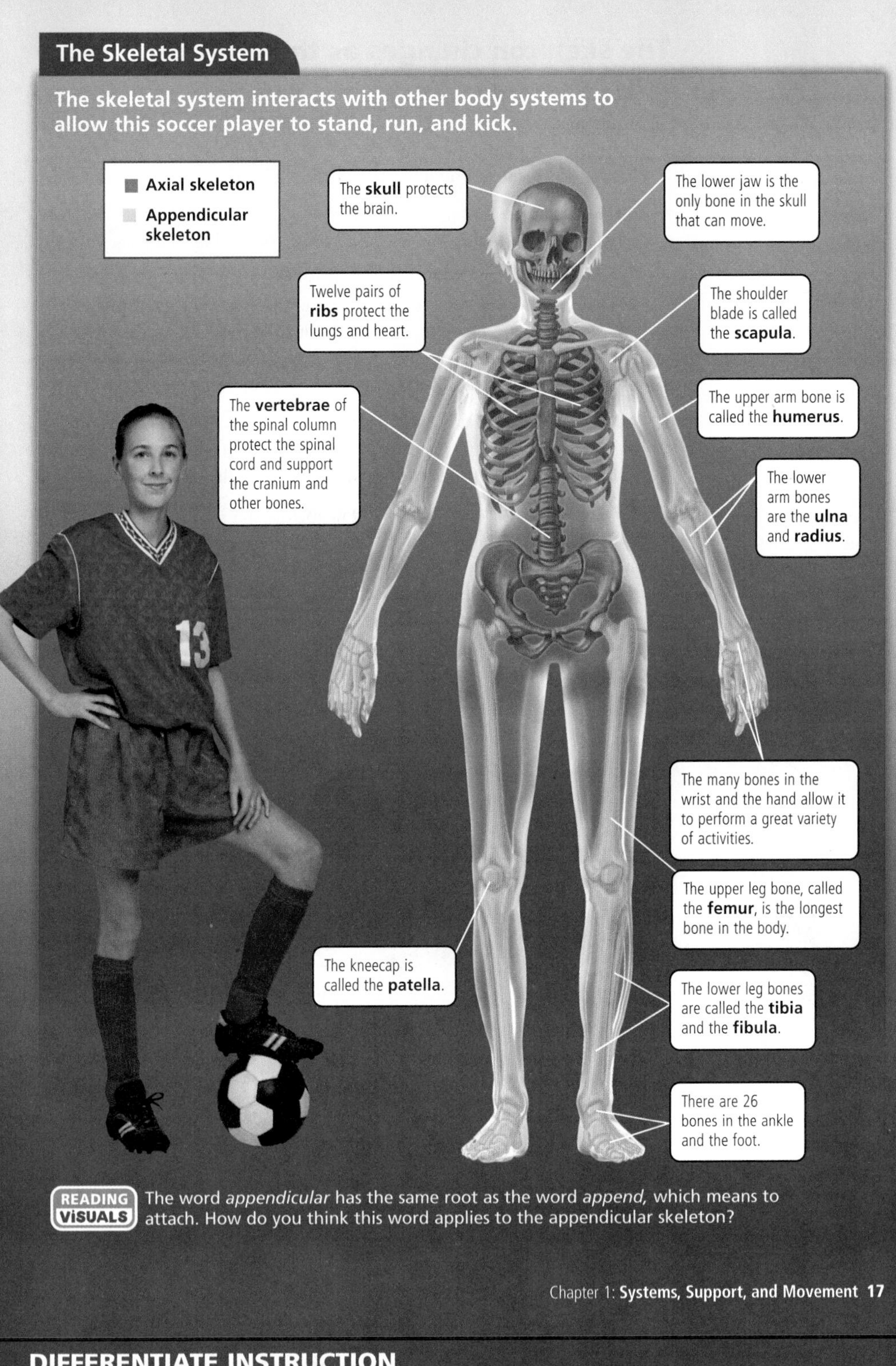

Axial skeleton
Appendicular skeleton
The **skull** protects the brain.
The lower jaw is the only bone in the skull that can move.
Twelve pairs of **ribs** protect the lungs and heart.
The shoulder blade is called the **scapula**.
The **vertebrae** of the spinal column protect the spinal cord and support the cranium and other bones.
The upper arm bone is called the **humerus**.
The lower arm bones are the **ulna** and **radius**.
13
The many bones in the wrist and the hand allow it to perform a great variety of activities.
The upper leg bone, called the **femur**, is the longest bone in the body.
The kneecap is called the **patella**.
The lower leg bones are called the **tibia** and the **fibula**.
There are 26 bones in the ankle and the foot.

READING VISUALS The word *appendicular* has the same root as the word *append,* which means to attach. How do you think this word applies to the appendicular skeleton?

DIFFERENTIATE INSTRUCTION

Advanced Have students research the skeletons of other animals, such as birds, and infer how differences in movement can result from differences in structure. They could draw and label diagrams and present their findings to the class in posters.

R Challenge and Extension, p. 30

Teach from Visuals

To help students interpret the skeleton diagram, point out that the pelvis bone is sometimes considered to be part of the axial and sometimes part of the appendicular skeleton. Ask:

- What functions do vertebrae perform? *They protect the spinal cord and support the cranium and other bones.*
- How many bones are in the ankle and foot? *26*
- Which division of the skeleton are these bones part of? *appendicular skeleton*
- What is the only bone in the skull that can move? *lower jaw*

EXPLORE the BIG idea

Revisit "How Many Bones Are in Your Hand?" on p. 7. Have students compare the numbers they found with the number of bones shown in the foot.

Arts Connection

Have students draw a skull from an actual model. Compare this activity with drawing the head of a live person. Drawing a skeleton or a part such as a skull can help an artist to understand what he or she is seeing when drawing from life. The artist comes to know where deep shadows and raised highlights should be. These can be harder to find in life.

Ongoing Assessment

Explain how the human skeleton is organized.

Ask: How is the body's framework, the skeleton, structured? *The axial skeleton runs top to bottom. The appendicular skeleton runs side to side.*

READING VISUALS *Answer: The appendicular skeleton attaches the arms and legs to the axial skeleton.*

History of Science

The skeleton has been studied at least since the time of the ancient Greeks. The physician Galen (129–c. 199 C.E.) dissected many animals, drawing logical conclusions about the shape and weight of certain bones. Galen produced many treatises on medicine. His medical writings were translated by Arab thinkers of the ninth century and were highly respected during the European Renaissance.

Integrate the Sciences

Calcium is a major component of bone tissue. It is a silvery white element, one of the alkaline-earth metals. Its atomic number is 20. Calcium is the fifth most abundant element in Earth's crust, but in nature it is always found as part of compounds.

- Marble, limestone, and chalk are calcium carbonate.
- Lime, cement, and mortar are calcium hydroxide.
- Teeth and bones are calcium hydroxyphosphate.

In the body, calcium is essential to the contraction of muscles, the transmission of nerve impulses, and the clotting of blood.

Social Studies Connection

Discuss with students how bone has been used to make tools, musical instruments, decorative inlays, and other objects in many cultures. Other examples include knife and gun handles, spear points, whale-bone sled runners and roof supports, scrimshaw, sewing needles, fishhooks, buttons, beads, and pendants. Keep a bulletin board or scrapbook in the classroom. Each student can contribute a page.

Real World Example

Deterioration of and loss of density in bone lead to a disease called osteoporosis. It is estimated that 10 million Americans have osteoporosis, and 34 million—55 percent of Americans 50 years of age or older—are at risk. Osteoporosis is responsible for more than 1.5 million fractures annually.

The skeleton changes as the body develops and ages.

MAIN IDEA WEB Make a web of the important terms and details about the main idea: *The skeleton changes as the body develops and ages.*

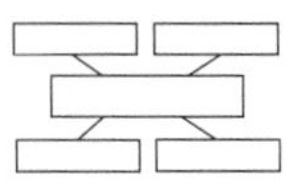

You will remember that bones are living tissue. During infancy and childhood, bones grow as the rest of the body grows. Bones become harder as they stop growing. In adulthood, bones continue to change.

Infancy The skull of a newborn is made up of several bones that have spaces between them. As the brain grows, the skull also grows. During the growth of the skull, the spaces between the bones close.

Childhood Bone growth occurs at areas called growth plates. These growth plates are made of cartilage, a flexible bone tissue. The length and shape of bones is determined by growth plates. Long bones grow at the ends of the bone surrounding growth plates.

Adolescence Toward the end of adolescence (AD-uhl-EHS-uhns) bones stop growing. The growth plate is the last portion of the bone to become hard. Once growth plates become hard, arms and legs stop growing.

REMINDER Density is the ratio of mass over volume. Bone density is a measure of the mass of a bone divided by the bone's volume.

Adulthood Even after bones stop growing, they go through cycles in which old bone is broken down and new bone is formed. As people age, more bone is broken down than is formed. This can lead to a decrease in bone mass, which causes a decrease in bone density. The strength of bones depends upon their density. As people age, their bone density may decrease. Bones that are less dense may break more easily. Many doctors recommend that adults over a certain age get regular bone density tests.

Test of Bone Density

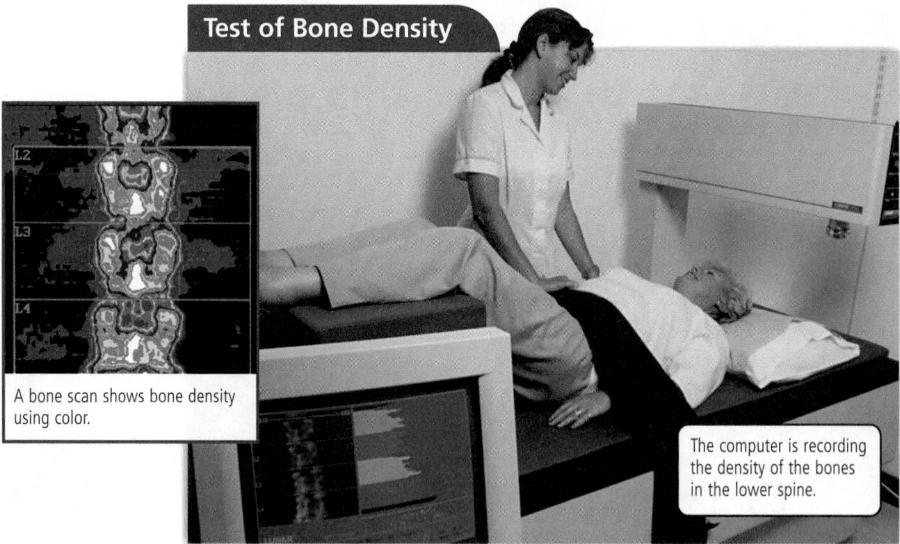

A bone scan shows bone density using color.

The computer is recording the density of the bones in the lower spine.

DIFFERENTIATE INSTRUCTION

More Reading Support

D Where does bone growth occur? *growth plates*

E When do your arms and legs stop growing? *toward the end of adolescence*

Additional Investigation To reinforce Section 1.2 learning goals, use the following full-period investigation:

R **Additional INVESTIGATION,** Bones and Calcium, A, B, & C, pp. 59–67, 305–306
(Advanced students should complete Levels B and C.)

Below Level Use pictures showing various bonelike parts of animals, such as shells, horns, and beaks. Tell whether each is or is not considered bone.

Joints connect parts of the skeletal system.

F A joint is a place at which two parts of the skeletal system meet. There are three types of joints: immovable, slightly movable, and freely movable.

Immovable and Slightly Movable Joints An immovable joint locks bones together like puzzle pieces. The bones of your skull are connected by immovable joints. Slightly movable joints are able to flex slightly. Your ribs are connected to your sternum by slightly movable joints.

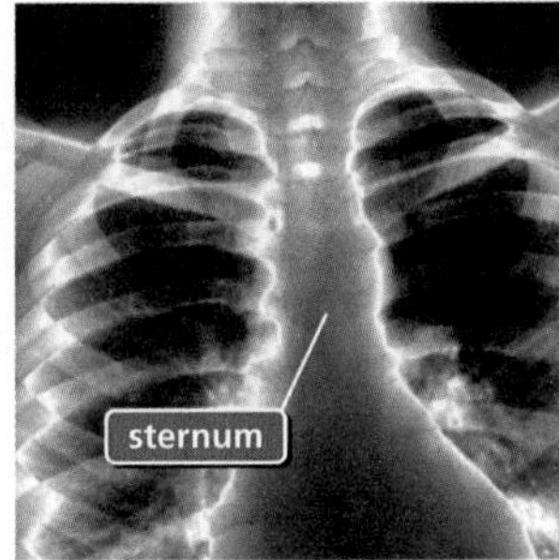

The sternum is an example of a slightly movable joint.

G **Freely Movable Joints** Freely movable joints allow your body to bend and to move. Tissues called ligaments hold the bones together at movable joints. Other structures inside the joint cushion the bones and keep them from rubbing together. The entire joint also is surrounded by connective tissue.

Movable joints can be classified by the type of movement they produce. Think about the movement of your arm when you eat an apple. Your arm moves up, then down, changing the angle between your upper and lower arms. This is angular movement. The joint that produces this movement is called a hinge joint.

INVESTIGATE Movable Joints

How can you move at joints?

PROCEDURE

1. Perform several activities that involve your joints. Twist at the waist. Bend from your waist to one side. Reach into the air with one arm. Open and close your mouth. Push a book across your desk. Lift the book.
2. Record each activity and write a note describing the motion that you feel at each joint.
3. Try to see how many different ways you can move at joints.

WHAT DO YOU THINK?

- How was the motion you felt similar for each activity? How was it different?
- Based on your observations, identify two or more ways that joints move.

CHALLENGE Draw a diagram showing how you think each joint moves. How might you classify different types of joints based upon the way they move?

SKILL FOCUS
Observing

MATERIALS
book

TIME
20 minutes

DIFFERENTIATE INSTRUCTION

More Reading Support

F What are the three types of joints? *immovable, slightly movable, and freely movable*

G What holds bones together at joints? *ligaments*

Advanced Osteoporosis is a decrease in bone mass that can put bones at risk of breaking. Have students research statistics on osteoporosis, along with the treatment and prevention. What methods of screening and diagnosis are in current use?

Have students who are interested in the makeup of bones read the following article:

 Challenge Reading, pp. 43–44

EXPLORE (the BIG idea)

Revisit "How Does It Move?" on p. 7. Have students explain how each type of joint moves.

INVESTIGATE Movable Joints

PURPOSE To make observations about how joints work

TIP *20 min.* Have students remove jackets and roll up sleeves to better observe joint movements.

INCLUSION Include students with restricted movement in this activity by having them serve as recorders and by having them analyze the movements of others.

WHAT DO YOU THINK? *Sample answer: Each motion had a limited range of movement. The types and ranges of movement differed.*

Sample answer: Different joints allow rotational movements, angular movements, and gliding movements.

CHALLENGE *Diagrams may vary, but should show an understanding of the types of joint movements.*

Datasheet, Movable Joints, p. 31

Technology Resources

Customize this student lab as needed or look for an alternative. Print rubrics to assess student lab reports.

 **Lab Generator CD-ROM**

Teaching with Technology

Students can further investigate movable joints by using a digital camera to record ranges of movement.

Ongoing Assessment

Describe how joints allow movement.

Ask: How are joints classified or grouped into categories?

Answer: Joints are classified by the types of movement they allow: freely movable, slightly movable, and immovable.

Ongoing Assessment

READING VISUALS *Answer: The elbow joint is shaped like a hinge allowing angular movement. The hip joint is shaped like a ball and socket allowing rotational movement.*

Reinforce 

Have students relate the section to the Big Idea.

 Reinforcing Key Concepts, p. 32

1.2 ASSESS & RETEACH

Assess

 Section 1.2 Quiz, p. 4

Reteach

On the board, draw an unlabeled diagram of a cross section of a bone (see p. 15) and an unlabeled diagram of a skeleton. Have students copy the diagrams. Have volunteers suggest labels for the diagrams. *Labels for the cross section should include* compact bone, spongy bone, blood vessels, *and* marrow. *Labels for the skeleton should include* axial skeleton, appendicular skeleton, hinge joint (elbow), ball-and-socket joint (hip), *and* names of individual bones.

Technology Resources

Have students visit **ClassZone.com** for reteaching of Key Concepts.

 CONTENT REVIEW

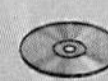 CONTENT REVIEW CD-ROM

Movable Joints

The joints in the elbow and hip allow different types of movement.

Angular movement (elbow)

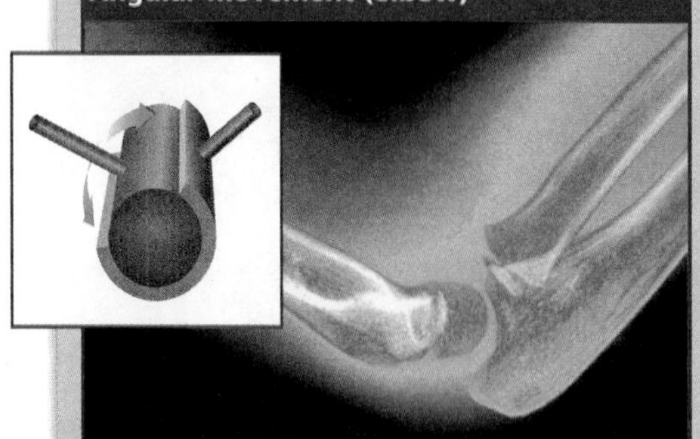

Rotational movement (hip)

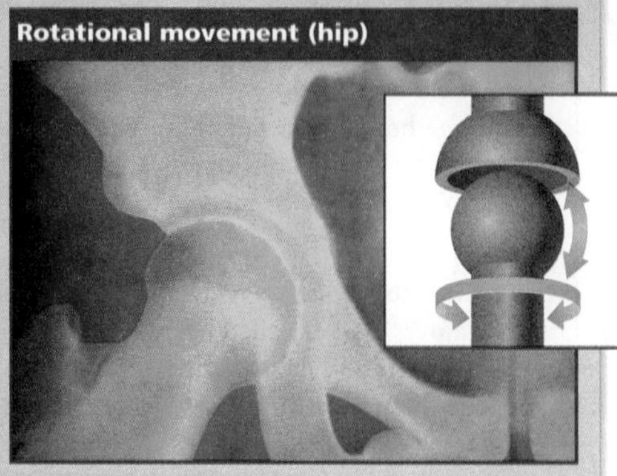

READING VISUALS INFER How do the structure and shape of each joint allow bones to move?

Your arm can also rotate from side to side, as it does when you turn a doorknob. Rotational movement like this is produced by a pivot joint in the elbow. You can also rotate your arm in a circle, like the motion of a softball pitcher winding up and releasing a ball. The joint in the shoulder that produces this type of rotational movement is called a ball-and-socket joint.

Joints also produce gliding movement. All joints glide, that is, one bone slides back and forth across another. In some cases, as with the joints in your backbone, a small gliding movement is the only movement the joint produces.

1.2 Review

KEY CONCEPTS

1. What are the functions of the two types of bone tissue?
2. What are the main divisions of the human skeleton?
3. Name three types of movement produced by movable joints and give an example of each.

CRITICAL THINKING

4. **Infer** What function do immovable joints in the skull perform? Think about the different stages of development in the human body.
5. **Analyze** Which type of movable joint allows the most movement? How does the joint's shape and structure contribute to this?

CHALLENGE

6. **Classify** The joints in your hand and wrist produce three different types of movement. Using your own wrist, classify the joint movement of the fingers, palm, and wrist. Support your answer.

E 20 Unit: Human Biology

ANSWERS

1. Compact bone gives a bone strength. The marrow inside spongy bone produces red blood cells.

2. axial skeleton and appendicular skeleton

3. Ball-and-socket (shoulder) allows movement in almost a full circle. Gliding (wrist) allows movement back and forth or up and down. Hinge (knee) allows movement in one plane.

4. allow rapid brain growth during infancy, then later protect the brain

5. ball-and-socket; ball shape allows movement in all directions

6. Fingers have angular movement, thumb has gliding, and wrist has angular and gliding movement.

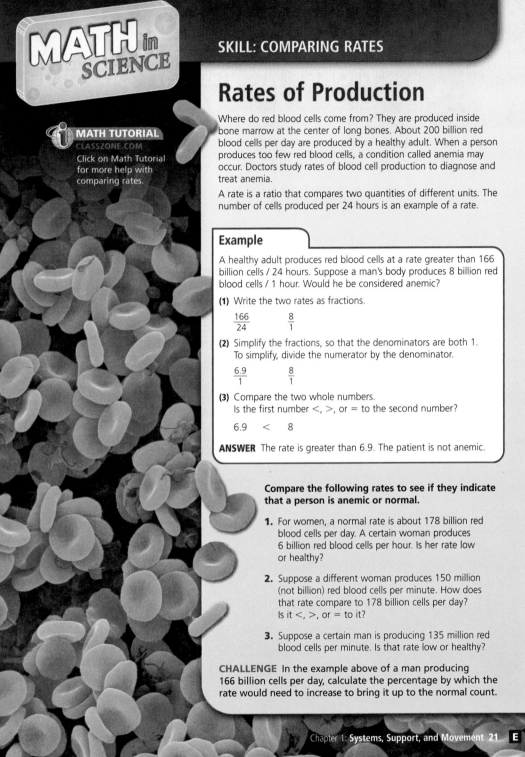

MATH in SCIENCE

SKILL: COMPARING RATES

Rates of Production

MATH TUTORIAL
CLASSZONE.COM
Click on Math Tutorial for more help with comparing rates.

Where do red blood cells come from? They are produced inside bone marrow at the center of long bones. About 200 billion red blood cells per day are produced by a healthy adult. When a person produces too few red blood cells, a condition called anemia may occur. Doctors study rates of blood cell production to diagnose and treat anemia.

A rate is a ratio that compares two quantities of different units. The number of cells produced per 24 hours is an example of a rate.

Example

A healthy adult produces red blood cells at a rate greater than 166 billion cells / 24 hours. Suppose a man's body produces 8 billion red blood cells / 1 hour. Would he be considered anemic?

(1) Write the two rates as fractions.

$\frac{166}{24}$ $\qquad$ $\frac{8}{1}$

(2) Simplify the fractions, so that the denominators are both 1. To simplify, divide the numerator by the denominator.

$\frac{6.9}{1}$ $\qquad$ $\frac{8}{1}$

(3) Compare the two whole numbers. Is the first number $<$, $>$, or $=$ to the second number?

$6.9 < 8$

ANSWER The rate is greater than 6.9. The patient is not anemic.

Compare the following rates to see if they indicate that a person is anemic or normal.

1. For women, a normal rate is about 178 billion red blood cells per day. A certain woman produces 6 billion red blood cells per hour. Is her rate low or healthy?
2. Suppose a different woman produces 150 million (not billion) red blood cells per minute. How does that rate compare to 178 billion cells per day? Is it $<$, $>$, or $=$ to it?
3. Suppose a certain man is producing 135 million red blood cells per minute. Is that rate low or healthy?

CHALLENGE **In the example above of a man producing 166 billion cells per day, calculate the percentage by which the rate would need to increase to bring it up to the normal count.**

ANSWERS

1. low

2. greater than

3. healthy

CHALLENGE 20%

MATH IN SCIENCE
Math Skills Practice for Science

Set Learning Goal

To compare rates of blood-cell production

Present the Science

It is important for the body to maintain an adequate number of red blood cells. Disorders such as iron or vitamin B12 deficiency and diseases such as cancer or leukemia can cause an underproduction of these vital cells. An underproduction of red blood cells is called anemia, and people suffering from it experience fatigue and weakness, shortness of breath, and paleness of skin, eyes, and nails.

Develop Number Sense

- Remind students that both decimals and fractions can represent rates. When a fraction shows a rate, the line may mean "out of."
- Review fraction-to-decimal conversion.

Close

There is a difference between the healthy rate (166 billion per day) and the normal rate (178 billion per day) for women. Ask students to explain this difference. *The normal rate is higher than the rate of cell production needed just to be healthy. That means that if a person's rate decreases due to temporary conditions, they can still be healthy.*

- Math Support, p. 48
- Math Practice, p. 49

Technology Resources

Students can visit **ClassZone.com** for practice in comparing rates.

MATH TUTORIAL

1.3 FOCUS

Set Learning Goals

Students will

- Explain the functions of muscles.
- Describe the different types of muscles and how they work.
- Explain how muscles grow and heal.
- Observe how muscles change with movement.

3-Minute Warm-Up

Display Transparency 5 or copy this exercise on the board:

Match the definitions to the correct terms.

Definitions

1. a bony structure that allows rotation *d*
2. the skeleton division containing the skull *b*
3. the part of a bone that produces red blood cells *e*

Terms

a. compact bone
b. axial skeleton
c. immovable joint
d. ball-and-socket joint
e. marrow

3-Minute Warm-Up, p. T5

1.3 MOTIVATE

EXPLORE Muscles

PURPOSE To observe how muscles change with movement

TIP *10 min.* Remind students to feel both the top and bottom of the leg during movement.

WHAT DO YOU THINK? *When the leg was straightened, the muscles on the top of the leg got thicker. When the leg was bent, the muscles on the bottom of the leg got thicker. Students' questions should relate to changes during movement of muscles.*

Ongoing Assessment

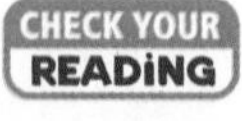

Answer: Muscles produce movement, keep body temperature stable, maintain posture.

KEY CONCEPT

1.3 The muscular system makes movement possible.

BEFORE, you learned

- There are different types of bone tissue
- The human skeleton has two separate divisions
- Joints function in several different ways

NOW, you will learn

- About the functions of muscles
- About the different types of muscles and how they work
- How muscles grow and heal

VOCABULARY

muscular system p. 23
skeletal muscle p. 24
voluntary muscle p. 24
smooth muscle p. 24
involuntary muscle p. 24
cardiac muscle p. 24

How do muscles change as you move?

PROCEDURE

1. Sit on a chair with your feet on the floor.
2. Place your hand around your leg. Straighten one leg as shown in the photograph.
3. Repeat step 2 several times.

WHAT DO YOU THINK?

- How did your muscles change during the activity?
- Record your observations.
- What questions do you have about the muscular system?

Muscles perform important functions.

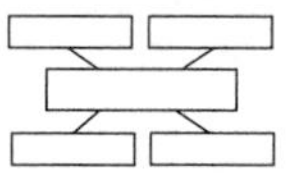

MAIN IDEA WEB
Make a web for the main idea: *Muscles perform important functions.*

Every movement of your body—from the beating of your heart to the movement of food down your throat, to the blinking of your eyes—occurs because of muscles. Some movements are under your control, and other movements seem to happen automatically. However, muscles do more than produce movement. They perform other functions as well. Keeping body temperature stable and maintaining posture are two additional functions of muscles.

What are three functions that muscles perform?

RESOURCES FOR DIFFERENTIATED INSTRUCTION

Below Level

UNIT RESOURCE BOOK
- Reading Study Guide A, pp. 35–36
- Decoding Support, p. 47

Advanced

UNIT RESOURCE BOOK
Challenge and Extension, p. 41

English Learners

UNIT RESOURCE BOOK
Spanish Reading Study Guide, pp. 39–40

- Audio Readings in Spanish
- Audio Readings (English)

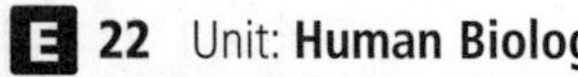
E 22 Unit: Human Biology

Movement

RESOURCE CENTER
CLASSZONE.COM
Discover more about muscles.

The **muscular system** works with the skeletal system to allow movement. Like all muscles, the muscles that produce movement are made up of individual cells called muscle fibers. These fibers contract and relax.

Most of the muscles involved in moving the body work in pairs. As they contract, muscles shorten, pulling against bones. It may surprise you to know that muscles do not push. Rather, a muscle on one side of a bone pulls in one direction, while another muscle relaxes. Muscles are attached to bones by stretchy connective tissue.

Maintaining Body Temperature

Earlier you read that processes within the body require certain conditions, such as temperature and the right amount of water and other materials. The balance of conditions is called homeostasis. One of the functions of the muscular system is related to homeostasis. Muscles function to maintain body temperature.

When muscles contract, they release heat. Without this heat from muscle contraction, the body could not maintain its normal temperature. You may have observed the way your muscles affect your body temperature when you shiver. The quick muscle contractions that occur when you shiver release heat and raise your body temperature.

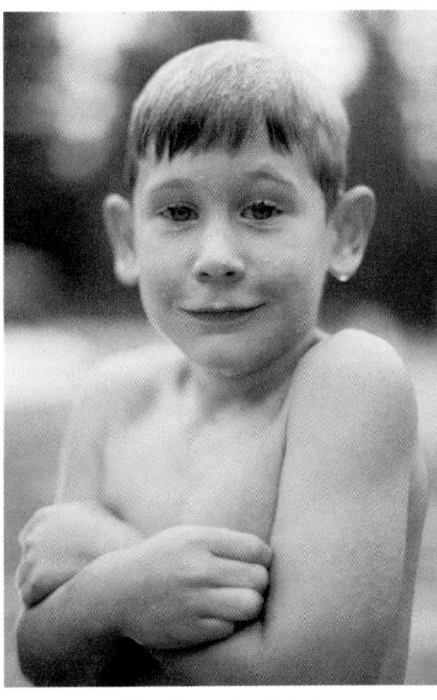
Muscles contract during shivering, raising body temperature.

CHECK YOUR READING How do muscles help maintain homeostasis?

Maintaining Posture

B Have you ever noticed that you stand up straight without thinking about it, even though gravity is pulling your body down? Most muscles in your body are always a little bit contracted. This tension, or muscle tone, is present even when you are sleeping. The muscles that maintain posture relax completely only when you are unconscious.

Try standing on the balls of your feet for a few moments, or on one leg. When you are trying to balance or hold one position for any length of time, you can feel different muscles contracting and relaxing. Your muscles make constant adjustments to keep you sitting or standing upright. You don't have to think about these tiny adjustments; they happen automatically.

1.3 INSTRUCT

Integrate the Sciences

Muscles atrophy when not used. A person's muscles may atrophy when illness confines the person to bed or when the person has to wear a cast. Muscles also atrophy in space, because there is no gravity for the muscles to resist.

In space, muscles can also change type. Without gravity, the nerve receptors in the muscles that usually carry information about gravitational load and muscle use change because of the lack of resistance. Research in this area is being conducted by NASA and is expected to benefit not only astronauts but also people with spinal-cord injuries and long-term illnesses.

Ongoing Assessment

Explain the functions of muscles.

Ask: How do muscles help maintain posture? *All muscles are always a little bit contracted to hold the body up against gravity.*

CHECK YOUR READING *Answer: They create heat to keep the body warm in a cool environment.*

DIFFERENTIATE INSTRUCTION

More Reading Support

A How do muscles work to maintain body temperature? *They generate heat when they contract.*

B What force pulling the body down do muscles work against? *gravity*

English Learners Students new to English may find the following science words challenging: *fiber, posture, expend, bundles, twitch, coordination.* Preview vocabulary by adding these words and their definitions to a science word wall.

Real World Example

Muscular dystrophy is a disease in which the muscles get progressively weaker and may stop working. Sometimes the protective membranes around the muscles are affected. Muscular dystrophy is a genetic disease that affects how the body makes proteins for muscle. Some current research is aimed at correcting the defective genes; other research is aimed at creating chemicals that act like the missing proteins.

Ongoing Assessment

Describe the different types of muscles and how they work.

Ask: What kind of bundles make up skeletal muscles, and what do they do? *Skeletal muscles are made of fast-twitch and slow-twitch bundles. Fast-twitch bundles can move quickly. Slow-twitch bundles move slowly.*

Answer: They move when you want them to.

Answer: You can consciously control skeletal muscles but not smooth muscles. Smooth muscle contracts slowly, whereas skeletal muscle contains both slow-twitch and fast-twitch bundles. Cardiac muscle is restricted to the heart; the other two can be found throughout the body. Like smooth muscle, cardiac muscle contracts slowly.

Your body has different types of muscle.

Your body has three types of muscle. All three types of muscle tissue share certain characteristics. For example, each type of muscle contracts and relaxes. Yet all three muscle types have different functions, and different types of muscle are found in different locations.

Skeletal Muscle

C

The muscles that are attached to your skeleton are called **skeletal muscles.** Skeletal muscle performs voluntary movement—that is, movement that you choose to make. Because they are involved in voluntary movement, skeletal muscles are also called **voluntary muscles.**

READING TIP The root of the word *voluntary* comes from the Latin root *vol-*, meaning "wish." In the word *involuntary* the prefix *in-* suggests the meaning "unwished for." *Involuntary movement* means movement you can't control.

Skeletal muscle, like all muscle, is made of long fibers. The fibers are made up of many smaller bundles, as a piece of yarn is made up of strands of wool. One type of bundle allows your muscles to move slowly. Those muscles are called slow-twitch muscles. Another type of bundle allows your muscles to move quickly. These are called fast-twitch muscles. If you were a sprinter, you would want to develop your fast-twitch muscles. If you were a long distance runner, you would develop your slow-twitch muscles.

What does it mean that skeletal muscles are voluntary muscles?

Smooth Muscle

Smooth muscle is found inside some organs, such as the intestines and the stomach. Smooth muscles perform automatic movement and are called **involuntary muscles.** In other words, smooth muscles work without your knowing it. You have no control over their movement. For example, smooth muscles line your stomach wall and push food through your digestive system. Smooth muscle fibers are not as long as skeletal muscle fibers. Also, unlike skeletal muscles, smooth muscles are not fast-twitch. Smooth muscles contract slowly.

VOCABULARY Remember to add four squares for *involuntary muscles* and *voluntary muscles* to your notebook. Note differences in the two diagrams.

Cardiac Muscle

D

Your heart is made of **cardiac muscle.** Like smooth muscle, cardiac muscle moves without conscious control. Each cardiac muscle cell has a branched shape. The cells of the heart connect in a chain. These chains form webs of layered tissue that allow cardiac cells to contract together and make the heart beat. Just like the smooth muscle cells, the cardiac muscle cells contract slowly.

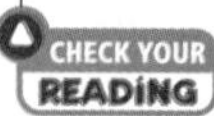

Compare and contrast the three types of muscle described: skeletal, smooth, and cardiac.

DIFFERENTIATE INSTRUCTION

More Reading Support

C Which muscles perform voluntary movements? *skeletal muscles*

D What kind of muscle makes up the heart? *cardiac muscle*

Advanced Have students research past fat-loss fads, such as vibrating belts and electronic muscle stimulators, and explain why these were ineffective. Have students design an exercise machine that they would enjoy and that would work to help people keep fit.

Challenge and Extension, p. 41

Muscle Tissue
The marchers in this band are using all three different types of muscle tissue.
250 ×
Cardiac muscle allows the hearts of the band members to pump blood as they march to the beat of the music.
150 ×
Smooth muscle in the lungs allows the band members to breathe as they play their instruments.
360 ×
Skeletal muscle moves the legs of these marchers.
READING VISUALS Which movements of these band members are voluntary, and which are involuntary?

Teach from Visuals

To help students see the science in the photograph of a band, ask:

- What are the voluntary muscles of the band members doing? *controlling their marching, allowing them to play instruments*
- What are involuntary muscles doing in each band member? *making the heart beat, breathing*
- Does everything you do involve all three types of muscle? Explain. *No; when you are asleep, you are not using skeletal muscle.*

Real World Example

A backpack is usually the best way for students to carry all their books and other school supplies. The strongest muscles in the body, the back and abdominal muscles, support the weight of the pack, which is designed to hang evenly. A backpack can overload the body, however, if it is too heavy or hangs unevenly. In such cases, a cart with wheels can help.

Ongoing Assessment

READING VISUALS *Answer: Their walking in time to music and playing instruments are voluntary. Their breathing and the beating of their hearts is involuntary.*

DIFFERENTIATE INSTRUCTION

Below Level Have students find pictures of people performing various movements and use them to make a poster. As captions, have the students name the muscles being used in the pictures.

Address Misconceptions

IDENTIFY Ask: How do muscles push on bones? If students attempt to explain, they may hold the misconception that muscles push and pull, whereas they can only pull.

CORRECT Point out that tendons are like string and that pushing on a string does not work. Explain that muscles and bones are arranged so that pulling the strings—the tendons—moves bones. You may want to include a physical demonstration by giving students lengths of string to manipulate.

REASSESS Ask: How do muscles move bones? *Muscles can only contract or relax. By contracting, they pull on the tendons.*

Technology Resources

Visit **ClassZone.com** for background on common student misconceptions.

 **MISCONCEPTION DATABASE**

Real World Example

Botox injections were first developed to treat problems that involved muscle spasms or rigidity. Examples of such problems are strabismus (crossed eyes) and cervical dystonia (rigidity of the neck muscles). "Botox" is the trade name for botulinum toxin A, a neurotoxin produced by the bacterium *Clostridium botulinum.* People who eat food contaminated by this bacterium can suffer fatal paralysis.

Teach from Visuals

To help students see the differences in the diagrams of muscles and bones, ask:

- In which diagram is the bottom muscle of the thigh contracted? *upper diagram*
- What does that do to the leg? *bends it*
- How does the other muscle in the pair look? *long and flat*

Skeletal muscles and tendons allow bones to move.

Skeletal muscles are attached to your bones by strong tissues called tendons. The tendons on the end of the muscle attach firmly to the bone. As the fibers in a muscle contract, they shorten and pull the tendon. The tendon, in turn, pulls the bone and makes it move.

You can feel your muscles moving your bones. Place your left arm, stretched out flat, in front of you on a table. Place the fingers of your right hand just above your left elbow. Bend your elbow and raise and lower your left arm. You are contracting your biceps. Can you feel the muscle pull on the tendon?

The dancers in the photograph are using many sets of muscles. The diagrams show how muscles and tendons work together to move bones. Muscles are shown in red. Notice how each muscle crosses a joint. Most skeletal muscles do. One end of the muscle attaches to one bone, crosses a joint, then attaches to a second bone. As the muscle contracts, it pulls on both bones. This pulling produces movement—in the case of these dancers, very exciting movement.

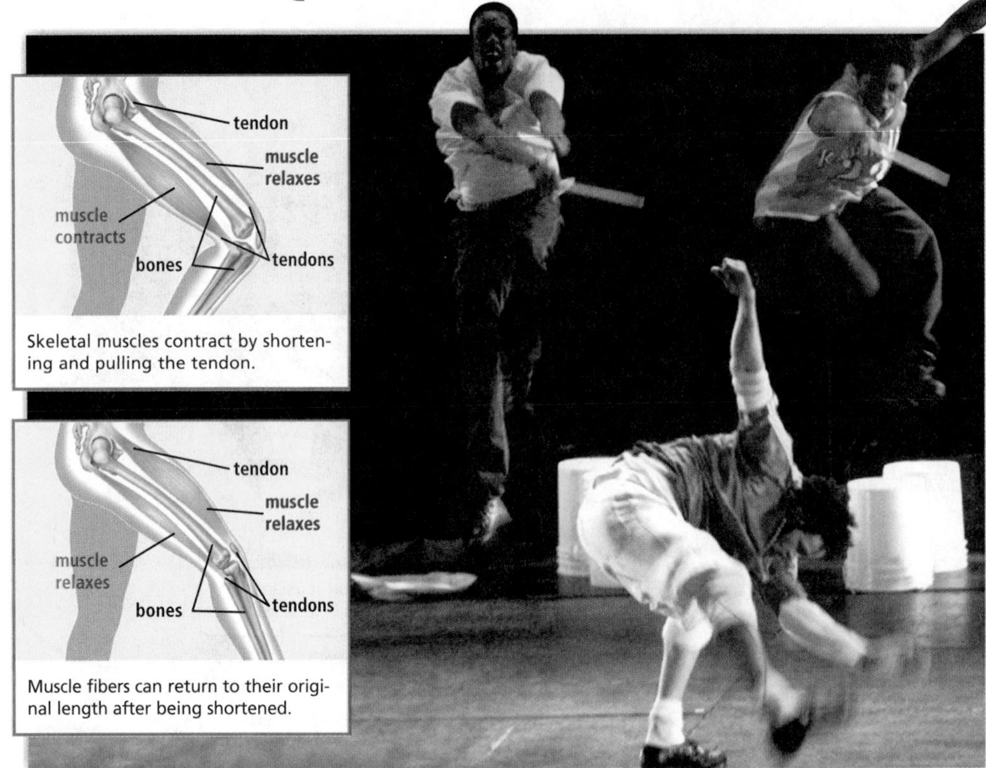

Skeletal muscles contract by shortening and pulling the tendon.

Muscle fibers can return to their original length after being shortened.

DIFFERENTIATE INSTRUCTION

 **More Reading Support**

E What attaches skeletal muscles to bone? *tendons*

F How do muscles produce movement? *They contract.*

Below Level Have students write and illustrate a pamphlet or picture storybook for younger students, explaining how to build and strengthen muscles. Alternatively, they may wish to write and illustrate a picture storybook about how muscles produce movement in the body.

Muscles grow and heal.

Developing Muscles An infant's muscles cannot do very much. A baby cannot lift its head, because the neck muscles are not strong enough to support it. For the first few months of life, a baby needs extra support, until the neck muscles grow strong and can hold up the baby's head.

The rest of the skeletal muscles also have to develop and strengthen. During infancy and childhood and into adolescence, humans develop muscular coordination and become more graceful in their movements. Coordination reaches its natural peak in adolescence but can be further improved by additional training.

Exercise and Muscles When you exercise regularly, your muscles get bigger. Muscles increase in size with exercise, because their cells reproduce more rapidly in response to the increased activity. Exercise also stimulates growth of individual muscle cells, making them larger.

You may have experienced sore muscles during or after exercising. During exercise, chemicals can build up in the muscles and make them cramp or ache. The muscle soreness you feel a day or so after exercise is caused by damage to the muscle fibers. The muscle fibers have been overstretched or torn. Such injuries take time to heal, because the body must remove injured cells, and new ones must form.

Stretching your muscles before exercise helps prevent injury.

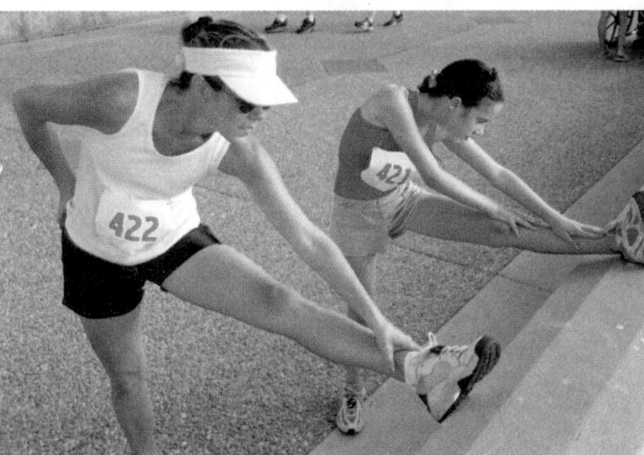

1.3 Review

KEY CONCEPTS

1. What are the three main functions of the muscular system?
2. Make a rough outline of a human body and label places where you could find each of the three types of muscles.
3. Explain why you may be sore after exercise.

CRITICAL THINKING

4. **Apply** You are exercising and you begin to feel hot. Explain what is happening in your muscles.
5. **Analyze** Describe what happens in your neck muscles when you nod your head.

CHALLENGE

6. **Infer** The digestive system breaks down food and transports materials. How are the short length and slow movement of smooth muscle tissues in the stomach and intestines related to the functions of these organs?

ANSWERS

1. movement, maintaining temperature, and posture

2. Outlines should label appropriate locations for skeletal, smooth, and cardiac muscles.

3. Muscle fibers can be overstretched or torn. Chemicals can build up in muscles during exercise.

4. Your muscles contract in exercise and expend energy that is converted to heat.

5. The muscles in front of your neck contract to pull your head forward. The muscles behind your spine contract to raise your head.

6. Sample answer: Long, powerful muscle fibers are not needed for involuntary muscle movements. Slow contractions make it possible for involuntary muscle movement to go unnoticed.

Ongoing Assessment

Explain how muscles grow and heal.

Ask: How do you build muscles safely? *Warm up and stretch before exercising, and exercise regularly.*

Reinforce the BIG idea

Have students relate the section to the Big Idea.

 Reinforcing Key Concepts, p. 42

1.3 ASSESS & RETEACH

Assess

 Section 1.3 Quiz, p. 5

Reteach

Write these headings on the board: "Functions," "Structure," "Movement," "Development." Have volunteers help you add under them entries that describe muscles. *Functions: movement, body temperature, posture; Structure: skeletal muscle, smooth muscle, cardiac muscle; Movement: contraction; Development: growth with age and exercise*

Technology Resources

Have students visit **ClassZone.com** for reteaching of Key Concepts.

 **CONTENT REVIEW**

 **CONTENT REVIEW CD-ROM**

CHAPTER INVESTIGATION

Focus

PURPOSE To examine how muscles and bones interact

OVERVIEW Students examine the muscles and bones of a chicken wing. Students then use their observations to

- describe the characteristics of muscles
- describe how the interaction of the muscles and bones in the chicken wing compares with the interactions of human muscles and bones

Lab Preparation

- Soak chicken wings in a bleach solution of 2 Tbsp bleach to 1 cup water for at least 5 minutes to kill any surface bacteria. Rinse them thoroughly in water and pat them dry. Store wings in a refrigerator until you are ready to use them.
- Prior to the investigation have students read through the investigation and prepare their data tables. Or you may wish to copy and distribute datasheets and rubrics.

 UNIT RESOURCE BOOK, pp. 50–58

 SCIENCE TOOLKIT, F14

Lab Management

- If necessary, help students locate the muscles, tendons, and ligaments in the chicken wings. Tendons are the tough, shiny white cords that attach the muscles to the bones.
- Arrange for proper disposal of wings at the conclusion of the lab.

SAFETY Caution students not to touch the chicken wings with their bare hands. Disinfect all work surfaces with bleach at the end of the lab.

INCLUSION Prior to the activity, dissect a wing, removing the muscles, tendons, and ligaments. Display the dissected parts of the wings with labels for students to use as a guide. Or, you might provide a full-page unlabeled outline of the wing for students to fill in with labels.

CHAPTER INVESTIGATION

A Closer Look at Muscles

OVERVIEW AND PURPOSE You use the muscles in your body to do a variety of things. Walking, talking, reading the words on this page, and scratching your head are all actions that require muscles. How do your muscles interact with your bones? In this investigation you will

- examine chicken wings to see how the muscles and the bones interact
- compare the movement of the chicken wing with the movement of your own bones and muscles

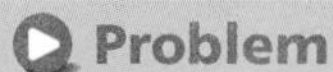

Problem

What are some characteristics of muscles?

Hypothesize

Write a hypothesis to propose how muscles interact with bones. Your hypothesis should take the form of an "If . . . , then . . . , because . . ." statement.

MATERIALS
- uncooked chicken wing and leg (soaked in bleach)
- paper towels
- dissection tray
- scissors

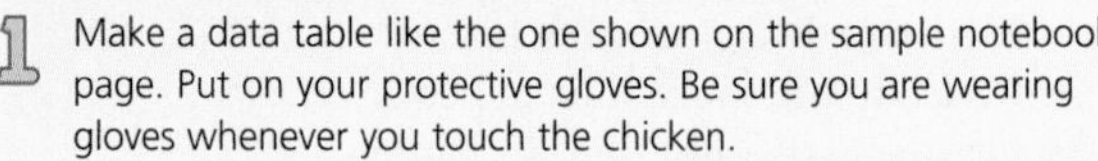

Procedure

1. Make a data table like the one shown on the sample notebook page. Put on your protective gloves. Be sure you are wearing gloves whenever you touch the chicken.
2. Obtain a chicken wing from your teacher. Rinse it in water and pat dry with a paper towel. Place it in the tray.
3. Extend the wing. In your notebook, draw a diagram of the extended wing. Be sure to include any visible external structures. Label the following on your diagram: lower limb, upper joint, and the wing tip.
4. Use scissors to remove the skin. Use caution so that you cut only through the skin. Peel back the skin and any fat so you can examine the muscles.

step 3

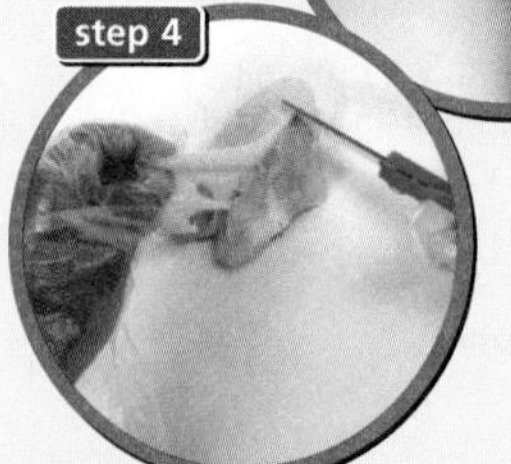

step 4

E 28 Unit: Human Biology

INVESTIGATION RESOURCES

 CHAPTER INVESTIGATION, A Closer Look at Muscles
- Level A, pp. 50–53
- Level B, pp. 54–57
- Level C, p. 58

Advanced students should complete Levels B & C.

 Writing a Lab Report, D12–13

Technology Resources

Customize this student lab as needed or look for an alternative. Print rubrics to assess student lab reports.

 **Lab Generator CD-ROM**

5 The muscles are the pink tissues that extend from one end of the bone to the other. Locate these in the upper wing and observe the way they move when you move the wing. Record your observations in your notebook.

6 Repeat this procedure for the muscles in the lower wing. In your notebook, draw a diagram of the muscles in the chicken wing.

7 There are also tendons in the chicken wing. These are the shiny white tissues at the end of the muscles. Add the tendons to your diagram.

8 Dispose of the chicken wing and parts according to your teacher's instructions. **Be sure to wash your hands well.**

Observe and Analyze

Write It Up

1. **RECORD** Write a brief description of how the bones and muscles work together to allow movement.
2. **EVALUATE** What difficulties, if any, did you encounter in carrying out this experiment?

Conclude

1. **INTERPRET** How does the chicken wing move when you bend it at the joint?
2. **OBSERVE** What happens when you pull on one of the wing muscles?
3. **COMPARE** Using your diagram of the chicken wing as an example, locate the same muscle groups in your own arm. How do they react when you bend your elbow?
4. **APPLY** What role do the tendons play in the movement of the muscles or bones?

INVESTIGATE Further

CHALLENGE Using scissors, carefully remove the muscles and the tendons from the bones. Next find the ligaments, which are located between the bones. Add these to your diagram. Describe how you think ligaments function.

A Closer Look at Muscles

Problem What are some characteristics of muscles?

Table 1. Observations

Draw your diagrams	Write your observations
Extended wing	Muscles in the upper wing
Muscles in the wing	Muscles in the lower wing

Observe and Analyze

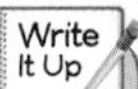

1. *Skeletal muscles are attached to bones by tendons. When muscles contract, the tendons pull on the bones, causing them to move.*
2. *Answers will vary. Some students may have had difficulty in distinguishing between ligaments and tendons.*

Conclude

1. *It moves with a back-and-forth motion.*
2. *A bone moves.*
3. *When the elbow bends, the biceps contracts and the triceps relaxes. When the arm straightens, the triceps contracts and the biceps relaxes.*
4. *Tendons attach the muscles to the bones and pull the bones when the muscles contract.*

INVESTIGATE Further

CHALLENGE Ligaments hold bones together at movable joints.

Post-Lab Discussion

- Ask students why pulling on a muscle in the chicken wing causes the wing to move. *Pulling on the muscle has the same effect as the muscle contracting; it pulls a tendon and thereby moves bones.*
- How are the chicken wing and a human arm similar? *Both have muscles, tendons, and bones that interact to cause movement.*
- Ask students what new questions arose, and discuss ways they could investigate those questions. *Students' questions should relate to the interaction of muscles with bones.*

BACK TO

the BIG idea

Have students name some of the organ systems of the human body. Then have them tell what each system does and how it may interact with other systems. *Sample answer: When you raise your hand, nerve tissue carries a message from your brain, and muscle tissue contracts to move the bones, made of connective and epithelial tissues.*

KEY CONCEPTS SUMMARY

SECTION 1.1
Ask students to describe the structures of tissues, organs, and organ systems. *Tissues are groups of similar cells that function together. Organs are made up of groups of tissues. Organ systems are made up of groups of organs.*

SECTION 1.2
Ask students to describe the main functions of the skeletal system. *The skeletal system provides a framework for the body that furnishes support, allows movement, and protects organ systems.*

SECTION 1.3
Ask students to identify organs in the body that contain smooth muscle and to explain why they contain that kind of muscle. *Smooth muscle is found in the lining of the stomach and the intestines. Since these organs perform involuntary movements, they need to be lined with smooth muscle.*

Review Concepts

- Big Idea Flow Chart, p. T1
- Chapter Outline, pp. T7–T8

1 Chapter Review

the BIG idea

The human body is made up of systems that work together to perform necessary functions.

KEY CONCEPTS SUMMARY

1.1 The human body is complex.

You can think of the body as having five levels of organization: cells, tissues, organs, organ systems, and the whole organism itself. The different systems of the human body work together to maintain homeostasis.

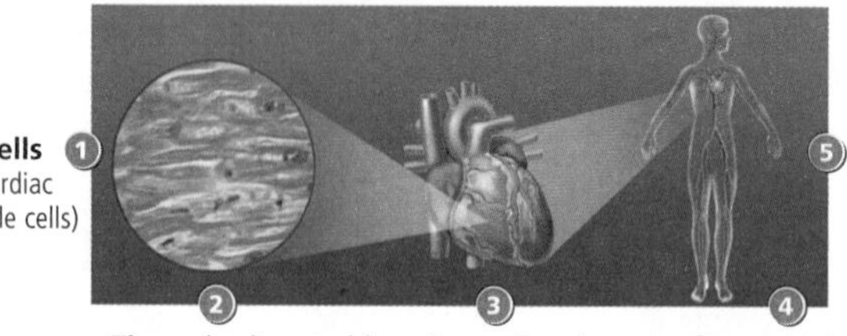

VOCABULARY
tissue p. 12
organ p. 12
organ system p. 12
homeostasis p. 12

1.2 The skeletal system provides support and protection.

Bones are living tissue. The skeleton is the body's framework and has two main divisions, the **axial skeleton** and the **appendicular skeleton**. Bones come together at joints.

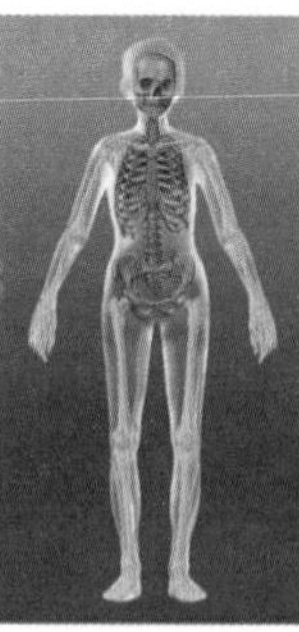

VOCABULARY
skeletal system p. 14
compact bone p. 14
spongy bone p. 15
axial skeleton p. 15
appendicular skeleton p. 16

1.3 The muscular system makes movement possible.

Types of muscle	Function
skeletal muscle, voluntary	moves bones, maintains posture, maintains body temperature
smooth muscle, involuntary	moves internal organs, such as the intestines
cardiac muscle, involuntary	pumps blood throughout the body

VOCABULARY
muscular system p. 24
skeletal muscle p. 24
voluntary muscle 24
smooth muscle p. 24
involuntary muscle p. 24
cardiac muscle p. 24

E 30 Unit: Human Biology

Technology Resources

Have students visit **ClassZone.com** or use the CD-ROM for a cumulative review of concepts.

CONTENT REVIEW

CONTENT REVIEW CD-ROM

Engage students in a whole-class interactive review of Key Concepts. Edit content as you wish.

POWER PRESENTATIONS

Reviewing Vocabulary

In one or two sentences describe how the vocabulary terms in each of the following pairs of words are related. Underline each vocabulary term in your answer.

1. cells, tissues
2. organs, organ systems
3. axial skeleton, appendicular skeleton
4. skeletal muscle, voluntary muscle
5. smooth muscle, involuntary muscle
6. compact bone, spongy bone

Reviewing Key Concepts

Multiple Choice *Choose the letter of the best answer.*

7. Which type of tissue carries electrical impulses from your brain?
   - **a.** epithelial tissue
   - **b.** muscle tissue
   - **c.** nerve tissue
   - **d** connective tissue
8. Connective tissue functions to provide
   - **a.** support and strength
   - **b.** messaging system
   - **c.** movement
   - **d.** heart muscle
9. Inside bone cells is a network made of
   - **a.** tendons
   - **b.** calcium
   - **c.** marrow
   - **d.** joints
10. The marrow produces
    - **a.** spongy bone
    - **b.** red blood cells
    - **c.** compact bone
    - **d.** calcium
11. Which bones make up the axial skeleton?
    - **a.** skull, shoulder blades, arm bones
    - **b.** skull, spinal column, leg bones
    - **c.** shoulder blades, spinal column, and hip bones
    - **d.** skull, spinal column, ribs
12. Bones of the skeleton connect to each other at

    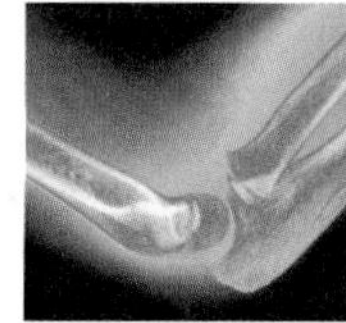

    - **a.** tendons
    - **b.** ligaments
    - **c.** joints
    - **d.** muscles
13. How do muscles contribute to homeostasis?
    - **a.** They keep parts of the body together.
    - **b.** They control the amount of water in the body.
    - **c.** They help you move.
    - **d.** They produce heat when they contract.
14. Cardiac muscle is found in the
    - **a.** heart
    - **b.** stomach
    - **c.** intestines
    - **d.** arms and legs
15. The stomach is made up of
    - **a.** cardiac muscle
    - **b.** skeletal muscle
    - **c.** smooth muscle
    - **d.** voluntary muscle

Short Answer *Write a short answer to each question.*

16. What is the difference between spongy bone and compact bone?
17. The root word *homeo* means "same," and the root word *stasis* means "to stay." How do these root words relate to the definition of *homeostasis*?
18. Hold the upper part of one arm between your elbow and shoulder with your opposite hand. Feel the muscles there. What happens to those muscles as you bend your arm?

ASSESSMENT RESOURCES

UNIT ASSESSMENT BOOK
- Chapter Test, Level A, pp. 6–9
- Chapter Test, Level B, pp. 10–13
- Chapter Test, Level C, pp. 14–17
- Alternative Assessment, pp. 18–19

SPANISH ASSESSMENT BOOK
Spanish Chapter Test, pp. 81–84

Technology Resources

Edit test items and answer choices.

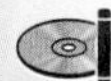 **Test Generator CD-ROM**

Visit **ClassZone.com** to extend test practice.

 **Test Practice**

Reviewing Vocabulary

Sample answers:

1. *Cells are the basic units of life. Tissues are made up of groups of similar cells that function together.*
2. *Organs are body structures that have specific functions. Organ systems are made up of groups of organs that perform related functions.*
3. *There are two main divisions of the human skeleton. The axial skeleton provides support and protection; the appendicular skeleton enables movement.*
4. *Skeletal muscle makes up the muscles attached to the skeleton. Because it produces voluntary movement, such a muscle is called voluntary muscle.*
5. *Smooth muscle is found inside body organs. It produces involuntary movement and is therefore called involuntary muscle.*
6. *Compact bone forms the hard outer layer of bones. Inside it is a strong, lightweight tissue called spongy bone.*

Reviewing Key Concepts

7. *c*
8. *a*
9. *b*
10. *b*
11. *d*
12. *c*
13. *d*
14. *a*
15. *c*
16. *Compact bone makes up a bone's outer layer. Spongy bone is less dense and is surrounded by compact bone.*
17. *Homeostasis is the ability of the body to maintain a stable internal condition—in other words, stay the same.*
18. *The muscle on the top of the arm contracts, and the muscle on the opposite side relaxes.*

Thinking Critically

19. Cells—muscle cell; tissues—bone; organs—heart; organ systems—digestive system

20. connective tissue; Blood flows through the body, connecting all organ systems.

21. The bones of the human skeleton contain calcium. The skeleton is hard, providing support and protecting internal organs. It is lightweight, and has joints, much like the hinge of the clam's shells. The difference is that a clam's shell is outside; bones are inside.

22. These are immovable joints. They lock together tissues that support and protect rather than move.

23. The arms and legs belong to the appendicular skeleton. They are attached by ball-and-socket joints, allowing a wide range of movement.

24. The skeleton provides support just like the framework of a house. Unlike the house, the skeleton is made of living tissues and is flexible.

25. The skeleton provides shape and support; it allows the body to move; it protects soft organs.

26. Bringing the toothbrush up involves rotational movement at the shoulder, pivotal of the wrist, angular of the elbow and fingers gripping. The small bones in the hand glide as the hand turns.

27. For movement to occur, muscles on one side of a bone must relax while muscles on the other side contract.

28. More effort means more fibers are contracted.

the BIG idea

29. Answers will depend on students' original answers.

30. Sample answer: <u>Bones</u> come together at <u>joints</u>. <u>Skeletal muscles</u> that are attached to the bones contract and relax, pulling the bones to move the body.

UNIT PROJECTS

Give students unit projects worksheets for their projects. Both directions and rubrics can be used as guides.

 Unit Projects, pp. 5–10

Thinking Critically

19. PROVIDE EXAMPLES What are the levels of organization of the human body from simple to most complex? Give an example of each.

20. CLASSIFY There are four types of tissue in the human body: epithelial, nerve, muscles, and connective. How would you classify blood? Explain your reasoning.

21. CONNECT A clam shell is made of a calcium compound. The material is hard, providing protection to the soft body of a clam. It is also lightweight. Describe three ways in which the human skeleton is similar to a seashell. What is one important way in which it is different?

Use the diagram below to answer the next two questions

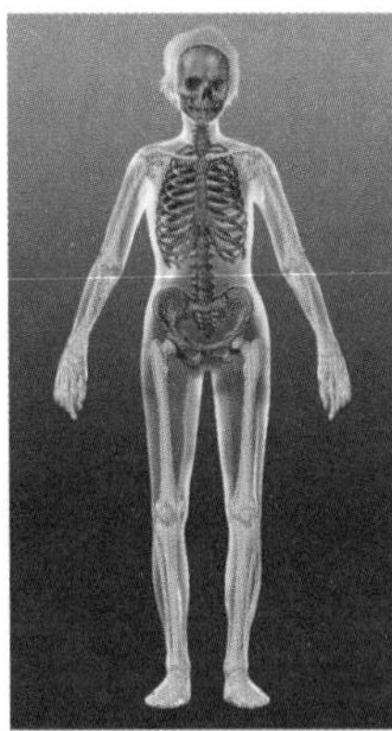

22. SYNTHESIZE Identify the type of joints that hold together the bones of the skull and sternum. How does this type of joint relate to the function of the skull and sternum?

23. SYNTHESIZE The human skeleton has two main divisions. Which skeleton do the arms and legs belong to? How do the joints that connect the arms to the shoulders and the legs to the hips relate to the function of this skeleton?

24. COMPARE AND CONTRAST How is the skeletal system of your body like the framework of a house or building? How is it different?

25. SUMMARIZE Describe three important functions of the skeleton.

26. APPLY The joints in the human body can be described as producing three types of movement. Relate these three types of movement to the action of brushing your teeth.

27. COMPARE AND CONTRAST When you stand, the muscles in you legs help to keep you balanced. Some of the muscles on both sides of your leg bones contract. How does this differ from how the muscles behave when you start to walk?

28. INFER Muscles are tissues that are made up of many muscle fibers. A muscle fiber can either be relaxed or contracted. Some movements you do require very little effort, like picking up a piece of paper. Others require a lot of effort, like picking up a book bag. How do you think a muscle produces the effort needed for a small task compared with a big task?

the BIG idea

29. INFER Look again at the picture on pages 6–7. Now that you have finished the chapter, how would you change or add details to your answer to the question on the photograph?

30. SUMMARIZE Write a paragraph explaining how skeletal muscles, bones, and joints work together to allow the body to move and be flexible. Underline the terms in your paragraph.

UNIT PROJECTS

If you are doing a unit project, make a folder for your project. Include in your folder a list of resources you will need, the date on which the project is due, and a schedule to track your progress. Begin gathering data.

MONITOR AND RETEACH

If students are having trouble applying the concepts in Chapter Review items 22 and 23, suggest that they review the illustration of the skeletal system on p. 17. Then have students create a concept map detailing the functions and structure of the skeleton.

Students may benefit from summarizing one or more sections of the chapter.

 Summarizing the Chapter, pp. 68–69

Standardized Test Practice

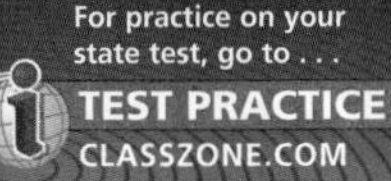

Interpreting Diagrams

The action of a muscle pulling on a bone can be compared to a simple machine called a lever. A lever is a rod that moves about a fixed point called the fulcrum. Effort at one end of the rod can move a load at the other end. In the human body, a muscle supplies the effort needed to move a bone—the lever. The joint is the fulcrum, and the load is the weight of the body part being moved. There are three types of levers, which are classified according to the position of the fulcrum, the effort, and the load.

Read the text and study the diagrams, and then choose the best answer for the questions that follow.

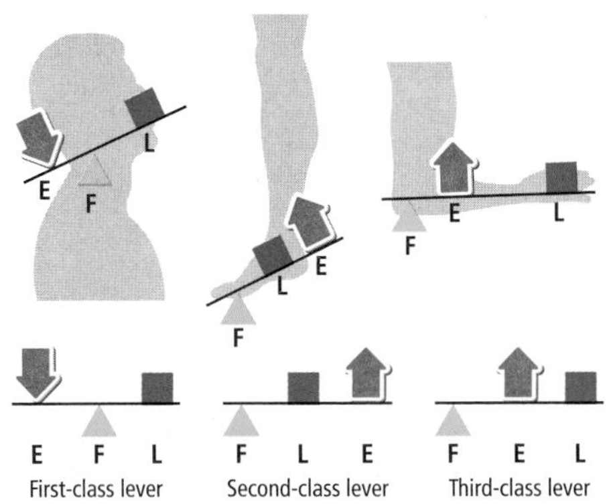

1. In a first-class lever
   - **a.** the load is at end of the lever opposite the fulcrum
   - **b.** the load is between the effort and the fulcrum
   - **c.** the fulcrum is between the load and the effort
   - **d.** the effort and load are on the same side

2. What is true of all levers?
   - **a.** The fulcrum must be located at the center of a lever.
   - **b.** The force of the load and effort point in the same direction.
   - **c.** The load and effort are on the same side of the fulcrum.
   - **d.** The force of the effort points in a direction opposite that of the load.

3. The fulcrum represents what structure in the human body?
   - **a.** a joint
   - **b.** a bone
   - **c.** a muscle
   - **d.** a part of the body

4. The main point of the diagram is to show
   - **a.** how bones work
   - **b.** that there are three types of levers and how they are classified
   - **c.** where to apply a force
   - **d.** the forces involved in moving parts of the body

Extended Response

Use the diagrams above and terms from the word box below to answer the next question. Underline each term you use in your answer.

fulcrum	load	effort	rod
bone	muscle	joint	

5. Suppose you had a heavy box to lift. Your first thought might be to bend over, stretch out your arms, and grab the box. Your body would be acting as a simple machine. Identify the type of lever this is and the parts of this machine.

6. A doctor would advise you not to lift a heavy object, like a box, simply by bending over and picking it up. That action puts too much strain on your back. It is better to bend your knees, hold the box close to your body, and then lift. How does this way of lifting change how you are using your body?

METACOGNITIVE ACTIVITY

Have students answer the following questions in their **Science Notebook:**

1. Are there any concepts relating to the skeletal and muscular systems that you are still confused about? Explain.
2. What did you learn in this unit that surprised you? Explain.
3. While researching your unit project, how did you prefer to do your research? Why?

Interpreting Diagrams

1. c 2. d 3. a 4. d

Extended Response

5. RUBRIC

4 points for a response that correctly answers the question and includes all of the following terms:

- fulcrum
- load
- bone
- muscle
- effort
- joint
- rod

Sample answer: Your body would be functioning as a third-class lever. The <u>bones</u> of your back act as the <u>rod</u> or lever, the <u>fulcrum</u> is the <u>joint</u> at the lower part of your back, and the <u>load</u> is the box. The <u>muscles</u> in your back provide the <u>effort</u>.

3 points correctly answers the question and includes five or six of the terms listed

2 points correctly answers the question and includes three or four of the terms

1 point correctly answers the question and includes one or two of the terms

6. RUBRIC

4 points for a response that correctly answers the question and includes all of the following points:

- the muscles of the legs are now involved as well.
- more muscles do the same work, so the effort is shared
- the fulcrum has changed
- now the legs are used as levers

Sample answer: By bending down next to the box to pick it up, you have changed the fulcrum to the joints in your legs. Because the legs are now levers, the leg muscles as well as the back muscles are involved in lifting the load. You are putting less strain on your back because your leg muscles are sharing the effort.

3 points includes three of the above points

2 points includes two of the above points

1 point includes one of the above points

CHAPTER

Absorption, Digestion, and Exchange

Life Science UNIFYING PRINCIPLES

PRINCIPLE 1
All living things share common characteristics.

PRINCIPLE 2
All living things share common needs.

PRINCIPLE 3
Living things meet their needs through interactions with the environment.

PRINCIPLE 4
The types and numbers of living things change over time.

Unit: Human Biology BIG IDEAS

CHAPTER 1
Systems, Support, and Movement
The body is made up of systems that work together to perform necessary functions.

CHAPTER 2
Absorption, Digestion, and Exchange
Systems in the body obtain and process materials and remove waste.

CHAPTER 3
Transport and Protection
Systems function to transport materials and to defend and protect the body.

CHAPTER 4
Control and Reproduction
The nervous and endocrine systems allow the body to respond to internal and external conditions.

CHAPTER 5
Growth, Development, and Health
The body develops and maintains itself over time.

CHAPTER 2 KEY CONCEPTS

SECTION 2.1

The respiratory system gets oxygen and removes carbon dioxide.
1. Your body needs oxygen.
2. Structures in the respiratory system function together.
3. The respiratory system is also involved in other activities.

SECTION 2.2

The digestive system breaks down food.
1. The body needs energy and materials.
2. The digestive system moves and breaks down food.
3. Materials are broken down as they move through the digestive tract.
4. Other organs aid digestion and absorption.

SECTION 2.3

The urinary system removes waste materials.
1. Life processes produce wastes.
2. The urinary system removes waste from the blood.
3. The kidneys act as filters.

T The Big Idea Flow Chart is available on p. T9 in the **UNIT TRANSPARENCY BOOK.**

Previewing Content

SECTION

The respiratory system gets oxygen and removes carbon dioxide.
pp. 37–44

1. Your body needs oxygen.
During a process called **cellular respiration,** the cells of the body use oxygen in chemical reactions in order to release energy. The respiratory system interacts with the environment and the digestive and circulatory systems to make cellular respiration possible. Carbon dioxide is a byproduct of cellular respiration.

2. Structures in the respiratory system function together.
- Air enters the body through the nose and mouth, then travels down the trachea and into the bronchial tubes that lead to the lungs.
- The exchange of oxygen and carbon dioxide with the circulatory system takes place in the alveoli, air sacs in the lungs.
- The diaphragm and involuntary muscles surrounding the ribs control the inhalation and exhalation of gases.

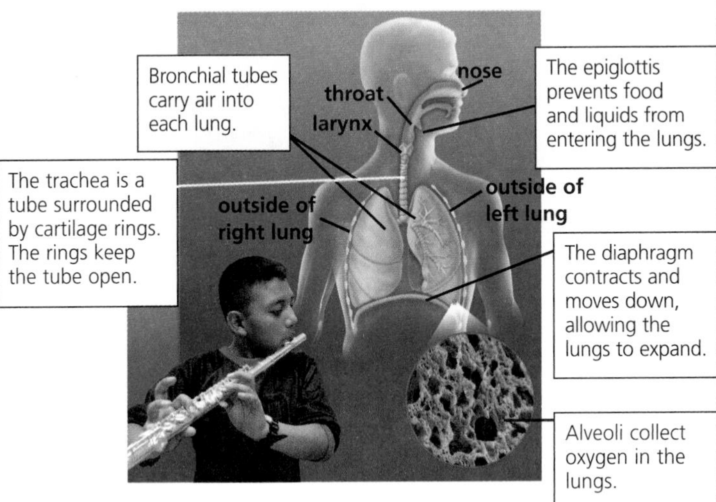

3. The respiratory system is also involved in other activities.
Before air passes into the trachea, it moves over the vocal cords of the larynx, allowing sounds to be produced. Water is removed from the respiratory system through the actions of hiccups, coughs, sneezes, and yawns.

SECTION

The digestive system breaks down food. pp. 45–51

1. The body needs energy and materials.
Nutrients, such as proteins, carbohydrates, fats, and water, are used by the body during the process of **digestion,** enabling the body to move, grow, and maintain homeostasis.

2. The digestive system moves and breaks down food.
Mechanical and chemical digestion both begin in the mouth. The action of physical digestion begins as teeth tear food and mash it. The action of chemical digestion begins as saliva changes starches into sugars. Food passes through the esophagus by the action of **peristalsis** and moves into the stomach.

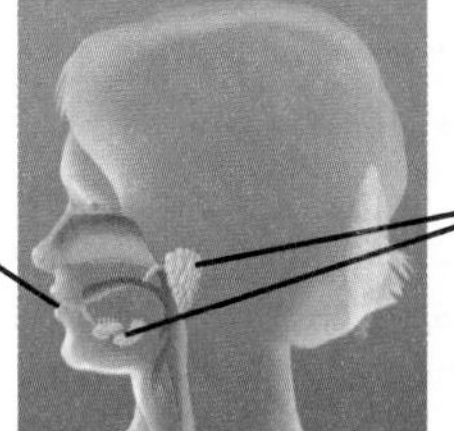

The mechanical stage of digestion begins when food is chewed in the mouth.
Salivary glands release saliva, which begins to chemically digest food.

The mouth gives a good example of the difference between mechanical digestion and chemical digestion, and shows how the two work together.

3. Materials are broken down as they move through the digestive tract.
- Muscle contractions in the esophagus allow food to be pushed through it to the stomach.
- In the stomach, mechanical digestion and chemical digestion continue, aided by acids in the stomach.
- Partially digested food moves into the small intestine, where digestion is completed. Nutrients are absorbed into the circulatory system through structures called villi.
- Water is absorbed into the body in the large intestine. Solid wastes pass out of the body through the rectum.

4. Other organs aid digestion and absorption.
- The liver secretes bile, which aids in the breakdown of fats.
- The gallbladder stores bile produced in the liver and secretes it into the small intestine.
- The pancreas aids in digestion of proteins, fats, and starch.

Common Misconceptions

MISCONCEPTION DATABASE
CLASSZONE.COM Background on student misconceptions

FOOD AND GROWTH Middle school students know food is related to growing, but they are not aware of how. They may think that food is only a source of energy, rather than also being a source of matter for growth. The energy gained from food is thought to be for movement or work rather than for all body functions.

This misconception is addressed on p. 46.

STOMACH LOCATION Students often think that the stomach is located much lower in the abdomen than it really is.

This misconception is addressed on p. 48.

SECTION

The urinary system removes waste materials. pp. 52–55

1. Life processes produce wastes.
Removal of wastes is essential in maintaining body functions. In addition to the urinary system, the digestive and respiratory systems, and the skin enable the body to remove wastes.

2. The urinary system removes waste from the blood.

- Kidneys filter materials out of the circulatory system. Some of these materials are eliminated as waste. Other materials are returned to the blood.
- From the kidneys, liquid waste material called **urine** travels down structures called ureters and into the bladder. There the urine is eliminated from the body through the urethra.

3. The kidneys act as filters.
Nephrons, which are looping tubes found inside the kidney, act as filters. Blood passes through a structure in the nephron called the glomerulus, where waste materials are filtered out. Once blood passes through the nephron, the waste material passes out of the kidney and into the ureters. The kidneys also regulate the amount of water in the body.

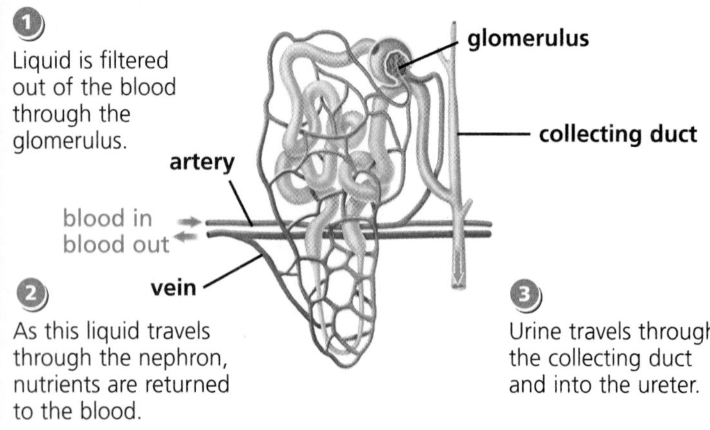

1
Liquid is filtered out of the blood through the glomerulus.
glomerulus
collecting duct
artery
blood in
blood out
vein
2
As this liquid travels through the nephron, nutrients are returned to the blood.
3
Urine travels through the collecting duct and into the ureter.

Point out to students that the urinary system is not "breaking down food to obtain nutrients" and so is not part of the digestive system.

Common Misconceptions

MISCONCEPTION DATABASE
CLASSZONE.COM Background on student misconceptions

URINARY BLADDER Many students think that the urinary bladder is a part of the digestive system when in fact it is part of the urinary system.

This misconception is addressed on p. 53.

Previewing Labs

Lab Generator CD-ROM
Edit these Pupil Edition labs and generate alternative labs.

EXPLORE the BIG idea

Mirror, Mirror, p. 35 Students observe, by exhaling on a mirror, that breath contains water.	**TIME** 10 minutes **MATERIALS** mirror
Water Everywhere, p. 35 Students determine how much liquid they consume in a 24-hour period.	**TIME** 10 minutes **MATERIALS** notebook
Internet Activity: Lung Movement, p. 35 Students observe a simulation of diaphragm and lung movement during respiration.	**TIME** 20 minutes **MATERIALS** computer with Internet access

SECTION 2.1

EXPLORE Breathing, p. 37 Students observe the movement of the rib cage during respiration.	**TIME** 10 minutes **MATERIALS** notebook
INVESTIGATE Lungs, p. 39 Students design a model of lungs to represent respiration and lung behavior.	**TIME** 15 minutes **MATERIALS** medium balloon, 1L plastic bottle with labels removed, duct tape

SECTION 2.2

EXPLORE Digestion, p. 45 Students explore how the digestive system breaks down fat.	**TIME** 15 minutes **MATERIALS** graduated cylinders, test tube with caps, test tube stand, 3 graduated cylinders, 5 mL vegetable oil, water, 5 mL dish detergent
INVESTIGATE Chemical Digestion, p. 47 Students model how saliva breaks down food.	**TIME** 25 minutes **MATERIALS** knife, cooked potato, droppers, 50 mL bacterial amylase solution (solution A), water

SECTION 2.3

EXPLORE Waste Removal, p. 52 Students demonstrate how the skin excretes wastes.	**TIME** 10 minutes **MATERIALS** plastic bag, masking tape, stopwatch
CHAPTER INVESTIGATION **Modeling a Kidney,** pp. 56–57 Students model the filtering process of kidneys to determine what types of materials are filtered by the kidneys.	**TIME** 40 minutes **MATERIALS** small funnel, fine filter paper, 10 g salt dissolved in 1000 mL water (solution A), 10 g glucose dissolved in 1000 mL water (solution B), 10 g albumin powder (or one egg white) dissolved in 1000 mL water (solution C), graduated cylinder, 100 mL beaker, glucose test strips, salinity test strips, protein test strips

R **Additional INVESTIGATION,** How Much Air Can Your Lungs Hold?, A, B, & C, pp. 118–126; Teacher Instructions, pp. 305–306

Previewing Chapter Resources

CHAPTER 2
Absorption, Digestion, and Exchange

INTEGRATED TECHNOLOGY

CLASSZONE.COM
- eEdition Plus
- EasyPlanner Plus
- Misconception Database
- Content Review
- Test Practice
- Resource Centers
- Visualization
- Internet Activity: Lung Movement
- Math Tutorial

SCILINKS.ORG

CD-ROMS
- eEdition
- EasyPlanner
- Power Presentations
- Content Review
- Lab Generator
- Test Generator

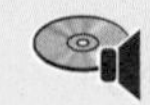
AUDIO CDS
- Audio Readings
- Audio Readings in Spanish

LABS AND ACTIVITIES

EXPLORE the Big Idea, p. 35
- Mirror, Mirror
- Water Everywhere
- Internet Activity: Lung Movement

UNIT RESOURCE BOOK
Unit Projects, pp. 5–10

Lab Generator CD-ROM
Generate customized labs.

SECTION

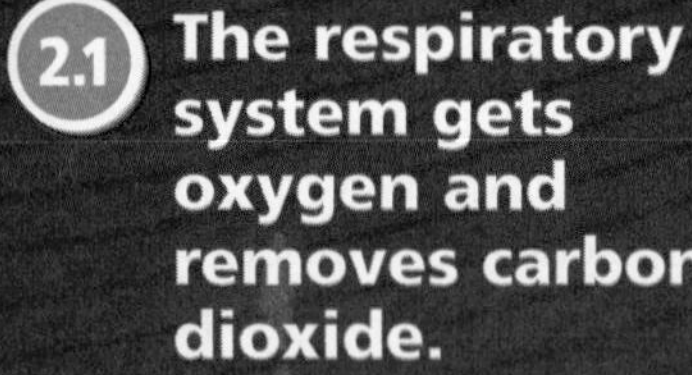
2.1 The respiratory system gets oxygen and removes carbon dioxide.
pp. 37–44

Time: 2 periods (1 block)

Lesson Plan, pp. 70–71

INTEGRATED TECHNOLOGY

RESOURCE CENTER, Respiratory System

UNIT TRANSPARENCY BOOK
- Big Idea Flow Chart, p. T9
- Daily Vocabulary Scaffolding, p. T10
- Note-Taking Model, p. T11
- 3-Minute Warm-Up, p. T12

LABS AND ACTIVITIES

- EXPLORE Breathing, p. 37
- INVESTIGATE Lungs, p. 39
- Science on the Job, p. 44

UNIT RESOURCE BOOK
- Datasheet, Lungs, p. 79
- ADDITIONAL Investigation, How Much Air Can Your Lungs Hold?, A, B, & C, pp. 118–126

SECTION

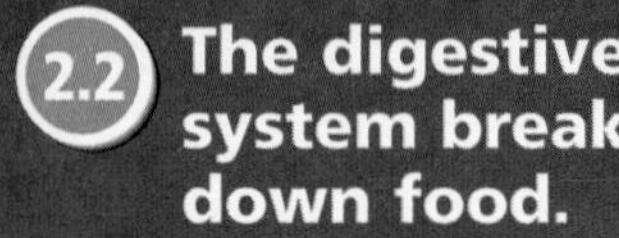
2.2 The digestive system breaks down food.
pp. 45–51

Time: 2 periods (1 block)

Lesson Plan, pp. 81–82

INTEGRATED TECHNOLOGY

- **VISUALIZATION,** Peristalsis
- **MATH TUTORIAL**

UNIT TRANSPARENCY BOOK
- Daily Vocabulary Scaffolding, p. T10
- 3-Minute Warm-Up, p. T12
- "Digestive System" Visual, p. T14

LABS AND ACTIVITIES

- EXPLORE Digestion, p. 45
- INVESTIGATE Chemical Digestion, p. 47
- Math in Science, p. 51

UNIT RESOURCE BOOK
- Datasheet, Chemical Digestion, p. 90
- Math Support, p. 107
- Math Practice, p. 108

SECTION

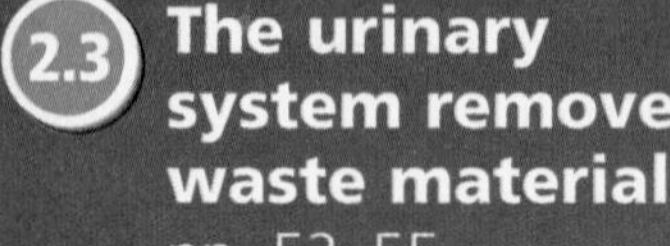
2.3 The urinary system removes waste materials.
pp. 52–55

Time: 4 periods (2 blocks)

Lesson Plan, pp. 92–93

INTEGRATED TECHNOLOGY

RESOURCE CENTER, Urinary System

UNIT TRANSPARENCY BOOK
- Big Idea Flow Chart, p. T9
- Daily Vocabulary Scaffolding, p. T10
- 3-Minute Warm-Up, p. T13
- Chapter Outline, pp. T15–T16

LABS AND ACTIVITIES

- EXPLORE Waste Removal, p. 52
- CHAPTER INVESTIGATION, Modeling a Kidney, pp. 56–57

UNIT RESOURCE BOOK
CHAPTER INVESTIGATION, Modeling a Kidney, A, B, & C, pp. 109–117

KEY TO ICONS

 CD/CD-ROM

INTERNET

 Teacher Edition

Pupil Edition

UNIT RESOURCE BOOK

 UNIT TRANSPARENCY BOOK

UNIT ASSESSMENT BOOK

 SPANISH ASSESSMENT BOOK

SCIENCE TOOLKIT

READING AND REINFORCEMENT

ASSESSMENT

STANDARDS

- Magnet Word Diagram, B24–25
- Outline, C43
- Daily Vocabulary Scaffolding, H1–8

UNIT RESOURCE BOOK
- Vocabulary Practice, pp. 104–105
- Decoding Support, p. 106
- Summarizing the Chapter, pp. 127–128

Audio Readings CD
Listen to Pupil Edition.

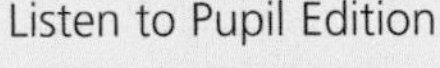

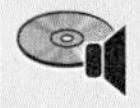

Audio Readings in Spanish CD
Listen to Pupil Edition in Spanish.

- Chapter Review, pp. 59–60
- Standardized Test Practice, p. 61

UNIT ASSESSMENT BOOK
- Diagnostic Test, pp. 20–21
- Chapter Test, A, B, & C, pp. 25–36
- Alternative Assessment, pp. 37–38

Spanish Chapter Test, pp. 85–88

Test Generator CD-ROM
Generate customized tests.

Lab Generator CD-ROM
Rubrics for Labs

National Standards
A.1–8, A.9.a–c, A.9.e–g, C.1.e, F.1.e, G.1.b

See p. 34 for the standards.

UNIT RESOURCE BOOK
- Reading Study Guide, A & B, pp. 72–75
- Spanish Reading Study Guide, pp. 76–77
- Challenge and Extension, p. 78
- Reinforcing Key Concepts, p. 80
- Challenge Reading, pp. 102–103

Ongoing Assessment, pp. 37–42

Section 2.1 Review, p. 43

UNIT ASSESSMENT BOOK
Section 2.1 Quiz, p. 22

National Standards
A.2–7, A.9.a–b, A.9.e–f, C.1.e, G.1.b

UNIT RESOURCE BOOK
- Reading Study Guide, A & B, pp. 83–86
- Spanish Reading Study Guide, pp. 87–88
- Challenge and Extension, p. 89
- Reinforcing Key Concepts, p. 91

Ongoing Assessment, pp. 46–48, 50

Section 2.2 Review, p. 50

UNIT ASSESSMENT BOOK
Section 2.2 Quiz, p. 23

National Standards
A.2–8, A.9.a–b, A.9.e–f, C.1.e, F.1.e, G.1.b

UNIT RESOURCE BOOK
- Reading Study Guide, A & B, pp. 94–97
- Spanish Reading Study Guide, pp. 98–99
- Challenge and Extension, p. 100
- Reinforcing Key Concepts, p. 101

Ongoing Assessment, pp. 52–55

Section 2.3 Review, p. 55

UNIT ASSESSMENT BOOK
Section 2.3 Quiz, p. 24

National Standards
A.1–7, A.9.a–b, A.9.e–g, C.1.e, G.1.b

Previewing Resources for Differentiated Instruction

CHAPTER INVESTIGATION

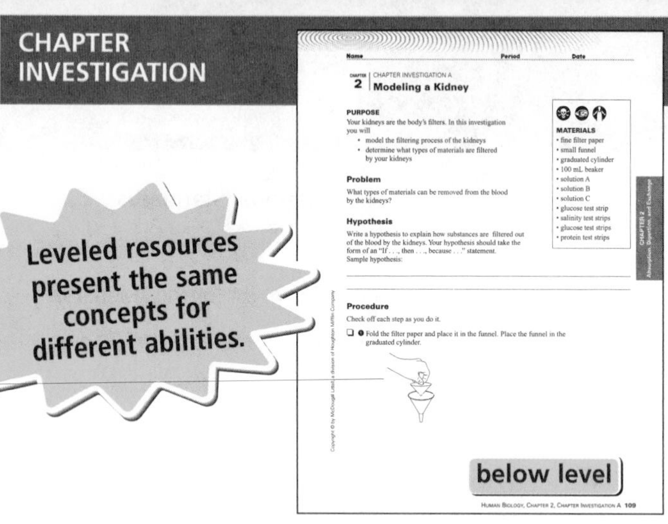

Leveled resources present the same concepts for different abilities.
Name Period Date
CHAPTER 2 | CHAPTER INVESTIGATION A
Modeling a Kidney
PURPOSE
Problem
Hypothesis
Procedure
MATERIALS
below level
HUMAN BIOLOGY, CHAPTER 2, CHAPTER INVESTIGATION A 109

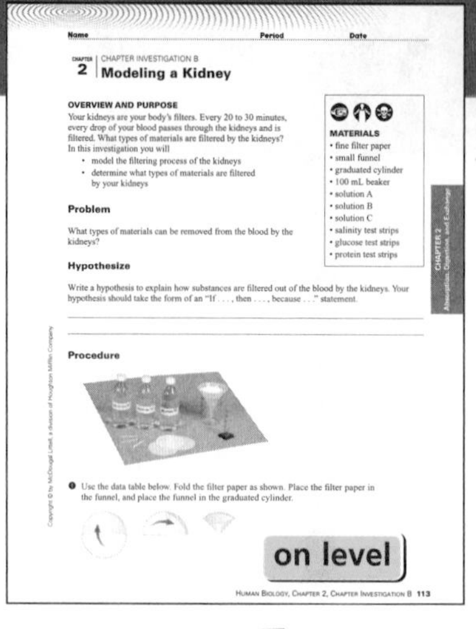

Name Period Date
CHAPTER 2 | CHAPTER INVESTIGATION B
Modeling a Kidney
OVERVIEW AND PURPOSE
Problem
Hypothesize
Procedure
MATERIALS
on level
HUMAN BIOLOGY, CHAPTER 2, CHAPTER INVESTIGATION B 113

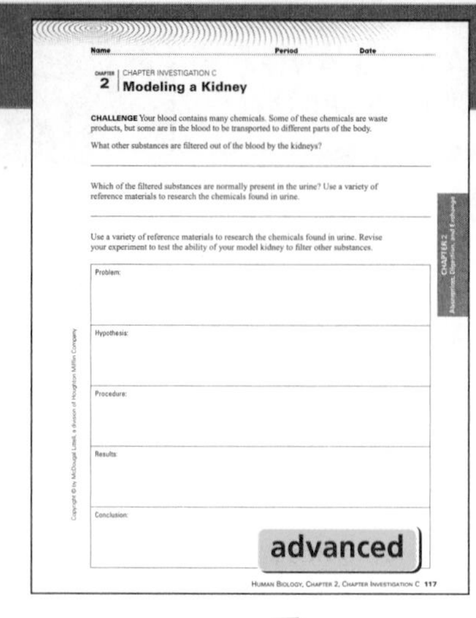

Name Period Date
CHAPTER 2 | CHAPTER INVESTIGATION C
Modeling a Kidney
Problem
Hypothesis
Procedure
Results
Conclusion
advanced
HUMAN BIOLOGY, CHAPTER 2, CHAPTER INVESTIGATION C 117

R UNIT RESOURCE BOOK, pp. 109–112 | **R** pp. 113–116 | **R** pp. 113–117

READING STUDY GUIDE

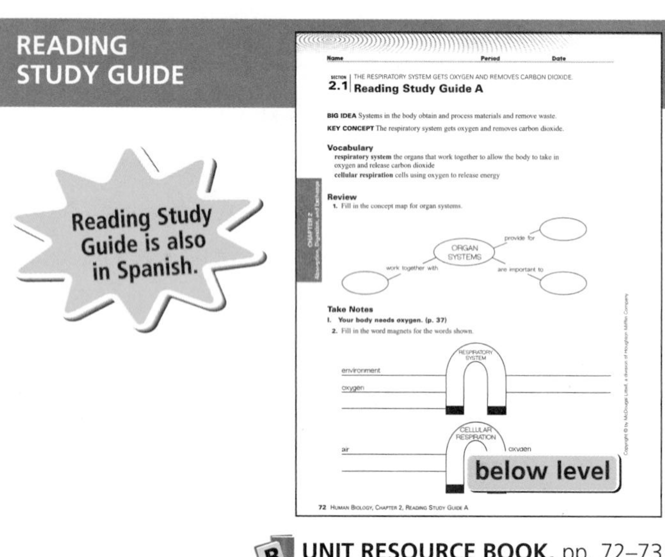

Reading Study Guide is also in Spanish.
Name Period Date
SECTION 2.1 | THE RESPIRATORY SYSTEM GETS OXYGEN AND REMOVES CARBON DIOXIDE.
Reading Study Guide A
Vocabulary
Review
Take Notes
below level
72 HUMAN BIOLOGY, CHAPTER 2, READING STUDY GUIDE A

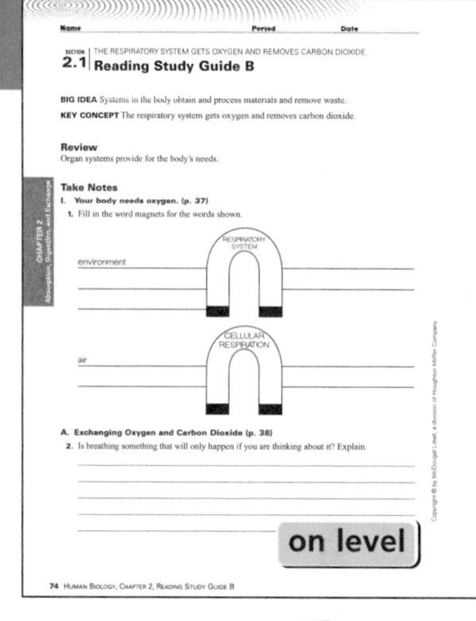

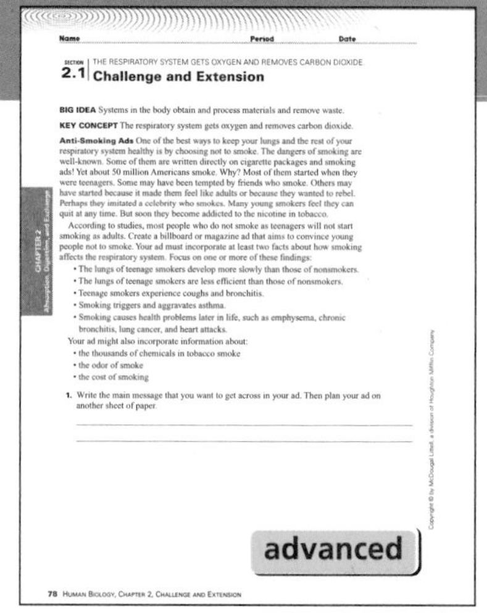

Name Period Date
SECTION 2.1 | THE RESPIRATORY SYSTEM GETS OXYGEN AND REMOVES CARBON DIOXIDE.
Challenge and Extension
advanced
78 HUMAN BIOLOGY, CHAPTER 2, CHALLENGE AND EXTENSION

R UNIT RESOURCE BOOK, pp. 72–73 | **R** pp. 74–75 | **R** p. 78

CHAPTER TEST

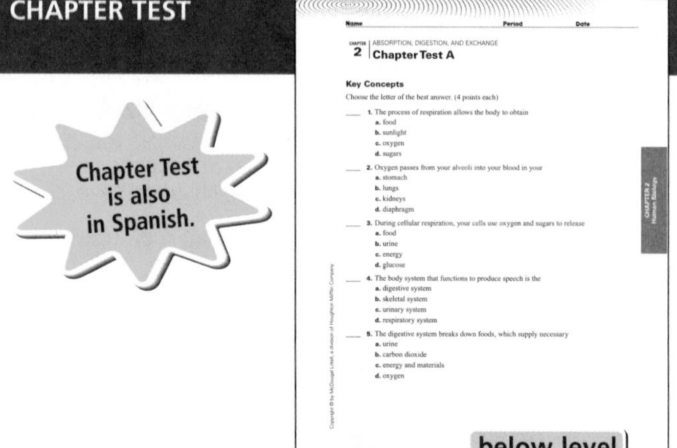

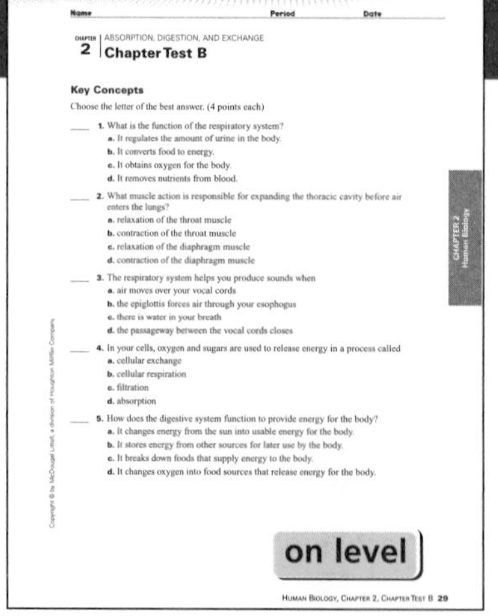

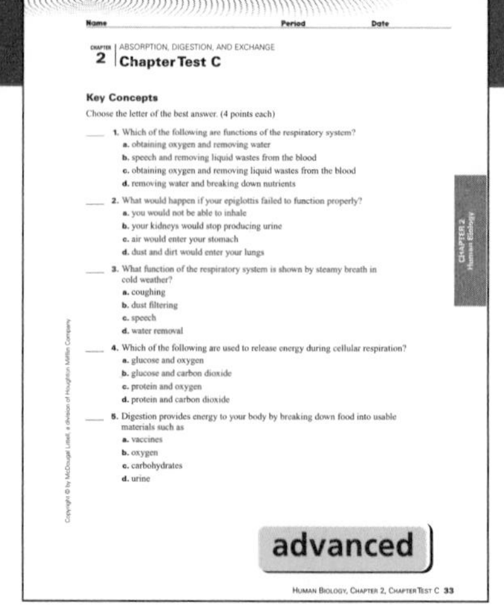

A UNIT ASSESSMENT BOOK, pp. 25–28 |  pp. 29–32 | 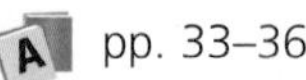 pp. 33–36

TECHNOLOGY

CLASSZONE
McDougal Littell Science
Life Science: Human Biology
Use these exciting animations, visuals, investigations, and links to enhance your knowledge of science.
There are two Resource Centers for this chapter.

McDOUGAL LITTELL
AUDIO READINGS
McDOUGAL LITTELL
LAB GENERATOR
Customize and edit labs with this easy-to-use CD-ROM
• Searchable database of all labs from the program
• Additional lab options
• Template for creating your own labs
• Rubrics and other resources
McDougal Littell Science

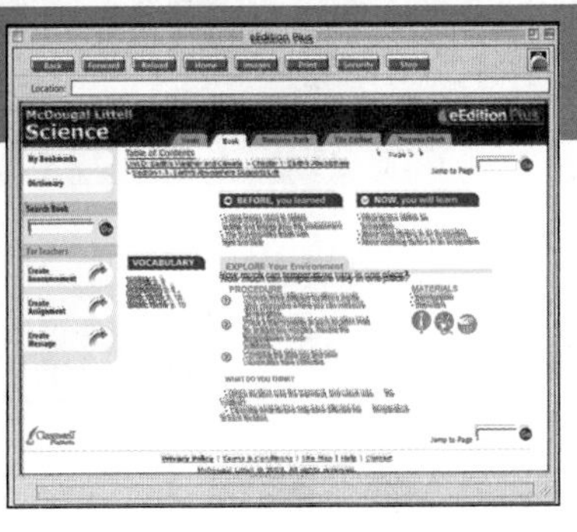

McDougal Littell Science
eEdition Plus

CLASSZONE.COM

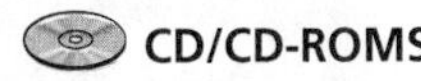

CD/CD-ROMS

CLASSZONE.COM

VISUAL CONTENT

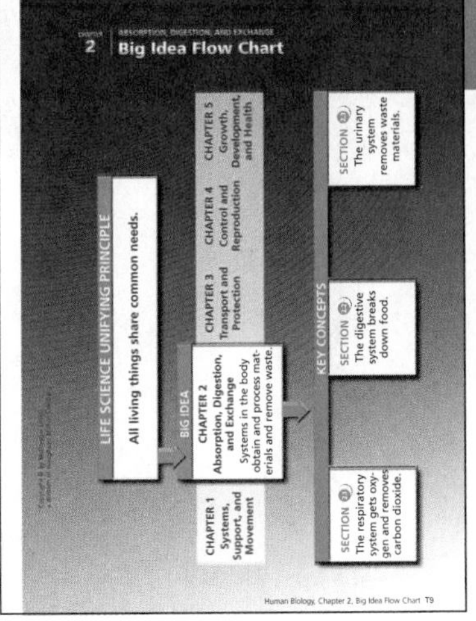

2 Big Idea Flow Chart
LIFE SCIENCE UNIFYING PRINCIPLE
All living things share common needs.
BIG IDEA
CHAPTER 2 Absorption, Digestion, and Exchange Systems in the body obtain and process materials and remove waste.
CHAPTER 1 Systems, Support, and Movement
CHAPTER 3 Transport and Protection
CHAPTER 4 Control and Reproduction
CHAPTER 5 Growth, Development, and Health
KEY CONCEPTS
SECTION The respiratory system gets oxygen and removes carbon dioxide.
SECTION The digestive system breaks down food.
SECTION The urinary system removes waste materials.

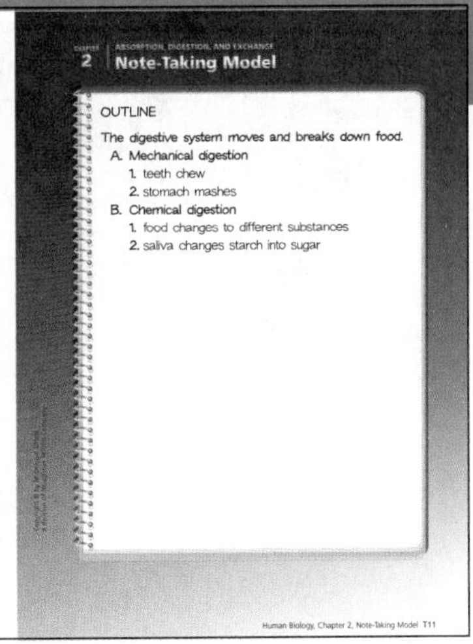

2 Note-Taking Model
OUTLINE
The digestive system moves and breaks down food.
A. Mechanical digestion
1. teeth chew
2. stomach mashes
B. Chemical digestion
1. food changes to different substances
2. saliva changes starch into sugar

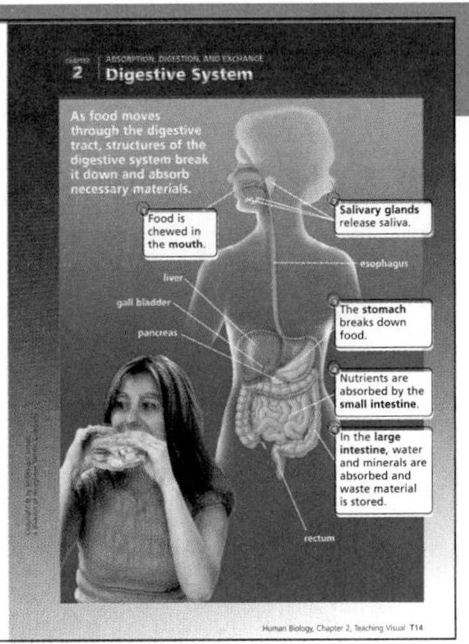

2 Digestive System
As food moves through the digestive tract, structures of the digestive system break it down and absorb necessary materials.
Food is chewed in the mouth.
Salivary glands release saliva.
esophagus
liver
gall bladder
pancreas
The stomach breaks down food.
Nutrients are absorbed by the small intestine.
In the large intestine, water and minerals are absorbed and waste material is stored.
rectum

UNIT TRANSPARENCY BOOK, p. T9

p. T11

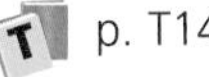

p. T14

MORE SUPPORT

Reinforcing Key Concepts for each section

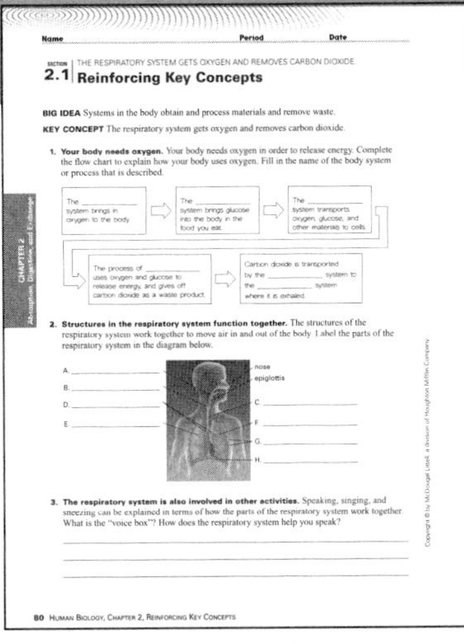

2.1 Reinforcing Key Concepts
BIG IDEA Systems in the body obtain and process materials and remove waste.
KEY CONCEPT The respiratory system gets oxygen and removes carbon dioxide.

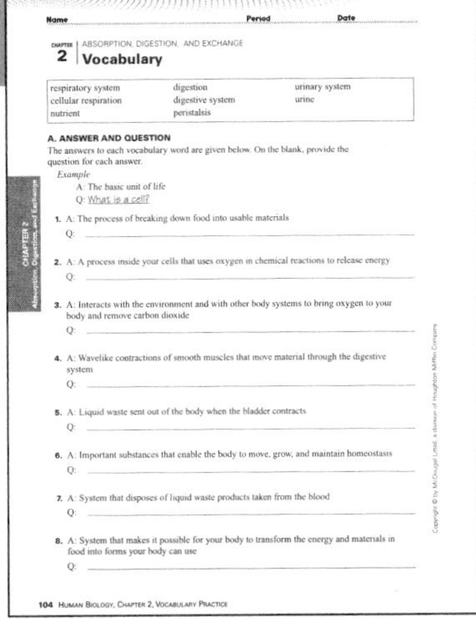

2 Vocabulary
A. ANSWER AND QUESTION

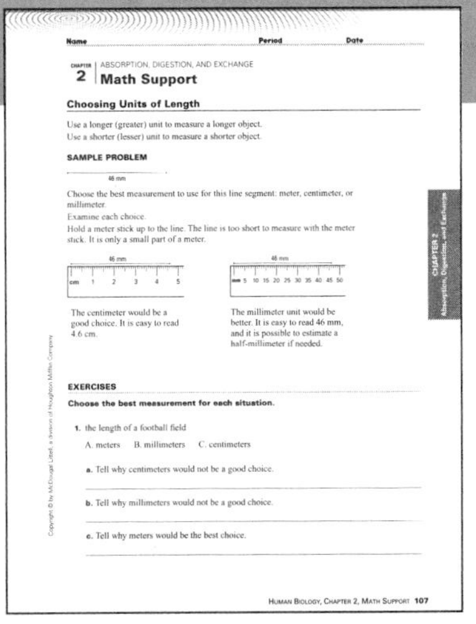

2 Math Support
Choosing Units of Length
SAMPLE PROBLEM
EXERCISES

UNIT RESOURCE BOOK, p. 80

pp. 104–105

p. 107

CHAPTER

Absorption, Digestion, and Exchange

INTRODUCE

the BIG idea

Have students look at the photograph and discuss how it shows materials the body needs to function.

- How do these materials get inside the body?
- How does your body manufacture waste materials? How do the waste materials leave the body?

National Science Education Standards

Content

C.1.e The human organism has systems for digestion, respiration, reproduction, circulation, excretion, movement, control, and coordination, and for the protection from disease. These systems interact with one another.

F.1.e Food provides energy and nutrients for growth and development. Nutrition requirements vary with body weight, age, sex, activity, and body functioning.

Process

A.1–8 Identify questions that can be answered through scientific investigations; design and conduct an investigation; use tools to gather and interpret data; use evidence to describe, predict, explain, model; think critically to make relationships between evidence and explanation; recognize different explanations and predictions; communicate scientific procedures and explanations; use mathematics.

A.9.a–c, A.9.e–g Understand scientific inquiry by using different investigations, methods, mathematics, and explanations based on logic, evidence, and skepticism. Data often results in new investigations.

G.1.b Science requires different abilities.

CHAPTER

Absorption, Digestion, and Exchange

the BIG idea

Systems in the body obtain and process materials and remove waste.

What materials does your body need to function properly?

Key Concepts

SECTION

2.1 The respiratory system gets oxygen and removes carbon dioxide.
Learn how the respiratory system functions.

SECTION

2.2 The digestive system breaks down food.
Learn how the digestive system provides cells with necessary materials.

SECTION

2.3 The urinary system removes waste materials.
Learn how the urinary system removes wastes.

Internet Preview

CLASSZONE.COM

Chapter 2 online resources: Content Review, two Visualizations, two Resource Centers, Math Tutorial, Test Practice.

INTERNET PREVIEW

CLASSZONE.COM For student use with the following pages:

Review and Practice
- Content Review, pp. 36, 58
- Math Tutorial: Measuring Lengths, p. 51
- Test Practice, p. 61

Activities and Resources
- Internet Activity: Lung Movement, p. 35
- Resource Centers: Respiratory System, p. 40; Urinary System, p. 54
- Visualization: Peristalsis, p. 46

Digestion **Code: MDL045**

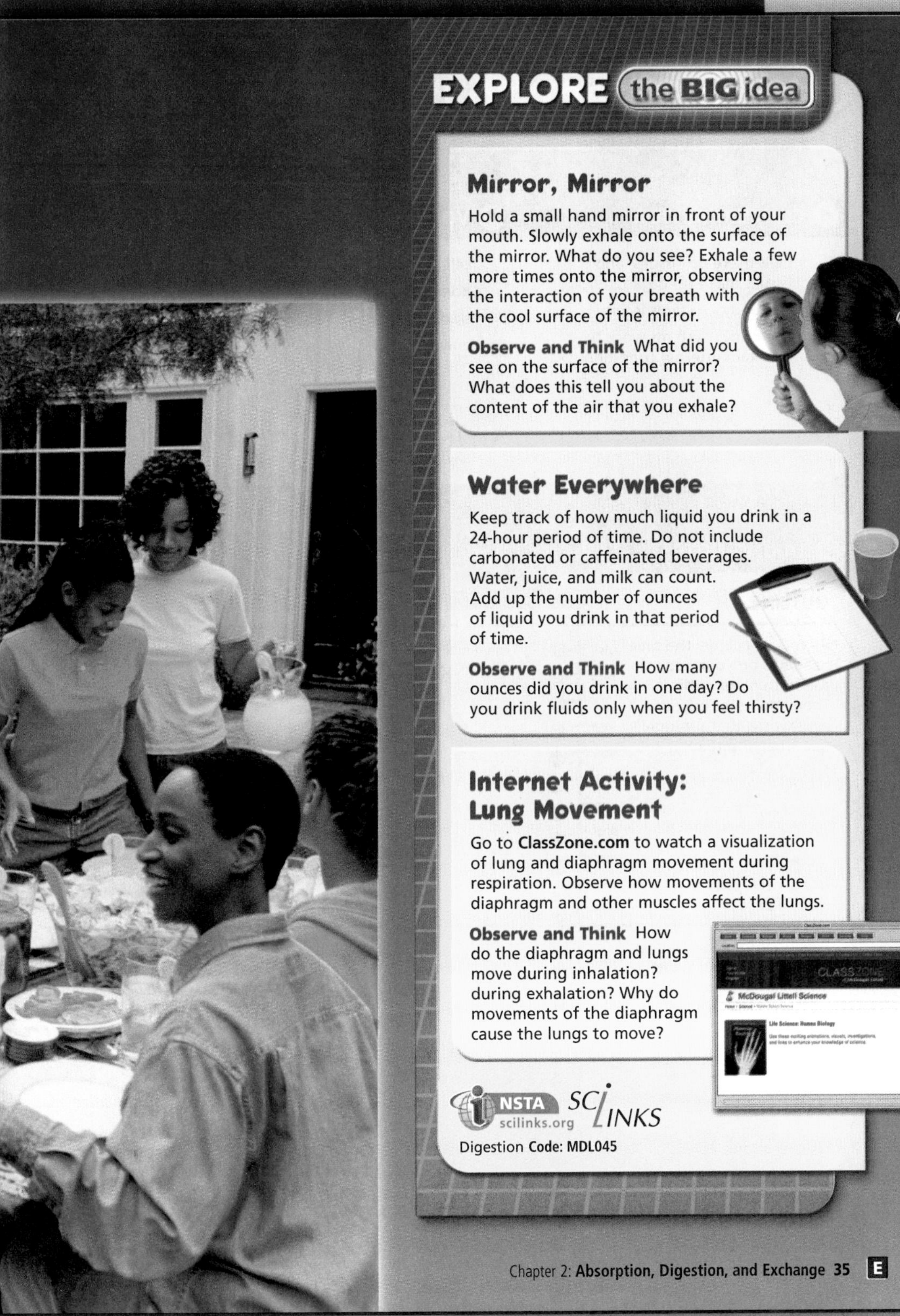

EXPLORE the BIG idea

Mirror, Mirror

Hold a small hand mirror in front of your mouth. Slowly exhale onto the surface of the mirror. What do you see? Exhale a few more times onto the mirror, observing the interaction of your breath with the cool surface of the mirror.

Observe and Think What did you see on the surface of the mirror? What does this tell you about the content of the air that you exhale?

Water Everywhere

Keep track of how much liquid you drink in a 24-hour period of time. Do not include carbonated or caffeinated beverages. Water, juice, and milk can count. Add up the number of ounces of liquid you drink in that period of time.

Observe and Think How many ounces did you drink in one day? Do you drink fluids only when you feel thirsty?

Internet Activity: Lung Movement

Go to **ClassZone.com** to watch a visualization of lung and diaphragm movement during respiration. Observe how movements of the diaphragm and other muscles affect the lungs.

Observe and Think How do the diaphragm and lungs move during inhalation? during exhalation? Why do movements of the diaphragm cause the lungs to move?

NSTA scilinks.org SCILINKS

Digestion Code: MDL045

EXPLORE the BIG idea

These inquiry-based activities are appropriate for use at home or as a supplement to classroom instruction.

Mirror, Mirror

PURPOSE To demonstrate to students that breath contains water.

TIP *10 min.* Caution students to be careful when using mirrors.

Answer: There was fog on the mirror. There is moisture in exhalation.

REVISIT after p. 43.

Water Everywhere

PURPOSE To determine how much liquid a student consumes each day.

TIP *10 min.* Supply students with a chart to record their data.

Answer: Answers will vary depending on how much liquid each student consumes.

REVISIT after p. 55.

Internet Activity: Lung Movement

PURPOSE To observe the movement of the lungs and diaphragm.

TIP *20 min.* Encourage students to feel their ribs as they watch the visualization.

Answer: During inhalation, the diaphragm contracts and moves downward, and the ribs expand. During exhalation, the diaphragm relaxes and moves upward, and the ribs contract. When the diaphragm contracts and moves downward, the thoracic cavity expands, putting pressure on the lungs to expand.

REVISIT after p. 40.

TEACHING WITH TECHNOLOGY

CBL and Probeware While students are investigating how the lungs work (p. 39), they can use a carbon dioxide sensor to measure the amount of carbon dioxide in their breath before and after exercise. Then they can use a graphing calculator or spreadsheet program to analyze the data.

Graphics Software Students could use drawing software or PowerPoint to create a flowchart that shows how the digestive and circulatory systems interact in transporting materials to the cells of the body (pp. 48–49).

PREPARE

CONCEPT REVIEW

Activate Prior Knowledge

- Display a picture of a fish, an insect, and a bird or mammal.
- Ask students how the animals obtain oxygen.
- Discuss how the various structures of spiracles, gills, and lungs enable the animals to interact with their environment.

TAKING NOTES

Outline

Encourage students to use the blue and red headings in their textbooks to form the basis of their outlines. The headings form the main ideas and subtopics of the outlines. Further details can be filled in with information found in the text and visuals.

Vocabulary Strategy

Students can include as many details as they want in a magnet diagram. Words and phrases that are associated with the vocabulary term are useful. The diagrams can become study aids as students read the chapter.

Vocabulary and Note-Taking Resources

- Vocabulary Practice, pp. 104–105
- Decoding Support, p. 106

- Daily Vocabulary Scaffolding, p. T10
- Note-Taking Model, p. T11

- Magnet Word Diagram, B24–25
- Outline, C43
- Daily Vocabulary Scaffolding, H1–8

CHAPTER 2

Getting Ready to Learn

CONCEPT REVIEW

- Cells make up tissues, and tissues make up organs.
- The body's systems interact.
- The body's systems work to maintain internal conditions.

VOCABULARY REVIEW

cell p. 10
homeostasis p. 12
smooth muscle p. 24
energy *See Glossary.*

CONTENT REVIEW
CLASSZONE.COM
Review concepts and vocabulary.

TAKING NOTES

OUTLINE

As you read, copy the blue headings on your paper in the form of an outline. Then add notes in your own words that summarize what you read.

VOCABULARY STRATEGY

Think about a vocabulary term as a **magnet word** diagram. Write the other terms or ideas related to that term around it.

See the Note-Taking Handbook on pages R45–R51.

SCIENCE NOTEBOOK

THE RESPIRATORY SYSTEM GETS OXYGEN AND REMOVES CARBON DIOXIDE.
A. Your body needs oxygen.
1. Oxygen is used to release energy
2. Oxygen is in air you breathe
B. Structures in the respiratory system function together
1. nose, throat, trachea
2. lungs

E 36 Unit: Human Biology

CHECK READINESS

Administer the Diagnostic Test to determine students' readiness for new science content and their mastery of requisite math skills.

 Diagnostic Test, pp. 20–21

Technology Resources

Students needing content and math skills should visit **ClassZone.com**.

- CONTENT REVIEW
- MATH TUTORIAL

 CONTENT REVIEW CD-ROM

KEY CONCEPT

2.1 The respiratory system gets oxygen and removes carbon dioxide.

BEFORE, you learned

- Cells, tissues, organs, and organ systems work together
- Organ systems provide for the body's needs
- Organ systems are important to the body's survival

NOW, you will learn

- About the structures of the respiratory system that function to exchange gases
- About the process of cellular respiration
- About other functions of the respiratory system

VOCABULARY

respiratory system p. 37
cellular respiration p. 38

EXPLORE Breathing

How do your ribs move when you breathe?

PROCEDURE

1. Place your hands on your ribs.
2. Breathe in and out several times, focusing on what happens when you inhale and exhale.
3. Record your observations in your notebook.

WHAT DO YOU THINK?

- What movement did you observe?
- Think about your observations. What questions do you have as a result of your observations?

Your body needs oxygen.

During the day, you eat and drink only a few times, but you breathe thousands of times. In fact, breathing is a sign of life. The body is able to store food and liquid, but it is unable to store very much oxygen. The **respiratory system** is the body system that functions to get oxygen from the environment and remove carbon dioxide and other waste products from your body. The respiratory system interacts with the environment and with other body systems.

The continuous process of moving and using oxygen involves mechanical movement and chemical reactions. Air is transported into your lungs by mechanical movements and oxygen is used during chemical reactions that release energy in your cells.

VOCABULARY Make a word magnet diagram for the term *respiration*.

CHECK YOUR READING What are the two main functions of your respiratory system?

RESOURCES FOR DIFFERENTIATED INSTRUCTION

Below Level

UNIT RESOURCE BOOK

- Reading Study Guide A, pp. 72–73
- Decoding Support, p. 106

Additional INVESTIGATION, How Much Air Can Your Lungs Hold?, A, B, & C, pp. 118–126; Teacher Instructions, pp. 305–306

Advanced

UNIT RESOURCE BOOK

- Challenge and Extension, p. 78
- Challenge Reading, pp. 102–103

English Learners

UNIT RESOURCE BOOK

Spanish Reading Study Guide, pp. 76–77

- Audio Readings in Spanish
- Audio Readings (English)

2.1 FOCUS

Set Learning Goals

Students will

- Describe the structures and functions of the respiratory system.
- Analyze the process of cellular respiration.
- Explain other functions of the respiratory system.
- Model how air moves in and out of the lungs.

3-Minute Warm-Up

Display Transparency 12 or copy this exercise on the board:

Draw a diagram that shows how cells, tissues, organs, and organ systems are related. Describe the relationship between cells and tissues, tissues and organs, and organs and organ systems.

The diagram could show four concentric circles, with cells in innermost circle and organ systems in outermost circle. Cells make up tissues; tissues make up organs; organs make up organ systems.

3-Minute Warm-Up, p. T12

2.1 MOTIVATE

EXPLORE Breathing

PURPOSE To observe the movement of the rib cage during respiration

TIP *10 min.* Be sure that students use a light touch and breathe deeply.

WHAT DO YOU THINK? *Sample answer: During inhalation, the chest expanded; during exhalation, the chest contracted. The ribs moved this way because of air rushing into the lungs. The lungs expand when you take in air. Students' questions should relate to effects of respiration on the body.*

Ongoing Assessment

CHECK YOUR READING *Answer: Respiratory system gets oxygen from environment and removes carbon dioxide and other wastes from body.*

2.1 INSTRUCT

Teach from Visuals

To help students interpret the photograph of the diver, ask:

- What kind of gas might be found in the air tank of the diver? *oxygen*
- How do you know? *Humans need oxygen to survive.*

History of Science

Before the mid-1940s, underwater sea exploration was limited to explorers who wore diving suits with air hoses attached to their helmets. In 1943, Jacques Cousteau, with the aid of Emile Gagnan, invented the Aqua Lung, which would eventually come to be known as scuba (self-contained underwater breathing apparatus). Scuba gear became commercially available in 1946. For the first time, divers could move about freely underwater for long periods of time without having to be attached to air hoses from ships above the ocean's surface.

Ongoing Assessment

CHECK YOUR READING *Answer: The air you breathe is made up of a mixture of gases including oxygen and nitrogen.*

Exchanging Oxygen and Carbon Dioxide

Like almost all living things, the human body needs oxygen to survive. Without oxygen, cells in the body die quickly. How does the oxygen you need get to your cells? Oxygen, along with other gases, enters the body when you inhale. Oxygen is then transported to cells throughout the body.

The air that you breathe contains only about 20 percent oxygen and less than 1 percent carbon dioxide. Almost 80 percent of air is nitrogen gas. The air that you exhale contains more carbon dioxide and less oxygen than the air that you inhale. It's important that you exhale carbon dioxide because high levels of it will damage, even destroy, cells.

In cells and tissues, proper levels of both oxygen and carbon dioxide are essential. Recall that systems in the body work together to maintain homeostasis. If levels of oxygen or carbon dioxide change, your brain or blood vessels signal the body to breathe faster or slower.

The photograph shows how someone underwater maintains proper levels of carbon dioxide and oxygen. The scuba diver needs to inhale oxygen from a tank. She removes carbon dioxide wastes with other gases when she exhales into the water. The bubbles you see in the water are formed when she exhales.

CHECK YOUR READING What gases are in the air that you breathe?

Gas Exchange

This scuba diver breathes the same mixture of gases present in air.

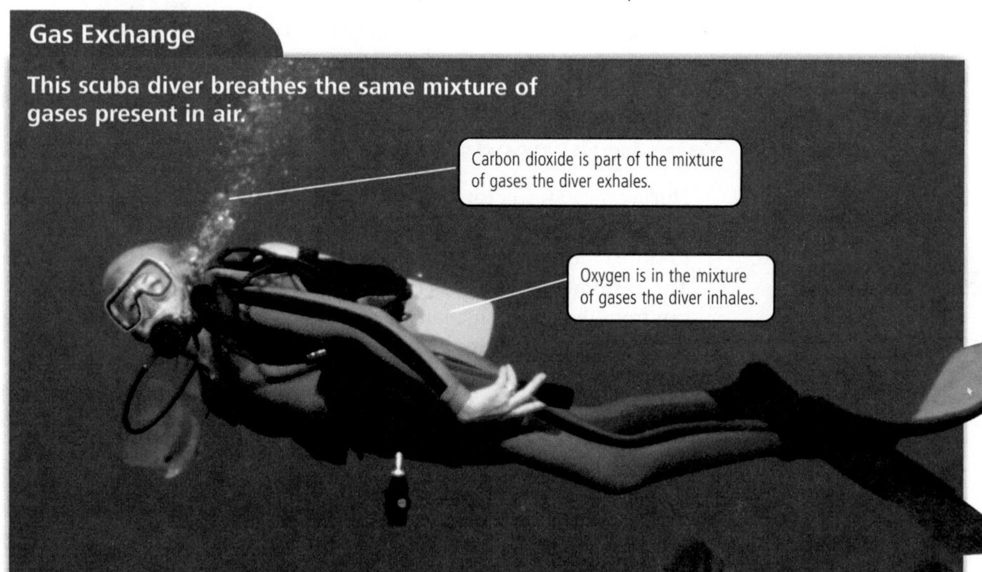

DIFFERENTIATE INSTRUCTION

More Reading Support

A What gas is necessary in air that you breathe? *oxygen*

B What gas is important to exhale so that cells are not damaged or destroyed? *carbon dioxide*

English Learners Have students write the definitions for the terms *cellular respiration, cilia,* and *alveoli* in their Science Word Dictionaries. Encourage them to add a sketch and a "reminder sentence" for each, based on their reading. Use the Chapter Review, p. 58, to help students focus on key concepts and vocabulary.

INVESTIGATE Lungs

How does air move in and out of lungs?

SKILL FOCUS
Modeling

MATERIALS
- one medium balloon
- 1-L clear plastic bottle with labels removed

TIME
15 minutes

PROCEDURE

1. Create a model of your lungs as shown. Insert an uninflated balloon into the top of the plastic bottle. While squeezing the bottle to force out some air, stretch the end of the balloon over the lip of the bottle. The balloon should still be open to the outside air. Tape the balloon in place with duct tape to make a tight seal
2. Release the bottle so that it expands back to its normal shape. Observe what happens to the balloon. Squeeze and release the bottle several times while observing the balloon. Record your observations.

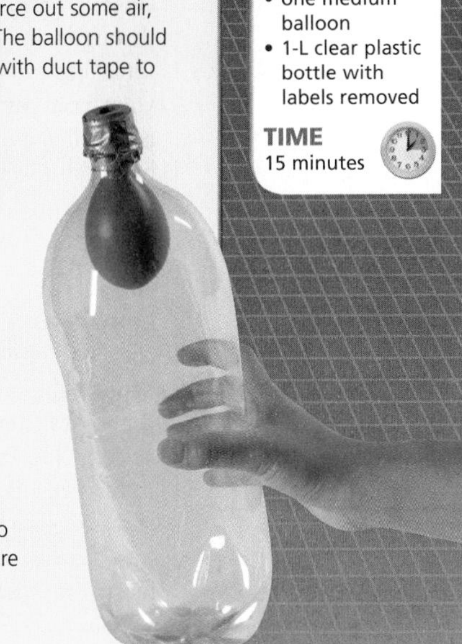

WHAT DO YOU THINK?
- Describe, in words, what happens when you squeeze and release the bottle.
- How do you think your lungs move when you inhale? when you exhale?

CHALLENGE Design an addition to your model that could represent a muscle called the diaphragm. What materials do you need? How would this work? Your teacher may be able to provide additional materials so you can test your model. Be sure to come up with a comprehensive list of materials as well as a specific diagram.l as a specific diagram.

Cellular Respiration

Inside your cells, a process called **cellular respiration** uses oxygen in chemical reactions that release energy. The respiratory system works with the digestive and circulatory systems to make cellular respiration possible. Cellular respiration requires glucose, or sugars, which you get from food, in addition to oxygen, which you get from breathing. These materials are transported to every cell in your body through blood vessels. You will learn more about the digestive and circulatory systems later in this unit.

? C

VOCABULARY
Add a magnet diagram for *cellular respiration* to your notebook. Include the word *energy* in your diagram.

During cellular respiration, your cells use oxygen and glucose to release energy. Carbon dioxide is a waste product of the process. Carbon dioxide must be removed from cells.

CHECK YOUR READING What three body systems are involved in cellular respiration?

DIFFERENTIATE INSTRUCTION

? More Reading Support

C How are materials from food transported throughout your body? *blood vessels*

Additional Investigation To reinforce Section 2.1 learning goals, use the following full-period investigation:

Additional INVESTIGATION, How Much Air Can Your Lungs Hold?, A, B, & C, pp. 118–126, 305–306

Advanced Have students draw a diagram or flow chart that shows the process of cellular respiration. Students should include in their diagrams the reaction between glucose and oxygen and both the resulting energy and the byproduct, carbon dioxide.

R Challenge and Extension, p. 78

Teaching with Technology

If a gas sensor is available, have students measure the amount of carbon dioxide exhaled before and after they exercise. Have them plot the data using a graphing calculator or spreadsheet program.

INVESTIGATE Lungs

PURPOSE To create a model showing how the lungs function

TIPS *15 min.*
- Have students stretch the balloon before putting it in the bottle.
- The balloon is best held on with duct tape. Be sure that the tape has made a tight seal.
- Make sure students do not squeeze the bottle too hard while putting in the balloon. The bottle should re-expand once the balloon is in place.

WHAT DO YOU THINK? *When the bottle is squeezed, the balloon deflates. When the bottle is released, the balloon expands slightly. When you inhale, your lungs expand. When you exhale, your lungs shrink.*

CHALLENGE *A diaphragm could be added by cutting off the bottom of the bottle and adding a stretchy material, such as a rubber glove, over the bottom. Pulling on the glove would make the balloon inflate. Releasing the glove would make the balloon deflate.*

Datasheet, Lungs, p. 79

Technology Resources

Customize this student lab as needed or look for an alternative. Print rubrics to assess student lab reports.

Lab Generator CD-ROM

Ongoing Assessment

Analyze the process of cellular respiration.

Ask: What gas is a waste product of cellular respiration? *carbon dioxide*

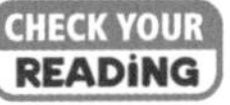
CHECK YOUR READING *Answer: The respiratory, circulatory, and digestive systems are involved in cellular respiration.*

Develop Critical Thinking

APPLY To help students understand the structure of the respiratory system, ask:

- By what criteria would biologists include an organ as part of the respiratory system? *Sample answer: The organ must be part of the group of organs that work together to get oxygen into the body and to release carbon dioxide from the body.*
- Have students describe why the organs of the respiratory system are similar to the branches and leaves of a tree. *Sample answer: The bronchial tubes are like the branches of a tree. They separate into smaller and smaller tubes until they end in alveoli. The alveoli are like the leaves of a tree.*

Integrate the Sciences

The contraction of the diaphragm and expansion of the ribs cause the capacity of the thoracic cavity to increase. Air pressure inside the thoracic cavity decreases because there is more room for the air. The difference in the air pressure inside the body and the air pressure outside the body causes air to rush into the lungs.

EXPLORE the BIG idea

Revisit "Internet Activity: Lung Movement" on p. 35. Have students use their own words to explain the movement of the diaphragm and ribs.

Ongoing Assessment

CHECK YOUR READING *Answer: Air inhaled through the nose or mouth enters the throat, passes through the trachea, and moves into the lungs via the bronchial tubes.*

CHECK YOUR READING *Answer: When you inhale, the diaphragm contracts and pulls downward, while other muscles draw the ribs outward.*

OUTLINE
Add *Structures in the respiratory system function together* to your outline. Be sure to include the six respiratory structures in your outline.

I. Main idea
 A. Supporting idea
  1. Detail
  2. Detail
 B. Supporting idea

Structures in the respiratory system function together.

The respiratory system is made up of many structures that allow you to move air in and out of your body, communicate, and keep out harmful materials.

Nose, Throat, and Trachea When you inhale, air enters your body through your nose or mouth. Inside your nose, tiny hairs called cilia filter dirt and other particles out of the air. Mucus, a sticky liquid in your nasal cavity, also helps filter air by trapping particles such as dirt and pollen as air passes by. The nasal cavity warms the air slightly before it moves down your throat toward a tubelike structure called the windpipe, or trachea (TRAY-kee-uh). A structure called the epiglottis (EHP-ih-GLAHT-ihs) keeps air from entering your stomach.

D

Lungs The lungs are two large organs located on either side of your heart. When you breathe, air enters the throat, passes through the trachea, and moves to the lungs through structures called bronchial tubes. Bronchial tubes branch throughout the lungs into smaller and smaller tubes. At the ends of the smallest tubes air enters tiny air sacs called alveoli. The walls of the alveoli are only one cell thick. In fact, one page in this book is much thicker than the walls of the alveoli. Oxygen passes from inside the alveoli through the thin walls and is dissolved into the blood. At the same time, carbon dioxide waste passes from the blood into the alveoli.

CHECK YOUR READING Through which structures does oxygen move into the lungs?

Ribs and Diaphragm If you put your hands on your ribs and take a deep breath, you can feel your ribs expand. The rib cage encloses a space inside your body called the thoracic (thu-RHAS-ihk) cavity. Some ribs are connected by cartilage to the breastbone or to each other, which makes the rib cage flexible. This flexibility allows the rib cage to expand when you breathe and make room for the lungs to expand and fill with air.

E

A large muscle called the diaphragm (DY-uh-FRAM) stretches across the floor of the thoracic cavity. When you inhale, your diaphragm contracts and pulls downward, which makes the thoracic cavity expand. This movement causes the lungs to push downward, filling the extra space. At the same time, other muscles draw the ribs outward and expand the lungs. Air rushes into the lungs, and inhalation is complete. When the diaphragm and other muscles relax, the process reverses and you exhale.

RESOURCE CENTER CLASSZONE.COM
Explore the respiratory system.

CHECK YOUR READING Describe how the diaphragm and the rib cage move.

DIFFERENTIATE INSTRUCTION

More Reading Support

D Where in the lungs are O_2 and CO_2 exchanged? *alveoli*

E What muscle stretches across the bottom of the thoracic cavity? *diaphragm*

Below Level Have students make a flow chart to show the path of oxygen as it travels through the respiratory system. Students should include the structures listed on p. 40—nose, throat, ribs, lungs, trachea, and diaphragm—and the terms *oxygen* and *carbon dioxide.*

Respiratory System

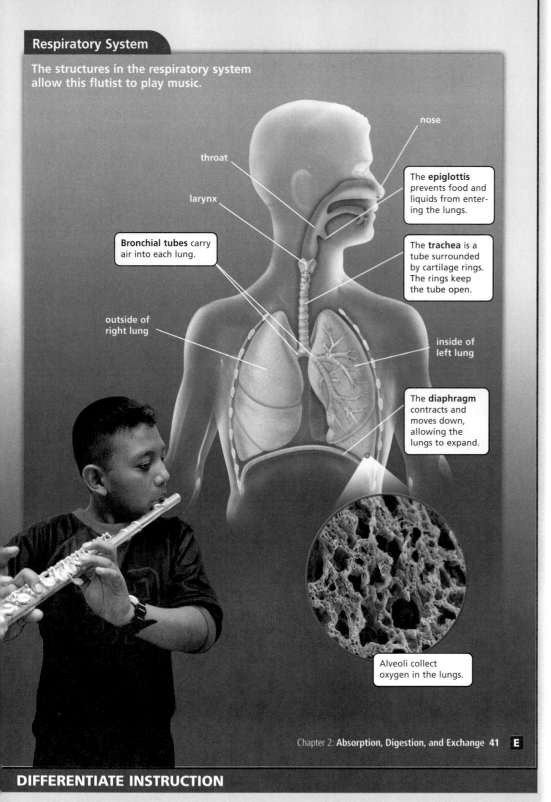

The structures in the respiratory system allow this flutist to play music.

nose

throat

larynx

The **epiglottis** prevents food and liquids from entering the lungs.

Bronchial tubes carry air into each lung.

The **trachea** is a tube surrounded by cartilage rings. The rings keep the tube open.

outside of right lung

inside of left lung

The **diaphragm** contracts and moves down, allowing the lungs to expand.

Alveoli collect oxygen in the lungs.

DIFFERENTIATE INSTRUCTION

Advanced Challenge students to time their breathing for one minute. Then have them calculate how many breaths they take in a year and how many in 18 years. Finally, ask students to determine how many breaths they have taken so far in their lives and how many they will have taken when they reach 80 years old.

Have students who are interested in the respiratory system read the following article:

 Challenge Reading, pp. 102–103

Teach from Visuals

To help students interpret the diagram of the respiratory system, ask:

- Which structures in the respiratory system are connected by the trachea? *The trachea connects the larynx and bronchial tubes.*
- Which structure directs air down the trachea? *the epiglottis*

Teacher Demo

Have students breathe normally for about 20 seconds. Then have them pause at the end of an exhalation. Ask them if they can gently push out more air before they breathe in again. Make sure no students have a health condition for which this would be dangerous. Students should discover that not all air is expelled from the lungs during exhalation.

Teach Difficult Concepts

Some students may have a hard time understanding that the diaphragm moves downward. Ask students to recall the Explore activity on p. 37. Have them describe again what they observed about the motion of the ribs as they breathed. Point out that the diaphragm pulls downward when they inhale, even though it feels as if the lungs and diaphragm go up.

Ongoing Assessment

Describe the structures and functions of the respiratory system.

Ask: How does the structure of the bronchial tubes allow air to pass into the lungs? *The bronchial tubes branch throughout the lungs into smaller and smaller tubes.*

Teacher Demo

Help students understand how the vocal cords produce sound.

- Provide pairs of students with different sizes of rubber bands. Invite each pair to the front of the classroom and have one student pull the rubber band taut. Have the other student pluck the rubber band to produce a sound. Relate the vibrating motion of the rubber band to how sound is produced by the vocal cords.
- Encourage students to place their fingers on their throats as they speak. They should be able to feel the vibration of the vocal chords.

Teach from Visuals

To help students understand how sound is produced by the vocal cords, ask:

- What path does air take as it produces sound? *Air from the lungs passes over the vocal cords. The air travels out over the tongue, and sound waves leave the mouth.*
- Is it possible to speak if you are holding your breath? Explain. *No; air is necessary for sounds to be produced.*

Ongoing Assessment

Describe other functions of the respiratory system.

Ask: How do the vocal cords produce sound? *Air from the lungs moves over the vocal cords, causing them to vibrate and produce sound.*

The respiratory system is also involved in other activities.

F In addition to providing oxygen and removing carbon dioxide, the respiratory system is involved in other activities of the body. Speaking and singing, along with actions such as sneezing, can be explained in terms of how the parts of the respiratory system work together.

Speech and Other Respiratory Movements

G If you place your hand on your throat and hum softly, you can feel your vocal cords vibrating. Air moving over your vocal cords allows you to produce sound, and the muscles in your throat, mouth, cheeks, and lips allow you to form sound into words. The vocal cords are folds of tissue in the larynx. The larynx, sometimes called the voice box, is a two-inch, tube-shaped organ about the length of your thumb, located in the neck, at the top of the trachea. When you speak, the vocal cords become tight, squeeze together, and force air from the lungs to move between them. The air causes the vocal cords to vibrate and produce sound.

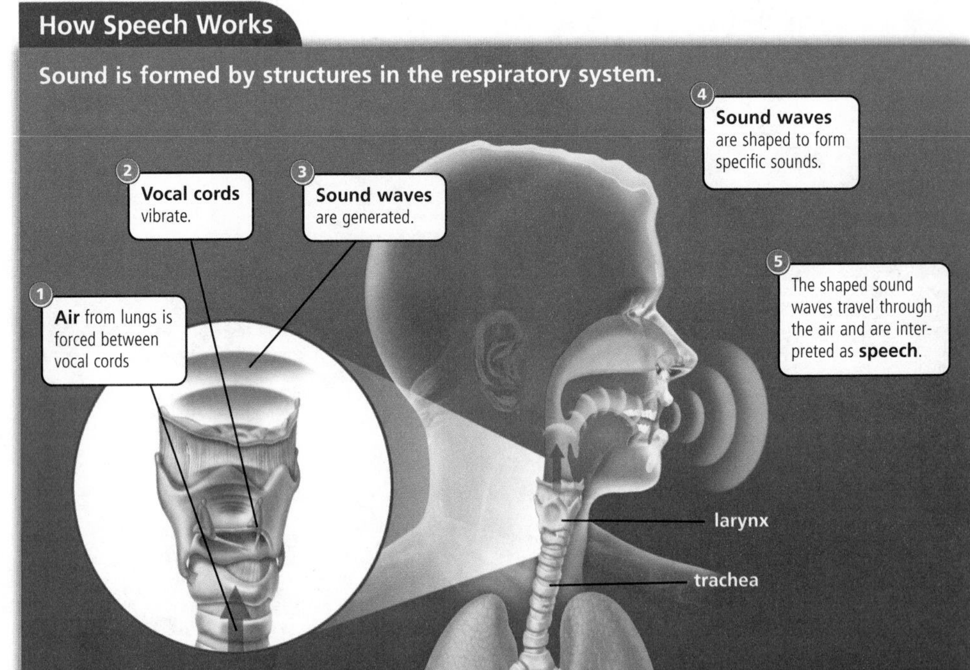
How Speech Works
Sound is formed by structures in the respiratory system.
1 Air from lungs is forced between vocal cords
2 Vocal cords vibrate.
3 Sound waves are generated.
4 Sound waves are shaped to form specific sounds.
5 The shaped sound waves travel through the air and are interpreted as speech.
larynx
trachea
lungs

DIFFERENTIATE INSTRUCTION

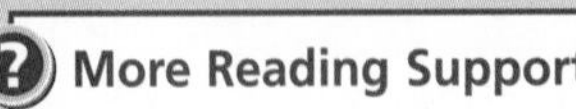

More Reading Support

F What are some activities the respiratory system is involved in? *speaking, singing, sneezing*

G Where are vocal cords located? *larynx, or voice box*

Below Level Students may confuse the terms *respiration* and *breathing*. Have them write the definitions on index cards to use as study guides. Remind students that respiration is the exchange of gases—and can happen at the level of cell, organ, or organism. Breathing is the movement of air in and out of the body.

E 42 Unit: Human Biology

Some movements of the respiratory system allow you to clear particles out of your nose and throat or to express emotion. The respiratory system is involved when you cough or sneeze. Sighing, yawning, laughing, and crying also involve the respiratory system.

Sighing and yawning both involve taking deep breaths. A sigh is a long breath followed by a shorter exhalation. A yawn is a long breath taken through a wide-open mouth. Laughing and crying are movements that are very similar to each other. In fact, sometimes it's difficult to see the difference between laughing and crying.

The respiratory system also allows you to hiccup. A hiccup is a sudden inhalation that makes the diaphragm contract. Several systems are involved when you hiccup. Air rushes into the throat, causing the diaphragm to contract. When the diaphragm contracts, the air passageway between the vocal cords closes. The closing of this passageway produces the sound of the hiccup. Hiccups can be caused by eating too fast, sudden temperature changes, and stress.

Water Removal

Hiccups, coughs, yawns, and all other respiratory movements, including speaking and breathing, release water from your body into the environment. Water is lost through sweat, urine, and exhalations of air. When it is cold enough outside, you can see your breath in the air. That is because the water vapor you exhale condenses into larger droplets when it moves from your warm body to the cold air.

Water leaves your body through your breath every time you exhale.

2.1 Review

KEY CONCEPTS

1. How is oxygen used by your body's cells?
2. What are the structures in the respiratory system and what do they do?
3. In addition to breathing, what functions does the respiratory system perform?

CRITICAL THINKING

4. **Sequence** List in order the steps that occur when you exhale.
5. **Compare and Contrast** How is the air you inhale different from the air you exhale?

CHALLENGE

6. **Hypothesize** Why do you think a person breathes more quickly when exercising?

ANSWERS

1. in the respiration process, to release energy

2. Nose, throat, trachea, and bronchial tubes carry air in and out of the lungs, where gas exchange occurs. Diaphragm and ribs move to cause lungs to expand.

3. also filters air and aids in speaking, coughing, sneezing

4. Diaphragm relaxes up to its original position, pushing lungs upward. At the same time, ribs move inward, collapsing lungs inward. Air moves out of lungs through bronchial tubes into trachea, then into throat and out nose or mouth.

5. Air you inhale contains more O_2 and less CO_2 than air you exhale.

6. Energy is expended when you exercise; rapid breathing brings more O_2 into the body, which can add more energy.

EXPLORE the BIG idea

Revisit "Mirror, Mirror" on p. 35 and the observations students made. How has their understanding changed?

Teach Difficult Concepts

Point out that water is released from the body through sweating and urination as well as through exhalation.

Reinforce the BIG idea

Have students relate the section to the Big Idea.

 Reinforcing Key Concepts, p. 80

2.1 ASSESS & RETEACH

Assess

 Section 2.1 Quiz, p. 22

Reteach

With a small group, write a comic strip frame by frame about how oxygen travels through the body. Prompt the group with questions such as:

- Where does oxygen go?
- Through what does oxygen pass?
- Where does oxygen end up?
- How does it leave the body?

Technology Resources

Have students visit **ClassZone.com** for reteaching of Key Concepts.

 **CONTENT REVIEW**

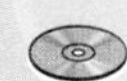 **CONTENT REVIEW CD-ROM**

SCIENCE ON THE JOB
Relevance of Science to Non-science Jobs

Set Learning Goal
To understand why yoga instructors need knowledge of the respiratory system

Present the Science
Remind students that the movement of the diaphragm and the rib cage controls breathing. As the diaphragm contracts and pulls downward, the rib cage expands, and air rushes into the lungs.

SAFETY Find out any health conditions that students may have that might prevent them from doing this activity. Students should not do deep inhalations or hold their breath if they are suffering from hypertension or heart problems. Students should not do deep exhalations if suffering from hypotension or depression. If students become tired, fatigued, or irritated, relax and stop the exercise. If students become hot, it is a sign of strain.

Discussion Questions
Ask: How is full-lung breathing different from the way a person normally breathes? *In full-lung breathing, the belly expands first, then the rib cage, then the upper chest. When a person breathes normally, the breathing does not occur in these stages.*

Ask: What are two functions of nostrils? *filtering dust before air enters lungs and warming air as it enters body*

Use the diagram for students to point out the diaphragm, and discuss how it moves during abdominal breathing.

Close
Ask: Why do yoga instructors need knowledge of the respiratory system? *Yoga instructors need to understand how breathing occurs so they can help their students to exercise and to breathe in healthy ways.*

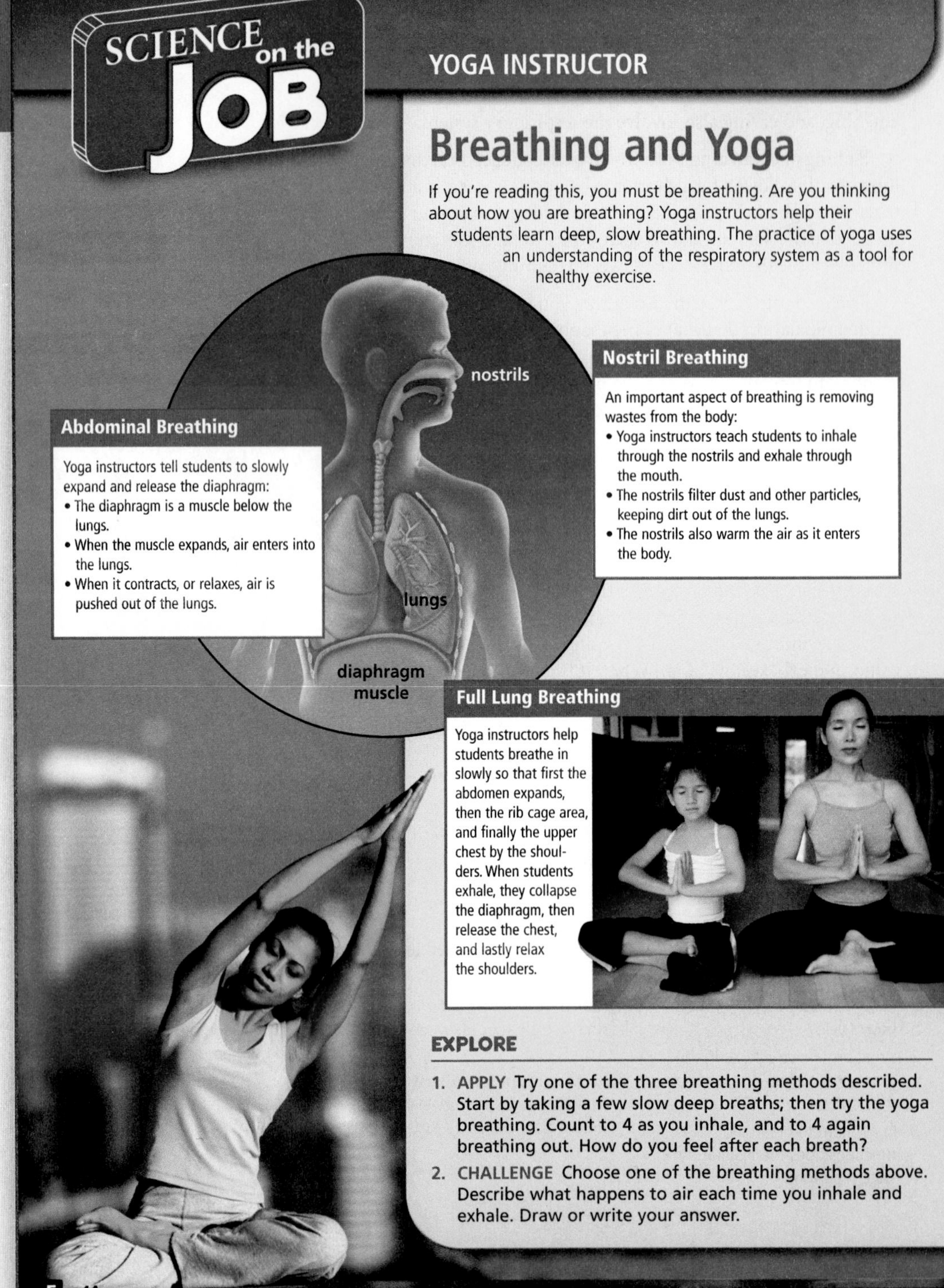

SCIENCE on the JOB
YOGA INSTRUCTOR
Breathing and Yoga
If you're reading this, you must be breathing. Are you thinking about how you are breathing? Yoga instructors help their students learn deep, slow breathing. The practice of yoga uses an understanding of the respiratory system as a tool for healthy exercise.
nostrils
lungs
diaphragm muscle
Abdominal Breathing
Yoga instructors tell students to slowly expand and release the diaphragm:
• The diaphragm is a muscle below the lungs.
• When the muscle expands, air enters into the lungs.
• When it contracts, or relaxes, air is pushed out of the lungs.
Nostril Breathing
An important aspect of breathing is removing wastes from the body:
• Yoga instructors teach students to inhale through the nostrils and exhale through the mouth.
• The nostrils filter dust and other particles, keeping dirt out of the lungs.
• The nostrils also warm the air as it enters the body.
Full Lung Breathing
Yoga instructors help students breathe in slowly so that first the abdomen expands, then the rib cage area, and finally the upper chest by the shoulders. When students exhale, they collapse the diaphragm, then release the chest, and lastly relax the shoulders.
EXPLORE
1. APPLY Try one of the three breathing methods described. Start by taking a few slow deep breaths; then try the yoga breathing. Count to 4 as you inhale, and to 4 again breathing out. How do you feel after each breath?
2. CHALLENGE Choose one of the breathing methods above. Describe what happens to air each time you inhale and exhale. Draw or write your answer.
E 44

EXPLORE
1. ***APPLY*** *Students may report that they feel relaxed. Others may feel energized. Some may find that their breathing is smoother.*
2. ***CHALLENGE*** *Sample answer: I take in breath through my nostrils. My lungs expand. My diaphragm descends. Then there is a slight pause before I exhale. When I exhale, my lungs deflate, my diaphragm lifts, and air passes out through my nostrils.*

KEY CONCEPT

2.2 The digestive system breaks down food.

BEFORE, you learned

- The respiratory system takes in oxygen and expels waste
- Oxygen is necessary for cellular respiration
- The respiratory system is involved in speech and water removal

NOW, you will learn

- About the role of digestion in providing energy and materials
- About the chemical and mechanical process of digestion
- How materials change as they move through the digestive system

VOCABULARY

nutrient p. 45
digestion p. 46
digestive system p. 46
peristalsis p. 46

EXPLORE Digestion

How does the digestive system break down fat?

PROCEDURE

1. Using a dropper, place 5 mL of water into a test tube. Add 5 mL of vegetable oil. Seal the test tube with a screw-on top. Shake the test tube for 10 seconds, then place it in a test tube stand. Record your observations.
2. Drop 5 mL of dish detergent into the test tube. Seal the tube. Shake the test tube for 10 seconds, then place in the stand. Observe the mixture for 2 minutes. Record your observations.

MATERIALS

- water
- dropper
- test tube
- vegetable oil
- test tube stand
- liquid dish detergent

WHAT DO YOU THINK?

- What effect does detergent have on the mixture of oil and water?
- How do you think your digestive system might break down fat?

OUTLINE
Remember to add *The body needs energy and materials* to your outline.

I. Main idea
  A. Supporting idea
    1. Detail
    2. Detail
  B. Supporting idea

The body needs energy and materials.

After not eating for a while, have you ever noticed how little energy you have to do the simplest things? You need food to provide energy for your body. You also need materials from food. Most of what you need comes from nutrients within food. **Nutrients** are important substances that enable the body to move, grow, and maintain homeostasis. Proteins, carbohydrates, fats, and water are some of the nutrients your body needs.

You might not think of water as a nutrient, but it is necessary for all living things. In fact, more than half of your body is made up of

2.2 FOCUS

Set Learning Goals

Students will

- Describe the role of digestion in providing energy for the body.
- Describe the process of digestion.
- Describe changes in materials as they move through the digestive system.
- Model how the stomach breaks down food.

3-Minute Warm-Up

Display Transparency 12 or copy this exercise on the board:

Decide if these statements are true. If not true, correct them.

1. The digestive system works with the respiratory system to make cellular respiration possible. *true*
2. Carbon dioxide is necessary for cellular respiration. *Oxygen is necessary for cellular respiration.*
3. The digestive system is involved in speech and water removal. *The digestive system is involved in digestion and water removal.*

3-Minute Warm-Up, p. T12

2.2 MOTIVATE

EXPLORE Digestion

PURPOSE To understand how the digestive system breaks down fat

TIPS *15 min.* Have students predict what they think will happen to the oil. If students work in groups, be sure they work out who is measuring and pouring ahead of time. A well-covered surface, and/or premeasured liquids will help students avoid spills. Smocks and goggles can also be used.

WHAT DO YOU THINK? *The detergent allows the oil to break down. The body breaks down fat into smaller pieces that can be absorbed.*

RESOURCES FOR DIFFERENTIATED INSTRUCTION

Below Level

UNIT RESOURCE BOOK
- Reading Study Guide A, pp. 83–84
- Decoding Support, p. 106

AUDIO CDS

Advanced

UNIT RESOURCE BOOK
Challenge and Extension, p. 89

English Learners

UNIT RESOURCE BOOK
Spanish Reading Study Guide, pp. 87–88

AUDIO CDS

- Audio Readings in Spanish
- Audio Readings (English)

2.2 INSTRUCT

Address Misconceptions

IDENTIFY Ask: What does food provide for the body? If students answer that food provides the body with energy, they may hold the misconception that food is only a source of energy rather than also being a source of matter for growth.

CORRECT Direct students to the definition of digestion on p. 46. Explain that digestion is involved in breaking down food into materials that cells use for growth and repair as well as for energy.

REASSESS Ask: What happens to food during the process of digestion? *Food is broken down into nutrients, used to provide energy for the body as well as material for growth and repair.*

Technology Resources

Visit **ClassZone.com** for background on common student misconceptions.

MISCONCEPTION DATABASE

Teach from Visuals

To help students interpret the sequence in the diagram, ask:

- What structures does the esophagus connect? *the mouth and the stomach*
- Why is the esophagus important in the process of digestion? *It transports food to the next organ of the digestive system.*

Teacher Demo

Model peristalsis by inserting a marble or bead into a piece of rubber tubing. Pinch the tubing so the marble is pushed down through the tubing. Ask:

- What does the tubing represent? *the esophagus*
- What does the marble represent? *food*

Ongoing Assessment

Describe the role of digestion in providing energy for the body.

In what 2 ways does the digestive system process food? *mechanically, chemically*

water. Protein is another essential nutrient; it is the material that the body uses for growth and repair. Cells in your body—such as those composing muscles, bones, and skin—are built of proteins. Carbohydrates are nutrients that provide cells with energy. Carbohydrates make up cellulose, which helps move materials through the digestive system. Another nutrient, fat, stores energy.

A

Before your body can use these nutrients, they must be broken into smaller substances. **Digestion** is the process of breaking down food into usable materials. Your digestive system transforms the energy and materials in food into forms your body can use.

The digestive system moves and breaks down food.

VISUALIZATION CLASSZONE.COM

Observe the process of peristalsis.

B

Your **digestive system** performs the complex jobs of moving and breaking down food. Material is moved through the digestive system by wavelike contractions of smooth muscles. This muscular action is called **peristalsis** (PEHR-ih-STAWL-sihs). Mucous glands throughout the system keep the material moist so it can be moved easily, and the muscles contract to push the material along. The muscles move food along in much the same way as you move toothpaste from the bottom of the tube with your thumbs. The body has complicated ways of moving food, and it also has complicated ways of breaking down food. The digestive system processes food in two ways: physically and chemically.

Peristalsis

The process of peristalsis moves food from the esophagus and into the stomach.

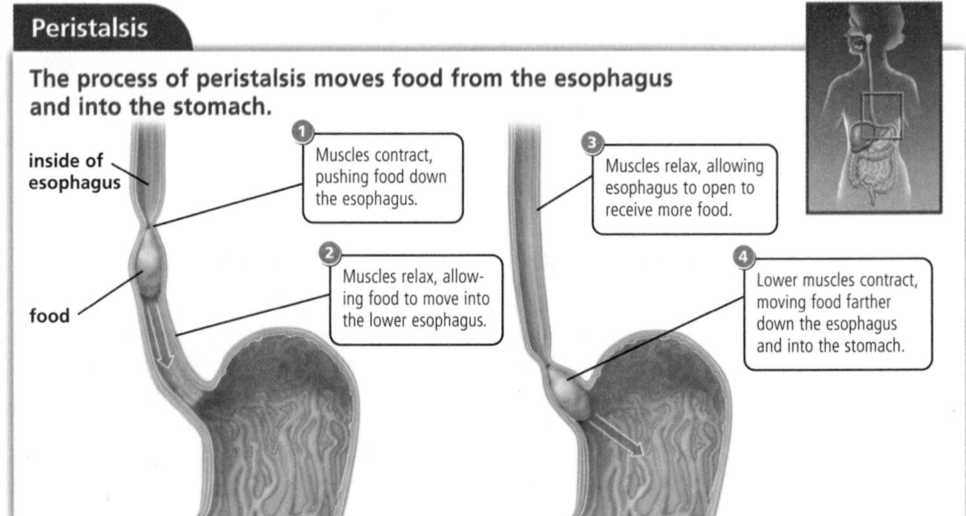

inside of esophagus
food
1 Muscles contract, pushing food down the esophagus.
2 Muscles relax, allowing food to move into the lower esophagus.
3 Muscles relax, allowing esophagus to open to receive more food.
4 Lower muscles contract, moving food farther down the esophagus and into the stomach.

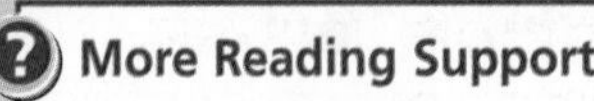

DIFFERENTIATE INSTRUCTION

More Reading Support

A What is digestion? *process of breaking down food into usable materials*

B What is peristalsis? *wavelike contractions of smooth muscles*

English Learners Have students write the definitions for *digestion, digestive system, nutrients, mechanical digestion,* and *chemical digestion* in their Science Word Dictionaries. Beside each word, students can sketch their favorite food being digested. The sketch should have labels and/or arrows to show each aspect of digestion.

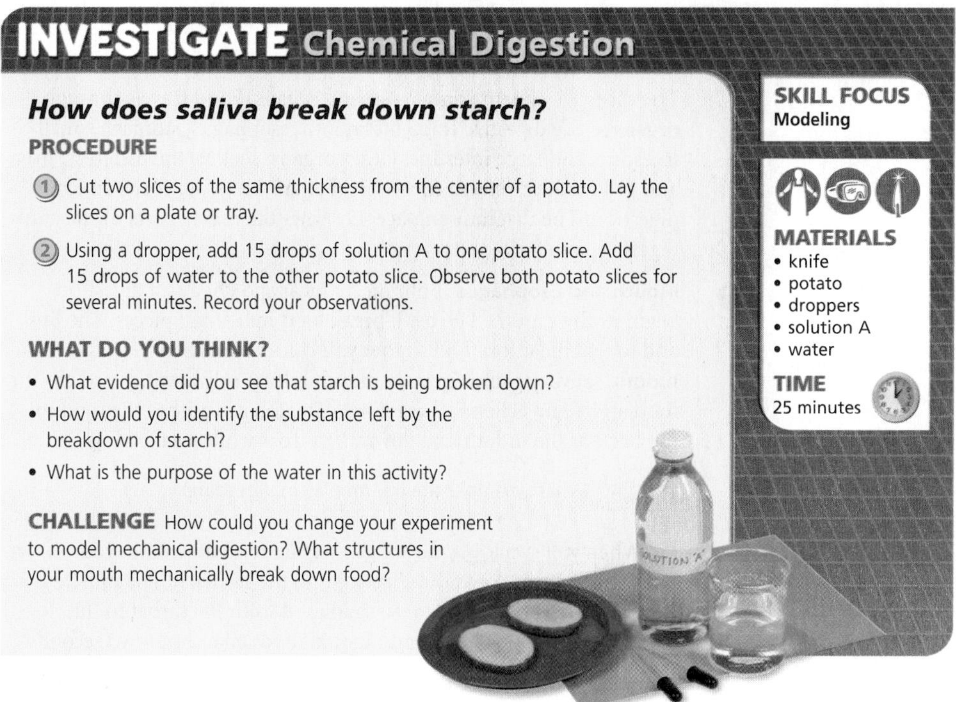

INVESTIGATE Chemical Digestion

How does saliva break down starch?

PROCEDURE

1. Cut two slices of the same thickness from the center of a potato. Lay the slices on a plate or tray.
2. Using a dropper, add 15 drops of solution A to one potato slice. Add 15 drops of water to the other potato slice. Observe both potato slices for several minutes. Record your observations.

WHAT DO YOU THINK?

- What evidence did you see that starch is being broken down?
- How would you identify the substance left by the breakdown of starch?
- What is the purpose of the water in this activity?

CHALLENGE How could you change your experiment to model mechanical digestion? What structures in your mouth mechanically break down food?

SKILL FOCUS
Modeling

MATERIALS
- knife
- potato
- droppers
- solution A
- water

TIME
25 minutes

Mechanical Digestion

C

Physical changes, which are sometimes called mechanical changes, break food into smaller pieces. Your teeth chew your food so you are able to swallow it. Infants without teeth need an adult to cut up or mash food for them. Otherwise they need soft food that they can swallow without chewing. Your stomach also breaks down food mechanically by mashing and pounding it during peristalsis.

Chemical Digestion

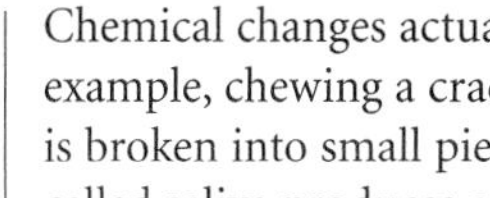
D

Chemical changes actually change food into different substances. For example, chewing a cracker produces a physical change—the cracker is broken into small pieces. At the same time, liquid in the mouth called saliva produces a chemical change—starches in the cracker are changed to sugars. If you chew a cracker, you may notice that after you have chewed it for a few seconds, it begins to taste sweet. The change in taste is a sign of a chemical reaction.

VOCABULARY
Don't forget to add magnet word diagrams for *digestion, digestive system,* and *peristalsis* to your notebook.

CHECK YOUR READING What are the two types of changes that take place during digestion?

DIFFERENTIATE INSTRUCTION

More Reading Support

C Describe mechanical digestion in the mouth. *Teeth chew the food into small pieces.*

D What substance in the mouth causes chemical changes? *saliva*

Advanced Have students find out more about the names and functions of the various types of teeth found in humans. Incisors bite and cut into food. Canines, or eye teeth, tear and break food into smaller pieces. Molars grind and crush food so it can be swallowed. Encourage students to report their findings with diagrams and pictures.

Challenge and Extension, p. 89

INVESTIGATE Chemical Digestion

PURPOSE To create a model of the breakdown of starch by saliva

TIPS *25 min.* Prior to the activity, cook potato slices until soft. Cooking breaks down the cell walls and releases starch for digestion. Amylase solution is available through science supply catalogs. Keep the solution refrigerated. You could do a control with uncooked potato and ask what a lack of a reaction suggests about eating raw potatoes.

WHAT DO YOU THINK? *A hole formed in the potato slice where solution A was dropped on it. The substance left by the breakdown of starch could be identified by adding iodine or by testing it on a glucose test strip. The water is a control.*

CHALLENGE *Mechanical digestion could be modeled by using a grater or masher to break the potato down into smaller pieces. In your mouth, the teeth mechanically break down food.*

Datasheet, Chemical Digestion, p. 90

Technology Resources

Customize this student lab as needed or look for an alternative. Print rubrics to assess student lab reports.

Lab Generator CD-ROM

Teach Difficult Concepts

The concept of chemical change in digestion is not something that students can see happening. To demonstrate, ask volunteers to chew on crackers. Explain that the saliva in the mouth causes the complex starch in the cracker to break down into simple sugar. Students should notice that the crackers begin to taste sweet. Explain that both starch and sugar are carbohydrates, but that starches have larger molecules made up of more atoms.

Ongoing Assessment

CHECK YOUR READING *Answer: Mechanical digestion breaks down food into smaller pieces, and chemical digestion changes food into different substances.*

History of Science

In 1822, an army surgeon named William Beaumont researched the stomach's function. Beaumont treated a patient who had received gunshot wounds. The man, Alexis St. Martin, had one wound that penetrated his stomach. Martin allowed Beaumont to perform experiments and make observations while the wound healed. Beaumont observed stomach secretions and collected samples from Martin's stomach in glass tubes. Beaumont reported the presence of acid in gastric juices, as well as the effects of different types of food.

Address Misconceptions

IDENTIFY Ask: Where is the stomach located in your body? If students answer that the stomach is in the lower abdomen, they may believe that the stomach is lower than it actually is.

CORRECT Point to your left side, just below your rib cage, and explain to students that this is the area where the stomach is located.

REASSESS Have students locate areas on their bodies to show the stomach, small intestine, and the ascending, transversing, and descending sections of the large intestine.

Technology Resources

Visit **ClassZone.com** for background on common student misconceptions.

MISCONCEPTION DATABASE

Ongoing Assessment

Describe the process of digestion.

Ask: Why are the structures called *villi* important in the process of digestion? *Villi allow nutrients to be absorbed by the blood and taken through the body.*

CHECK YOUR READING *Answer: The mouth begins digestion by mechanically breaking down food with teeth and chemically breaking it down with saliva.*

Answer: Mechanical digestion occurs primarily in the mouth and stomach.

Materials are broken down as they move through the digestive tract.

The digestive system contains several organs. Food travels through organs in the digestive tract: the mouth, esophagus, stomach, small intestine, and large intestine. Other organs, such as the pancreas, liver, and gall bladder, release chemicals that are necessary for chemical digestion. The diagram on page 49 shows the major parts of the entire digestive system.

READING TIP

As you read about the digestive tract, look at the structures on page 49.

Mouth and Esophagus Both mechanical and chemical digestion begin in the mouth. The teeth break food into small pieces. The lips and tongue position food so that you can chew. When food is in your mouth, salivary glands in your mouth release saliva, which softens the food and begins chemical reactions. The tongue pushes the food to the back of the mouth and down the throat while swallowing. E

What part does the mouth play in digestion?

When you swallow, your tongue pushes food down into your throat. Food then travels down the esophagus to the stomach. The muscle contractions of peristalsis move solid food from the throat to the stomach in about eight seconds. Liquid foods take about two seconds.

Stomach Strong muscles in the stomach further mix and mash food particles. The stomach also uses chemicals to break down food. Some of the chemicals made by the stomach are acids. These acids are so strong that they could eat through the stomach itself. To prevent this, the stomach's lining is replaced about every three days.

Small Intestine Partially digested food moves from the stomach to the small intestine. There, chemicals released by the pancreas, liver, and gallbladder break down nutrients. Most of the nutrients broken down in digestion are absorbed in the small intestine. F

Villi allow broken-down nutrients to be absorbed into your bloodstream.

Structures called villi are found throughout the small intestine. These structures contain folds that absorb nutrients from proteins, carbohydrates, and fats. Once absorbed by the villi, nutrients are transported by the circulatory system around the body. You will read more about the circulatory system in Chapter 3.

Large Intestine In the large intestine, water and some other nutrients are absorbed from the digested material. Most of the solid material then remaining is waste material, which is compacted and stored. Eventually it is eliminated through the rectum.

Where in your digestive system does mechanical digestion occur?

DIFFERENTIATE INSTRUCTION

More Reading Support

E Where does chemical digestion begin? *the mouth*

F Once partially digested food leaves the stomach, where does it go? *into the small intestine*

Advanced Have students find out what happens when people choke on pieces of food. Students can make a poster showing how to recognize that a person is choking and how to assist if a choking situation occurs.

Digestive System

As food moves through the digestive tract, structures of the digestive system break it down and absorb necessary materials.

DIFFERENTIATE INSTRUCTION

English Learners English learners may not be familiar with the various uses of the phrase "break down." Commonly, "break down" is used to mean something has broken, either physically (a car) or figuratively (negotiations). In this section, the meaning of "break down" is more complex. It indicates the taking in of food and breaking it into energy, usable materials, and waste materials.

Below Level Have students make a flow chart that shows the path of food as it travels through the digestive system. Students should include the structures listed on p. 48 and the term *villi* in their flow charts.

Teach from Visuals

To help students interpret the diagram of the digestive system, ask:

- What structures in the digestive system are connected by the esophagus? *The esophagus connects the mouth with the stomach.*
- How is the small intestine different from the large intestine? *The small intestine is a coiled tube that sits inside the sections of the large intestine.*

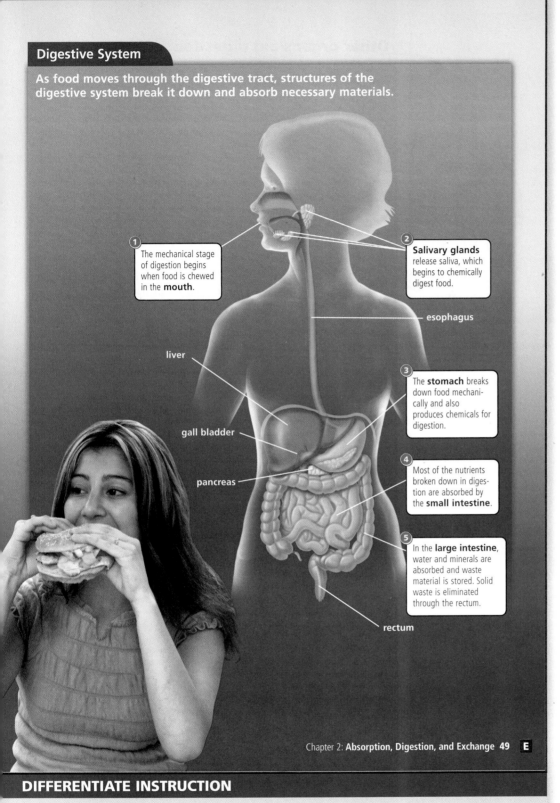

The visual "Digestive System" is available as T14 in the Unit Transparency Book.

Teach Difficult Concepts

Many students confuse the trachea with the esophagus. Have students review the diagram of the respiratory system on p. 41, then compare it with the diagram of the digestive system. Point out that the trachea and the esophagus are two different structures connected with two different systems in the body: The trachea helps transport air to the lungs in the respiratory system; the esophagus helps transport food to the stomach in the digestive system.

Metacognitive Strategy

Challenge students to devise a mnemonic device to remember the order of organs through which materials pass in the digestive tract.

Teaching with Technology

Use graphics software to create a flow chart of the process of digestion. Students can show how the digestive and circulatory systems are related.

Ongoing Assessment

Describe changes in materials as they move through the digestive system.

Ask: How does the sandwich in the picture change as it moves through a person's body? *The mouth softens it and breaks it apart. In the esophagus and stomach it breaks down chemically. As it moves through the intestines, it breaks into nutrients such as molecules of carbohydrates, proteins, and fats.*

Ongoing Assessment

CHECK YOUR READING *Answer: The pancreas produces chemicals that lower acidity in the small intestine and help break down proteins, fats, and starch.*

Reinforce the BIG idea

Have students relate the section to the Big Idea.

 Reinforcing Key Concepts, p. 91

2.2 ASSESS & RETEACH

Assess

 Section 2.2 Quiz, p. 23

Reteach

Write these terms on the board: *mouth, esophagus, stomach, small intestine, large intestine, liver, pancreas, gall bladder.* For each term, have students tell (a) whether it is part of the digestive tract, (b) whether mechanical or chemical digestion occurs there, and (c) what the organ's function is in the digestion process.

Technology Resources

Have students visit **ClassZone.com** for reteaching of Key Concepts.

 CONTENT REVIEW

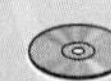 CONTENT REVIEW CD-ROM

Other organs aid digestion and absorption.

The digestive organs not in the digestive tract—the liver, gallbladder, and pancreas—also play crucial roles in your body. Although food does not move through them, all three of these organs aid in chemical digestion by producing or concentrating important chemicals.

Liver The liver—the largest internal organ of the body—is located in your abdomen, just above your stomach. Although you can survive losing a portion of your liver, it is an important organ. The liver filters blood, cleansing it of harmful substances, and stores unneeded nutrients for later use in the body. It produces a golden yellow substance called bile, which is able to break down fats, much like to the way soap breaks down oils. The liver also breaks down medicines and produces important proteins, such as those that help clot blood if you get a cut.

Bile is transferred from the liver to the gallbladder and small intestines through the bile duct.

Gallbladder The gallbladder is a tiny pear-shaped sac connected to the liver. Bile produced in the the liver is stored and concentrated in the gallbladder. The bile is then secreted into the small intestine.

Pancreas Located between the stomach and the small intestine, the pancreas produces chemicals that are needed as materials move between the two. The pancreas quickly lowers the acidity in the small intestine and breaks down proteins, fats, and starch. The chemicals produced by the pancreas are extremely important for digesting and absorbing food substances. Without these chemicals, you could die of starvation, even with plenty of food in your system. Your body would not be able to process and use the food for energy without the pancreas.

 How does the pancreas aid in digestion?

2.2 Review

KEY CONCEPTS

1. List three of the functions of the digestive system.
2. Give one example each of mechanical digestion and chemical digestion.
3. How does your stomach process food?

CRITICAL THINKING

4. **Apply** Does an antacid deal with physical or chemical digestion?
5. **Apply** You have just swallowed a bite of apple. Describe what happens as the apple moves through your digestive system. Include information about what happens to the material in the apple.

CHALLENGE

6. **Compare and Contrast** Describe the roles of the large and the small intestines. How are they similar? How are they different?

ANSWERS

1. Sample answers: breaking food into small pieces; absorbing nutrients; absorbing water; processing waste products

2. Sample answer: bread is broken in small pieces by chewing; saliva changes starches into sugars (chemical)

3. Muscles mix and mash food. Acid breaks down food.

4. An antacid deals with chemical digestion.

5. Answers should include all basic steps and structures in digestion and mention materials in the apple absorbed by intestines, and waste expelled.

6. Both absorb nutrients. The small intestine absorbs more nutrients; the large absorbs water and some nutrients, and stores waste.

MATH in SCIENCE

SKILL: CHOOSING UNITS OF LENGTH

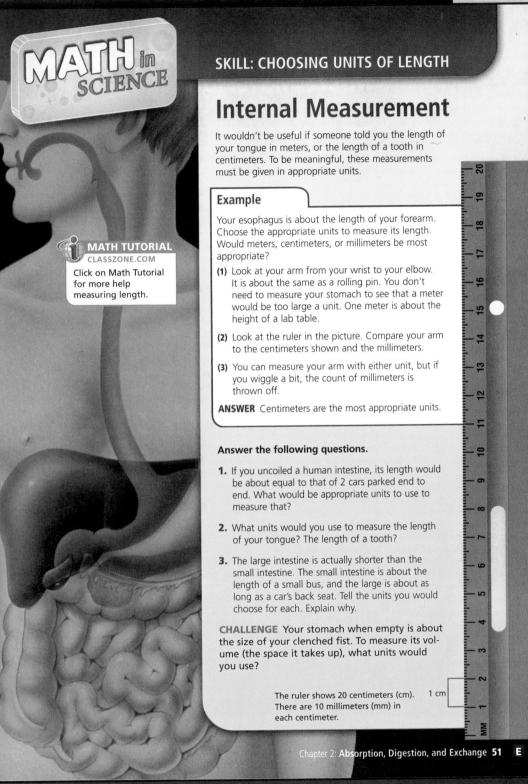

Internal Measurement

It wouldn't be useful if someone told you the length of your tongue in meters, or the length of a tooth in centimeters. To be meaningful, these measurements must be given in appropriate units.

Example

Your esophagus is about the length of your forearm. Choose the appropriate units to measure its length. Would meters, centimeters, or millimeters be most appropriate?

(1) Look at your arm from your wrist to your elbow. It is about the same as a rolling pin. You don't need to measure your stomach to see that a meter would be too large a unit. One meter is about the height of a lab table.

(2) Look at the ruler in the picture. Compare your arm to the centimeters shown and the millimeters.

(3) You can measure your arm with either unit, but if you wiggle a bit, the count of millimeters is thrown off.

ANSWER Centimeters are the most appropriate units.

MATH TUTORIAL
CLASSZONE.COM
Click on Math Tutorial for more help measuring length.

Answer the following questions.

1. If you uncoiled a human intestine, its length would be about equal to that of 2 cars parked end to end. What would be appropriate units to use to measure that?

2. What units would you use to measure the length of your tongue? The length of a tooth?

3. The large intestine is actually shorter than the small intestine. The small intestine is about the length of a small bus, and the large is about as long as a car's back seat. Tell the units you would choose for each. Explain why.

CHALLENGE Your stomach when empty is about the size of your clenched fist. To measure its volume (the space it takes up), what units would you use?

The ruler shows 20 centimeters (cm). There are 10 millimeters (mm) in each centimeter.

ANSWERS

1. *meters*
2. *centimeters, millimeters*
3. *meters for both, because you can compare both vehicles without changing units*
4. ***CHALLENGE*** *The size of my fist would be measured in cubic centimeters. Volume is length · width · height: Centimeters are appropriate for a stomach's length, width, or height.* $cm \cdot cm \cdot cm = cm^3$.

MATH IN SCIENCE
Math Skills Practice for Science

Set Learning Goal

To choose appropriate units for measuring different lengths

Present the Science

Remind students of the differences between the small and large intestines. Point out that the small intestine is actually much longer than the large intestine, but has a smaller diameter.

Develop Measurement Skills

Review metric units of measure: millimeters, centimeters, and meters. Display a meter stick and have students note the differences among the three units of measure.

DIFFERENTIATION TIP Have students use metric rulers to measure parts of the body, for example, the length of a fingernail, a finger, a hand, and an arm, and the height of the whole body. Have them determine which metric unit is right for which body part.

Close

Ask students how they could estimate the length or the surface area of other internal organs. *Use estimation, use formulas for area, use comparison to other organs or to benchmarks, such as a dinner plate or orange skin.*

- Math Support, p. 107
- Math Practice, p. 108

Technology Resources

Students can visit **ClassZone.com** for help with measuring length.

MATH TUTORIAL

2.3 FOCUS

Set Learning Goals

Students will

- Recognize that different body systems remove waste.
- Discuss why kidneys are important organs.
- Analyze the role of the kidneys in homeostasis.
- Model the process of kidney filtration.

3-Minute Warm-Up

Display Transparency 13 or copy this exercise on the board:

Match each definition to the correct term. Show the letter of the correct answer.

Definitions

1. the organ that secretes bile *e*
2. the organ in which acids break down food *b*
3. the organ in which villi absorb nutrients into the bloodstream *c*

Terms

a. mouth
b. stomach
c. small intestine
d. large intestine
e. liver
f. pancreas

3-Minute Warm-Up, p. T13

2.3 MOTIVATE

EXPLORE Waste Removal

PURPOSE To observe how the skin helps to get rid of waste

TIP *10 min.* Have students share their observations in small groups.

WHAT DO YOU THINK? *The bag gets some fog on the inside. It demonstrates that the skin releases moisture through sweat.*

Ongoing Assessment

CHECK YOUR READING *Answer: through the urinary system (eliminates liquid waste), respiratory system (removes water vapor and waste gases), digestive system (removes solid waste from food), and skin (releases waste through sweat glands)*

KEY CONCEPT

2.3 The urinary system removes waste materials.

BEFORE, you learned

- The digestive system breaks down food
- Organs in the digestive system have different roles

NOW, you will learn

- How different body systems remove different types of waste
- Why the kidneys are important organs
- About the role of the kidney in homeostasis

VOCABULARY

urinary system p. 53
urine p. 53

EXPLORE Waste Removal

How does the skin get rid of body waste?

PROCEDURE

1. Place a plastic bag over the hand you do not use for writing and tape it loosely around your wrist.
2. Leave the bag on for five minutes. Write down the changes you see in conditions within the bag.

MATERIALS

- plastic bag
- tape
- stopwatch

WHAT DO YOU THINK?

- What do you see happen to the bag?
- How does what you observe help explain the body's method of waste removal?

Life processes produce wastes.

OUTLINE
Add *Life processes produce wastes* to your outline. Include four ways the body disposes of waste products.

I. Main idea
  A. Supporting idea
    1. Detail
    2. Detail
  B. Supporting idea

You have read that the respiratory system and the digestive system provide the body with energy and materials necessary for important processes. During these processes waste materials are produced. The removal of these wastes is essential for the continuing function of body systems. Several systems in your body remove wastes.

- The urinary system disposes of liquid waste products removed from the blood.
- The respiratory system disposes of water vapor and waste gases from the blood.
- The digestive system disposes of solid waste products from food.
- The skin releases wastes through sweat glands.

What are four ways the body disposes of waste products?

RESOURCES FOR DIFFERENTIATED INSTRUCTION

Below Level

UNIT RESOURCE BOOK

- Reading Study Guide A, pp. 94–95
- Decoding Support, p. 106

Advanced

UNIT RESOURCE BOOK

Challenge and Extension, p. 100

English Learners

UNIT RESOURCE BOOK

Spanish Reading Study Guide, pp. 98–99

- Audio Readings in Spanish
- Audio Readings (English)

The urinary system removes waste from the blood.

If you have observed an aquarium, you have seen a filter at work. Water moves through the filter, which removes waste materials from the water. Just as the filter in a fish tank removes wastes from the water, structures in your urinary system filter wastes from your blood.

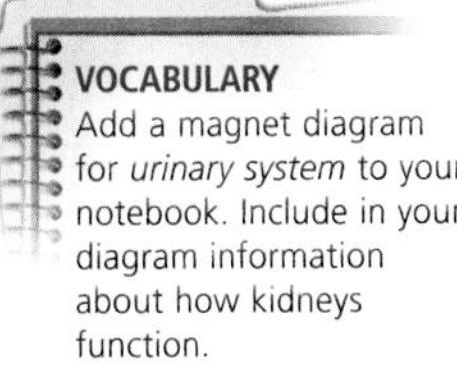

VOCABULARY
Add a magnet diagram for *urinary system* to your notebook. Include in your diagram information about how kidneys function.

As shown in the diagram, the **urinary system** contains several structures. The kidneys are two organs located high up and toward the rear of the abdomen, one on each side of the spine. Kidneys function much as the filter in the fish tank does. In fact, the kidneys are often called the body's filters. Materials travel in your blood to the kidneys. There, some substances are removed, and others are returned to the blood. A

After the kidneys filter chemical waste from the blood, the liquid travels down two tubes called ureters (yu-REE-tuhrz). The ureters bring the waste to the bladder, a storage sac with a wall of smooth muscle. The lower neck of the bladder leads into the urethra, a tube that carries the liquid waste outside the body. Voluntary muscles at one end of the bladder allow a person to hold the urethra closed until he or she is ready to release the muscles. At that time, the bladder contracts and sends the liquid waste, or **urine,** out of the body. B

Urinary System

The urinary system transports wastes out of the body.

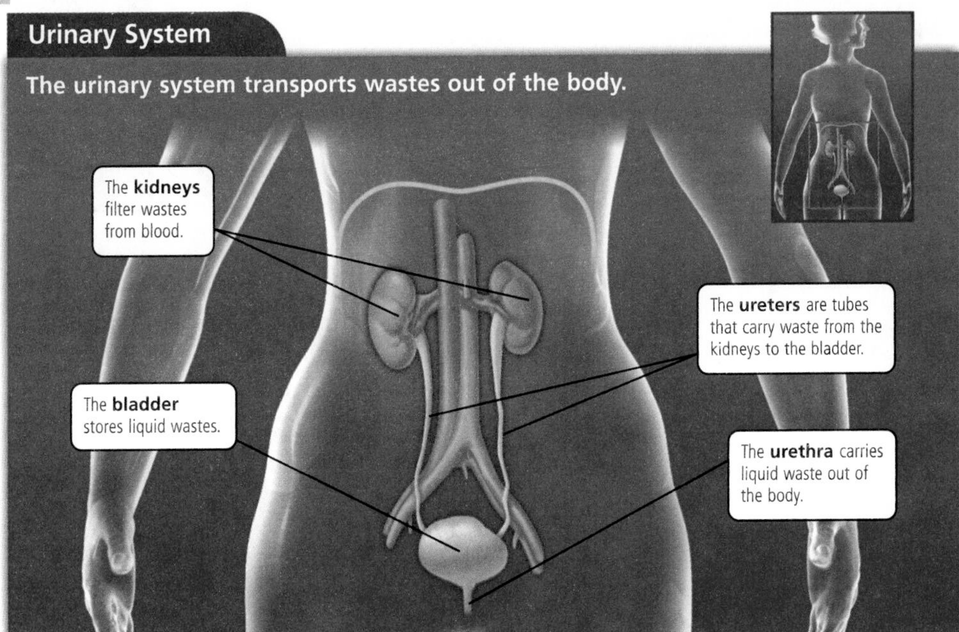

The **kidneys** filter wastes from blood.
The **ureters** are tubes that carry waste from the kidneys to the bladder.
The **bladder** stores liquid wastes.
The **urethra** carries liquid waste out of the body.

2.3 INSTRUCT

Real World Example

People with compromised kidney function often undergo dialysis, a process in which a patient's blood is removed, purified, and returned to the bloodstream. This process is accomplished with an artificial kidney, called a hemodialyzer. Blood is directed from an artery into the dialyzer. There, it flows along the surface of a membrane where impurities are filtered out. The blood then passes through a trap that removes clots and bubbles and is returned to the bloodstream through a vein, usually in the patient's forearm.

Teach from Visuals

To help students interpret the diagram of the urinary system, ask:

- Which structures make up the urinary system? *kidneys, bladder, ureters, and urethra*
- Through which structure is waste eliminated from the body? *the urethra*

Address Misconceptions

IDENTIFY Ask: The urinary bladder is a part of which body system? Students may hold the misconception that the bladder is part of the digestive system.

CORRECT Have students study the diagram of the urinary system, noting the position of the urinary bladder.

REASSESS Ask students to create a two-column chart, naming the organs of the digestive and respiratory systems.

Technology Resources

Visit **ClassZone.com** for background on common student misconceptions.

 **MISCONCEPTION DATABASE**

Ongoing Assessment

Recognize that different body systems remove waste.

Ask: What are the four body systems that remove waste from the body? *urinary, digestive, respiratory, skin*

DIFFERENTIATE INSTRUCTION

More Reading Support

A Which organs are often called the body's filters? *the kidneys*

B What is the term for the liquid waste that collects in the bladder? *urine*

English Learners Provide students with four index cards. Have them write a different term on the front of each card: *kidney, bladder, urethra, ureter.* On the back of each card, students can draw a picture of the structure and/or write a description of its function. Students can use the cards as they read.

Teach Difficult Concepts

Students may not understand where in the body cavity the kidneys are located. Point to the location of the kidneys on your body (on the back in the lower section of the rib cage) and have students do the same.

Teacher Demo

Some students may not know what a filter is or how it works. To demonstrate this concept, obtain a coffee filter, a glass beaker, a funnel, and some muddy water. Pour the muddy water through the filter and funnel into the beaker. Have students compare the original water with the filtrate. Ask: How is this demonstration similar to the function of the kidneys? *The kidneys filter waste substances out of the blood.*

Teach from Visuals

To help students interpret the diagram of the nephron, ask:

- What structure is made up of a cluster of capillaries? *the glomerulus*
- What is the purpose of the nephron? *The nephron removes waste from the blood and produces urine.*
- What happens to the nutrients and water that are filtered out in the nephron? *Some nutrients are returned to the blood. Some travel into the collecting duct and out through the ureter as urine.*

Ongoing Assessment

Explain why the kidneys are important organs.

Ask: Why is it important that blood be filtered by the kidneys? *Waste materials are filtered out of the blood. If they stayed in the blood, they might damage cells and organs.*

Find out more about the urinary system.

The kidneys act as filters.

At any moment, about one quarter of the blood leaving your heart is headed toward your kidneys to be filtered. The kidneys, which are about as long as your index finger—only 10 centimeters (3.9 in.) long—filter all the blood in your body many times a day.

The Nephron

C

Inside each kidney are approximately one million looping tubes called nephrons. The nephron regulates the makeup of the blood.

1. Fluid is filtered from the blood into the nephron through a structure called the glomerulus (gloh-MEHR-yuh-luhs). Filtered blood leaves the glomerulus and circulates around the tubes that make up the nephron.
2. As the filtered fluid passes through the nephron, some nutrients are absorbed back into the blood surrounding the tubes. Some water is also filtered out in the glomerulus, but most water is returned to the blood.
3. Waste products travel to the end of the nephron into the collecting duct. The remaining liquid, now called urine, passes out of the kidney and into the ureters.

D

Kidney Function

Kidneys filter waste materials out of the blood.

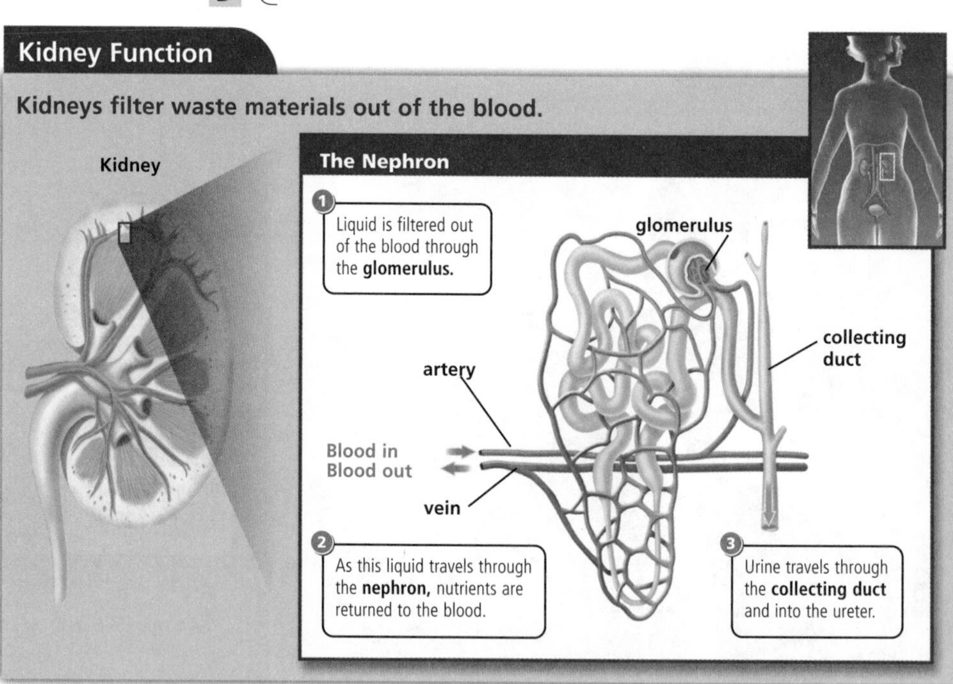

DIFFERENTIATE INSTRUCTION

More Reading Support

C What are the names of the looping tubes that are found inside kidneys? *nephrons*

D What product travels to the end of the nephron and into the ureter? *urine*

Advanced Many English idioms use the names of organs or terms related to organs to describe different emotions or problems that are not related to the function of the organ. For example, we might say that someone "dies of a broken heart." Ask students to research different idioms related to respiratory, digestive, or other organs. Have students prepare a chart to share with the class. *Sample answers: breathless in love, strong of heart, weak-stomached, lily-livered, heartache*

 Challenge and Extension, p. 100

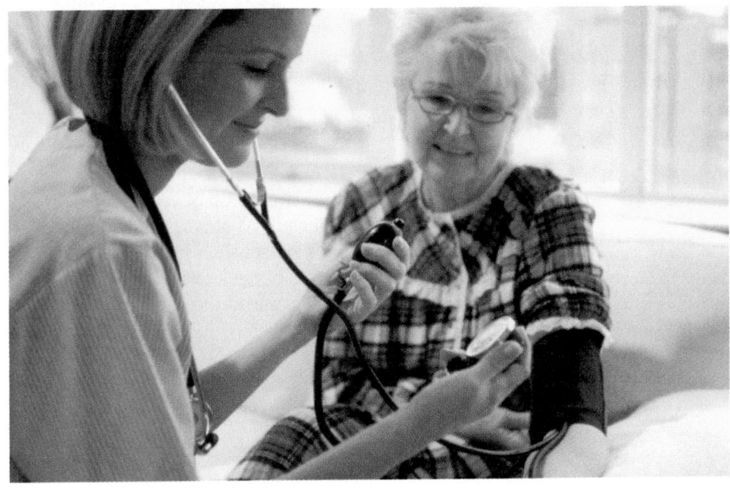

The amount of water in your body affects your blood pressure. Excess water increases blood pressure.

Water Balance

The kidneys not only remove wastes from blood, they also regulate the amount of water in the body. You read in Chapter 1 about the importance of homeostasis—a stable environment within your body. The amount of water in your cells affects homeostasis. If your body contains too much water, parts of your body may swell. Having too little water interferes with cell processes.

About one liter of water leaves the body every day. The kidneys control the amount of water that leaves the body in urine. Depending on how much water your body uses, the kidneys produce urine with more or less water.

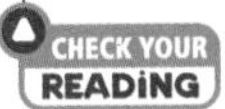
CHECK YOUR READING How do your kidneys regulate the amount of water in your body?

2.3 Review

KEY CONCEPTS

1. Describe the four organ systems that remove wastes and explain how each removes waste.
2. Describe the function of four organs in the urinary system.
3. Describe homeostasis and explain why the kidney is important to homeostasis.

CRITICAL THINKING

4. **Connect** Make a word web with the term *kidney* in the center. Add details about kidney function to the web.

CHALLENGE

5. **Synthesize** Explain why you may become thirsty on a hot day. Include the term *homeostasis* in your explanation.

ANSWERS

1. urinary system—removes liquid waste from the blood; respiratory—removes water vapor and waste gases; digestive—removes solid waste; sweat glands—remove waste water

2. Kidneys filter waste from blood. Waste leaves kidneys as urine via ureters to bladder, where stored. Urethra carries urine out of body.

3. Homeostasis is the process of maintaining internal conditions. The kidney helps control the water level in the body.

4. Webs should include the kidney's function as a filter.

5. If you sweat a lot, you will lose fluids. If too little water is in the body, the kidney releases a chemical that makes you thirsty. Drinking fluids helps restore homeostasis.

EXPLORE the BIG idea

Revisit "Water Everywhere" on p. 35. Ask what might happen if a person drinks more than 2 L of water in a day. *The person may produce more urine.*

Ongoing Assessment

Explain the role of the kidneys in homeostasis.

Ask: What might happen if the kidneys were unable to eliminate water from the body? *The body would become swollen.*

CHECK YOUR READING *Answer: The kidneys produce urine with varying amounts of water, depending on how much water the body uses.*

Reinforce the BIG idea

Have students relate the section to the Big Idea.

 Reinforcing Key Concepts, p. 101

2.3 ASSESS & RETEACH

Assess

 Section 2.3 Quiz, p. 24

Reteach

Write the following systems on the board. Have students describe how each system acts to remove wastes from the body.

- respiratory *removes water vapor and gaseous waste from the blood*
- digestive *removes solid waste*
- urinary *removes liquid waste from the blood*
- skin *removes waste via sweat*

Technology Resources

Have students visit **ClassZone.com** for reteaching of Key Concepts.

 **CONTENT REVIEW**

CHAPTER INVESTIGATION

Focus

PURPOSE To demonstrate the process of kidney filtration

OVERVIEW Students will use filters and solutions to determine the following:

- how the filtering process of the kidneys can be modeled
- what types of materials are filtered by the kidneys

Lab Preparation

- Prepare solution A: Add 10 g of table salt to 1000 mL of water. Stir until dissolved.
- Prepare solution B: Add 10 g of glucose to 1000 mL of water. Stir until dissolved.
- Prepare solution C: Add 10 g of albumin powder to 1000 mL of water. Stir gently until dissolved. Can substitute the white of one egg if albumin powder is not available.
- Prior to the investigation, have students read through the investigation and prepare their data tables. Or you may wish to copy and distribute datasheets and rubrics.

 UNIT RESOURCE BOOK, pp. 109–117

 SCIENCE TOOLKIT, F14

Lab Management

- Model use of test strips before distributing them.
- Have students work in pairs, one student pouring and testing the solutions, the other student recording the data.
- Have students rinse the funnel and beaker between each testing.
- **SAFETY** Caution students not to touch the solutions or test strips. Make sure they wash their hands when the lab is completed.
- **INCLUSION** To help students new to English or students who have varying abilities, have all students work in cooperative groups where there is a Reader, Leader, Recorder, and Reporter.

CHAPTER INVESTIGATION

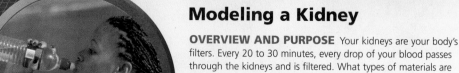

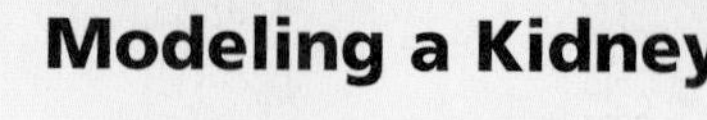

Modeling a Kidney

OVERVIEW AND PURPOSE Your kidneys are your body's filters. Every 20 to 30 minutes, every drop of your blood passes through the kidneys and is filtered. What types of materials are filtered by the kidneys? In this investigation you will

- model the filtering process of the kidneys
- determine what types of materials are filtered by your kidneys

Problem

What types of materials can be removed from the blood by the kidneys?

Hypothesize

Write a hypothesis to explain how substances are filtered out of the blood by the kidneys. Your hypothesis should take the form of an "If . . . , then . . . , because . . ." statement.

Procedure

1. Make a data table like the one shown on the sample notebook page. Fold the filter paper as shown. Place the filter paper in the funnel, and place the funnel in the graduated cylinder.

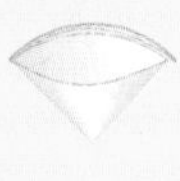

2. Pour 20 mL of solution A into a beaker. Test the solution for salt concentration using a test strip for salinity. Record the results in your notebook. Slowly pour the solution into the funnel. Wait for it all to drip through the filter paper.

MATERIALS

- fine filter paper
- small funnel
- graduated cylinder
- 100 mL beaker
- solution A
- solution B
- solution C
- salinity test strips
- glucose test strips
- protein test strips

E 56 Unit: Human Biology

INVESTIGATION RESOURCES

 CHAPTER INVESTIGATION, Modeling a Kidney

- Level A, pp. 109–112
- Level B, pp. 113–116
- Level C, p. 117

Advanced students should complete Levels B & C.

 Writing a Lab Report, D12–13

Technology Resources

Customize this student lab as needed or look for an alternative. Print rubrics to assess student lab reports.

 **Lab Generator CD-ROM**

3. Test the filtered liquid for salt concentration again. Record the results.

4. Repeat steps 1, 2, and 3 for solution B using glucose test strips. Record the results in your notebook.

5. Repeat steps 1, 2, and 3 for solution C using protein test strips. Record the results in your notebook.

Observe and Analyze

Write It Up

1. **RECORD** Be sure your data table is complete.
2. **OBSERVE** What substances were present in solutions A, B, and C?
3. **IDENTIFY VARIABLES** Identify the variables and constants in the experiment. List them in your notebook.

Conclude

1. **COMPARE AND CONTRAST** In what ways does your model function like a kidney? How is your model not like a kidney?
2. **INTERPRET** Which materials were able to pass through the filter and which could not?
3. **INFER** What materials end up in the urine? How might materials be filtered out of the blood but not appear in the urine?
4. **APPLY** How is a filtering device useful in your body?

INVESTIGATE Further

CHALLENGE Your blood contains many chemicals. Some of these chemicals are waste products, but some are in the blood to be transported to different parts of the body. What other substances are filtered out of the blood by the kidneys? Which of the filtered substances are normally present in the urine? Use a variety of reference materials to research the chemicals found in urine. Revise your experiment to test the ability of your model kidney to filter other substances.

Modeling a Kidney

Table 1. Test-strip results

	Before filtering	After filtering
Solution A		
Solution B		
Solution C		

Observe and Analyze

1. *Check to make sure data table is complete.*
2. *Solution A had salt, solution B had glucose (sugar), and solution C had protein.*
3. *The variables in the experiment were the different types of solutions. The constants were the filter paper and the amount of solution that was filtered.*

Conclude

1. *The model acts like a kidney in that it filters some, but not all, substances out of the blood. The model is not like a kidney because some substances that are passed through the filter would be reabsorbed into the blood.*
2. *Salt and glucose were able to pass through the filter; protein could not.*
3. *The urine usually contains salt but not glucose. Glucose is reabsorbed into the blood before the urine leaves the kidney.*
4. *A filtering device could perform the functions of the kidneys in removing specific substances from the blood.*

INVESTIGATE Further

CHALLENGE Sample answer: Other substances filtered out of the blood by the kidney include urea, uric acid, calcium, potassium, bicarbonate. All of these substances are normally found in urine. Substances not normally found in urine include protein, glucose, nitrite, hemoglobin, ketones, bilirubin.

Post-Lab Discussion

- Discuss why some substances passed through the filter paper and why others were unable to. *Pores allowed smaller particles to pass through, whereas larger particles were unable to pass through.*
- Discuss how this lab simulates the functions of the kidneys. *Kidneys filter waste out of the blood and return other materials to the blood.*
- Ask: After observing your results, what would you want to study further? What question arise? How would you experiment to find answers?

BACK TO

the BIG idea

For each of the systems discussed in the chapter, have students choose one of the structures in the system and explain how that structure functions within its system to process materials and remove wastes.

KEY CONCEPTS SUMMARY

SECTION 2.1

Have students look at the picture of the respiratory system and describe the path of oxygen as it travels into the lungs. Then have them describe what happens to the oxygen. *Oxygen is inhaled through the nose or mouth, then passes through the larynx, trachea, and bronchial tubes into the lungs. Once inside the lungs, the oxygen travels into smaller vessels, called capillaries, where gas exchange with the circulatory system takes place.*

SECTION 2.2

Ask students to describe the difference between mechanical and chemical digestion. Then have them examine the chart and name the structures of the digestive system in which chemical digestion takes place. *Mechanical digestion physically breaks food into smaller pieces. Chemical digestion changes food into different substances. Chemical digestion takes place in the mouth, stomach, and small intestine.*

SECTION 2.3

Ask: What systems are involved in waste removal? *respiratory, urinary, digestive, (and skin)*

Review Concepts

- Big Idea Flow Chart, p. T9
- Chapter Outline, pp. T15–T16

2 Chapter Review

the BIG idea

Systems in the body obtain and process materials and remove waste.

KEY CONCEPTS SUMMARY

2.1 The respiratory system gets oxygen and removes carbon dioxide.

- Your body needs oxygen
- Structures in the respiratory system function together
- Your respiratory system is involved in other functions

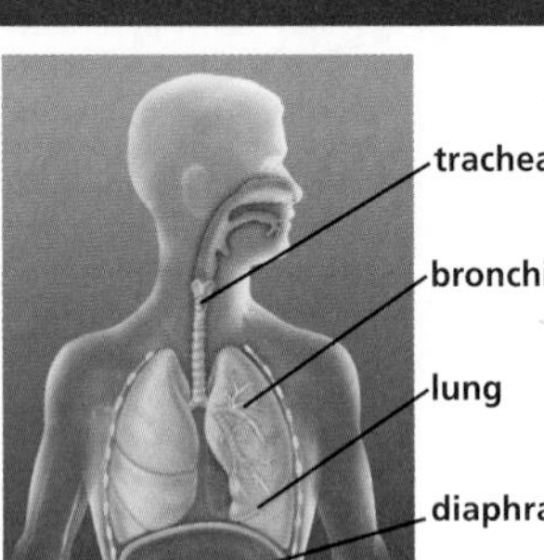

VOCABULARY
respiratory system p. 37
cellular respiration p. 38

2.2 The digestive system breaks down food.

Structure	Function
Mouth	chemical and mechanical digestion
Esophagus	movement of food by peristalsis from mouth to stomach
Stomach	chemical and mechanical digestion; absorption of broken-down nutrients
Small intestine	chemical digestion; absorption of broken-down nutrients
Large intestine	absorption of water and broken-down nutrients, elimination of wastes

VOCABULARY
nutrient p. 45
digestion p. 46
digestive system p. 46
peristalsis p. 46

2.3 The urinary system removes waste materials.

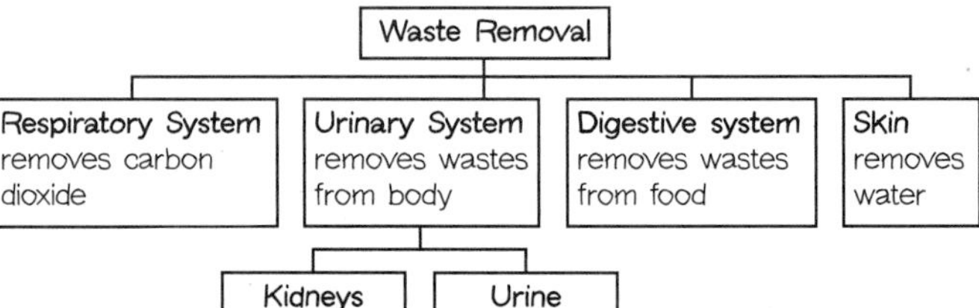

VOCABULARY
urinary system p. 53
urine p. 53

Technology Resources

Have students visit **ClassZone.com** or use the CD-ROM for a cumulative review of concepts.

 CONTENT REVIEW

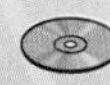 CONTENT REVIEW CD-ROM

Engage students in a whole-class interactive review of Key Concepts. Edit content as you wish.

 POWER PRESENTATIONS

Reviewing Vocabulary

Copy the chart below and write the definition for each word. Use the meaning of the word's root to help you.

Word	Root meaning	Definition
EXAMPLE: rib cage	to arch over	bones enclosing the internal organs of the body
1. respiration	to breathe	
2. nutrient	to nourish	
3. digestion	to separate	
4. urine	to moisten, to flow	

Reviewing Key Concepts

Multiple Choice *Choose the letter of the best answer.*

5. Which system brings oxygen to your body and removes carbon dioxide?
- **a.** digestive system
- **b.** urinary system
- **c.** respiratory system
- **d.** muscular system

6. Which body structure in the throat keeps air from entering the stomach?
- **a.** trachea
- **b.** epiglottis
- **c.** lungs
- **d.** alveoli

7. Oxygen and carbon dioxide are exchanged through structures in the lungs called
- **a.** bronchial tubes
- **b.** alveoli
- **c.** cartilage
- **d.** villi

8. Carbon dioxide is a waste product that is formed during which process?
- **a.** cellular respiration
- **b.** peristalsis
- **c.** urination
- **d.** circulation

9. Carbohydrates are nutrients that
- **a.** make up most of the human body
- **b.** make up cell membranes
- **c.** enable cells to grow and repair themselves
- **d.** provide cells with energy

10. Which is *not* a function of the digestive system?
- **a.** absorb water from food
- **b.** absorb nutrients from food
- **c.** filter wastes from blood
- **d.** break down food

11. Which is an example of a physical change?
- **a.** teeth grind cracker into smaller pieces
- **b.** liquids in mouth change starches to sugars
- **c.** bile breaks down fats
- **d.** stomach breaks down proteins

12. Where in the digestive system is water absorbed?
- **a.** small intestine
- **b.** stomach
- **c.** large intestine
- **d.** esophagus

13. Chemical waste is filtered from the body in which structure?
- **a.** alveoli
- **b.** kidney
- **c.** stomach
- **d.** villi

14. The kidneys control the amount of
- **a.** oxygen that enters the blood
- **b.** water that is absorbed by the body
- **c.** urine that leaves the body
- **d.** water that leaves the body

Reviewing Vocabulary

Sample answers:

1. *process taking place in cells by which energy is released*
2. *important substance in food that enables the body to grow and maintain homeostasis*
3. *the process of breaking down food into absorbable materials*
4. *liquid waste from the body*

Reviewing Key Concepts

5. c
6. b
7. b
8. a
9. d
10. c
11. a
12. c
13. b
14. c

(Answers to items that appear on p. 60.)

15. Students' sketches should show the thoracic cavity increasing in volume.
16. energy, water, and carbon dioxide
17. mouth, esophagus, stomach, small intestine, large intestine, rectum
18. filtering liquid waste from body

ASSESSMENT RESOURCES

UNIT ASSESSMENT BOOK
- Chapter Test A, pp. 25–28
- Chapter Test B, pp. 29–32
- Chapter Test C, pp. 33–36
- Alternative Assessment, pp. 37–38

SPANISH ASSESSMENT BOOK
Spanish Chapter Test, pp. 85–88

Technology Resources

Edit test items and answer choices.

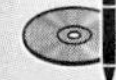 **Test Generator CD-ROM**

Visit **ClassZone.com** to extend test practice.

 **Test Practice**

(Answers for items 15–18 appear on p. 59.)

Thinking Critically

19. *Air moves into lungs through tubes that branch into smaller tubes and end in alveoli. O_2 moves through the walls of the alveoli and is dissolved into blood. CO_2 moves out of blood and into alveoli to be breathed out.*
20. *O_2 and sugars combine to release energy. Respiratory system provides the O_2 needed. Digestive system provides sugars.*
21. *Coughing and sneezing help remove particles from the respiratory system.*
22. *Water is produced in respiration. When air is exhaled, the water vapor condenses on the glass surface.*
23. *Mechanical digestion occurs in the mouth and stomach and physically changes food. Chemical digestion breaks chemical bonds.*
24. *Partially digested food would go straight into the small intestine. No; the small intestine is where nutrients are absorbed.*
25. *Kidneys regulate water in the body. The athlete will release water through urine, sweat, and water vapor in breath.*
26. *Nutrients are filtered back into bloodstream.*

the BIG idea

27. *The body needs oxygen, water, and nutrients.*
28. *The respiratory system allows O_2 to enter lungs. There, O_2 is exchanged with CO_2 from circulatory system, and CO_2 is exhaled. In the digestive system, nutrients and water are absorbed into circulatory system. In urinary system, blood passes through the kidneys. Nutrients are returned and liquid waste is eliminated.*

UNIT PROJECTS

Collect schedules, materials lists, and questions. Be sure dates and materials are obtainable, questions, focused.

 Unit Projects, pp. 5–10

Short Answer *Write a short answer to each question.*

15. Draw a sketch that shows how the thoracic cavity changes as the diaphragm contracts and pulls downward.
16. What are two products that are released into the body as a result of cellular respiration?
17. Through which organs does food pass as it travels through the digestive system?
18. What is the function of the urinary system?

Thinking Critically

19. **SUMMARIZE** Describe how gas exchange takes place inside the lungs.
20. **SYNTHESIZE** Summarize what happens during cellular respiration. Explain how the digestive system and the respiratory system are involved.
21. **ANALYZE** When there is a lot of dust or pollen in the air, people may cough and sneeze. What function of the respiratory system is involved?
22. **INFER** When you exhale onto a glass surface, the surface becomes cloudy with a thin film of moisture. Explain why this happens.
23. **COMPARE AND CONTRAST** Where does mechanical digestion take place? How is it different from chemical digestion?
24. **PREDICT** People with stomach disease often have their entire stomachs removed and are able to live normally. Explain how this is possible. Would a person be able to live normally without the small intestine? Explain your answer.
25. **APPLY** An athlete drinks a liter of water before a basketball game and continues to drink water during the game. Describe how the athlete's body is able to maintain homeostasis during the course of the game.
26. **INTERPRET** Use the diagram of the nephron shown below to describe what happens to the blood as it travels through the vessels surrounding the nephron.

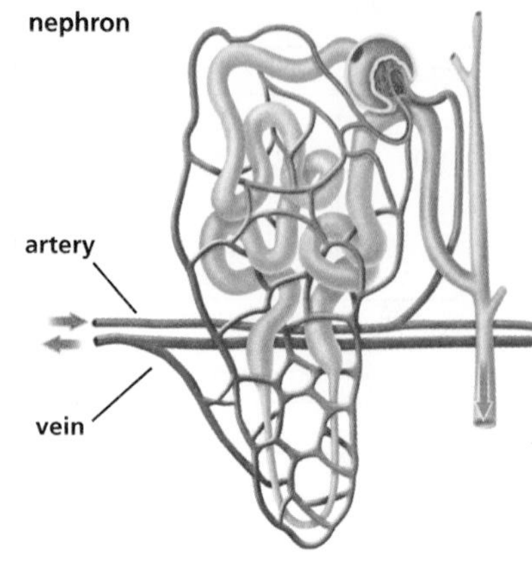

the BIG idea

27. **INFER** Look again at the picture on pages 34–35. Now that you have finished the chapter, how would you change or add details to your answer to the question on the photograph?
28. **SYNTHESIZE** Write a paragraph explaining how the respiratory system, the digestive system, and the urinary system work together with the circulatory system to eliminate waste materials from the body. Underline these terms in your paragraph.

UNIT PROJECTS

Check your schedule for your unit project. How are you doing? Be sure that you've placed data or notes from your research in your project folder.

MONITOR AND RETEACH

If students are having trouble applying the concepts in Chapter Review items 19, 20, and 28, suggest that they create a flow chart that shows how the respiratory, digestive, and urinary systems are connected. Students can refer to the diagrams on pp. 41, 49, and 53 to help them with their graphics.

Students may benefit from summarizing one or more sections of the chapter.

 Summarizing the Chapter, pp. 127–128

Standardized Test Practice

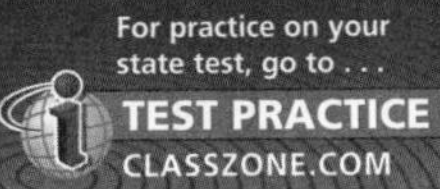

Analyzing Data

The bar graph below shows respiration rates.

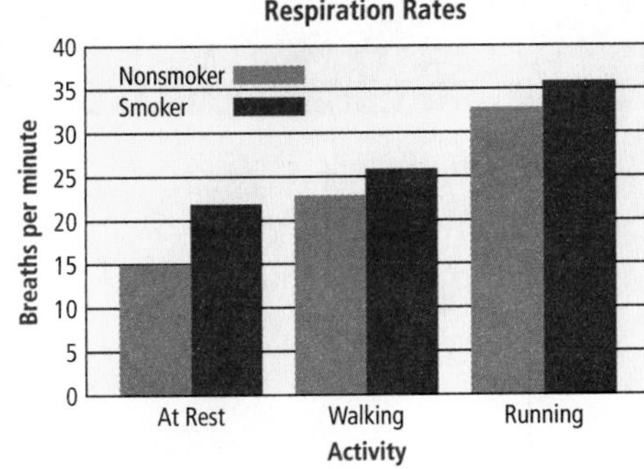

Use the graph to answer the questions below.

1. What is the best title for this graph?
   - **a.** Respiration Rates of Smokers and Nonsmokers
   - **b.** Cigarettes Smoked During Exercise
   - **c.** Activities Performed by Smokers and Nonsmokers
   - **d.** Blood Pressure Levels of Smokers and Nonsmokers.

2. How many breaths per minute were taken by a nonsmoker at rest?
   - **a.** 15 breaths per minute
   - **b.** 22 breaths per minute
   - **c.** 23 breaths per minute
   - **d.** 33 breaths per minute

3. For the nonsmokers, by how much did the respiration rate increase between resting and running?
   - **a.** 15 breaths per minute
   - **b.** 18 breaths per minute
   - **c.** 23 breaths per minute
   - **d.** 33 breaths per minute

4. Which statement is *not* true?
   - **a.** The nonsmoker at rest took more breaths per minute than the smoker at rest.
   - **b.** The nonsmoker took more breaths per minute running than walking.
   - **c.** The smoker took more breaths per minute than the nonsmoker while walking.
   - **d.** The nonsmoker took fewer breaths per minute than the smoker while running.

5. Which statement is the most logical conclusion to draw from the data in the chart?
   - **a.** Smoking has no effect on respiration rate.
   - **b.** Increased activity has no effect on respiration rate.
   - **c.** There is no difference in the respiration rates between the smoker and the nonsmoker.
   - **d.** Smoking and activity cause an increase in respiration rate.

Extended Response

6. Tar, which is a harmful substance found in tobacco smoke, coats the lining of the lungs over time. Based on the information in the graph and what you know about the respiratory system, write a paragraph describing how smoking cigarettes affects the functioning of the respiratory system.

7. Ads for cigarettes and other tobacco products have been banned from television. However, they still appear in newspapers and magazines. These ads make tobacco use look glamorous and exciting. Using your knowledge of the respiratory system, design an ad that discourages the use of tobacco products. Create a slogan that will help people remember how tobacco affects the health of the respiratory system.

METACOGNITIVE ACTIVITY

Have students answer the following questions in their **Science Notebook:**

1. What are some strategies that helped you learn about the structures and functions of the respiratory system?
2. What questions do you still have about how the respiratory system works together with the circulatory system?
3. How did your unit project help you gain a better understanding of the systems studied in this chapter?

Analyzing Data

1. a   *3. b*   *5. d*

2. a   *4. a*

Extended Response

6. RUBRIC

4 points for a response that correctly answers the question and uses the following terms accurately:

- bronchial tubes
- tar
- oxygen
- alveoli

Sample answer: Tar coats the lining of the passageways and the bronchial tubes that lead into the lungs. As the amount of tar increases, the smoker no longer is able to keep the passageways of the lungs clear of dust and other harmful particles. Less oxygen is exchanged in the alveoli, and the smoker's rate of respiration increases at rest as well as in other types of activities.

3 points correctly answers the question and uses three terms accurately

2 points correctly answers the question and uses two terms accurately

1 point correctly answers the question and uses one term accurately

7. RUBRIC

4 points for a response in which all requirements of the task are included. *Students' ads should include the following:*

- *knowledge of respiratory system structures (mentions lungs, alveoli, cilia, and/or bronchial tubes)*
- *affect of tobacco on the structures of the respiratory system*
- *respiratory diseases associated with the respiratory system (such as cancer or emphysema)*
- *slogan that describes the effects of smoking (Sample slogan: Today's smoke is tomorrow's choke.)*

3 points response fulfills three of the above items

2 points response fulfills two of the above items

1 point response fulfills one of the above items

CHAPTER

Transport and Protection

Life Science
UNIFYING PRINCIPLES

PRINCIPLE 1
All living things share common characteristics.

PRINCIPLE 2
All living things share common needs.

PRINCIPLE 3
Living things meet their needs through interactions with the environment.

PRINCIPLE 4
The types and numbers of living things change over time.

Unit: Human Biology
BIG IDEAS

CHAPTER 1
Systems, Support, and Movement
The human body is made up of systems that work together to perform necessary functions.

CHAPTER 2
Absorption, Digestion, and Exchange
Systems in the body obtain and process materials and remove waste.

CHAPTER 3
Transport and Protection
Systems function to transport materials and to defend and protect the body.

CHAPTER 4
Control and Reproduction
The nervous and endocrine systems allow the body to respond to internal and external conditions.

CHAPTER 5
Growth, Development, and Health
The body develops and maintains itself over time.

CHAPTER 3
KEY CONCEPTS

SECTION 3.1

The circulatory system transports materials.

1. The circulatory system works with other body systems.
2. Structures in the circulatory system function together.
3. Blood exerts pressure on blood vessels.
4. There are four different blood types.

SECTION 3.2

The immune system defends the body.

1. Many systems defend the body from harmful materials.
2. The immune system has response structures.
3. The immune system responds to attack.
4. Most diseases can be prevented or treated.

SECTION 3.3

The integumentary system shields the body.

1. Skin performs important functions.
2. The structure of skin is complex.
3. The skin grows and heals.

T The Big Idea Flow Chart is available on p. T17 in the **UNIT TRANSPARENCY BOOK.**

Previewing Content

SECTION

The circulatory system transports materials. pp. 65–73

1. **The circulatory system works with other body systems.**
The **circulatory system** transports digested materials from the digestive system to cells of the body. It also transports oxygen to cells of the body and removes carbon dioxide and other wastes through the **blood.**

2. **Structures in the circulatory system function together.**
The human heart is divided into four chambers. It pumps oxygen-poor blood from the right ventricle to the lungs. Oxygen-rich blood is pumped from the left ventricle to the rest of the body. The left atrium receives oxygen-rich blood from the lungs, and the right atrium receives oxygen-poor blood from the rest of the body.

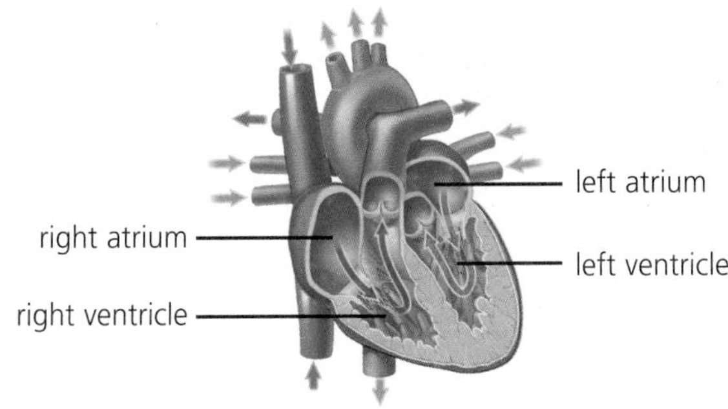

You may wish to leave this simple diagram on the board to emphasize that the heart is not shaped like a valentine and that blood flows both out from and into the heart.

- Blood is a tissue made up of plasma, **red blood cells,** white blood cells, and platelets. Red blood cells carry oxygen to the other cells of the body.
- Blood vessels include **arteries, veins,** and **capillaries.** Most arteries carry blood away from the heart, and most veins carry blood toward the heart. Exchanges of gases and other materi-als take place in capillaries.

3. **Blood exerts pressure on blood vessels.**
Blood pressure is the force that blood exerts on the arteries, veins, and capillaries. High or low blood pressure can cause serious medical conditions.

4. **There are four different blood types.**
The four blood types are A, B, AB, and O and are determined by antigens on red blood cells.

SECTION

The immune system defends the body. pp. 74–82

1. **Many systems defend the body from harmful materials.**
The integumentary, respiratory, and digestive systems are the first lines of defense against **pathogens,** disease-causing agents that may enter the body.

2. **The immune system has response structures.**
White blood cells are produced in the bone marrow, thymus gland, spleen, and lymph nodes. They are specialized to destroy foreign organisms. The lymphatic system transports white blood cells and antibodies throughout the body.

3. **The immune system responds to attack.**
Inflammation and fever are two immune responses in the body's defense against pathogens. Histamines are released to raise the temperature of tissues and increase blood flow to an infected area. Fever occurs when many tissues produce histamines. T cells and B cells are two kinds of white blood cells. B cells produce **antibodies** that attack foreign materials.

The diagram below shows the immune responses of T cells and B cells.

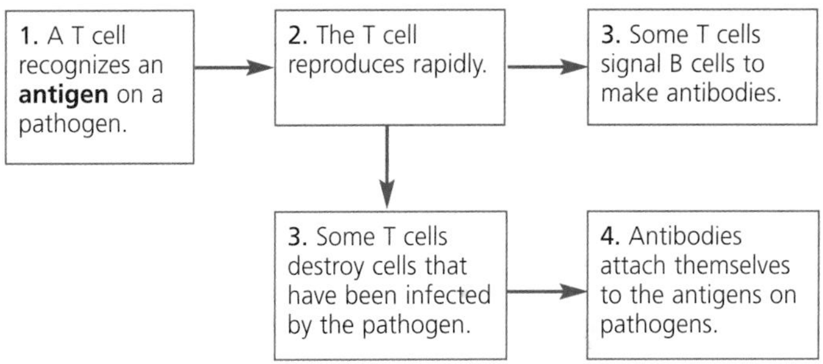

1. A T cell recognizes an **antigen** on a pathogen.
2. The T cell reproduces rapidly.
3. Some T cells signal B cells to make antibodies.
3. Some T cells destroy cells that have been infected by the pathogen.
4. Antibodies attach themselves to the antigens on pathogens.

- Active immunity is developed when the body makes its own antibodies after an illness. **Vaccines** provide a form of active immunity. Passive immunity occurs in newborns that inherit antibodies from their mothers.

4. **Most diseases can be prevented or treated.**
   - Vaccines provide a form of active immunity that can prevent some diseases.
   - **Antibiotics** and other medicines are used to treat diseases.

Common Misconceptions

MISCONCEPTION DATABASE
CLASSZONE.COM Background on student misconceptions

FUNCTION OF THE HEART Many students realize that the heart pumps blood to the extremities, but they are not aware that the blood also returns to the heart. Students often confuse its pump function with functions such as cleaning, filtering, and manufac-turing blood.

 This misconception is addressed on p. 66.

CIRCULATION AND RESPIRATION Students of all ages hold the misconception that a separate system of tubes carries air to the heart and other organs. Emphasize instead that although oxygen enters the body through the trachea and lungs, it is carried to cells in the blood.

 This misconception is addressed on p. 67.

Previewing Content

SECTION

The integumentary system shields the body. pp. 83–89

1. Skin performs important functions.
The body's integumentary system, which includes hair and nails, repels water, guards against infection, helps maintain homeostasis, and senses the environment.

2. The structure of skin is complex.
Human skin is composed of two layers, the **epidermis** and the **dermis.**

- The epidermis forms a thick, protective outer layer made up of many protein fibers.
- The dermis contains sweat and oil glands, sensory receptors, hair, blood vessels, and muscles.
- Nails grow from epidermal cells in nail beds.

3. The skin grows and heals.

- Growth occurs at the base of the epidermis. This layer completely replaces itself every two to four weeks.
- Serious injuries to the skin include burns, especially sunburn. Repeated burning increases a person's chances of getting skin cancer.

Explain to students that many layers and structures make up the skin they see on their hands.

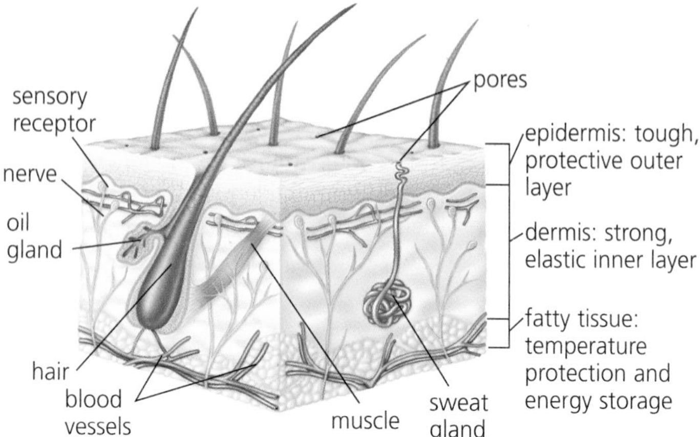

Common Misconceptions

MISCONCEPTION DATABASE
CLASSZONE.COM Background on student misconceptions

CAUSES OF ILLNESS Many students think that all illnesses are caused by germs and that germs enter only through the mouth during eating. In fact, pathogens can enter the body through a cut in the skin or through the lungs during breathing.

This misconception is addressed on p. 75.

Previewing Labs

Lab Generator CD-ROM
Edit these Pupil Edition labs and generate alternative labs.

EXPLORE the BIG idea

Lab	Time and Materials
Blood Pressure, p. 63 Students use balloons partially filled with air to simulate the action of blood pressure.	**TIME** 10 minutes **MATERIALS** a few balloons
Wet Fingers, p. 63 Students observe how the skin senses changes in the environment.	**TIME** 10 minutes **MATERIALS** cup, water
Internet Activity: Heart Pumping, p. 63 Students observe how the heart pumps blood to understand that blood circulates.	**TIME** 20 minutes **MATERIALS** computer with Internet access

SECTION 3.1

Lab	Time and Materials
EXPLORE The Circulatory System, p. 65 Students take their pulse to measure how fast the heart beats.	**TIME** 10 minutes **MATERIALS** stopwatch
CHAPTER INVESTIGATION, Heart Rate and Exercise, pp. 72–73 Students calculate resting, maximum, and target heart rates and examine the effects of exercise on heart rate.	**TIME** 40 minutes **MATERIALS** notebook, stopwatch, calculator, graph paper

SECTION 3.2

Lab	Time and Materials
EXPLORE Membranes, p. 74 Students model working membranes to understand how the body keeps out foreign particles.	**TIME** 15 minutes **MATERIALS** white cloth, zippered sandwich bag, large bowl, water, food coloring, small pin
INVESTIGATE Antibodies, p. 79 Students make models to investigate how antibodies stop pathogens from spreading.	**TIME** 15 minutes **MATERIALS** plastic containers with lids, index cards

SECTION 3.3

Lab	Time and Materials
EXPLORE The Skin, p. 83 Students remove the skin from apples to explore the functions of the skin.	**TIME** 10 minutes **MATERIALS** vegetable peeler, apple
INVESTIGATE Skin Protection, p. 85 Students observe how oil protects the skin.	**TIME** 10 minutes **MATERIALS** cotton ball, rubbing alcohol, dropper, water

R **Additional INVESTIGATION,** Listen to Your Heart, A, B, & C, pp. 177–185; Teacher Instructions, pp. 305–306

Previewing Chapter Resources

CHAPTER 3
Transport and Protection

INTEGRATED TECHNOLOGY

CLASSZONE.COM
- eEdition Plus
- EasyPlanner Plus
- Misconception Database
- Content Review
- Test Practice
- Resource Centers
- Internet Activity: Heart Pumping
- Math Tutorial

SCILINKS.ORG

SCILINKS

CD-ROMS
- eEdition
- EasyPlanner
- Power Presentations
- Content Review
- Lab Generator
- Test Generator

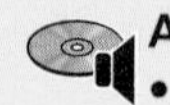
AUDIO CDS
- Audio Readings
- Audio Readings in Spanish

LABS AND ACTIVITIES

EXPLORE the Big Idea, p. 63
- Blood Pressure
- Wet Fingers
- Internet Activity: Heart Pumping

UNIT RESOURCE BOOK
Unit Projects, pp. 5–10

Lab Generator CD-ROM
Generate customized labs.

SECTION

3.1 The circulatory system transports materials.
pp. 65–73

Time: 3 periods (1.5 blocks)

R Lesson Plan, pp. 129–130

RESOURCE CENTERS, The Circulatory System, Blood Types

UNIT TRANSPARENCY BOOK
- Big Idea Flow Chart, p. T17
- Daily Vocabulary Scaffolding, p. T18
- Note-Taking Model, p. T19
- 3-Minute Warm-Up, p. T20
- "The Heart" Visual, p. T22

- EXPLORE The Circulatory System, p. 65
- CHAPTER INVESTIGATION, Heart Rate and Exercise, pp. 72–73

UNIT RESOURCE BOOK
- CHAPTER INVESTIGATION, Heart Rate and Exercise, A, B, & C, pp. 168–176
- Additional INVESTIGATION, Listen to Your Heart, A, B, & C, pp. 177–185

SECTION

3.2 The immune system defends the body.
pp. 74–82

Time: 2 periods (1 block)

R Lesson Plan, pp. 139–140

- **RESOURCE CENTER,** Lymphatic System
- **MATH TUTORIAL**

UNIT TRANSPARENCY BOOK
- Daily Vocabulary Scaffolding, p. T18
- 3-Minute Warm-Up, p. T20

- EXPLORE Membranes, p. 74
- INVESTIGATE Antibodies, p. 79
- Math in Science, p. 82

UNIT RESOURCE BOOK
- Datasheet, Antibodies, p. 148
- Math Support, p. 166
- Math Practice, p. 167

SECTION

3.3 The integumentary system shields the body.
pp. 83–89

Time: 3 periods (1.5 blocks)

R Lesson Plan, pp. 150–151

- **RESOURCE CENTER,** Structure of Skin
- **VISUALIZATION,** How Skin Heals

UNIT TRANSPARENCY BOOK
- Big Idea Flow Chart, p. T17
- Daily Vocabulary Scaffolding, p. T18
- 3-Minute Warm-Up, p. T21
- Chapter Outline, pp. T23–T24

- EXPLORE The Skin, p. 83
- INVESTIGATE Skin Protection, p. 85
- Extreme Science, p. 89

UNIT RESOURCE BOOK
Datasheet, Skin Protection, p. 159

KEY TO ICONS 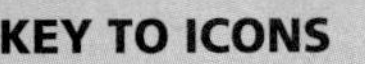

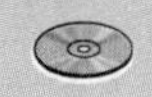 CD/CD-ROM

INTERNET

 Teacher Edition

Pupil Edition

 UNIT TRANSPARENCY BOOK

UNIT RESOURCE BOOK

UNIT ASSESSMENT BOOK

 SPANISH ASSESSMENT BOOK

SCIENCE TOOLKIT

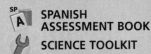

READING AND REINFORCEMENT

- Frame Game, B26–27
- Main Idea and Detail Notes, C37
- Daily Vocabulary Scaffolding, H1–8

 **UNIT RESOURCE BOOK**

- Vocabulary Practice, pp. 163–164
- Decoding Support, p. 165
- Summarizing the Chapter, pp. 186–187

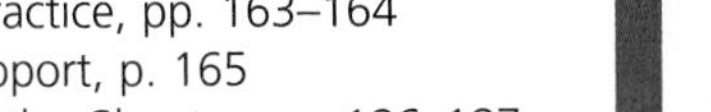

 **Audio Readings CD**
Listen to Pupil Edition.

 **Audio Readings in Spanish CD**
Listen to Pupil Edition in Spanish.

ASSESSMENT

- Chapter Review, pp. 90–91
- Standardized Test Practice, p. 93

UNIT ASSESSMENT BOOK

- Diagnostic Test, pp. 39–40
- Chapter Test, A, B, & C, pp. 44–55
- Alternative Assessment, pp. 56–57

Spanish Chapter Test, pp. 73–76

 **Test Generator CD-ROM**
Generate customized tests.

 **Lab Generator CD-ROM**
Rubrics for Labs

STANDARDS

National Standards
A.2–8, A.9.a–c, A.9.e–f, C.1.d–f, G.1.b

See p. 62 for the standards.

 **UNIT RESOURCE BOOK**

- Reading Study Guide, A & B, pp. 131–134
- Spanish Reading Study Guide, pp. 135–136
- Challenge and Extension, p. 137
- Reinforcing Key Concepts, p. 138

 Ongoing Assessment, pp. 66–67, 69–71

 Section 3.1 Review, p. 71

 **UNIT ASSESSMENT BOOK**
Section 3.1 Quiz, p. 41

National Standards
A.2–8, A.9.a–c, A.9.e–f, C.1.d–e, G.1.b

 **UNIT RESOURCE BOOK**

- Reading Study Guide, A & B, pp. 141–144
- Spanish Reading Study Guide, pp. 145–146
- Challenge and Extension, p. 147
- Reinforcing Key Concepts, p. 149

 Ongoing Assessment, pp. 74–81

 Section 3.2 Review, p. 81

 **UNIT ASSESSMENT BOOK**
Section 3.2 Quiz, p. 42

National Standards
A.2–7, A.9.a–b, A.9.e–f, C.1.d–f, G.1.b

 **UNIT RESOURCE BOOK**

- Reading Study Guide, A & B, pp. 152–155
- Spanish Reading Study Guide, pp. 156–157
- Challenge and Extension, p. 158
- Reinforcing Key Concepts, p. 160
- Challenge Reading, pp. 161–162

 Ongoing Assessment, pp. 83, 85–88

 Section 3.3 Review, p. 88

 **UNIT ASSESSMENT BOOK**
Section 3.3 Quiz, p. 43

National Standards
A.2–7, A.9.a–b, A.9.e–f, C.1.e–f, G.1.b

Previewing Resources for Differentiated Instruction

CHAPTER INVESTIGATION

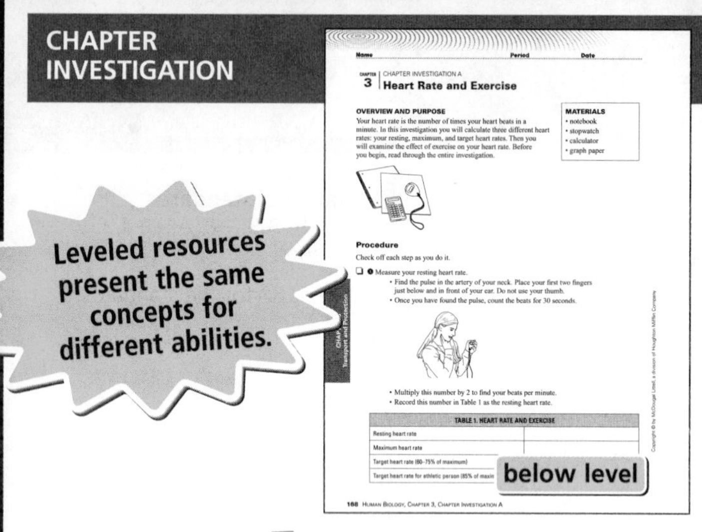

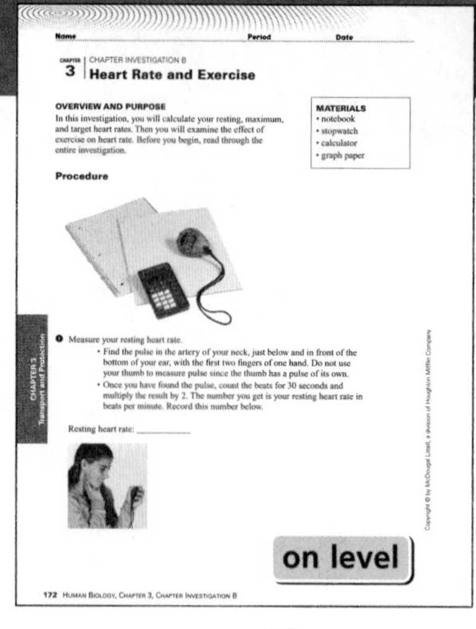

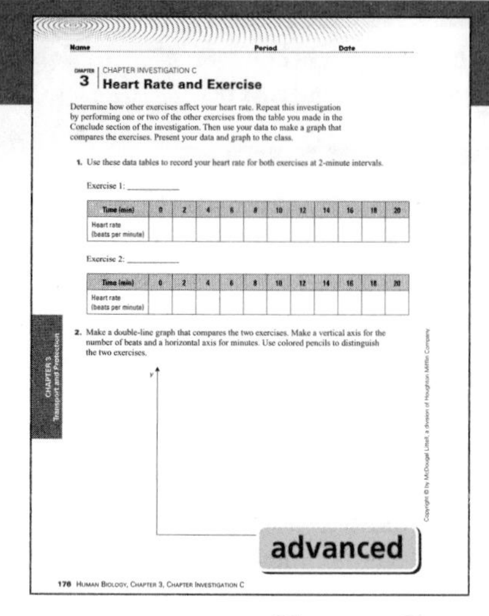

UNIT RESOURCE BOOK, pp. 168–171 | pp. 172–175 | pp. 172–176

READING STUDY GUIDE

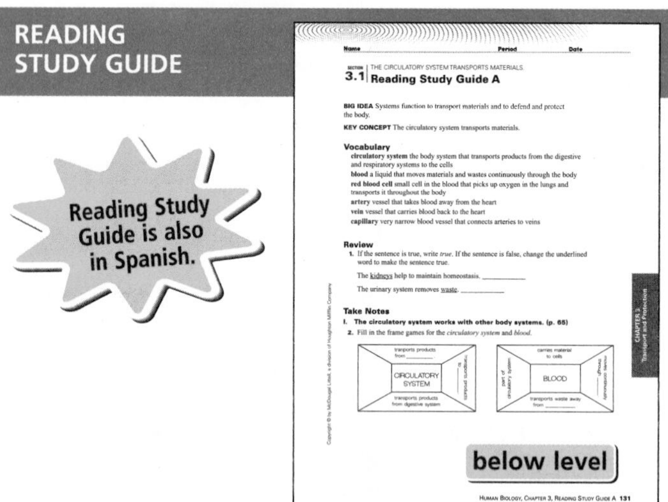

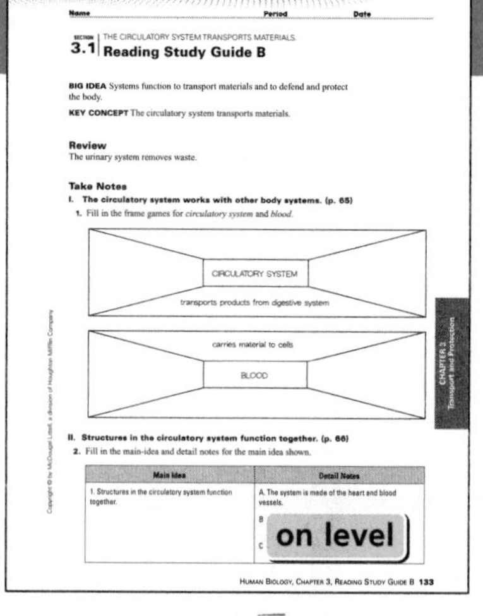

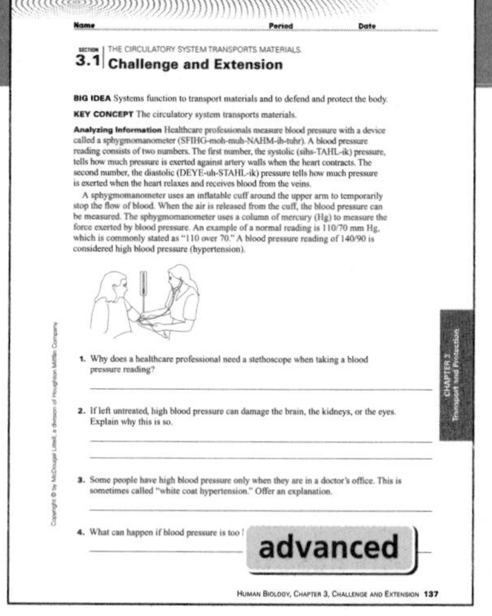

UNIT RESOURCE BOOK, pp. 131–132 | pp. 133–134 | p. 137

CHAPTER TEST

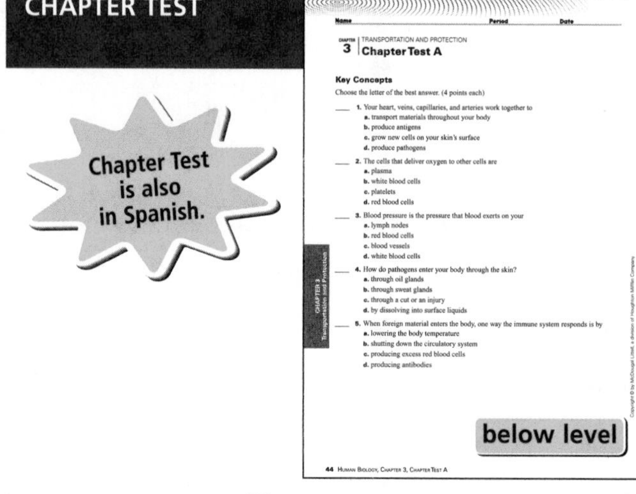

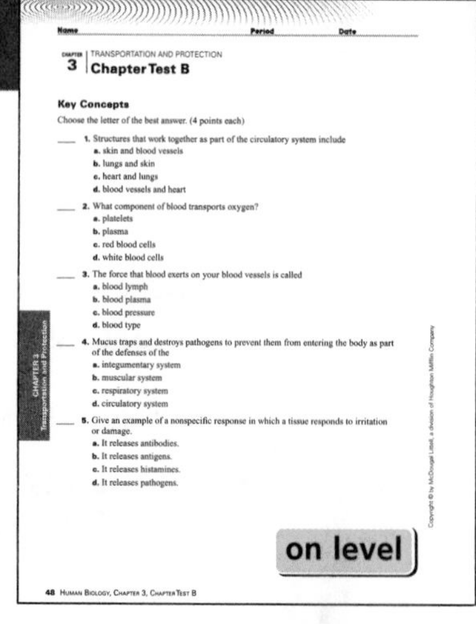

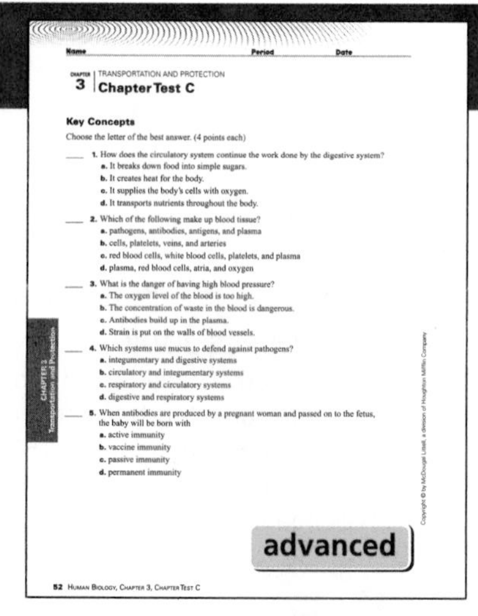

UNIT ASSESSMENT BOOK, pp. 44–47 | pp. 48–51 | 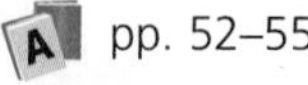 pp. 52–55

TECHNOLOGY

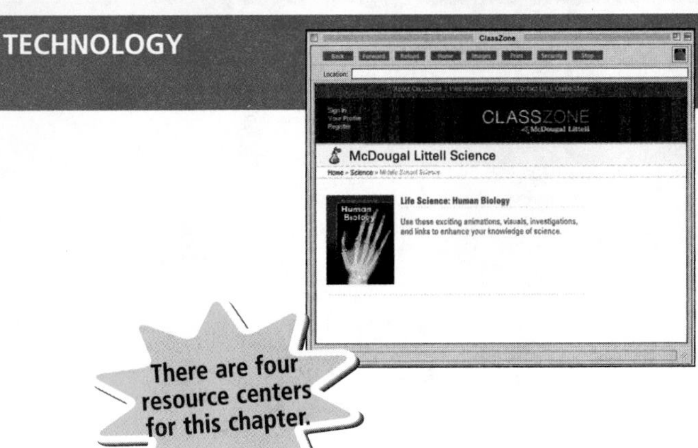

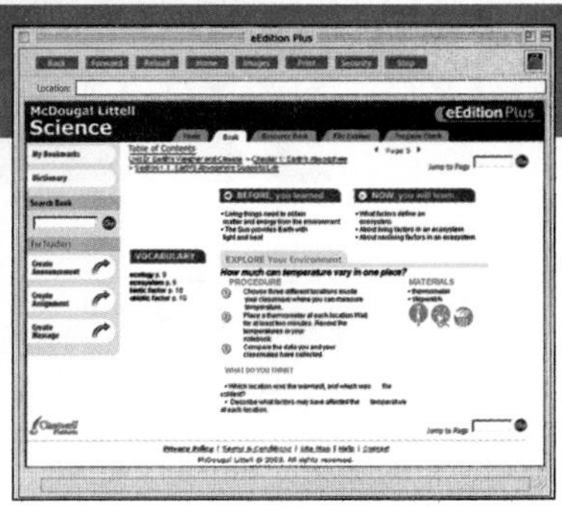

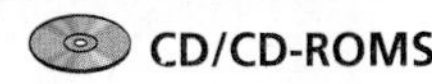

VISUAL CONTENT

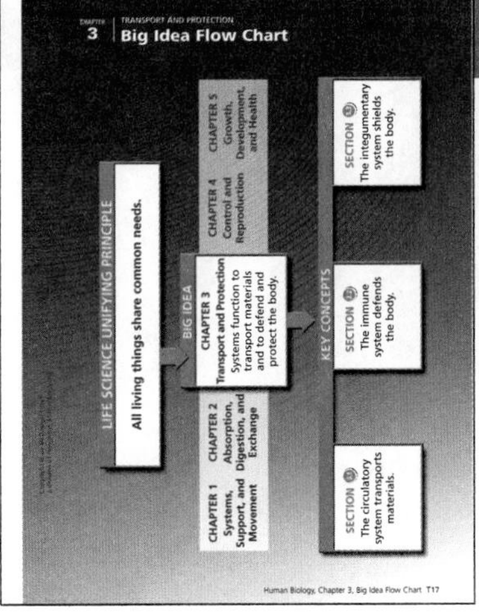

UNIT TRANSPARENCY BOOK, p. T17

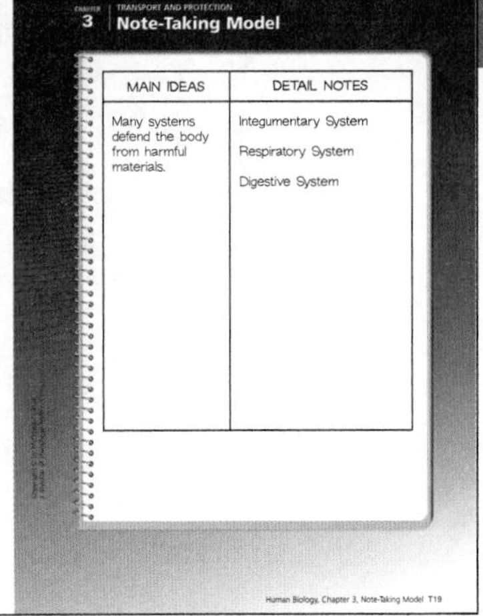

p. T19

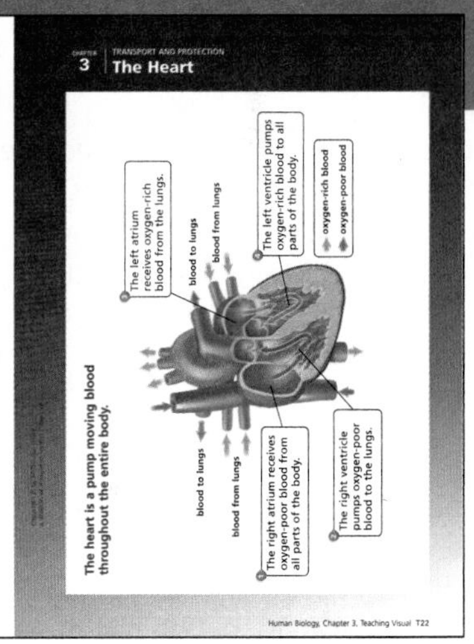

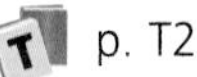
p. T22

MORE SUPPORT

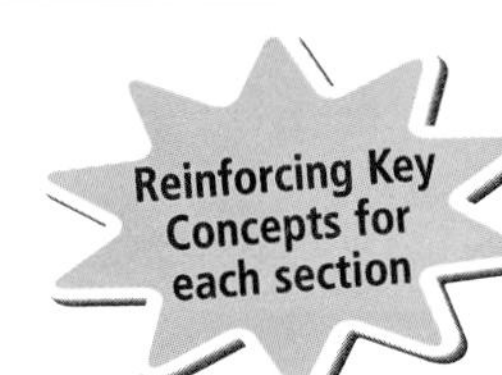

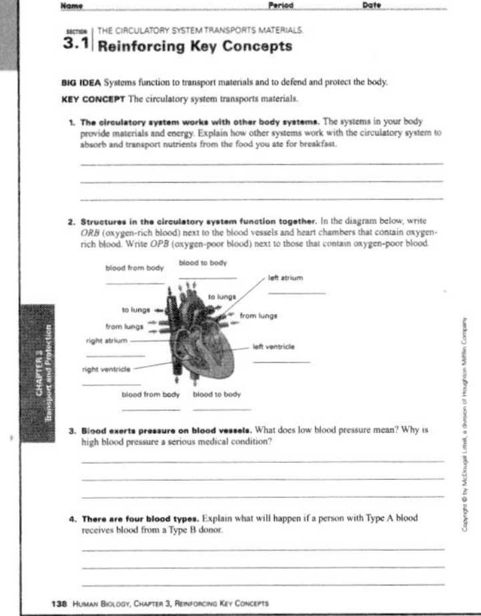

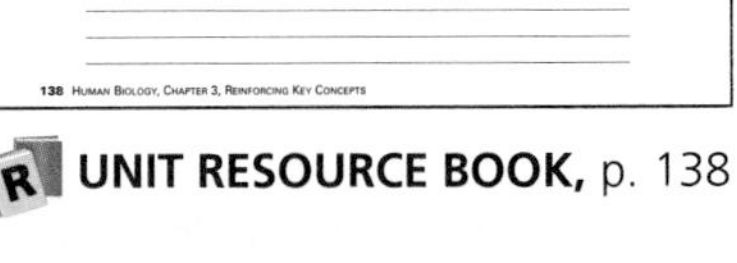

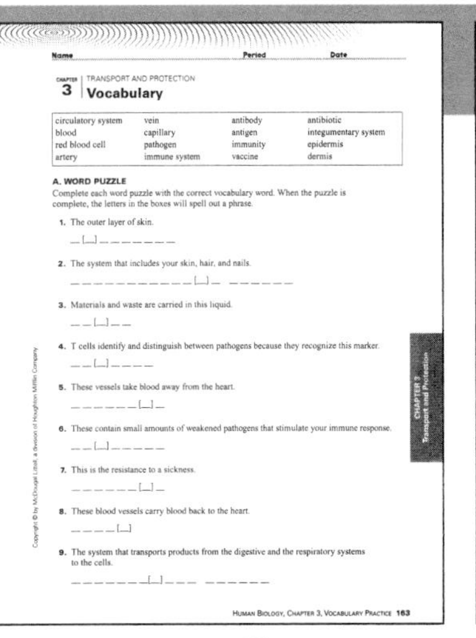

pp. 163–164

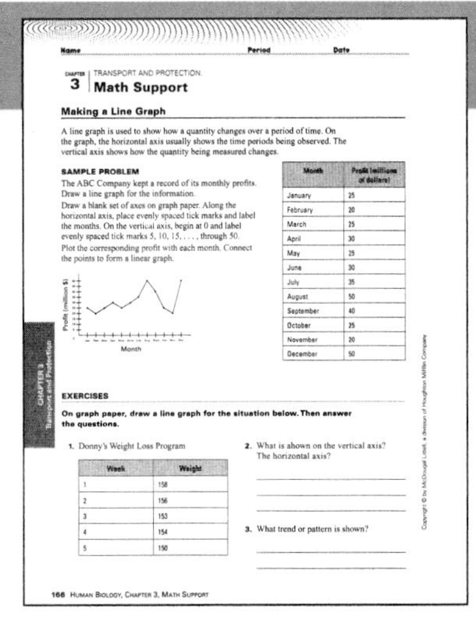

p. 166

CHAPTER

Transport and Protection

INTRODUCE

the BIG idea

Have students look at the photograph of red blood cells and discuss how the question in the box links to the Big Idea. Ask:

- How does the circulatory system act like a transportation system?
- Why are red blood cells key in transporting material to and from cells?

National Science Education Standards

Content

C.1.d Specialized cells perform specialized functions in multicellular organisms. Groups of specialized cells cooperate to form a tissue. Different tissues in turn form larger functional units, called organs. Each type of cell, tissue, and organ has a distinct structure and set of functions.

C.1.e The human organism has systems for digestion, respiration, reproduction, circulation, excretion, movement, control and coordination, and protection from disease. These systems interact.

C.1.f Disease is a breakdown in structures or functions. Some diseases are the result of intrinsic failures. Others are the result of damage by infection.

Process

A.2–8 Design and conduct an investigation; use tools to gather and interpret data; use evidence to describe, predict, explain, model; think critically to make relationships between evidence and explanation; recognize different explanations and predictions; communicate scientific procedures and explanations; use mathematics.

A.9.a–c, A.9.e–f Understand scientific inquiry by using different investigations, methods, mathematics, explanations based on logic, evidence, and skepticism.

G.1.b Science requires different abilities.

CHAPTER

Transport and Protection

the BIG idea

Systems function to transport materials and to defend and protect the body.

Red blood cells travel through a blood vessel. How do you think blood carries materials around your body?

Key Concepts

SECTION

The circulatory system transports materials.
Learn how materials move through blood vessels.

SECTION

The immune system defends the body.
Learn about the body's defenses and responses to foreign materials.

SECTION

The integumentary system shields the body.
Learn about the structure of skin and how it protects the body.

Internet Preview

CLASSZONE.COM

Chapter 3 online resources: Content Review, two Visualizations, four Resource Centers, Math Tutorial, Test Practice

INTERNET PREVIEW

CLASSZONE.COM For student use with the following pages:

Review and Practice
- Content Review, pp. 64, 90
- Math Tutorial, Making a Line Graph, p. 82
- Test Practice, p. 93

Activities and Resources
- Internet Activity: Heart Pumping, p. 63
- Resource Centers: Circulatory System, p. 66; Blood Types, p. 71; Lymphatic System, p. 76; Skin, p. 85
- Visualization, p. 87

Immune System
Code: MDL046

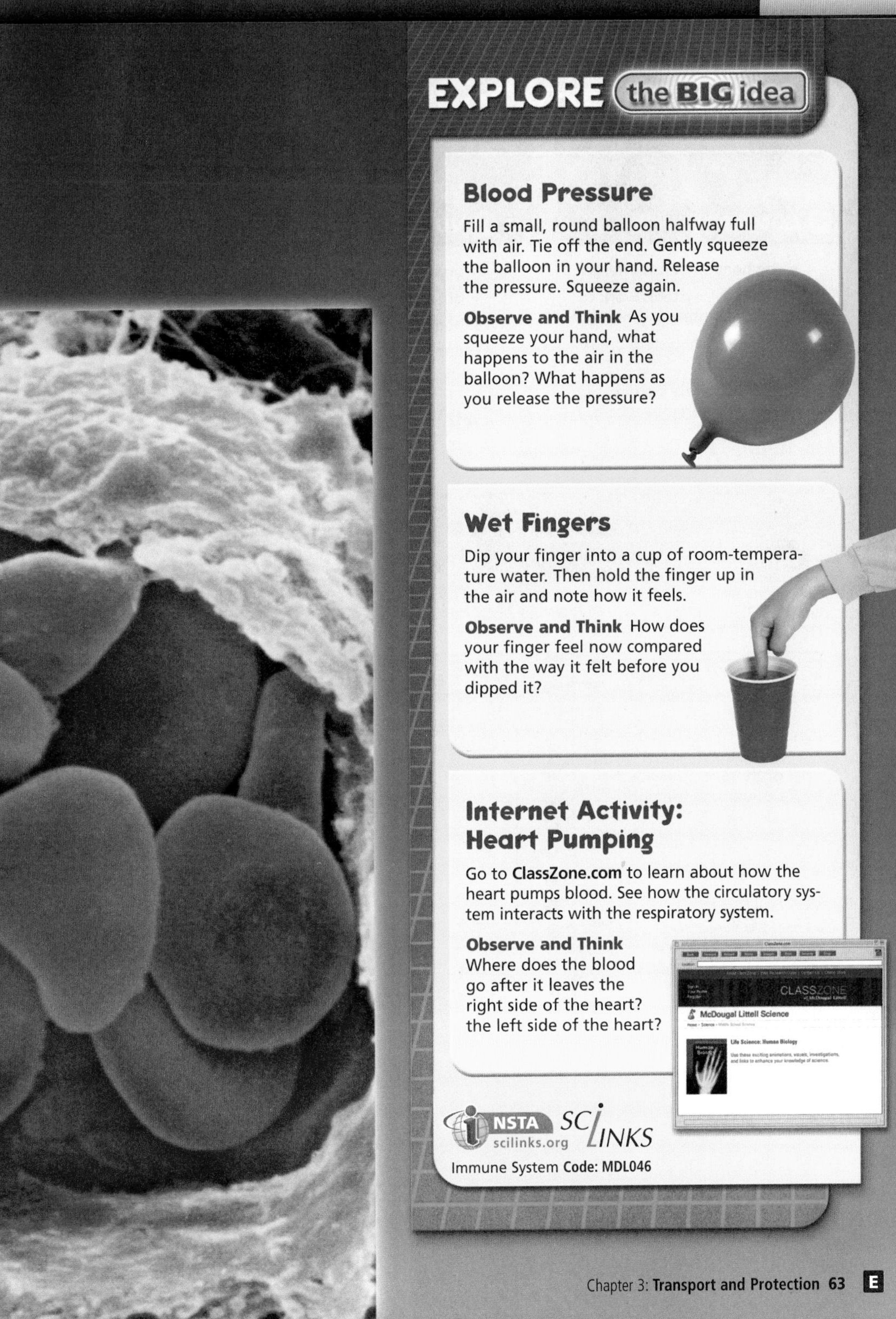

EXPLORE the BIG idea

Blood Pressure

Fill a small, round balloon halfway full with air. Tie off the end. Gently squeeze the balloon in your hand. Release the pressure. Squeeze again.

Observe and Think As you squeeze your hand, what happens to the air in the balloon? What happens as you release the pressure?

Wet Fingers

Dip your finger into a cup of room-temperature water. Then hold the finger up in the air and note how it feels.

Observe and Think How does your finger feel now compared with the way it felt before you dipped it?

Internet Activity: Heart Pumping

Go to **ClassZone.com** to learn about how the heart pumps blood. See how the circulatory system interacts with the respiratory system.

Observe and Think Where does the blood go after it leaves the right side of the heart? the left side of the heart?

NSTA scilinks.org SCILINKS

Immune System **Code: MDL046**

TEACHING WITH TECHNOLOGY

CBL and Probeware Students can use a heart rate sensor to measure heart rates before and after mild exercise in the Chapter Investigation on pp. 72–73. Students can use graphing calculators or spreadsheet software to collect, display, and analyze the data.

CBL and Probeware Students can use a skin surface temperature probe to measure the change in skin temperature caused by placing a cold pack or a heat pack on their arms. Relate this measurable change to the change in sensation. Students will learn about sensory receptors in skin on p. 86.

EXPLORE the BIG idea

These inquiry-based activities are appropriate for use at home or as a supplement to classroom instruction.

Blood Pressure

PURPOSE To simulate the action of blood pressure.

TIP *10 min.* Students can try the same activity with water instead of air in the balloon if they are careful not to break the balloon.

Answer: The air inside the balloon moves to one area as the balloon is squeezed. That area of the balloon is stretched by the pressure. When the hand is released, the air once again fills the balloon evenly.

REVISIT after p. 70.

Wet Fingers

PURPOSE To observe how the skin senses changes in the environment.

TIPS *10 min.* You might encourage students to try their other hand or a different temperature of water to compare the observations. A bowl or pan may be used instead of a cup.

Answer: The finger feels colder when exposed to air while wet.

REVISIT after p. 85.

Internet Activity: Heart Pumping

PURPOSE To observe how the heart pumps blood.

TIP *20 min.* Use student observations to introduce the interaction between the circulatory and respiratory systems.

Answer: Blood from the right side of the heart goes to the lungs. Blood from the left side goes to the rest of the body.

REVISIT after p. 66.

PREPARE

CONCEPT REVIEW

Activate Prior Knowledge

- Ask students to describe how oxygen reaches the cells of the body. *It is exchanged in the alveoli of the lungs and carried by the blood vessels.*
- Discuss how digested food reaches the cells of the body. *Nutrients are absorbed by villi in the intestines and carried by the blood vessels.*

TAKING NOTES

Main Idea and Detail Notes

Remind students that the blue headings in the text should be the main ideas in their graphic organizers. Students can find details in the text as well as in the visuals. Encourage students to compare their notes.

Vocabulary Strategy

The frame game diagrams help students organize facts, descriptions, words, and pictures in a systematic manner. The diagrams can be useful study devices when students look back over their notes.

Students can turn a diagram into a game by leaving the center space blank and asking other students to guess at what the frame is holding.

Vocabulary and Note-Taking Resources

- Vocabulary Practice, pp. 163–164
- Decoding Support, p. 165

- Daily Vocabulary Scaffolding, p. T18
- Note-Taking Model, p. T19

- Frame Game, B26–27
- Main Idea and Detail Notes, C37
- Daily Vocabulary Scaffolding, H1–8

CHAPTER 3

Getting Ready to Learn

CONCEPT REVIEW

- The body's systems interact.
- The body's systems work to maintain internal conditions.
- The digestive system breaks down food.
- The respiratory system gets oxygen and removes carbon dioxide.

VOCABULARY REVIEW

organ p. 11
organ system p. 12
homeostasis p. 12
nutrient p. 45

Review concepts and vocabulary.

TAKING NOTES

MAIN IDEA AND DETAIL NOTES

Make a two-column chart. Write the main ideas, such as those in the blue headings, in the column on the left. Write details about each of those main heads in the column on the right.

SCIENCE NOTEBOOK

MAIN IDEAS	DETAIL NOTES
1. The circulatory system works with other body systems.	1. Transports materials from digestive and respiratory systems to cells 2. Blood is fluid that carries materials and wastes 3. Blood is always moving through the body 4. Blood delivers oxygen and takes away carbon dioxide

VOCABULARY STRATEGY

Write each new vocabulary term in the center of a **frame game** diagram. Decide what information to frame it with. Use examples, descriptions, parts, sentences that use the term in context, or pictures. You can change the frame to fit each term.

carries material to cells
moves continuously through body
BLOOD
carries waste away from cells
circulatory system

See the Note-Taking Handbook on pages R45–R51.

E 64 Unit: Human Biology

CHECK READINESS

Administer the Diagnostic Test to determine students' readiness for new science content and their mastery of requisite math skills.

Diagnostic Test, pp. 39–40

Technology Resources

Students needing content and math skills should visit **ClassZone.com**.

- **CONTENT REVIEW**
- **MATH TUTORIAL**

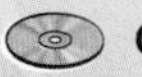
CONTENT REVIEW CD-ROM

KEY CONCEPT

The circulatory system transports materials.

BEFORE, you learned
- The urinary system removes waste
- The kidneys play a role in homeostasis

NOW, you will learn
- How different structures of the circulatory system work together
- About the structure and function of blood
- What blood pressure is and why it is important

VOCABULARY

circulatory system p. 65
blood p. 65
red blood cell p. 67
artery p. 69
vein p. 69
capillary p. 69

EXPLORE The Circulatory System

How fast does your heart beat?

PROCEDURE

1. Hold out your left hand with your palm facing up.
2. Place the first two fingers of your right hand on your left wrist below your thumb. Move your fingertips slightly until you can feel your pulse.
3. Use the stopwatch to determine how many times your heart beats in one minute.

MATERIALS
stopwatch

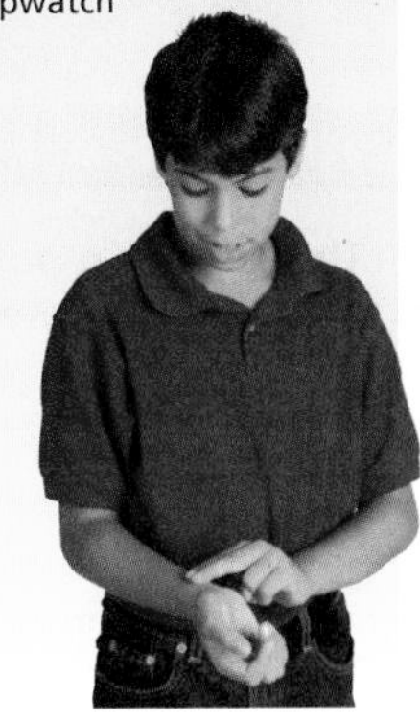

WHAT DO YOU THINK?
- How many times did your heart beat?
- What do you think you would find if you took your pulse after exercising?

The circulatory system works with other body systems.

VOCABULARY
Add a frame game diagram for the term *circulatory system* to your notebook.

You have read that the systems in your body provide materials and energy. The digestive system breaks down food and nutrients, and the respiratory system provides the oxygen that cells need to release energy. Another system, called the **circulatory system,** transports products from the digestive and the respiratory systems to the cells.

Materials and wastes are carried in a fluid called **blood**. Blood moves continuously through the body, delivering oxygen and other materials to cells and removing carbon dioxide and other wastes from cells.

3.1 FOCUS

Set Learning Goals

Students will
- Describe how the different structures of the circulatory system work together.
- Describe blood's components and functions.
- Explain what blood pressure is and why it is important.
- Determine how heart rate is affected by exercise.

3-Minute Warm-Up

Display Transparency 20 or copy this exercise on the board:

Decide which of these statements are true. If not true, correct them.

1. Blood transports digested food particles to cells of the body. *true*
2. Kidneys filter urine and send it out of the body through the esophagus. *Kidneys filter blood and send urine out of the body through the urethra.*
3. The respiratory system is involved in the exhalation of oxygen from the lungs. *The respiratory system is involved in the exhalation of carbon dioxide from the lungs.*

 3-Minute Warm-Up, p. T20

3.1 MOTIVATE

EXPLORE The Circulatory System

PURPOSE To calculate heart rate

TIPS *10 min.* Remind students not to use their thumb while taking their pulse, since the thumb has its own strong pulse. Encourage students to try other pulse points, such as at a carotid artery in the neck.

WHAT DO YOU THINK? *Sample answer: 72. Pulse rate would increase after exercise.*

RESOURCES FOR DIFFERENTIATED INSTRUCTION

Below Level

UNIT RESOURCE BOOK
- Reading Study Guide A, pp. 131–132
- Decoding Support, p. 165

 **AUDIO CDS**

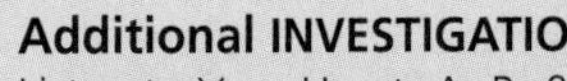

Additional INVESTIGATION, Listen to Your Heart, A, B, & C, pp. 177–185; Teacher Instructions, pp. 305–306

Advanced

UNIT RESOURCE BOOK
Challenge and Extension, p. 137

English Learners

UNIT RESOURCE BOOK
Spanish Reading Study Guide, pp. 135–136

 **AUDIO CDS**
- Audio Readings in Spanish
- Audio Readings (English)

3.1 INSTRUCT

Address Misconceptions

IDENTIFY Ask students to describe how the heart functions. They may understand that the heart is a pump but may omit that blood returns to the heart.

CORRECT Create a model of a heart with half of a pear. Refer to the heart illustration on p. 66. Scoop out the flesh to represent the four chambers. Use color-coded tubes to represent the main artery and vein. Explain how blood flows through the chambers of the heart and blood vessels.

REASSESS Where do the inflows come from? *from the lungs and from all other parts of the body*

Technology Resources

Visit **ClassZone.com** for background on common student misconceptions.

MISCONCEPTION DATABASE

Teach from Visuals

Explain that the cross section is a frontal view of the heart. What is on the heart's left is on the right side in the diagram. Ask:

- Which chambers contain oxygen-rich blood? *left atrium and ventricle*
- Why are the walls of the ventricles thicker than those of the atria? *ventricles: pumping chambers that pump blood; atria: filling chambers that simply receive blood*

EXPLORE the BIG idea

Revisit "Internet Activity: Heart Pumping" on p. 63.

Ongoing Assessment

Describe how the different structures of the circulatory system work together.

Ask: How do the heart and blood vessels provide the body with nutrients? *The heart pumps blood through blood vessels to all body parts.*

READING VISUALS *Answers: right side pumps blood to lungs; left side pumps blood to body*

Structures in the circulatory system function together.

Find out more about the circulatory system.

In order to provide the essential nutrients and other materials that your cells need, your blood must keep moving through your body. The circulatory system, which is made up of the heart and blood vessels, allows blood to flow to all parts of the body. The circulatory system works with other systems to provide the body with this continuous flow of life-giving blood. A

The Heart

The heart is the organ that pushes blood throughout the circulatory system. The human heart actually functions as two pumps—one pump on the right side and one on the left side. The right side of the heart pumps blood to the lungs, and the left side pumps blood to the rest of the body. The lungs receive oxygen when you inhale and remove carbon dioxide when you exhale. Inside the lungs, the respiratory system interacts with the circulatory system. B

The Heart

The heart is a pump moving blood throughout the entire body.

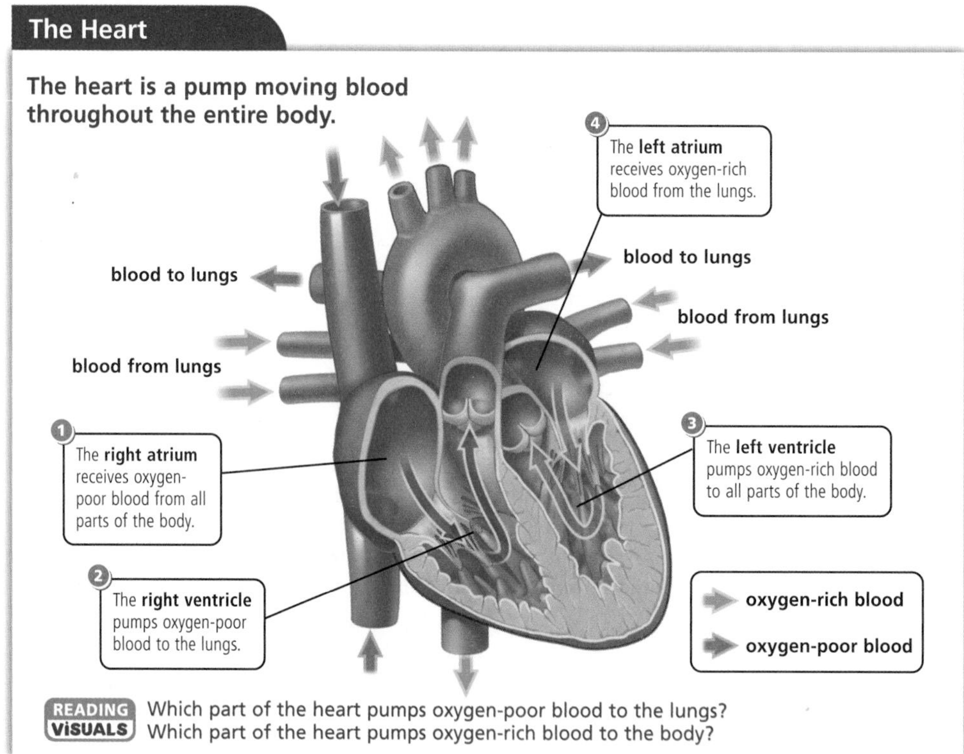
blood to lungs
blood from lungs
blood to lungs
blood from lungs
1 The right atrium receives oxygen-poor blood from all parts of the body.
2 The right ventricle pumps oxygen-poor blood to the lungs.
3 The left ventricle pumps oxygen-rich blood to all parts of the body.
4 The left atrium receives oxygen-rich blood from the lungs.
oxygen-rich blood
oxygen-poor blood

READING VISUALS Which part of the heart pumps oxygen-poor blood to the lungs? Which part of the heart pumps oxygen-rich blood to the body?

DIFFERENTIATE INSTRUCTION

More Reading Support

A The heart and what other structures make up the circulatory system? *blood vessels*

B To which organs does the right side of the heart pump blood? *the lungs*

English Learners This section of the chapter contains a variety of introductory clauses and phrases. Have students find the commas that separate them from the rest of the sentences. Then help them locate the subject of each sentence. Example: "In order to provide the essential nutrients and other materials that your cells need," p. 66. The subject, "blood," follows the clause.

Each side of the heart is divided into two areas called chambers. Oxygen-poor blood, which is, blood from the body with less oxygen, flows to the right side of your heart, into a filling chamber called the right atrium. With each heartbeat, blood flows into a pumping chamber, the right ventricle, and then into the lungs, where it releases carbon dioxide waste and absorbs oxygen.

After picking up oxygen, blood is pushed back to the heart, filling another chamber, which is called the left atrium. Blood moves from the left atrium to the left ventricle, a pumping chamber, and again begins its trip out to the rest of the body. Both oxygen-poor blood and oxygen-rich blood are red. However, oxygen-rich blood is a much brighter and lighter shade of red than is oxygen-poor blood. The diagram on page 66 shows oxygen-poor blood in blue, so that you can tell where in the circulatory system oxygen-poor and oxygen-rich blood are found.

Summarize the way blood moves through the heart. Remember, a summary contains only the most important information.

Blood

The oxygen that your cells need in order to release energy must be present in blood to travel through your body. Blood is a tissue made up of plasma, red blood cells, white blood cells, and platelets. About 60 percent of blood is plasma, a fluid that contains proteins, glucose, hormones, gases, and other substances dissolved in water.

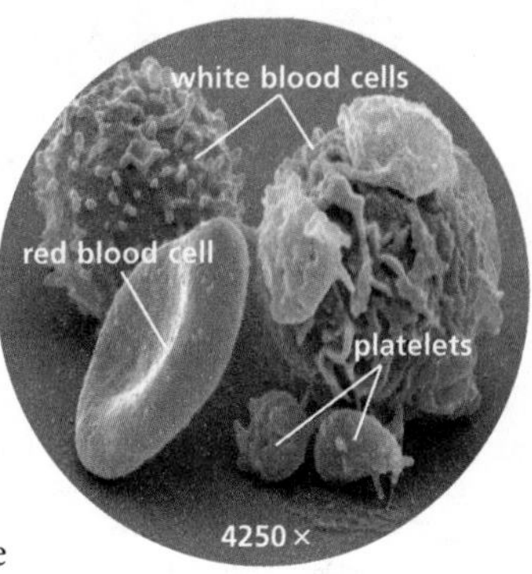

Blood is made mostly of red blood cells, white blood cells, and platelets.

White blood cells help your body fight infection by attacking disease-causing organisms. **Red blood cells** are more numerous than white blood cells and have a different function. They pick up oxygen in the lungs and transport it throughout the body. As red blood cells travel through the circulatory system, they deliver oxygen to other cells.

Platelets are large cell fragments that help form blood clots when a blood vessel is injured. You know what a blood clot is if you've observed a cut or a scrape. The scab that forms around a cut or scrape is made of clotted blood. After an injury such as a cut, platelets nearby begin to enlarge and become sticky. They stick to the injured area of the blood vessels and release chemicals that result in blood clotting. Blood clotting keeps blood vessels from losing too much blood.

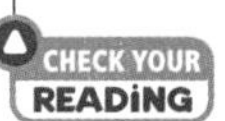

What are the four components that make up blood?

DIFFERENTIATE INSTRUCTION

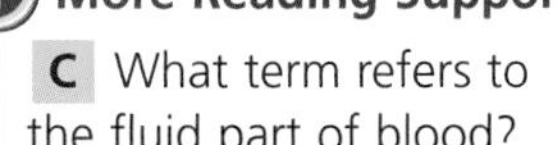

More Reading Support

C What term refers to the fluid part of blood? *plasma*

D Which cells carry oxygen to other cells of the body? *red blood cells*

Advanced Invite students to create a poster of the different ways to keep the heart healthy. They should include items such as the right foods to eat, the role of exercise in strengthening and improving heart function, and unhealthy behaviors, such as smoking, drug use, and excessive drinking.

R Challenge and Extension, p. 137

Address Misconceptions

IDENTIFY Ask students how oxygen gets to the heart and other parts of the body. Students may hold the misconception that a separate system of tubes carries air to the heart and other body structures.

CORRECT Ask students to study the diagram of the heart on p. 66 and make a flow chart to show how oxygen is transported from the lungs to the heart and from the heart to other parts of the body.

REASSESS Ask: Where does the oxygen in "oxygen-rich blood" come from? *It is breathed into the lungs and enters the bloodstream there.*

Technology Resources

Visit **ClassZone.com** for background on common student misconceptions.

MISCONCEPTION DATABASE

Teach Difficult Concepts

Students may think that blood is simply a liquid as opposed to a tissue containing cells. Ask students to describe the four components of blood. Explain that blood is a fluid tissue, a group of cells working together to perform a function.

Ongoing Assessment

Describe blood's components and functions.

Ask: Why are platelets important? *Platelets cause blood clotting and prevent blood from being lost at a cut.*

CHECK YOUR READING *Sample answer: Oxygen-poor blood flows from the body to the right atrium of the heart. The heart pumps it into the right ventricle, then into the lungs, where it releases CO_2 and picks up O_2. Then the blood flows into the left atrium of the heart and is pumped to the left ventricle. From there it is sent out to the body, and the cycle repeats.*

Answer: Blood is made of plasma, red blood cells, white blood cells, and platelets.

Teach from Visuals

To help students interpret the diagram of the circulatory system, ask: What organ pumps blood throughout the entire body? *the heart*

Point out that the left side of the diagram represents the right side of the person shown, who is depicted facing the reader.

History of Science

An artificial heart-lung machine powers blood circulation and oxygenates blood outside a patient's body. The first successful open-heart surgery using such a device was performed in 1953 by the American surgeon Dr. John H. Gibbon, Jr. In 1967 Dr. Christiaan Barnard, a South African surgeon, led a team of 20 surgeons in the first successful heart transplant. The patient, Louis Washkansky, died 18 days later from complications of double pneumonia and immune-suppressant drugs.

Metacognitive Strategy

Have students devise memory strategies or mnemonic devices to remember what happens to the blood as it flows through the heart and lungs to the rest of the body. For example, they might equate "vent" in *ventricle* with *sent* to remember that blood flows out from those chambers and equate "ate" in *atrium* with *gate* to think of going in. Similarly, they can relate *vein* to *vain* to recall "poor" and relate *artery* to *barter* to recall "rich" or perhaps to a highway to recall "traffic."

Mathematics Connection

Have students take their pulse by holding two fingers at the base of the inside of the wrist or just below the jawbone on the neck. Students can compile data on classmates' heartbeats per minute and display the data in a bar graph. Challenge students to find how many times the heart beats in one day, in one month, in one year, and in an average lifetime.

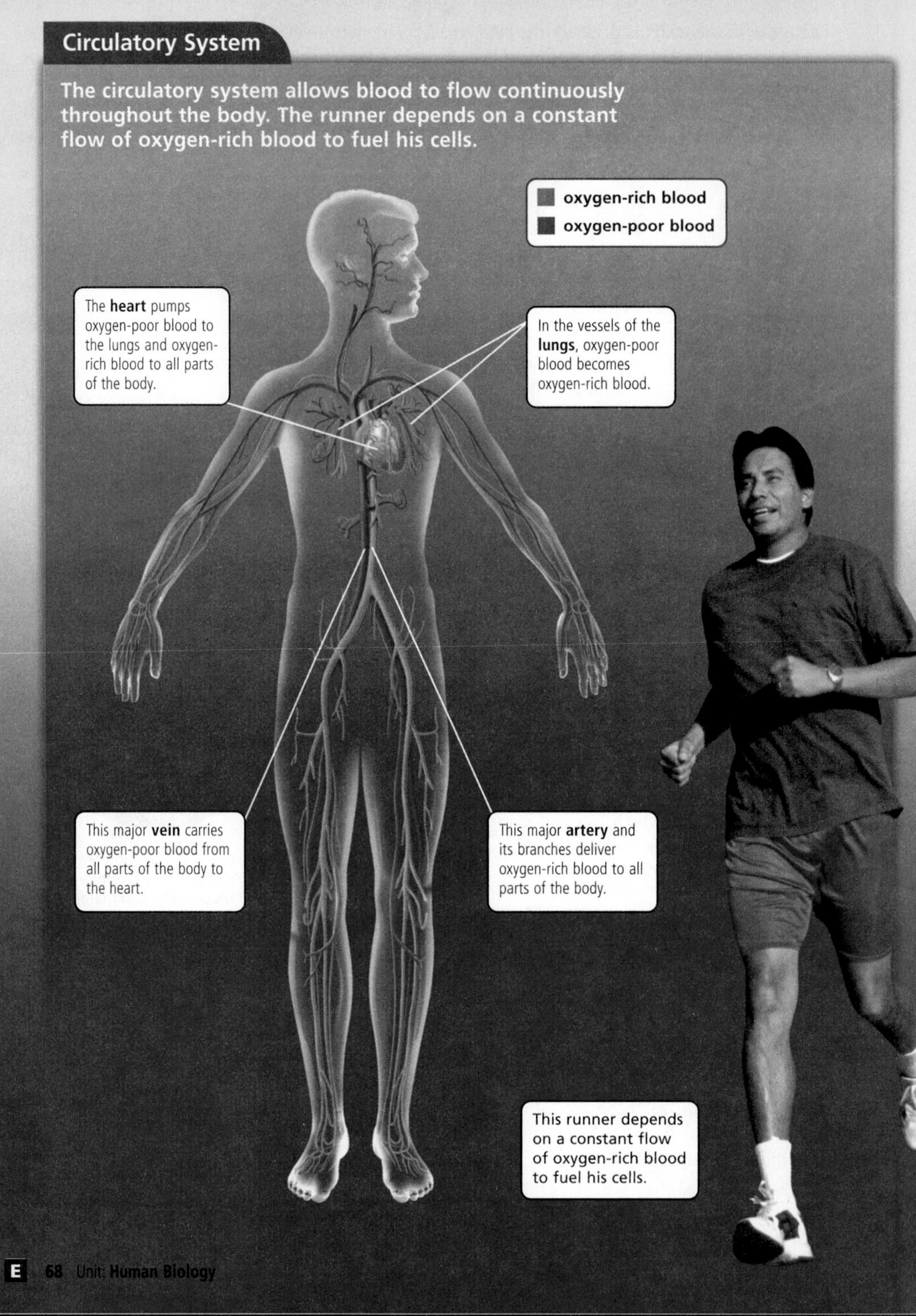

Circulatory System
The circulatory system allows blood to flow continuously throughout the body. The runner depends on a constant flow of oxygen-rich blood to fuel his cells.
oxygen-rich blood
oxygen-poor blood
The heart pumps oxygen-poor blood to the lungs and oxygen-rich blood to all parts of the body.
In the vessels of the lungs, oxygen-poor blood becomes oxygen-rich blood.
This major vein carries oxygen-poor blood from all parts of the body to the heart.
This major artery and its branches deliver oxygen-rich blood to all parts of the body.
This runner depends on a constant flow of oxygen-rich blood to fuel his cells.
E 68 Unit: Human Biology

DIFFERENTIATE INSTRUCTION

English Learners Encourage students whose knowledge of English is limited to use the visual in a focused, interactive way. Give them tracing paper to trace the overall path of the circulatory system, using red and blue pencils or markers. Have them label the heart, the lungs, a vein, and an artery and make a key indicating the meanings of red and blue.

As blood travels through blood vessels, some fluid is lost. This fluid, called lymph, is collected in lymph vessels and returned to veins and arteries. As you will read in the next section, lymph and lymph vessels are associated with your immune system. Sometimes scientists refer to the lymph and lymph vessels as the lymphatic system. The lymphatic system helps you fight disease.

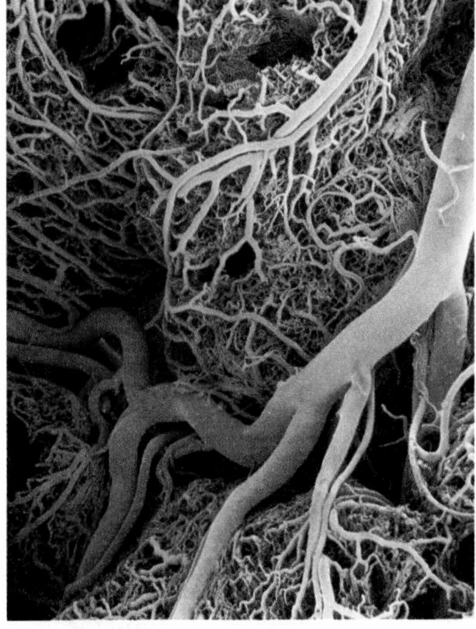

Arteries, capillaries, and veins form a complex web to carry blood to all the cells in the body (30×).

Blood Vessels

Blood moves through a network of structures called blood vessels. Blood vessels are tube-shaped structures that are similar to flexible drinking straws. The structure of blood vessels suits them for particular functions. **Arteries**, which are the vessels that take blood away from the heart, have strong walls. An artery wall is thick and elastic and can handle the tremendous force produced when the heart pumps. **Veins** are blood vessels that carry blood back to the heart. The walls of veins are thinner than those of arteries. However, veins are generally of greater diameter than are arteries.

Most arteries carry oxygen-rich blood away from the heart, and most veins carry oxygen-poor blood back to the heart. However, the pulmonary blood vessels are exceptions. Oxygen-poor blood travels through the two pulmonary arteries, one of which goes to each lung. The two pulmonary veins carry oxygen-rich blood from the lungs to the heart.

E

Veins and arteries branch off into very narrow blood vessels called capillaries. **Capillaries** connect arteries with veins. Through capillaries materials are exchanged between blood and tissues. Oxygen and materials from nutrients move from the blood cells of arteries to the body's tissues through tiny openings in the capillary walls. Waste materials and carbon dioxide move from the tissues' cells through the capillary walls and into the blood in the veins.

F

CHECK YOUR READING Compare and contrast arteries, veins, and capillaries.

Blood exerts pressure on blood vessels.

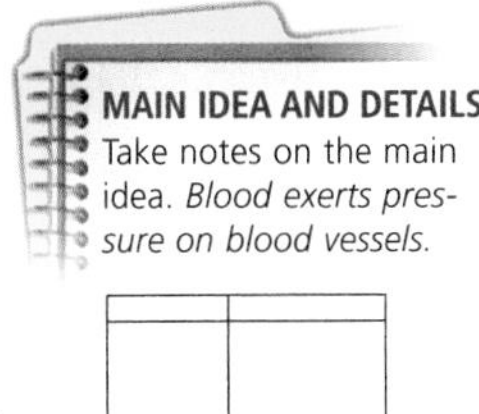

MAIN IDEA AND DETAILS Take notes on the main idea. *Blood exerts pressure on blood vessels.*

As you have read, the contractions of the heart push blood through blood vessels. The force produced when the heart contracts travels through the blood and exerts pressure on the blood vessels. This force is called blood pressure. To get an idea of how force affects pressure,

Teacher Demo

Ask each student to hold one hand above their head and one hand down by their side for about one minute. Have them compare the two hands. The fingertips of the hand held down will be redder, indicating that blood flowed down. The blood flowed out of the hand held above their head.

Teach Difficult Concepts

To help students understand the difference between arteries and veins, ask:

- How do you know that the blood in most arteries is oxygen-rich? *All arteries carry blood away from the heart. Most of the blood traveling away from the heart is carrying oxygen to the rest of the body.*
- How do you know that the blood in most veins is oxygen-poor? *All veins carry blood toward the heart. Most of the blood traveling toward the heart is coming from the rest of the body, where it has become oxygen-poor.*

Ongoing Assessment

CHECK YOUR READING *Answer: Arteries have strong walls and take blood away from the heart. Veins are blood vessels that carry blood back to the heart. They have thinner walls than arteries. Capillaries connect arteries and veins.*

DIFFERENTIATE INSTRUCTION

More Reading Support

E Which vessels carry blood from the heart to the lungs? *pulmonary arteries*

F In which vessels does exchange take place? *capillaries*

Below Level Have students make flow charts to show the flow of blood through the body. They should include the words *arteries, veins,* and *capillaries* in their diagrams. Once you have reviewed the diagrams for accuracy, have students add the flow of blood to and from the lungs through the pulmonary arteries and veins.

Additional Investigation To reinforce Section 1 learning goals, use the following full-period investigation:

R Additional Investigation, Listen to Your Heart, A, B, & C, pp. 177–185, pp. 305–306

EXPLORE (the BIG idea)

Revisit "Blood Pressure" on p. 63. Ask: How does this activity illustrate blood pressure? *The force against the balloon is like the pressure of blood against the walls of arteries, veins, and capillaries.*

Teach from Visuals

To help students interpret the diagram of blood pressure, ask: Would you expect blood pressure to be higher in arteries or in veins? Explain. *Arteries, because the blood is being pumped by the heart through the arteries. Arteries are closest to the impact of the pump.*

Teach Difficult Concepts

Some students may not understand the difference between pulse rate and blood pressure. Explain that pulse rate is a measure of the frequency of heartbeats, whereas blood pressure is a measure of the pressure that is exerted on the arteries and veins by the pumping action of the heart.

Health Connection

High blood pressure, or hypertension, is a serious condition that affects about one in four adults. It causes stroke and heart and kidney failure. In most cases, the cause of high blood pressure is unknown. However, there are many life choices to help reduce the risk:

- Have blood pressure checked regularly.
- Avoid excessive salt intake.
- Eat a healthy, low-fat diet.
- Exercise regularly.
- Reduce emotional stress; learn stress-reduction techniques.

Ongoing Assessment

Explain what blood pressure is and why is it important.

Ask: Why would a person with high blood pressure have certain health risks? *High blood pressure causes strain on the arteries and veins and can weaken them or cause the heart to work harder than it should.*

consider a sealed plastic bag filled with water. If you push down at the center of the bag, you can see the water push against the sides.

The heart pushes blood in a similar way, exerting pressure on the arteries, veins, and capillaries in the circulatory system. It is important to maintain healthy blood pressure so that materials in blood get to all parts of your body. If blood pressure is too low, some of the cells will not get oxygen and other materials. On the other hand, if blood pressure is too high, the force may weaken the blood vessels and require the heart to work harder to push blood through the blood vessels. High blood pressure is a serious medical condition, but it can be treated.

? G

The circulatory system can be considered as two smaller systems: one, the pulmonary system, moves blood to the lungs; the other, the systemic system, moves blood to the rest of the body. Blood pressure is measured in the systemic part of the circulatory system.

You can think of blood pressure as the pressure that blood exerts on the walls of your arteries when the heart contracts. Health professionals measure blood pressure indirectly with a device called a sphygmomanometer (SFIHG-moh-muh-NAHM-ih-tuhr).

Blood pressure is expressed with two numbers—one number over another number. The first number refers to the pressure in the arteries when the heart contracts. The second number refers to the pressure in the arteries when the heart relaxes and receives blood from the veins.

Blood Pressure

Blood pressure allows materials to travel to all parts of your body.

DIFFERENTIATE INSTRUCTION

More Reading Support

G What might happen if a person's blood pressure is too low? *Some cells will not get oxygen or other materials.*

English Learners English learners may be familiar with the *If . . . then* sentence construction. But oftentimes *then* is omitted from a sentence, and cause and effect are to be inferred. Help students recognize *If . . . then* sentences that do not contain *then.* Here are two examples from page 70:

"If you push down at the center of the bag, you can see the water push against the sides."

"If blood pressure is too low, some of the cells will not get oxygen and other materials."

Blood comes in four different types

Each red blood cell has special structures on its surface called antigens. There are two types of antigens, A and B. If your blood cells have A antigens, you have type A blood. If you have B antigens on your blood cells, you have type B blood. Some people have both antigens on their blood cells; they are said to have type AB blood. Some people have neither antigen and have type O blood. In the United States, O is the most common blood type, and AB is the least common.

Antibodies are present in the blood's liquid plasma. The antibodies will react with antigens of another blood type. Someone with type A blood has antibodies that react with type B blood; a person with type B blood has anti–type-A antibodies. Type O people have both types of antibodies; type AB people have none. Thus, people with type O blood are known as "universal donors," while those with AB blood are "universal recipients."

Knowing blood type is important, because one person can donate blood to another. However the blood types between donor and receiver must be compatible. If someone receives a blood of a type that triggers the antibodies in the plasma, his or her blood may clot. Such clotting can be fatal. The diagram shows which blood types are compatible. A person with blood type A can receive blood only from a person who has blood type O. Blood type O is said to be compatible with type A.

CHECK YOUR READING Why is it important to know your blood type?

Blood Type Compatibility

Blood Type	Can Donate Blood To	Can Receive Blood From
A	A, AB	A, O
B	B, AB	B, O
AB	AB	A, B, AB, O
O	A, B, AB, O	O

People can donate blood to others.

RESOURCE CENTER CLASSZONE.COM

Learn more about blood types.

3.1 Review

KEY CONCEPTS

1. What are the functions of the two sides of the heart?
2. What is the primary function of red blood cells?
3. Why can both high and low blood pressure be a problem?

CRITICAL THINKING

4. **Apply** List three examples of the circulatory system working with another system in your body.
5. **Compare and Contrast** Explain why blood pressure is expressed with two numbers.

CHALLENGE

6. **Identify Cause and Effect** You can feel the speed at which your heart is pumping by pressing two fingers to the inside of your wrist. This is your pulse. If you run for a few minutes, your pulse rate is faster for a little while, then it slows down again. Why did your pulse rate speed up and slow down?

ANSWERS

1. right side: pumps oxygen-poor blood to the lungs; left side: pumps oxygen-rich blood to the body

2. to carry oxygen to other cells in the body

3. too high: blood vessels can be weakened; too low: some cells won't get oxygen

4. Sample answer: Circulatory system works with muscular system (exercise), with respiratory system (breathing), and with digestive system (eating).

5. first number: pressure in the arteries when the heart contracts; second number: pressure in the arteries when the heart relaxes

6. The pulse rate sped up and slowed down because the body uses more energy when running. To produce extra energy, the body needs more oxygen, so the heart pumps faster to push more blood. Once the body stops running, the extra energy is not needed, and the pumping slows down.

Ongoing Assessment

CHECK YOUR READING *Answer: A person receiving blood may die if the blood is of a type not compatible with his or her own.*

Reinforce the BIG idea

Have students relate the section to the Big Idea.

 Reinforcing Key Concepts, p. 138

3.1 ASSESS & RETEACH

Assess

 Section 3.1 Quiz, p. 41

Reteach

Put a 4-column chart on the board for the four parts of blood; red blood cells, white blood cells, plasma, and platelets. Have students brainstorm a list of words associated with each, such as *oxygen* for red blood cells, *fluid* for plasma, and *scab* for platelets.

Then ask students to write comic strips about a day in the life of a blood cell. Students' comic strips should show the journey of the blood cell from the heart to the lungs and other parts of the body. They should name the type of cell and the fluid it travels in. Encourage them to have other characters in the comics. They should refer to diagrams in Section 3.1 and may trace them if they find doing so helpful.

Technology Resources

Have students visit **ClassZone.com** for reteaching of Key Concepts.

 CONTENT REVIEW

 CONTENT REVIEW CD-ROM

CHAPTER INVESTIGATION

Focus

PURPOSE To determine how heart rate is affected by exercise

OVERVIEW In this investigation, students will

- calculate resting, maximum, and target heart rates
- create and analyze graphs of heart rates during and after exercise
- determine a cause of exercise's effect on the heart

Lab Preparation

- Ask volunteers to demonstrate correct use of the stopwatch.
- Inquire about health exemptions for the exercise part of the activity.
- Prior to the investigation have students read through the investigation and prepare their data tables. Or copy and distribute datasheets and rubrics.

UNIT RESOURCE BOOK, pp. 168–176

SCIENCE TOOLKIT, F15

Lab Management

- Have students work in pairs. One partner records data, the other exercises.
- If students are able to find their pulse quickly, they can count beats for 15 seconds and multiply by four.

SAFETY Caution students not to exercise beyond their target rate. If students have difficulty breathing, chest discomfort, or feelings of dizziness, they should stop exercising.

INCLUSION Prepare blank graphs for students in advance. The graphs should be titled "Heart Rate During and After Exercise;" the x-axis should be labeled "Time" and the y-axis labeled "Heart rate (beats per minute)."

INCLUSION Students with mobility limitations can time and/or record data. They may perform modified or assisted exercise.

CHAPTER INVESTIGATION

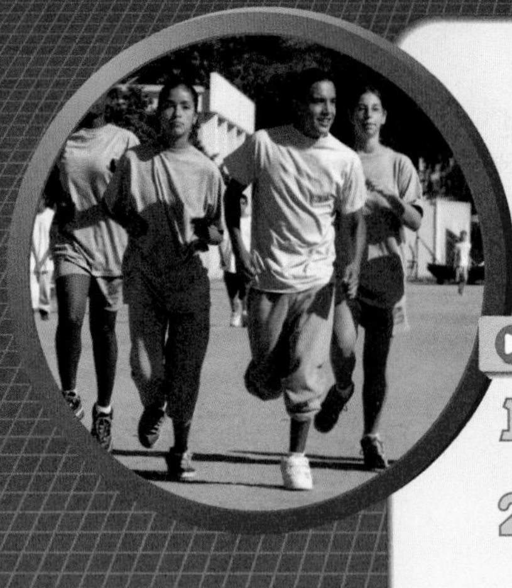

Heart Rate and Exercise

OVERVIEW AND PURPOSE In this activity, you will calculate your resting, maximum, and target heart rates. Then you will examine the effect of exercise on heart rate. Before you begin, read through the entire investigation.

MATERIALS
- notebook
- stopwatch
- calculator
- graph paper

Procedure

1. Make a data table like the one shown on the sample notebook page.

2. Measure your resting heart rate. Find the pulse in the artery of your neck, just below and in front of the bottom of your ear, with the first two fingers of one hand. Do not use your thumb to measure pulse since the thumb has a pulse of its own. Once you have found the pulse, count the beats for 30 seconds and multiply the result by 2. The number you get is your resting heart rate in beats per minute. Record this number in your notebook.

3. Calculate your maximum heart rate by subtracting your age from 220. Record this number in your notebook. Your target heart rate should be 60 to 75 percent of your maximum heart rate. Calculate and record this range in your notebook.

4. Someone who is very athletic or has been exercising regularly for 6 months or more can safely exercise up to 85 percent of his or her maximum heart rate. Calculate and record this rate in your notebook.

5. Observe how quickly you reach your target heart rate during exercise. Begin by running in place at an intensity that makes you breathe harder but does not make you breathless. As with any exercise, remember that if you experience difficulty breathing, dizziness, or chest discomfort, stop exercising immediately.

E 72 Unit: Human Biology

INVESTIGATION RESOURCES

CHAPTER INVESTIGATION, Heart Rate and Exercise
- Level A, pp. 168–171
- Level B, pp. 172–175
- Level C, p. 176

Advanced students should complete Levels B & C.

Writing a Lab Report, D12–13

Technology Resources

Customize this student lab as needed or look for an alternative. Print rubrics to assess student lab reports.

Lab Generator CD-ROM

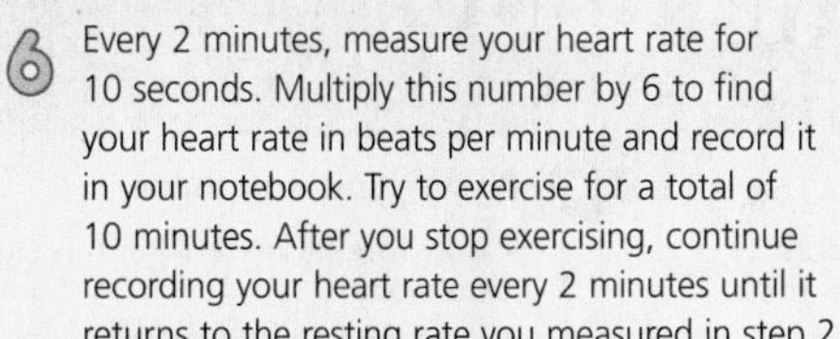

6 Every 2 minutes, measure your heart rate for 10 seconds. Multiply this number by 6 to find your heart rate in beats per minute and record it in your notebook. Try to exercise for a total of 10 minutes. After you stop exercising, continue recording your heart rate every 2 minutes until it returns to the resting rate you measured in step 2.

Observe and Analyze

Write It Up

1. **GRAPH DATA** Make a line graph of your heart rate during and after the exercise. Graph the values in beats per minute versus time in minutes. Your graph should start at your resting heart rate and continue until your heart rate has returned to its resting rate. Using a colored pencil, shade in the area that represents your target heart-rate range.
2. **ANALYZE DATA** How many minutes of exercising were needed for you to reach your target heart rate of 60 to 75 percent of maximum? Did your heart rate go over your target range?
3. **INTERPRET DATA** How many minutes after you stopped exercising did it take for your heart rate to return to its resting rate? Why do you think your heart rate did not return to its resting rate immediately after you stopped exercising?

Conclude

Write It Up

1. **INFER** Why do you think that heart rate increases during exercise?
2. **IDENTIFY** What other body systems are affected when the heart rate increases?
3. **PREDICT** Why do you think that target heart rate changes with age?
4. **CLASSIFY** Create a table comparing the intensity of different types of exercise, such as walking, skating, bicycling, weight lifting, and any others you might enjoy.

INVESTIGATE Further

CHALLENGE Determine how other exercises affect your heart rate. Repeat this investigation by performing one or two of the other exercises from your table. Present your data, with a graph, to the class.

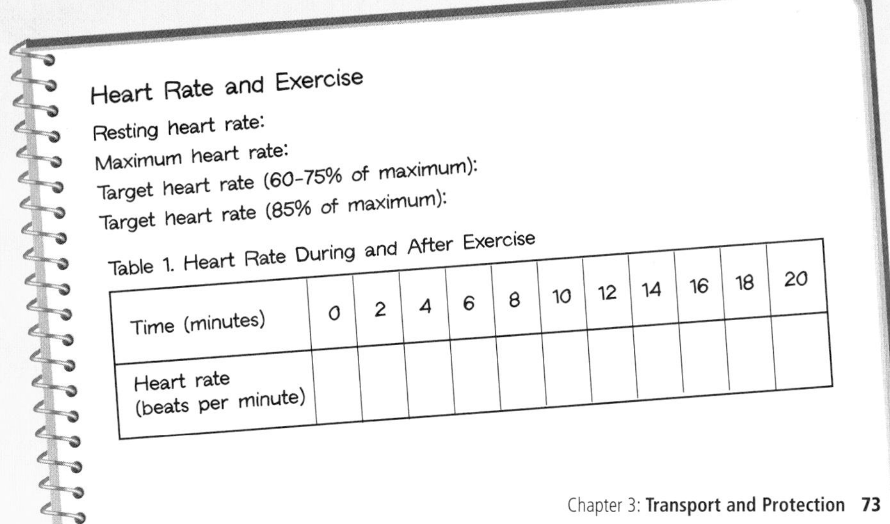

Heart Rate and Exercise

Resting heart rate:

Maximum heart rate:

Target heart rate (60–75% of maximum):

Target heart rate (85% of maximum):

Table 1. Heart Rate During and After Exercise

Time (minutes)	0	2	4	6	8	10	12	14	16	18	20
Heart rate (beats per minute)											

Observe and Analyze

Write It Up

1. *Graphs will vary but should show an increase in heart rate during exercise and a return to normal when exercise stops. Target heart-rate range should be similar for all students.*
2. *Answers will vary. Heart rate should reach the target range within 5 minutes. Students should try not to exercise above their target heart rate.*
3. *Answers will vary. Heart rate should return to resting rate within 10 minutes after exercise stops. Heart rate remains elevated for a few minutes after exercise stops because it takes time for the body to restore homeostasis.*

Conclude

Write It Up

1. *Heart rate increases during exercise to pump more oxygen-rich blood to the body cells. The body cells use oxygen at a higher rate during exercise to produce energy.*
2. *The respiratory system is also affected when the heart rate increases, because the breathing rate also increases. The nervous system increases the secretions of sweat glands to get rid of excess heat.*
3. *The heart's ability to pump blood and deliver oxygen decreases as it ages.*
4. *Tables will vary. Exercises such as bicycling are more intense than exercises such as walking or weightlifting.*

INVESTIGATE Further

CHALLENGE Graphs will vary. Exercises of higher intensity will cause the heart to reach its target rate faster than exercises of lower intensity. After higher-intensity exercises, the heart will also take longer to return to its resting rate than it will after lower intensity exercises.

Post-Lab Discussion

- Discuss the connection between heart rate and blood pressure. When heart rate increases, so does blood pressure. Well-conditioned athletes have low resting pulses, and therefore low blood pressure.
- Discuss the health risks of high blood pressure. Point out that proper diet, avoidance of alcohol and tobacco, and exercise can help keep blood pressure at a normal level.

Teaching with Technology

If probeware for monitoring heart rate is available, have students use it to record their heart rates before and after mild exercise. Make sure students do not have any health or heart conditions before allowing them to participate in the activity. They can use a graphing calculator to show the data visually.

3.2 FOCUS

Set Learning Goals

Students will

- Describe how foreign material enters the body.
- Explain how the immune system responds to foreign material.
- Describe ways that the body can become immune to disease.
- Make models to show how antibodies stop pathogens from spreading.

3-Minute Warm-Up

Display Transparency 20 or copy this exercise on the board:

Match the definitions to the correct terms.

Definitions

1. blood cells that carry oxygen *d*
2. blood vessels that carry blood to the heart *e*
3. system for obtaining oxygen *f*

Terms

a. white blood cells
b. arteries
c. systemic
d. red blood cells
e. veins
f. respiratory

 3-Minute Warm-Up, p. T20

3.2 MOTIVATE

EXPLORE Membranes

PURPOSE To understand how the body keeps out foreign particles

TIP *15 min.* Have students start by predicting what will happen. Warn students to use caution handling pins. Provide a pin cushion if possible.

WHAT DO YOU THINK? *When the bag has even a small puncture, it cannot protect the cloth, and the cloth is stained.*

Ongoing Assessment

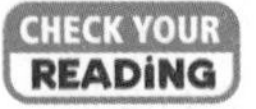

Answer: The integumentary, respiratory, and digestive systems are the first line of defense against pathogens.

KEY CONCEPT

3.2 The immune system defends the body.

BEFORE, you learned

- The circulatory system works with other systems to fuel the body cells
- Structures in the circulatory system work together
- Blood pressure allows materials to reach all parts of the body

NOW, you will learn

- How foreign material enters the body
- How the immune system responds to foreign material
- Ways that the body can become immune to a disease

VOCABULARY

pathogen p. 74
immune system p. 75
antibody p. 75
antigen p. 78
immunity p. 80
vaccine p. 80
antibiotic p. 81

EXPLORE Membranes

How does the body keep out foreign particles?

PROCEDURE

1. Place a white cloth into a sandwich bag and seal it. Fill a bowl with water and stir in several drops of food coloring.
2. Submerge the sandwich bag in the water. After five minutes, remove the bag and note the condition of the cloth.
3. Puncture the bag with a pin. Put the bag back in the water for five minutes. Remove the bag and note the condition of the cloth.

WHAT DO YOU THINK?

- How does a puncture in the bag affect its ability to protect the cloth?

MATERIALS

- white cloth
- zippered sandwich bag
- large bowl
- water
- food coloring
- small pin

Many systems defend the body from harmful materials.

MAIN IDEA AND DETAILS Add the main idea *Many systems defend the body from harmful materials* to your chart along with detail notes.

You might not realize it, but you come into contact with harmful substances constantly. Your body has ways to defend itself, and that is why you don't even notice. One of the body's best defenses is to keep foreign materials from entering in the first place. The integumentary (ihn-TEHG-yu-MEHN-tuh-ree), respiratory, and digestive systems are the first line of defense against **pathogens,** or disease-causing agents. Pathogens can enter through your skin, the air you breathe, and even the food you eat or liquids you drink.

Which systems are your first line of defense against pathogens?

RESOURCES FOR DIFFERENTIATED INSTRUCTION

Below Level

UNIT RESOURCE BOOK

- Reading Study Guide A, pp. 141–142
- Decoding Support, p. 165

Advanced

UNIT RESOURCE BOOK

Challenge and Extension, p. 147

English Learners

UNIT RESOURCE BOOK

Spanish Reading Study Guide, pp. 145–146

- Audio Readings in Spanish
- Audio Readings (English)

Integumentary System Defenses Most of the time, your skin functions as a barrier between you and the outside world. The physical barrier the skin forms is just one obstacle for pathogens. The growth of pathogens on your eyes can be slowed by other substances contained in tears. The millions of bacteria cells that live on the skin can also kill pathogens. The only way most pathogens can enter the body through the skin is through a cut. The circulatory system is then able to help defend the body because blood contains cells that destroy pathogens.

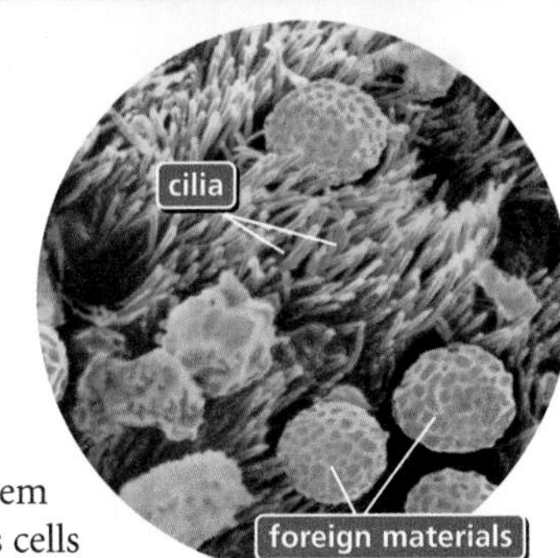

Cilia are hairlike protrusions that trap materials entering your respiratory system (600×).

Respiratory System Defenses Sneezing and coughing are two ways the respiratory system defends the body from harmful substances.  A Cilia and mucus also protect the body. Cilia are tiny, hairlike protrusions in the nose and the lungs that trap dust particles present in the air. Mucus is a thick and slippery substance found in the nose, throat, and lungs. Like the cilia, mucus traps dirt and other particles. Mucus contains substances similar to those in tears that can slow the growth of pathogens.

Digestive System Defenses Some foreign materials manage to enter your digestive system, but many are destroyed by saliva, mucus, enzymes, and stomach acids. Saliva in your mouth helps kill bacteria. Mucus protects the digestive organs by coating them. Pathogens can also be destroyed by enzymes produced in the liver and pancreas or by the acids in the stomach.

The immune system has response structures.

Sometimes foreign materials do manage to get past the first line of defense. When this happens, the body relies on the **immune system** to respond. This system functions in several ways:

- Tissues in the bone marrow, the thymus gland, the spleen, and the lymph nodes produce white blood cells, which are specialized cells that function to destroy foreign organisms.
- Some white blood cells produce a nonspecific response to injury or infection that is the second line of defense.
- B Some white blood cells produce proteins called **antibodies,** which are part of a specific immune response to foreign materials.

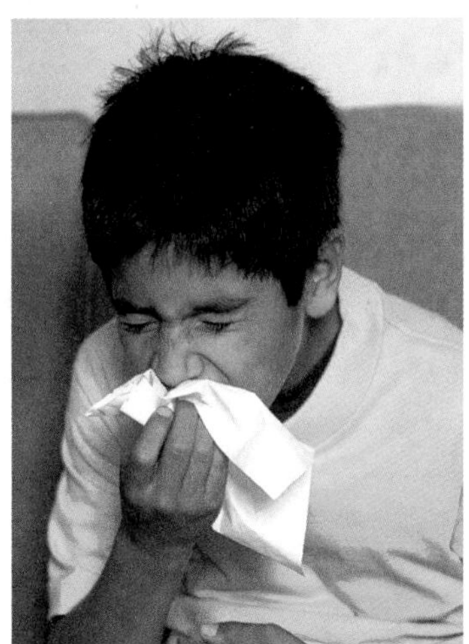
Sneezing helps to expel harmful substances from the body.

3.2 INSTRUCT

Address Misconceptions

IDENTIFY Ask students to describe what causes illness. Some students might hold the misconception that all illnesses are caused by germs and that germs enter through the mouth during eating.

CORRECT Have students make a graphic organizer that shows the body's first line of defense. Students should note the various ways that foreign materials can enter the body.

REASSESS Ask students to extend the diagram to show possible effects of materials such as irritants, bacteria, viruses, and parasites entering the body through each system.

Technology Resources

Visit **ClassZone.com** for background on common student misconceptions.

 **MISCONCEPTION DATABASE**

Teacher Demo

Demonstrate the presence of bacteria on a person's skin. Have a volunteer run his or her fingers across agar in a petri dish. Cover and seal the dish, and place it in a warm place for several days. Students should note that bacteria from the student's hand has grown on the agar. Ask: How does the skin protect the body from these bacteria? *The skin is a protective barrier that prevents bacteria from entering the body.*

Note: This demo is presented as a full-period investigation in Chapter 5.

Ongoing Assessment

Describe how foreign material enters the body.

Ask: How can foreign materials enter through the skin? *If there is a cut in the skin, pathogens may be able to enter the body through it.*

DIFFERENTIATE INSTRUCTION

More Reading Support

A What is the function of cilia? *Cilia trap foreign particles.*

B What proteins are produced by white blood cells? *antibodies*

English Learners English learners may be sensitive to the use of the term *foreign* with regard to human social situations. Point out that *foreign material* is a scientific and medical term and that *foreign* does not correctly describe people who travel or relocate from one country to another. In such a case *immigrant, visitor, guest,* or *newly arrived resident* is preferable and kinder.

Metacognitive Strategy

To help students understand the difference between the circulatory system and the lymphatic system, have them create a Venn diagram comparing and contrasting the two systems. Make sure that students understand that the lymphatic system, like the circulatory system, also contains a network of vessels but that it has no pump.

Teach Difficult Concepts

Some students may not understand the connection between the lymphatic and circulatory systems. The lymphatic system collects fluids that escape from the blood and returns them to the circulatory system. Relate these systems to other real-world systems that have main and secondary or supportive relationships, such as highway systems and computer networks.

Ongoing Assessment

CHECK YOUR READING *Answer: The lymphatic system transports white blood cells throughout the body, so that when they are needed in a part of the body, they are ready to attack.*

White Blood Cells

The immune system has specialized cells called white blood cells. There are five major types of white blood cells. The number of white blood cells in the blood can increase during an immune response. These cells travel through the circulatory system and the lymphatic system to an injured or infected area of the body. White blood cells leave the blood vessels and travel into the damaged tissue where the immune response takes place.

The Lymphatic System

Learn more about the lymphatic system.

C

The lymphatic system transports pathogen-fighting white blood cells throughout the body, much as the circulatory system does. The two systems are very similar. The lymphatic system carries lymph, and the circulatory system carries blood. Both fluids transport similar materials, such as white blood cells.

Lymph is the fluid left behind by the circulatory system. It moves through lymph vessels, which are similar to blood vessels. However, the lymphatic system has no pump like the heart to move fluid. Lymph drifts through the lymph vessels when your skeletal muscles contract or by gravity when your body changes position. As it moves, it passes through lymph nodes, which filter out pathogens and store white blood cells and antibodies. Because they filter out pathogens, infections are often fought in your lymph nodes, causing them to swell when you get sick.

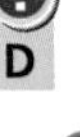
D

CHECK YOUR READING How does the lymphatic system help the immune system?

The immune system responds to attack.

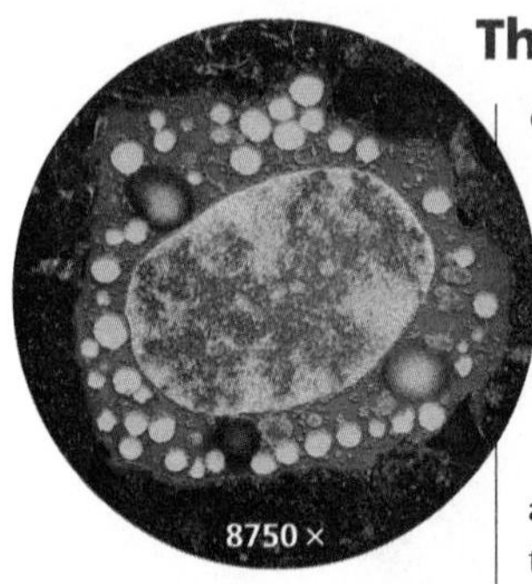

The mast cell above is an important part of the immune system.

Certain illnesses can cause you to cough, sneeze, or have a fever. These signs of illness, or symptoms, make you uncomfortable when you are sick. But in fact, most symptoms are the result of the immune system's response to disease.

The immune system responds in two ways. The white blood cells that first respond to the site of injury or infection attack pathogens in a nonspecific response. These cells attack pathogens by engulfing them and also produce chemicals that help other white blood cells work better. The second part of the response is very specific to the types of pathogens invading the body. These white blood cells produce antibodies specific to each pathogen and provide your body with immunity.

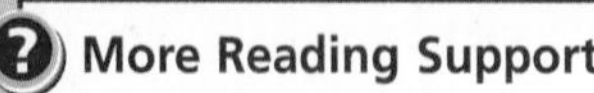

DIFFERENTIATE INSTRUCTION

More Reading Support

C What fluid does the lymphatic system transport? *lymph*

D What causes lymph nodes to swell? *the fight against infection carried out in them*

English Learners Have students write the definitions of *pathogen, antibody, antigen, immunity,* and *vaccine* in their Science Word Dictionaries. In the section review on p. 81, students are asked to make two different charts in response to questions. Help English learners by showing them a model of a type of chart that might work for each question.

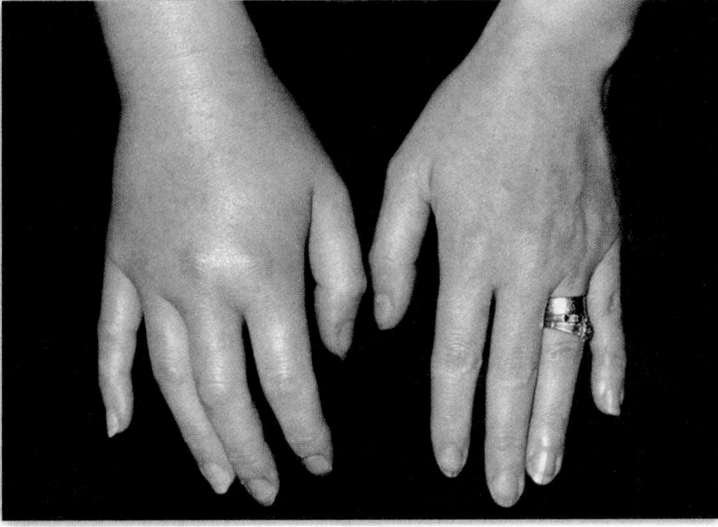

Wasp stings cause an immediate immune response. The area of the sting swells up and increases in temperature while your body battles the injury.

Nonspecific Response

Swelling, redness, and heat are some of the symptoms that tell you that a cut or scrape has become infected by foreign materials. They are all signs of inflammation, your body's first defense reaction against injuries and infections.

When tissue becomes irritated or damaged, it releases large amounts of histamine (HIHS-tuh-meen). Histamine raises the temperature of the tissues and increases blood flow to the area. Increased blood flow, which makes the injured area appear red, allows antibodies and white blood cells to arrive more quickly for battle. Higher temperatures improve the speed and power of white blood cells. Some pathogens cannot tolerate heat, so they grow weaker. The swelling caused by the production of histamine can be a small price to pay for this chemical's important work.

Histamine may also be produced when an illness affects more than one area of your body. In these cases, many tissues produce histamine. As a result, the temperature of your whole body rises. Any temperature above 37 degrees Celsius (98.6°F) is considered a fever, but only temperatures hot enough to damage tissues are dangerous. Trying to lower a high fever with medication is advisable in order to avoid tissue damage. When you have a small fever, lowering your body temperature might make you more comfortable, but it will not affect how long you stay sick.

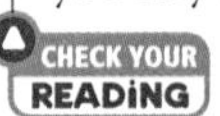

CHECK YOUR READING What causes a fever when you are sick?

Teach from Visuals

To help students interpret the photograph showing an immune response, ask: How can you tell that an inflammation has occurred in this person's hand? *Swelling and redness (and an increase in temperature) are signs that an inflammatory response has occurred in the body.*

Real World Example

Influenza is probably one of the last remaining pandemic infections in humans. It is an infectious disease of the respiratory tract caused by any of three types of influenza virus (A, B, and C), with many different strains. Influenza pandemics are periodically caused by new strains of the type-A virus. Most notably, the pandemic of 1918 to 1919 is estimated to have killed about 25 million people worldwide. More recent influenza pandemics include the Asian flu of 1957 and the Hong Kong flu of 1968.

Ongoing Assessment

CHECK YOUR READING *Answer: Many tissues produce histamine, which raises the body's temperature to help fight illness.*

DIFFERENTIATE INSTRUCTION

More Reading Support

E What is the body's first defense reaction against infection? *inflammation*

F How does histamine cause redness? *by increasing blood flow*

Below Level Have students make trading cards with questions and answers about how the immune system protects the body. Allow students time to trade their cards and answer the questions.

Teach Difficult Concepts

Make sure students understand the difference between pathogens and antibodies. Explain that pathogens are foreign materials that can enter the body and cause disease. Antibodies are produced by the body to fight against pathogens. Ask: How does the body produce antibodies? *They are produced by white blood cells.*

Teach from Visuals

To help students interpret the diagram of immune response, ask:

- What are two types of white blood cells? *T cells and B cells*
- What is the role of antibodies in the immune response? *Antibodies attach themselves to antigens on the pathogens.*
- What are antigens? *Antigens are markers that are carried by pathogens.*

Ongoing Assessment

Explain how the immune system responds to foreign material.

Ask: What is the difference between the nonspecific and specific immune responses? *During a nonspecific response to any injury or infection, histamine is released by tissues to cause inflammation. During a specific immune response, antibodies attach to antigens causing pathogens to be contained and eliminated.*

Specific Response

If a pathogen is not destroyed by the nonspecific response of the immune system, then a specific immune response occurs. T cells and B cells are the two major types of white blood cells that produce the immune response. They have different roles.

T Cells T cells identify and distinguish between pathogens because they recognize a marker called an **antigen,** which is carried by each pathogen. Each pathogen has a different type of antigen that only one type of T cell can recognize. When a T cell finds a pathogen with an antigen it recognizes, it begins to reproduce rapidly, making a large army of T cells that are also capable of identifying the antigen. Some T cells then attack cells that have been infected by the pathogen.

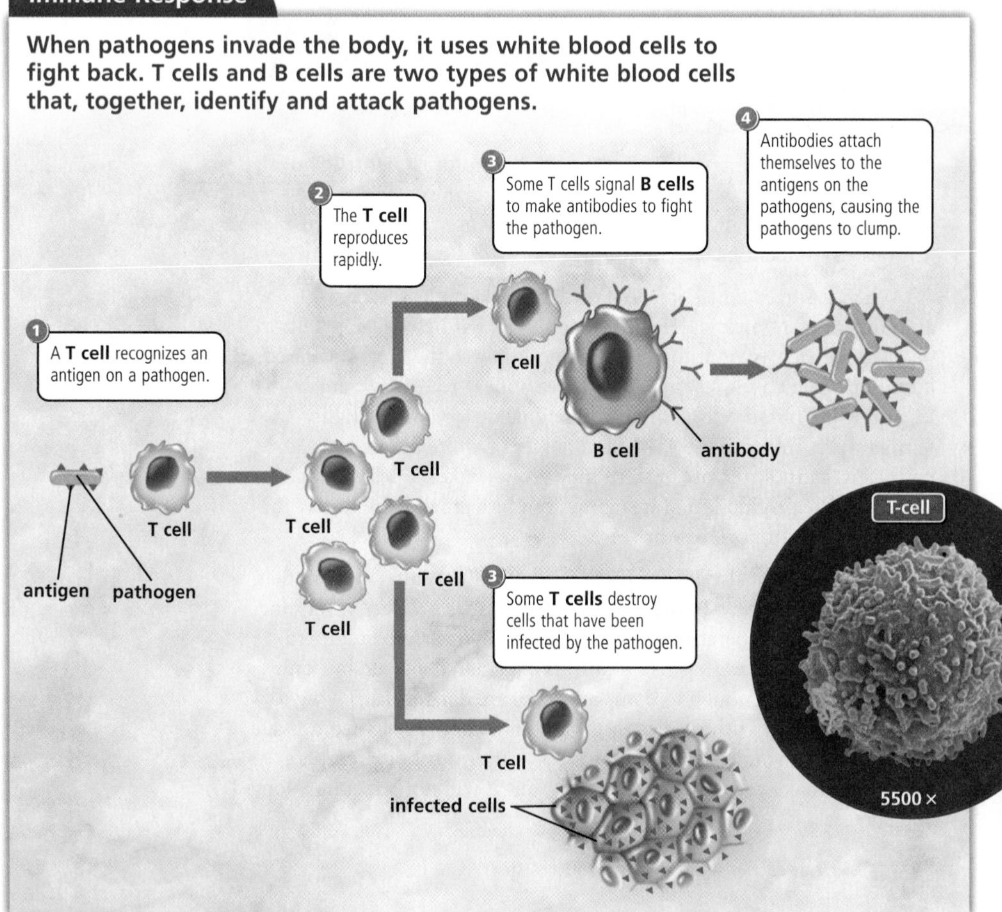

Immune Response
When pathogens invade the body, it uses white blood cells to fight back. T cells and B cells are two types of white blood cells that, together, identify and attack pathogens.
1 A T cell recognizes an antigen on a pathogen.
antigen
pathogen
T cell
2 The T cell reproduces rapidly.
T cell
T cell
T cell
T cell
3 Some T cells signal B cells to make antibodies to fight the pathogen.
T cell
B cell
antibody
4 Antibodies attach themselves to the antigens on the pathogens, causing the pathogens to clump.
3 Some T cells destroy cells that have been infected by the pathogen.
T cell
infected cells
T-cell
5500 ×

DIFFERENTIATE INSTRUCTION

Below Level Have students draw a chart with two columns headed "B Cells" and "T Cells." Students can write the functions of each, along with the cells' reactions during immune response. Encourage students to draw pictures to illustrate their notes.

INVESTIGATE Antibodies

How do antibodies stop pathogens from spreading?

SKILL FOCUS
Making models

MATERIALS
- plastic containers with lids
- index cards

TIME
15 minutes

PROCEDURE

1. Your teacher will hand out plastic lids, each labeled with the name of a different pathogen. You will see plastic containers spread throughout the room. There is one container in the room with the same label as your lid.
2. At the signal, find the plastic container with the pathogen that has the same label as your lid and wait in place for the teacher to tell you to stop. If you still haven't found the matching container when time is called, your model pathogen has spread.
3. If your pathogen has spread, write its name on the board.

WHAT DO YOU THINK?

Which pathogens spread?

- What do you think the lid and container represent? Why?
- How do antibodies identify pathogens?

CHALLENGE Why do you think it is important for your body to identify pathogens?

B Cells After the T cells recognize an antigen, B cells are signaled to make antibodies to destroy the pathogen. Once a B cell finds the antigen, it releases antibodies, which attach themselves to the antigen. This causes pathogens to clump, making them unable to cause damage. In some cases, antigens and antibodies mark pathogens for attack by T cells.

READING TIP
Antigen and *antibody* are words that look very similar. Antigens are markers on pathogens. Antibodies fight pathogens.

Some of the antibodies remain in the body as a form of immune system memory. If the same type of foreign material appears, the system can respond much more quickly, because B cells do not have to produce a new antibody. Structures in the lymphatic system called lymph nodes store masses of white blood cells and antibodies until a large number are needed at one location.

G

CHECK YOUR READING Why is it important for the body to store antibodies?

DIFFERENTIATE INSTRUCTION

More Reading Support

G Which structures in the lymphatic system store masses of white blood cells? *lymph nodes*

Advanced Challenge students to create a board game called Invaders from the Outside World. Students should set up the game to model the immune response illustrated on p. 78. The goal of the game is to "collect" enough T cells and B cells to destroy the invading pathogens from the outside world. Allow students class time to play their game.

R Challenge and Extension, p. 147

INVESTIGATE Antibodies

PURPOSE To create a model to show how antibodies keep pathogens from spreading

TIP *15 min.* Prior to the activity, prepare the plastic lids and containers with the names of various pathogens such as *E. coli, streptococcus,* influenza, cold virus (*rhinovirus*), or chicken pox (*varicella*).

WHAT DO YOU THINK? *Students' answers should include any pathogen for which they did not find the container that matched their lid. The lid represents an antibody. The container represents the corresponding pathogen. Antibodies identify pathogens by markers called antigens that are carried by the pathogens.*

CHALLENGE *When the body is able to identify a pathogen, B cells are signaled to make antibodies that destroy the pathogen and prevent it from spreading.*

Datasheet, Antibodies, p. 148

Technology Resources

Customize this student lab as needed or look for an alternative. Print rubrics to assess student lab reports.

Lab Generator CD-ROM

Develop Critical Thinking

ANALYZE Explain to students that HIV is a virus that destroys the body's T cells. Ask: How might HIV eventually affect the ability of the body to defend itself against pathogens? *Since the production of T cells is reduced, the body will lose its ability to fight disease.*

Remind students that HIV spreads only through exchange of internal fluids, such as blood.

Ongoing Assessment

Answer: so that antibodies will be available when they are needed to destroy pathogens

History of Science

The first vaccination is attributed to the British physician Edward Jenner. In 1796 Jenner used the cowpox virus to create a vaccine against smallpox, a disease caused by a similar virus. Jenner had observed that dairymaids who had contracted cowpox, a relatively mild disease, did not contract smallpox when exposed to it. In 1881 Louis Pasteur contributed to the further development of vaccines, immunizing sheep against anthrax by injecting them with forms of the bacteria that cause the disease. Since Pasteur's time, vaccinations have been developed for diphtheria (1891), polio (1955), and measles (1963).

Develop Critical Thinking

APPLY Have students explain the difference between active immunity and passive immunity in their own words. Ask: Why are young schoolchildren more susceptible to flu and colds than older children? *Young children may not have had the chance to build up active immunity and to develop antibodies.*

Ongoing Assessment

Describe ways that the body can become immune to disease.

Ask: What is a vaccine? *A vaccine contains small amounts of weakened pathogens that enable the body to develop an active immunity against the pathogens.*

CHECK YOUR READING *Answer: In active immunity, a body makes its own antibodies. In passive immunity, a body does not develop immunity on its own, as in the case of a baby who inherits immunity from its mother.*

Development of Immunity

COMPARE A doctor gives a girl a vaccination. Is getting a vaccination an example of passive or active immunity?

After your body has won out against a specific pathogen, antibodies designed to fight that pathogen remain in your system. If the same pathogen were to attack again, your immune system would almost certainly destroy it before you became ill. This resistance to a sickness is called **immunity.** (H)

Immunity takes two forms: passive and active. When babies are first born, they have only the immune defenses given to them by their mothers. They have not had the chance to develop antibodies on their own. This type of immunity is called passive immunity. Antibodies are not produced by the person's own body but given to the body from another source. Babies must develop their own antibodies after a few months. (I)

You have active immunity whenever your body makes its own antibodies. Your body will again fight against any specific pathogen you have developed antibodies against. For example, it is most unlikely that you will get the same cold twice.

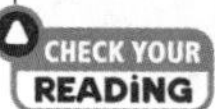

CHECK YOUR READING What is the difference between active and passive immunity?

Most diseases can be prevented or treated.

Given enough time, your body will fight off most diseases. However, some infections can cause significant and lasting damage before they are defeated by the body's defenses. Other infections are so strong that the immune system cannot fight them. Medical advances in the prevention and treatment of diseases have reduced the risks of serious illness.

Vaccination

Another way to develop an immunity is to receive a **vaccine.** Vaccines contain small amounts of weakened pathogens that stimulate your immune response. Your B cells are called into action to create antibodies as if you were fighting the real illness. The pathogens are usually weakened, so that you will not get sick, yet they still enable your body to develop an active immunity.

DIFFERENTIATE INSTRUCTION

More Reading Support

H What term refers to a resistance to an illness? *immunity*

I Is a baby more likely to have passive immunity or active immunity? *passive immunity*

Advanced Challenge students to find out about the kinds of vaccines they received as young children. Ask them to compare the kinds of vaccines they received with those that were administered to schoolchildren in the 1950s, 1960s, and 1970s. How have the vaccines changed? What might account for the changes?

Today we have vaccines for many common pathogens. Most children who are vaccinated will not get many diseases that their great grandparents, grandparents, and even parents had. Vaccinations can be administered by injection or by mouth. Babies are not the only ones who get them, either. You can be vaccinated at any age.

CHECK YOUR READING Why don't vaccinations usually make you sick?

Treatment

Not all diseases can be prevented, but many of them can be treated. In some cases, treatments can only reduce the symptoms of the disease while the immune system fights the disease-causing pathogens. Other treatments attack the pathogens directly.

In some cases, treatment can only prevent further damage to body tissues by a pathogen that cannot be cured or defeated by the immune system. The way in which a disease is treated depends on what pathogen causes it. Most bacterial infections can be treated with antibiotics. **Antibiotics** are medicines that block the growth and reproduction of bacteria. You may have taken antibiotics when you have had a disease such as strep throat or an ear infection.

Types of Pathogens

Disease	Pathogen
Colds, chicken pox, hepatitis, AIDS, influenza, mumps, measles, rabies	virus
Food poisoning, strep throat, tetanus, tuberculosis, acne, ulcers, Lyme disease	bacteria
Athlete's foot, thrush, ringworm	fungus
Malaria, parasitic pneumonia, pinworm, lice, scabies	parasites

3.2 Review

KEY CONCEPTS

1. Make a chart showing three ways that foreign material enters the body and how the immune system defends against each type of attack.
2. What are white blood cells and what is their function in the body?
3. What are two ways to develop immunity?

CRITICAL THINKING

4. **Compare and Contrast** Make a chart comparing B cells and T cells. Include an explanation of the function of antibodies.
5. **Apply** Describe how your immune system responds when you scrape your knee.

CHALLENGE

6. **Hypothesize** Explain why, even if a person recovers from a cold, that person could get a cold again.

ANSWERS

1. Charts will vary but should include that foreign materials can enter the body through a wound, during eating or drinking, or during breathing.

2. White blood cells are types of blood cell that destroy foreign materials.

3. through illness and through vaccination

4. Charts will vary but should include that B cells release antibodies, whereas T cells recognize markers called antigens.

5. Answers should show an understanding of the process of inflammation.

6. Students should speculate that different colds are caused by different foreign materials, since getting one cold does not give you immunity against another cold.

Real World Example

Although the measles vaccine has decreased the incidence of measles in the United States, measles remains a health threat in many developing countries. In 2002 the World Health Organization reported that there were 26 million cases of measles worldwide, with 745,000 deaths for the year 2001. Measles results in childhood death from complications like pneumonia, diarrhea, and malnutrition. Measles is also the major cause of preventable blindness in the world.

Ongoing Assessment

CHECK YOUR READING *Answer: You receive a very small amount of a weakened pathogen in a vaccine, which is not usually enough to make you sick.*

Reinforce the BIG idea

Have students relate the section to the Big Idea.

 Reinforcing Key Concepts, p. 149

3.2 ASSESS & RETEACH

Assess

 Section 3.2 Quiz, p. 42

Reteach

Write the following terms on the board: *white blood cells, T cells, B cells, antibodies, antigens, fever, inflammation, mucus, cilia, lymph nodes, vaccine.* For each term have students tell whether it names a part of the body's first line of defense or a function of the immune system. Then have students describe how each item either keeps pathogens from entering the body or fights to destroy or remove pathogens that have entered the body.

Technology Resources

Have students visit **ClassZone.com** for reteaching of Key Concepts.

 **CONTENT REVIEW**

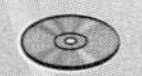 **CONTENT REVIEW CD-ROM**

MATH IN SCIENCE
Math Skills Practice for Science

Set Learning Goal

To use graphing to explore and represent pollen-count data

Present the Science

Allergies occur when allergens like pollen and dust enter the body. The allergens cause the production of histamine, which results in sneezing, runny nose, and watery eyes.

Develop Graphing Skills

Review the structure of a graph:

- The horizontal axis is the *x*-axis.
- The vertical axis is the *y*-axis.
- The *x*- and *y*-axes intersect at a point, 0.

Line graphs show a relationship between sets of data. Line graphs are particularly effective at showing data that change over time. Double line graphs present two independent sets of data that can be compared.

DIFFERENTIATION TIP A rate is a special kind of ratio in which the denominator is 1. The rate shown in the table is number of grains per 1 cubic meter.

Close

Ask students why labels are important on a graph. *Sample answer: to show exactly what data are measured and how they are being compared*

Suggest that students obtain local pollen counts over a number of days. They can graph the data and post the graph as a community service.

- Math Support, p. 166
- Math Practice, p. 167

Technology Resources

Students can visit **ClassZone.com** to practice graphing skills.

MATH TUTORIAL

SKILL: MAKING A LINE GRAPH

Pollen Counts

Every year, sometime between July and October, in nearly every state in the United States, the air will fill with ragweed pollen. For a person who has a pollen allergy, these months blur with tears. Linn County, Iowa, takes weekly counts of ragweed and non-ragweed pollen.

MATH TUTORIAL
CLASSZONE.COM
Click on Math Tutorial for more help making line graphs.

Weekly Pollen Counts, Linn County, Iowa

	Jul. 29	Aug. 5	Aug. 12	Aug. 19	Aug. 26	Sept. 2	Sept. 9	Sept. 16	Sept. 23	Sept. 30	Oct. 7
Ragweed (Grain/m^3)	0	9	10	250	130	240	140	25	20	75	0
Non-Ragweed (Grain/m^3)	10	45	15	50	100	50	40	10	20	25	0

Example

A line graph of the data will show the pattern of increase and decrease of ragweed pollen in the air.

(1) Begin with a quadrant with horizontal and vertical axes.

(2) Mark the weekly dates at even intervals on the horizontal axis.

(3) Starting at 0 on the vertical axis, mark even intervals of 50 units.

(4) Graph each point. Connect the points with line segments.

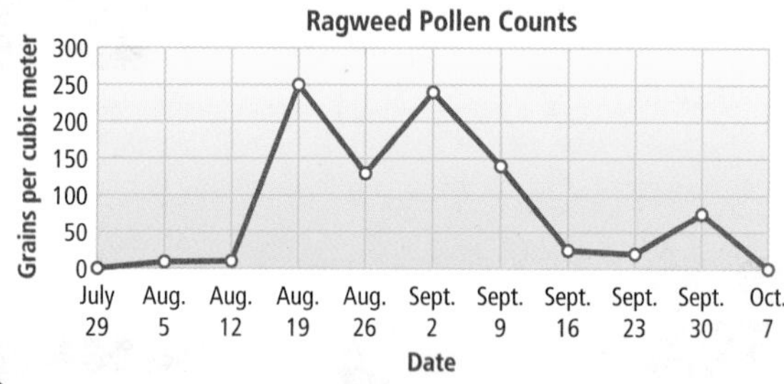

The pollen of *Ambrosia artemisiifolia* (common ragweed) sets off a sneeze.

Complete and present your graph as directed below.

1. Use graph paper to make your own line graph of the non-ragweed pollen in Linn County.
2. Write some questions that can be answered by comparing the two graphs. Trade questions with a partner.
3. Which weeks have the highest pollen counts in Linn County?

CHALLENGE Try making a double line graph combining both sets of data in one graph.

ANSWERS

1. Graphs should show an accurate portrayal of the data, with x- and y-axes labeled and a title.

2. Sample answers: Which day had the most pollen? Which had the least? If you have ragweed allergies, which day is the safest to be outside? Which is the worst day for people with allergies?

3. August 19–September 9

CHALLENGE *Graphs should show an accurate portrayal of the data, with the x- and y-axes labeled. It should have a key to the lines and a title.*

KEY CONCEPT

3.3 The integumentary system shields the body.

BEFORE, you learned

- The body is defended from harmful materials
- Response structures fight disease
- The immune system responds in many ways to illness

NOW, you will learn

- About the functions of the skin
- How the skin helps protect the body
- How the skin grows and heals

VOCABULARY

integumentary system p. 83
epidermis p. 84
dermis p. 84

EXPLORE The Skin

What are the functions of skin?

PROCEDURE

1. Using a vegetable peeler, remove the skin from an apple. Take notes on the characteristics of the apple's peeled surface. Include observations on its color, moisture level, and texture.
2. Place the apple on a dry surface. After fifteen minutes, note any changes in its characteristics.

MATERIALS

- vegetable peeler
- apple

WHAT DO YOU THINK?

- What is the function of an apple's skin? What does it prevent?
- What does this experiment suggest about how skin might function in the human body?

Skin performs important functions.

MAIN IDEA AND DETAILS Start a two-column chart with the main idea *Skin performs important functions.* Add detail notes about those functions.

Just as an apple's skin protects the fruit inside, your skin protects the rest of your body. Made up of flat sheets of cells, your skin protects the inside of your body from harmful materials outside. The skin is part of your body's **integumentary system** (ihn-TEHG-yu-MEHN-tuh-ree), which also includes your hair and nails.

Your skin fulfills several vital functions:

- Skin repels water.
- Skin guards against infection.
- Skin helps maintain homeostasis.
- Skin senses the environment.

RESOURCES FOR DIFFERENTIATED INSTRUCTION

Below Level

UNIT RESOURCE BOOK

- Reading Study Guide A, pp. 152–153
- Decoding Support, p. 165

AUDIO CDS

Advanced

UNIT RESOURCE BOOK

- Challenge and Extension, p. 158
- Challenge Reading, pp. 161–162

English Learners

UNIT RESOURCE BOOK

Spanish Reading Study Guide, pp. 156–157

AUDIO CDS

- Audio Readings in Spanish
- Audio Readings (English)

3.3 FOCUS

Set Learning Goals

Students will

- Describe the functions of the skin.
- Analyze how the skin protects the body.
- Explain how the skin grows and heals.
- Observe how oil protects the skin.

3-Minute Warm-Up

Display Transparency 21 or copy this exercise on the board:

Complete the chart describing the body's lines of defense against foreign materials.

Body System	Defense(s)
Integumentary System	*skin as barrier*
Respiratory System	*mucus, cilia, coughing, sneezing*
Digestive System	*saliva, mucus, enzymes, acids*
Circulatory System	*white blood cells, antibodies*

 3-Minute Warm-Up, p. T21

3.3 MOTIVATE

EXPLORE The Skin

PURPOSE To understand the functions of skin

TIPS *10 min.* Have students work in groups or pairs. Caution them to avoid touching the sharp edge of the peeler.

WHAT DO YOU THINK? *The apple's skin keeps the apple from turning brown, decaying, and drying out. Skin keeps the inside of the body moist and healthy.*

Ongoing Assessment

Describe the functions of the skin.

Ask: What are the four functions of the integumentary system? *repels water, guards against infection, helps maintain homeostasis, senses the environment*

3.3 INSTRUCT

Teach from Visuals

To help students interpret the diagram of skin structure, ask:

- What structures are found in the dermis layer of the skin? *sensory receptors, nerves, oil glands, hair, blood vessels, muscles, and sweat glands*
- How does the skin provide insulation for the body? *Fatty tissue below the dermis provides protection against extremes in temperature.*
- Which layer of skin protects the body from foreign materials? *epidermis*

Metacognitive Strategy

Many students have trouble remembering the difference between the epidermis and dermis layers of the skin. Challenge students to write a rhyme that will help them remember the difference between the two layers.

Develop Critical Thinking

ANALYZE Ask students how the structure of the skin helps the body maintain homeostasis. *The fatty layer of tissue helps regulate body temperature; sweat glands help cool the body through evaporation.*

Language Arts Connection

The word *epidermis* is derived from the Greek prefix *epi-,* which means "above" or "upon" and the Greek word *derma*, which means "skin." Have students use a dictionary to find the meanings of Greek prefixes such as *neo-*, *peri-*, and *exo-*, which are often used in life science terminology. Ask a few students to list the prefixes and meanings on the board. Challenge others to discover words that contain the prefixes and to add them to the list.

Skin Structure

When you look at your hand, you only see the outer layer of skin. The skin has many structures to protect your body.

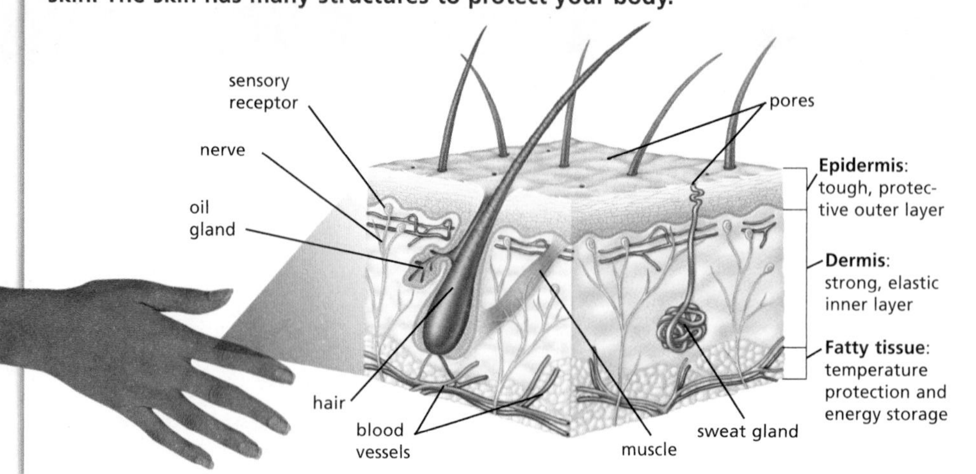

The structure of skin is complex.

Have you ever looked closely at your skin? Your skin is more complex than it might at first seem. It does more than just cover your body. The skin is made up of many structures, which perform many different jobs.

Dermis and Epidermis

VOCABULARY
Add frame games for *epidermis* and *dermis* to your notebook.

As you can see in the diagram above, human skin is composed of two layers: an outer layer, called the **epidermis,** and an inner layer, called the **dermis.** The cells of the epidermis contain many protein fibers that give the skin tough, protective qualities. These cells are formed in the deepest part of the epidermis. Skin cells move upward slowly as new cells form below them. Above new cells, older cells rub off. The surface cells in the epidermis are dead but form a thick, waterproof layer about 30 cells deep.

The dermis, the inner layer of skin, is made of tissue that is strong and elastic. The structure of the dermis allows it to change shape instead of tear when it moves against surfaces. The dermis is rich in blood vessels, which supply oxygen and nutrients to the skin's living cells. Just beneath the dermis lies a layer of fatty tissue. This layer protects the body from extremes in temperature, and it stores energy for future use. Also in the dermis are structures that have special functions, including sweat and oil glands, hair, nails, and sensory receptors.

DIFFERENTIATE INSTRUCTION

More Reading Support

A What gives the epidermis its protective qualities? *protein fibers*

B What is formed by the dead surface cells? *a thick, waterproof layer*

English Learners English learners may be unfamiliar with the use of prefixes and suffixes in English. Discuss the terms *dermis* and *epidermis.* Then ask students to look up the prefix *epi-* in the dictionary. Ask students how adding this prefix changes the word *dermis.* Also discuss with students the different words related to *sense* found in this section, such as *sensory, sensor, sensation,* and *sensitive.*

Sweat and Oil Glands

RESOURCE CENTER
CLASSZONE.COM

Explore the structure of skin.

Deep within the dermis are structures that help maintain your body's internal environment. Sweat glands help control body temperature, and oil glands protect the skin by keeping it moist. Both types of glands open to the surface through tiny openings in the skin called pores. Pores allow important substances to pass to the skin's surface. Pores can become clogged with dirt and oil. Keeping the skin clean can prevent blockages.

Sweat glands, which are present almost everywhere on the body's surface, help maintain homeostasis. When you become too warm, the sweat glands secrete sweat, a fluid that is 99% water. This fluid travels from the sweat glands, through the pores, and onto the skin's surface. You probably know already about evaporation. Evaporation is the process by which a liquid becomes a gas. During evaporation, heat is released. Thus, sweating cools the skin's surface and the body.

Like sweat glands, oil glands are present almost everywhere on the body. They secrete an oil that moistens skin and hair and keeps them from becoming dry. Skin oils add flexibility and provide part, but not all, of the skin's waterproofing.

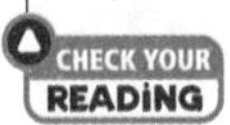

CHECK YOUR READING What are the functions of oil glands?

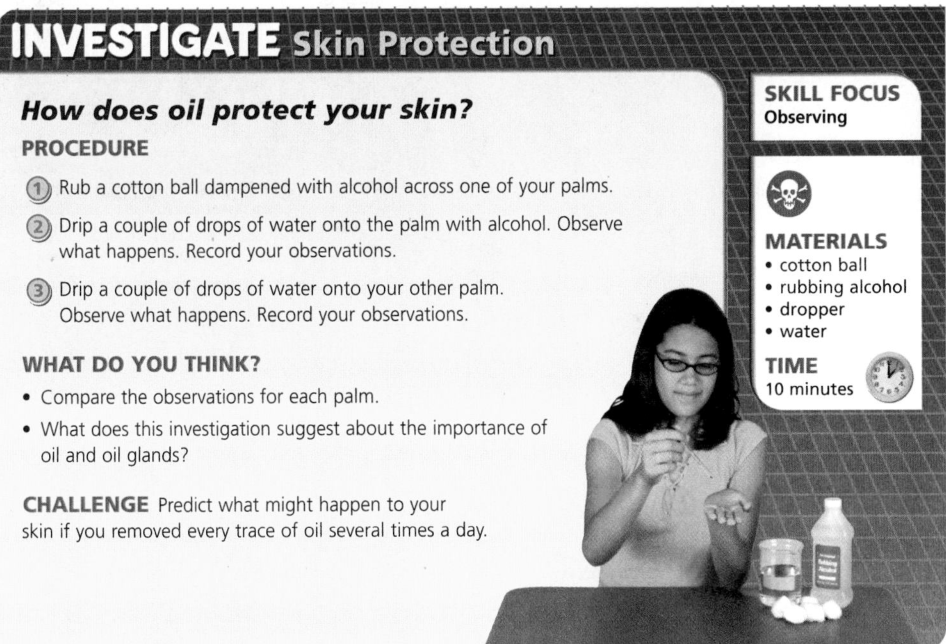

INVESTIGATE Skin Protection

How does oil protect your skin?

PROCEDURE

1. Rub a cotton ball dampened with alcohol across one of your palms.
2. Drip a couple of drops of water onto the palm with alcohol. Observe what happens. Record your observations.
3. Drip a couple of drops of water onto your other palm. Observe what happens. Record your observations.

WHAT DO YOU THINK?

- Compare the observations for each palm.
- What does this investigation suggest about the importance of oil and oil glands?

CHALLENGE Predict what might happen to your skin if you removed every trace of oil several times a day.

SKILL FOCUS
Observing

MATERIALS
- cotton ball
- rubbing alcohol
- dropper
- water

TIME
10 minutes

INVESTIGATE Skin Protection

PURPOSE To observe how oil protects the skin

TIPS *10 min.* Keep the room well ventilated. Caution students not to inhale fumes from the alcohol. You may wish to have petroleum jelly or baby oil available for students whose skin oils are minimal.

WHAT DO YOU THINK? *On the hand without the alcohol, the water formed small beads. On the hand rubbed with alcohol, the water spread thinly over the skin and seemed to sink into the skin. The oil must act as a barrier on the skin. Oil helps keep liquids that don't belong in the skin from sinking into the skin.*

CHALLENGE *The skin would become dry and could absorb foreign materials.*

 Datasheet, Skin Protection, p. 159

Technology Resources

Customize this student lab as needed or look for an alternative. Print rubrics to assess student lab reports.

 **Lab Generator CD-ROM**

EXPLORE the BIG idea

Revisit "Wet Fingers" on p. 63. Ask students how this activity demonstrates how evaporation of sweat helps to cool the body.

Ongoing Assessment

Analyze how the skin protects the body.

Ask: What are two types of glands that help the body maintain homeostasis? *oil glands and sweat glands*

CHECK YOUR READING *Answer: Oil glands help to moisten the skin and hair and keep them from drying out.*

DIFFERENTIATE INSTRUCTION

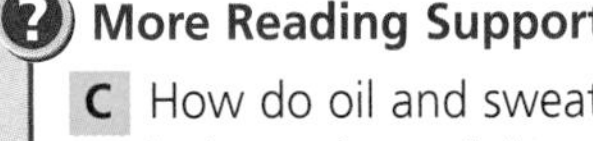

More Reading Support

C How do oil and sweat reach the surface of the skin? *through pores*

D How does sweat cool the body? *through evaporation*

Alternative Assessment Have students create a 3-D cross section of the skin. They can use materials such as plastic foam, pipe cleaners, yarn, beads, and wires. Ask them to add labels and captions about the three ways skin helps the body.

Teach from Visuals

Have students compare the magnification photographs of hair and nails. Ask:

- Which magnification shows a greater area? *hair*
- Which image is blown up the most? *nails*
- What do hair and nails have in common with the cells on the surface of the skin? *They are made up of dead cells.*

Teaching with Technology

If skin surface temperature probeware is available, have students use it to measure the change in skin temperature caused by placing a cold pack or a heat pack on their arms. They can use a graphing calculator to show the data visually.

Real World Example

Most students will have experienced getting goosebumps on their skin when they are cold. Explain that the bumps that appear on the surface of the skin are the result of the contraction of tiny muscles at the base of each hair, making the hair stand up.

Integrate the Sciences

Electrical engineers are trying to improve skin for robots. This is a tough problem because robot skin must be elastic but also able to conduct electricity. If sensors and metal wires are used to conduct electricity, they break when the skin stretches. One solution might be to implant broad metal strips that are corrugated. The corrugated metal is stretchy; it can be flattened out or compressed and continue to conduct electricity. Scientists' goal is to make a robot that can "sense" its environment.

Ongoing Assessment

CHECK YOUR READING *Answer: Hair protects your skin and keeps you warm. Nails protect the tips of the fingers and toes from injury.*

CHECK YOUR READING *Answer: Sensory receptors sense pressure, temperature, pain, touch, and vibration.*

Hair and Nails

In addition to your skin, your integumentary system includes your hair and nails. Many cells in your hair and nails are actually dead but continue to perform important functions.

The hair on your head helps your body in many ways. When you are outside, it shields your head from the Sun. In cold weather, it traps heat close to your head to keep you warmer. Your body hair works the same way, but it is much less effective at protecting your skin and keeping you warm.

E Fingernails and toenails protect the tips of the fingers and toes from injury. Both are made of epidermal cells that are thick and tough. They grow from the nail bed, which continues to manufacture cells as the cells that form the nail bed bond together and grow.

CHECK YOUR READING What are the functions of hair and nails?

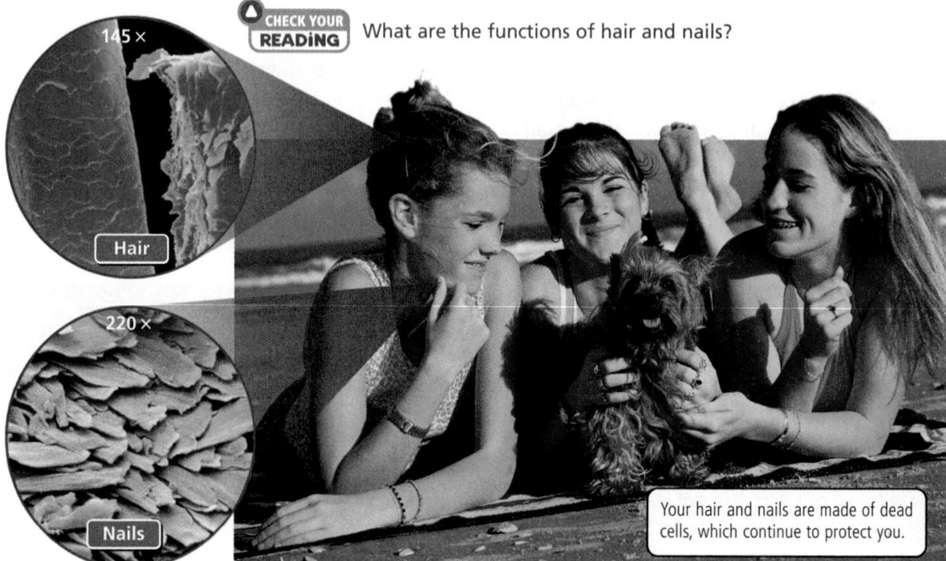

Your hair and nails are made of dead cells, which continue to protect you.

Sensory Receptors

F How does your body know when you are touching something too hot or too cold? You get that information from sensory receptors attached to the nerves. These receptors are actually part of the nervous system, but they are located in your skin. Your skin contains receptors that sense pressure, temperature, pain, touch, and vibration. These sensors help protect the body. For example, temperature receptors sense when an object is hot. If it is too hot and you touch it, pain receptors send signals to your brain telling you that you have been burned.

CHECK YOUR READING What are the five types of sensory receptors in skin?

DIFFERENTIATE INSTRUCTION

More Reading Support

E What are fingernails and toenails made of? *epidermal cells from the nail beds*

F Sensory receptors belong to which body system? *the nervous system*

Advanced Invite interested students to investigate the different types of fingerprints and the way forensic scientists use fingerprinting in helping to solve crimes. Students should research the limitations of using fingerprinting in forensics. Encourage students to make impressions of their own fingerprints for display.

 Challenge and Extension, p. 158

The skin grows and heals.

As a person grows, skin also grows. As you have noticed if you have ever had a bruise or a cut, your skin is capable of healing. Skin can often repair itself after injury or illness.

Growth

As your bones grow, you get taller. As your muscles develop, your arms and legs become thicker. Through all your body's growth and change, your skin has to grow, too.

Most of the growth of your skin occurs at the base of the epidermis, just above the dermis. The cells there grow and divide to form new cells, constantly replacing older epidermal cells as they die and are brushed off during daily activity. Cells are lost from the skin's surface all the time: every 2 to 4 weeks, your skin surface is entirely new. In fact, a percentage of household dust is actually dead skin cells.

Healing Skin

Small injuries to the skin heal by themselves over time.

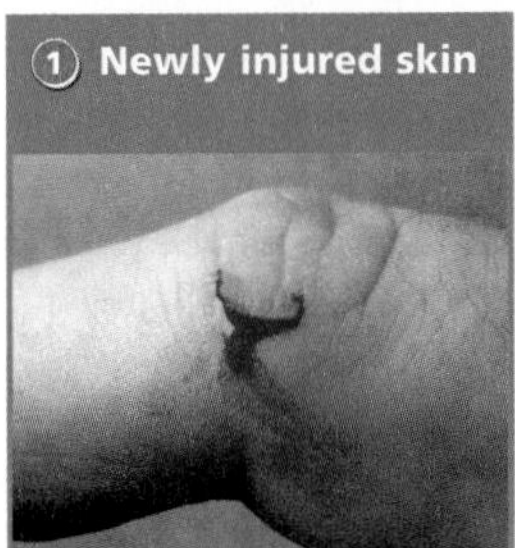

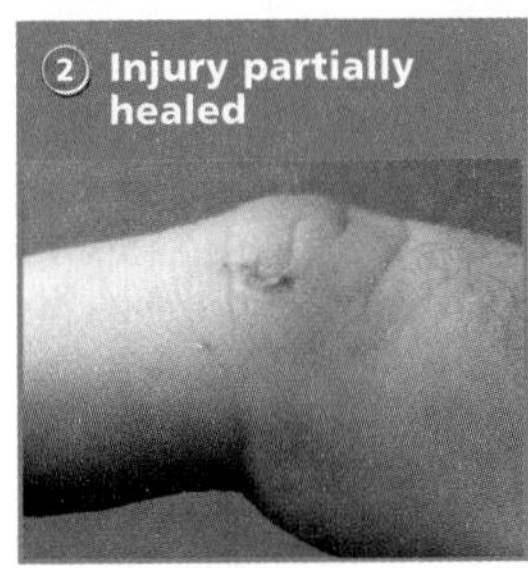

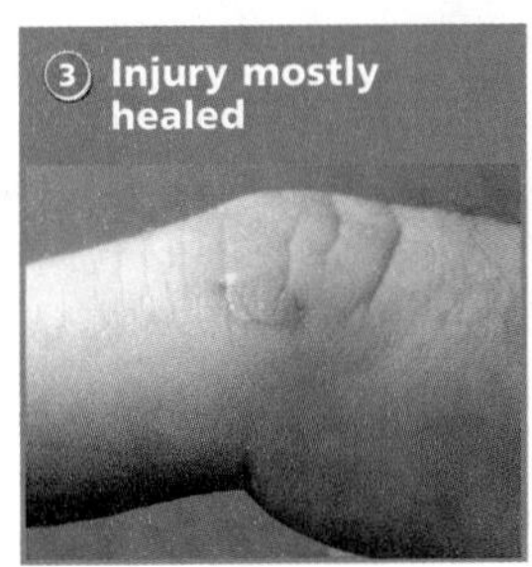

READING VISUALS How do you think small injuries to the skin heal?

Injuries and Healing

You have probably experienced some injuries to your skin, such as blisters, burns, cuts, and bruises. Most such injuries result from the skin's contact with the outside world, such as a concrete sidewalk. In simple injuries, the skin can usually repair itself.

Explore how the skin heals.

Burns can be serious injuries. They can be caused by heat, electricity, radiation, or certain chemicals. In mild cases—those of first-degree burns—skin merely becomes red, and the burn heals in a day or two. In severe cases—those of second-degree and third-degree burns—the body loses fluids, and death can result from fluid loss, infection, and other complications.

DIFFERENTIATE INSTRUCTION

More Reading Support

G How much time does it take for the surface of the skin to replace itself? *two to four weeks*

Below Level Have students create a poster showing the various ways to keep the skin safe from the Sun. Students may include a chart of the UV Index, showing the various dangers at each level in the index. Encourage students to check a local newspaper and report on the daily UV levels. Allow class time for students to present their findings.

Advanced Have students who are interested in the effects of tanning on the skin read the following article.

R Challenge Reading, pp. 161–162

Teach from Visuals

Discuss the photographic sequence of healing skin. At the site of the cut on the finger, three things happen:

- Platelets help the blood clot, or thicken and dry, to form a scab.
- New skin cells grow and cover the area as the scab falls off.
- The new skin and surrounding skin continue to generate new cells while the dead cells form a protective, seamless outer layer.

Integrate the Sciences

The ozone layer is about 6–30 miles (10–50 km) above the surface of Earth. Human activity has caused a thinning of this layer, resulting in higher levels of ultraviolet (UV) radiation. The increase in radiation in turn has elevated the incidence of skin cancers and cataracts. The National Weather Service and the Environmental Protection Agency have developed the UV Index, which predicts daily levels of UV radiation on a 0 to 10+ scale.

Health Connection

About 2.4 million burn injuries to the skin occur each year. Of those, between 8000 and 12,000 result in death. The most frequent place where burns occur in the home is the kitchen. Burns are the most common cause of accidental death for children under the age of 14.

Ongoing Assessment

Explain how the skin grows and heals.

Ask: How are new skin cells formed? *Cells are formed at the base of the epidermis. As new cells are produced, the layer of older epidermal cells is replaced and brushed off during normal daily activity.*

READING VISUALS *Answer: The skin cells at the edge of the injury begin to divide to replace the skin cells that were damaged. Eventually the new skin cells fill in the injured area.*

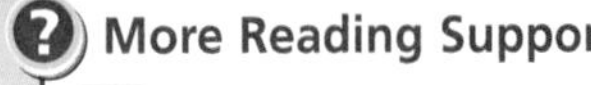

Ongoing Assessment

 *Answer: Severe heat and severe cold can damage the skin.*

Reinforce 

Have students relate the section to the Big Idea.

 Reinforcing Key Concepts, p. 160

3.3 ASSESS & RETEACH

Assess

Section 3.3 Quiz, p. 43

Reteach

Write the following notes on the board. For each, have students tell whether it pertains to the dermis or epidermis layer of the skin.

1. contains protein fibers *epidermis*
2. contains nerves and sweat glands *dermis*
3. rich in blood vessels *dermis*
4. forms a waterproof covering *epidermis*
5. Most skin growth occurs at its base. *epidermis*

Then have partners create a Venn diagram comparing the functions of the dermis and the epidermis.

dermis—insulates; cushions; contains glands, vessels, and nerves
both—senses external environment; protects
epidermis—heals surface wounds; releases oil and sweat to surface; acts as shield

Technology Resources

Have students visit **ClassZone.com** for reteaching of Key Concepts.

 CONTENT REVIEW

 CONTENT REVIEW CD-ROM

Sunburns are usually minor first-degree burns, but that does not mean they cannot be serious. Rays from the Sun can burn and blister the skin much as a hot object can. Repeated burning can increase the chance of skin cancer. Specialized cells in the skin absorb the Sun's ultraviolet rays and help prevent tissue damage. These cells produce the skin pigment melanin when exposed to the Sun. The amount of melanin in your skin determines how dark your skin is.

Severe cold can damage skin as well. Skin exposed to cold weather can get frostbite, a condition in which the cells are damaged by freezing. Mild frostbite often heals just as well as a minor cut. In extreme cases, frostbitten limbs become diseased and have to be amputated.

Wearing sunblock when you are outside protects your skin from harmful rays from the Sun.

 What types of weather can damage your skin?

Protection

Your skin is constantly losing old cells and gaining new cells. Although your skin is always changing, it is still important to take care of it.

- Good nutrition supplies materials the skin uses to maintain and repair itself. By drinking water, you help your body, and thus your skin, to remain moist and able to replace lost cells.
- Appropriate coverings, such as sunblock in summer and warm clothes in winter, can protect the skin from weather damage.
- Skin also needs to be kept clean. Many harmful bacteria cannot enter the body through healthy skin, but they should be washed off regularly. This prevents them from multiplying and then entering the body through small cuts or scrapes.

3.3 Review

KEY CONCEPTS

1. List four functions of the skin.
2. How do the epidermis and dermis protect the body?
3. Make your own diagram with *How skin grows and repairs itself* at the center. Around the center, write at least five facts about skin growth and healing.

CRITICAL THINKING

4. **Apply** Give three examples from everyday life of sensory receptors in your skin reacting to changes in your environment.
5. **Connect** Describe a situation in which sensory receptors could be critical to survival.

 **CHALLENGE**

6. **Infer** Exposure to sunlight may increase the number of freckles on a person's skin. Explain the connection between sunlight, melanin, and freckles.

ANSWERS

1. It repels water, guards against infection, helps maintain homeostasis, and senses the environment.

2. The epidermis keeps harmful materials out because it contains tough protein fibers. The dermis helps because it is strong elastic tissue that resists tearing.

3. Diagrams will vary but should include five facts about skin growth and healing.

4. Answers should show how sensory receptors respond to the environment.

5. Answers should show the role of sensory receptors in survival.

6. Melanin concentrated in a small place on the skin that is surrounded by skin with less melanin produces a freckle. Exposure to sunlight increases melanin levels in some spots, which become visible as freckles.

A spray-on polymer creates an artificial outer skin to help heal surface wounds on an arm.

GROWING NEW SKIN

Artificial Skin

Skin acts like a barrier, keeping our insides in and infections out. Nobody can survive without skin. But when a large amount of skin is severely damaged, the body cannot work fast enough to replace it. In some cases there isn't enough undamaged skin left on the body for transplanting. Using skin from another person risks introducing infections or rejection by the body. The answer? Artificial skin.

Here's the Skinny

To make artificial skin, scientists start with cells in a tiny skin sample. Cells from infants are used because infant skin-cell molecules are still developing, and scientists can manipulate the molecules to avoid transplant rejection. The cells from just one small sample of skin can be grown into enough artificial skin to cover 15 basketball courts. Before artificial skin, badly burned victims didn't have much chance to live. Today, 96 out of 100 burn victims survive.

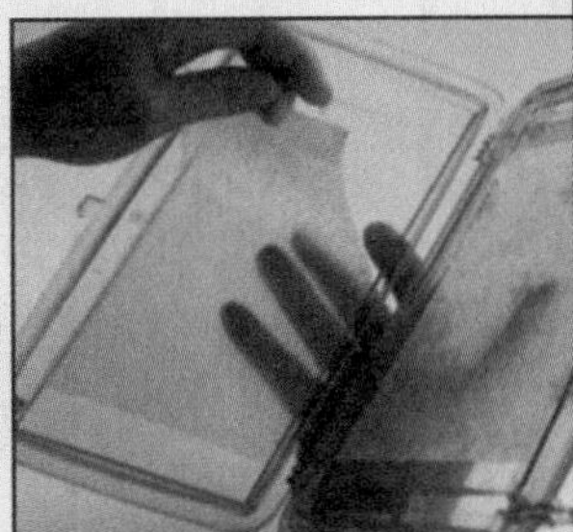
A surgeon lifts a layer of artificial skin. The skin is so thin, a newspaper could be read behind it.

What's Next?

- Scientists are hoping to be able to grow organs using this technology. Someday artificially grown livers, kidneys, and hearts may take the place of transplants and mechanical devices.
- A self-repairing plastic skin that knits itself back together when cracked has been developed. It may someday be used to create organs or even self-repairing rocket and spacecraft parts.
- Artificial polymer "skin" for robots is being developed to help robots do delicate work such as microsurgery or space exploration.

Robot designer David Hanson has developed the K-bot, a lifelike face that uses 24 motors to create expressions.

EXPLORE

1. COMPARE AND CONTRAST Detail the advantages and disadvantages of skin transplanted from another place on the body and artificial skin.
2. CHALLENGE Artificial skin is being considered for applications beyond those originally envisioned. Research and present a new potential application.

EXPLORE

1. *COMPARE AND CONTRAST Sometimes there is not enough skin to transplant. Transplants from another person can cause infections or rejections. Using artificial skin saves lives. Artificial skin is expensive to produce. Artificial skin can shorten hospital and rehabilitation time.*
2. *CHALLENGE Accept responses that are coherent and research-based.*

EXTREME SCIENCE

Fun and Motivating Science

Set Learning Goal

To understand the uses and future potential uses of artificial skin

Present the Science

Treatment for burns consists of removing the burned tissue and covering the dermis with skin grafts from other parts of the body. For people with large areas of burns, this is not always possible. Artificial skin has made it possible for surgeons to cover burned areas with a temporary synthetic epidermis. Cells from the dermis layer form a "lattice" with the artificial layer, eventually producing a new layer of epidermis.

Discussion Questions

Ask students to brainstorm a "top 10 list" for burn prevention.

Ask: Do you think burns are more severe if they are limited to the epidermis or if they affect the dermis as well? Explain. *Burns are likely to be more severe if they reach down to the dermis, because such burns go deeper into the skin, where live tissue and blood vessels are found.*

Close

Ask: What difficulties do you think scientists might encounter in using the technology of artificial skin for other purposes? *Accept responses that cite qualities of real skin and the difficulties of reproducing it.*

Technology Resources

Students can visit **ClassZone.com** to for more links about artificial skin.

 **RESOURCE CENTER**

BACK TO

the BIG idea

Have students explain how the words *transport, defend,* and *protect* convey how the circulatory and immune systems are related. *The circulatory system transports materials throughout the body; some of these materials are blood cells that defend and protect the body from foreign materials.*

KEY CONCEPTS SUMMARY

SECTION 3.1
Ask students to look at the picture of the heart and identify which chamber of the heart receives oxygen-rich blood. Where does this blood come from? *The left atrium receives oxygen-rich blood from the lungs.*

SECTION 3.2
Ask students to describe the structures and responses of the immune system. *white blood cells, antibodies; inflammation, fever, and development of immunity*

SECTION 3.3
Ask students which layer of the skin protects the body from foreign materials. How does it accomplish this? *The epidermis, the outer layer of skin, prevents foreign materials from entering the body. This layer is made up of tough, waterproof cells.*

Review Concepts

- Big Idea Flow Chart, p. T17
- Chapter Outline, pp. T23–T24

3 Chapter Review

the BIG idea

Systems function to transport materials and to defend and protect the body.

KEY CONCEPTS SUMMARY

3.1 The circulatory system transports materials.

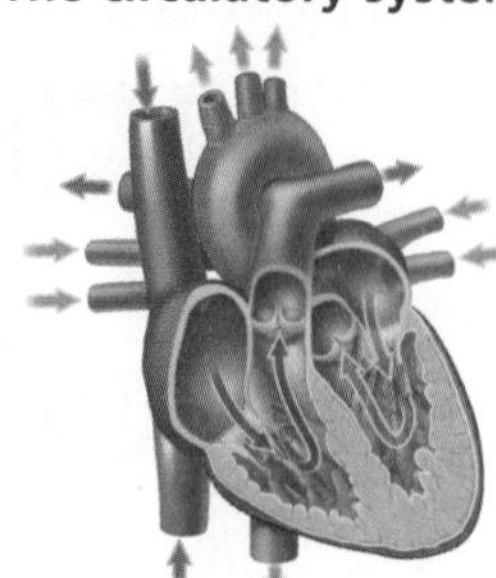

The heart, blood vessels, and blood of the circulatory system work together to transport materials from the digestive and respiratory systems to all cells. The blood exerts pressure on the walls of the blood vessels and keeps the blood moving around the body.

VOCABULARY
circulatory system p. 65
blood p. 65
red blood cell p. 67
artery p. 69
vein p. 69
capillary p. 69

3.2 The immune system defends the body.

The immune system defends the body from pathogens. White blood cells identify and attack pathogens that find their way inside the body. The immune system responds to attack with inflammation, fever, and development of immunity.

Types of Pathogens

Disease	Pathogen
colds, chicken pox, hepatitis, AIDS, influenza, mumps, measles, rabies	virus
food poisoning, strep throat, tetanus, tuberculosis, acne, ulcers, Lyme disease	bacteria
athlete's foot, thrush, ring worm	fungus
malaria, parasitic pneumonia, pinworm, lice, scabies	parasites

VOCABULARY
pathogen p. 74
immune system p. 75
antibody p. 75
antigen p. 78
immunity p. 80
vaccine p. 80
antibiotic p. 81

3.3 The integumentary system shields the body.

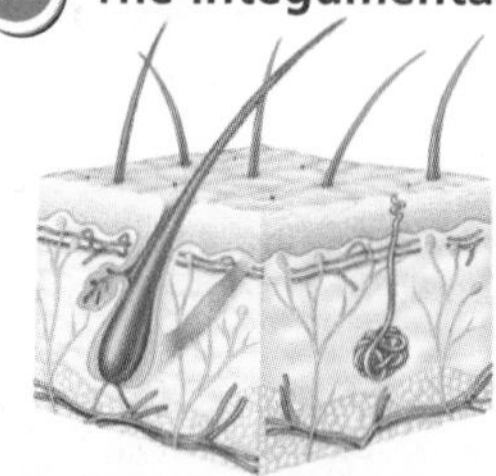

The skin protects the body from harmful materials in the environment, and allows you to sense temperature, pain, touch, and vibration. In most cases the skin is able to heal itself after injury.

VOCABULARY
integumentary system p. 83
epidermis p. 84
dermis p. 84

Technology Resources

Have students visit **ClassZone.com** or use the CD-ROM for a cumulative review of concepts.

 CONTENT REVIEW

 CONTENT REVIEW CD-ROM

Engage students in a whole-class interactive review of Key Concepts. Edit content as you wish.

 POWER PRESENTATIONS

Reviewing Vocabulary

Draw a word triangle for each of the terms below. Write a term and its definition in the bottom section. In the middle section, write a sentence in which you use the term correctly. In the top section, draw a small picture to illustrate the term.

The epidermis forms a thick waterproof layer that protects the body.

epidermis: the outer layer of human skin

1. capillary

2. blood

3. dermis

4. antigen

Write a sentence describing the relationship between each pair of terms.

5. pathogen, antibody

6. artery, vein

7. immunity, vaccine

Reviewing Key Concepts

Multiple Choice *Choose the letter of the best answer.*

8. Which chamber of the heart pumps oxygen-poor blood into the lungs?
- **a.** right atrium
- **b.** right ventricle
- **c.** left atrium
- **d.** left ventricle

9. Which blood structures carry blood back to the heart?
- **a.** veins
- **b.** capillaries
- **c.** arteries
- **d.** platelets

10. The structures in the blood that carry oxygen to the cells of the body are the
- **a.** plasma
- **b.** platelets
- **c.** white blood cells
- **d.** red blood cells

11. High blood pressure is unhealthy because it
- **a.** does not exert enough pressure on your arteries
- **b.** causes your heart to work harder
- **c.** does not allow enough oxygen to get to the cells in your body
- **d.** causes your veins to collapse

12. Which category of pathogens causes strep throat?
- **a.** virus
- **b.** bacteria
- **c.** fungus
- **d.** parasite

13. Which of the following is a function of white blood cells?
- **a.** destroying foreign organisms
- **b.** providing your body with nutrients
- **c.** carrying oxygen to the body's cells
- **d.** forming a blood clot

14. Which makes up the integumentary system?
- **a.** a network of nerves
- **b.** white blood cells and antibodies
- **c.** the brain and spinal cord
- **d.** the skin, hair, and nails

15. Which structure is found in the epidermis layer of the skin?
- **a.** pores
- **b.** sweat glands
- **c.** surface cells
- **d.** oil glands

16. The layer of fatty tissue below the dermis protects the body from
- **a.** cold temperatures
- **b.** bacteria
- **c.** sunburn
- **d.** infection

Short Answer *Write a short answer to each question.*

17. What are platelets? Where are they found?

18. What are antibodies? Where are they found?

19. What special structures are found in the dermis layer of the skin?

ASSESSMENT RESOURCES

UNIT ASSESSMENT BOOK
- Chapter Test A, pp. 44–47
- Chapter Test B, pp. 48–51
- Chapter Test C, pp. 52–55
- Alternative Assessment, pp. 56–57

SPANISH ASSESSMENT BOOK
Spanish Chapter Test, pp. 73–76

Technology Resources

Edit test items and answer choices.

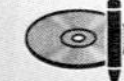

Test Generator CD-ROM

Visit **ClassZone.com** to extend test practice.

Test Practice

Reviewing Vocabulary

1. *Sample answer: Capillaries connect arteries with veins. The exchange of oxygen and carbon dioxide takes place in the capillaries.*
2. *Sample answer: Blood is a fluid that carries materials and wastes through the body. Blood is made up of plasma, red blood cells, white blood cells, and platelets.*
3. *The dermis is the inner layer of the skin. Sweat glands and oil glands are found in the dermis.*
4. *Antigens are markers carried by pathogens. T cells recognize antigens on pathogens.*
5. *Certain white blood cells produce antibodies, which attack and kill pathogens in the body.*
6. *Arteries are blood vessels that carry blood away from the heart; veins carry blood back from the body to the heart.*
7. *Vaccines produce immunity, or resistance to diseases caused by viruses or bacteria.*

Reviewing Key Concepts

8. b	*11. b*	*14. d*
9. a	*12. b*	*15. a*
10. d	*13. a*	*16. a*

17. *Platelets are cell fragments in the blood, which help form blood clots.*
18. *Antibodies attack specific foreign materials. They are found in the circulatory and lymphatic systems.*
19. *sweat and oil glands, hair, nails, and sensory receptors*

(Answers to items that appear on p. 92)

Thinking Critically

20. *Atria receive blood; ventricles pump blood. Ventricles have thicker walls for the pumping of blood.*
21. *Pumping action pushes blood through the arteries, so valves are not needed.*
22. *Sweat, tears, and mucus contain substances that attack foreign materials. Tiny hairs in the nose and lungs trap particles and prevent them from passing through membranes.*

(Answers for items 20–22 appear on p. 91.)

Thinking Critically

23. T cells identify and destroy pathogens in the blood and the lymph. As these cells are destroyed, the body is less able to protect itself against infection and disease.

24. Histamines raise the temperature of and increase blood flow to the injured area. White blood cells and antibodies move to the injured area.

25. Dead cells that make up the surface form a thick waterproof layer that prevents bacteria from entering the body.

26. Sweat glands: release fluid to cool skin; oil glands: secrete oils that help repel water, kill bacteria; hair: traps heat and keeps body warm.

27. Melanin is a pigment that determines how dark a person's skin is. People with darker skin are less likely to get sunburns. Repeated sunburns can increase a person's chances of getting skin cancer.

28–33. For answers, see p. 90.

the BIG idea

34. Answers should include blood pressure and functions of the heart and circulatory system.

35. The skin of the <u>integumentary system</u> forms a waterproof barrier that protects against foreign substances. Sweat and skin oil and mucus in the nose and throat trap harmful foreign substances and slow the growth of pathogens. White blood cells and antibodies of the <u>immune system</u> work together to kill invading bacteria.

UNIT PROJECTS

Check to make sure students are working on their projects. Check schedules and work in progress.

 Unit Projects, pp. 5–10

Thinking Critically

20. COMPARE AND CONTRAST How do the functions of the atria and ventricles of the heart differ? How are they alike? Use this diagram of the heart as a guide.

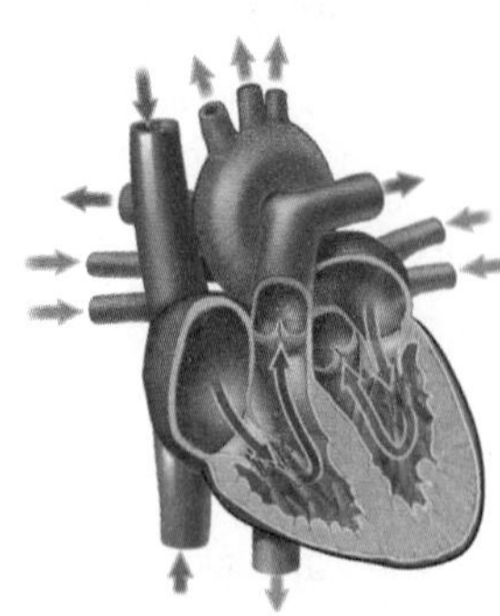

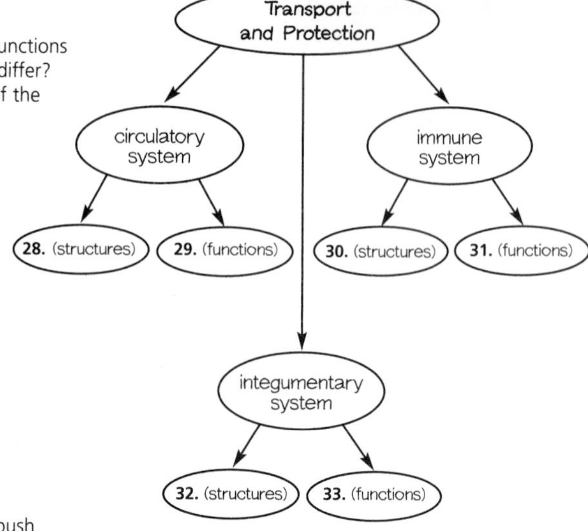

21. APPLY Veins have one-way valves that push the blood back to the heart. Most arteries do not have valves. Explain how these structures help the circulatory system function.

22. PROVIDE EXAMPLES Describe three structures in the body that help prevent harmful foreign substances from entering the body.

23. IDENTIFY CAUSE HIV is a virus that attacks and destroys the body's T cells. Why is a person who is infected with the HIV virus more susceptible to infection and disease?

24. APPLY You fall and scrape your knee. How does the production of histamines aid the healing of this injury?

25. ANALYZE Describe how the structure of the epidermis helps protect the body from disease.

25. SYNTHESIZE Explain how sweat glands, oil glands, and hair help your body maintain homeostasis.

27. HYPOTHESIZE People with greater concentrations of melanin in their skin are less likely to get skin cancer than people who have lesser concentrations of melanin. Write a hypothesis explaining why this is so.

the BIG idea

34. INFER Look again at the picture on pages 62–63. Now that you have finished the chapter, how would you change or add details to your answer to the question on the photograph?

35. SYNTHESIZE Write a paragraph explaining how the integumentary system and the immune system work together to help your body maintain its homeostasis. Underline these terms in your paragraph.

UNIT PROJECTS

If you need to create graphs or other visuals for your project, be sure you have grid paper, poster board, markers, and other supplies.

MONITOR AND RETEACH

If students have trouble applying the concepts in Chapter Review items 22, 25, and 26, suggest they review the diagram on p. 84. Have students create concept maps for the following:

- structures in the body that prevent foreign materials from entering
- structures in the skin that protect the body from disease
- ways in which structures in the skin help the body maintain homeostasis

Students may benefit from summarizing sections of the chapter.

 Summarizing the Chapter, pp. 186–187

Standardized Test Practice

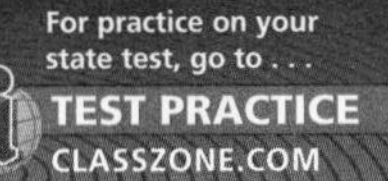

Analyzing Data

Choose the letter of the best answer.

This chart shows the amount of time a person can stay in the sun without burning, based on skin type and use of a sunscreen with the SPF shown.

Recommended Sun Protection Factors (SPF)

Skin Type	1 hr	2hr	3 hr	4hr	5hr
Very Fair/ Sensitive	15	30	30	45	45
Fair/Sensitive	15	15	30	30	45
Fair	15	15	15	30	30
Medium	8	8	15	15	30
Dark	4	8	8	15	15

1. What is the least SPF that a person with very fair skin should use while exposed to the sun?
   - **a.** 8
   - **b.** 15
   - **c.** 30
   - **d.** 45
2. If a person with a medium skin type is exposed to the sun for 5 hours, which SPF should be used?
   - **a.** 4
   - **b.** 8
   - **c.** 15
   - **d.** 30
3. Which skin type requires SPF 30 for three hours of sun exposure?
   - **a.** fair/sensitive
   - **b.** fair
   - **c.** medium
   - **d.** dark
4. Based on the data in the chart, which statement is a reasonable conclusion?
   - **a.** People with a fair skin type are less prone to UV damage than those with a dark skin type.
   - **b.** The darker the skin type, the more SPF protection a person needs.
   - **c.** A person with a medium skin type does not need as much SPF protection as a person with a fair skin type.
   - **d.** If exposure to the sun is longer, then a person needs a higher SPF for protection.
5. If a person normally burns after 10 minutes with no protection, an SPF 2 would protect that person for 20 minutes. How long would the same person be protected with SPF 15?
   - **a.** 1 hour
   - **b.** $1\frac{1}{2}$ hours
   - **c.** 2 hours
   - **d.** $2\frac{1}{2}$ hours

Extended Response

6. UV index levels are often broadcast with daily weather reports. A UV index of 0 to 2 indicates that it would take an average person about 60 minutes to burn. A UV index level of 10 indicates that it would take the average person about 10 minutes to burn. Write a paragraph describing some variable conditions that would affect this rate. Include both environmental as well as conditions that would apply to an individual.

7. Sun protection factors are numbers on a scale that rate the effectiveness of sunscreen. Without the use of sunscreen, UV rays from the Sun can cause sunburns. People who spend time in the sun without protection, or who get repeated burns are at a higher risk of developing deadly forms of skin cancer. Based on the information in the table and your knowledge of the layers of the skin, design a brochure encouraging people to protect their skin from the sun. Include in your brochure the harmful effects on your skin and ways to protect your skin from harmful UV rays.

METACOGNITIVE ACTIVITY

Have students answer the following questions in their **Science Notebook:**

1. What interests you most about the circulatory system?
2. What were you surprised to find out about the immune or integumentary system?
3. How have you accomplished the objectives for your unit project?

Analyzing Data

1. b *3. a* *5. d*

2. d *4. d*

Extended Response

6. RUBRIC

4 points for a response that correctly identifies both environmental and individual factors and includes these terms:

- SPF
- melanin

Sample answer: UV index levels may be affected by environmental conditions such as the time of day, the weather, the amount of cloud coverage, the season, and the location. Conditions that affect an individual would include the fairness or darkness of the skin, the amount of melanin present in the skin, the amount of time spent in the Sun, and the protection used, such as clothing or sunscreen with a particular SPF.

3 points correctly answers the question and includes a description of environmental factors and individual factors

2 points correctly answers the question and includes a description of either environmental factors or individual factors

1 point correctly identifies at least one environmental or individual factor

7. RUBRIC

4 points for a brochure that includes all of the following:

- harmful effects of UV rays
- ways to protect skin from UV rays
- a description of SPF rating
- a description of differences in the skin of individuals

Sample answer: Overexposure to UV rays can eventually lead to skin cancers. Use protection based on your skin type and the length of time you are exposed to the Sun. In addition, stay out of the Sun during peak hours from 10:00 A.M. to 2:00 P.M., wear hats, sunglasses, and cover-ups. Sunscreens are given SPF ratings according to how long they will provide protection from harmful UV rays. People with fair skin should use a sunscreen with an SPF of at least 45 during prolonged exposure to the Sun.

3 points includes at least three of the elements listed above

2 points includes at least two of the elements listed above

1 point includes at least one of the elements listed above

FOCUS

Set Learning Goals

Students will

- Examine technological advances that allow observation of the inner workings of the human body.
- Learn about some of the applications of the new technologies.
- Chart the different types of medical imaging and explain how they work.

National Science Education Standards

A.9.a–g Understandings About Scientific Inquiry

E.6.a–c Understandings About Science and Technology

F.5.a–e, F.5.g Science and Technology in Society

G.1.a–b Science as a Human Endeavor

G.2.a Nature of Science

G.3.a–c History of Science

INSTRUCT

Point out that the top half of the timeline depicts scientific progress in observing the inner workings of the human body. The bottom half of the timeline provides examples of technology and their application to that study. The gaps represent blocks of time that have been omitted.

Integrate the Sciences

Wilhelm Roentgen studied mechanical engineering and taught physics at several universities. When he discovered X-rays, he was studying the passage of electric current through gas. For his discovery he won the Nobel Prize in Physics in 1901.

TIMELINES in Science

SEEING INSIDE the Body

What began as a chance accident in a darkened room was only the beginning. Today, technology allows people to produce clear and complete pictures of the human body. From X-rays to ultrasound to the latest computerized scans, accidental discoveries have enabled us to study and diagnose the inner workings of the human body.

Being able to see inside the body without cutting it open would have seemed unthinkable in the early 1890s. But within a year of the discovery of the X-ray in 1895, doctors were using technology to see through flesh to bones. In the time since then, techniques for making images have advanced to allow doctors to see soft tissue, muscle, and even to see how body systems work in real time. Many modern employ X-ray technology, while others employ sound waves or magnetic fields.

1895

Accidental X-Ray Shows Bones

Working alone in a darkened lab to study electric currents passing through vacuum tubes, William Conrad Roentgen sees a mysterious light. He puts his hand between the tubes and a screen, and an image appears on the screen—a skeletal hand! He names his discovery the X-ray, since the images are produced by rays behaving like none known before them. Roentgen uses photographic paper to take the first X-ray picture, his wife's hand.

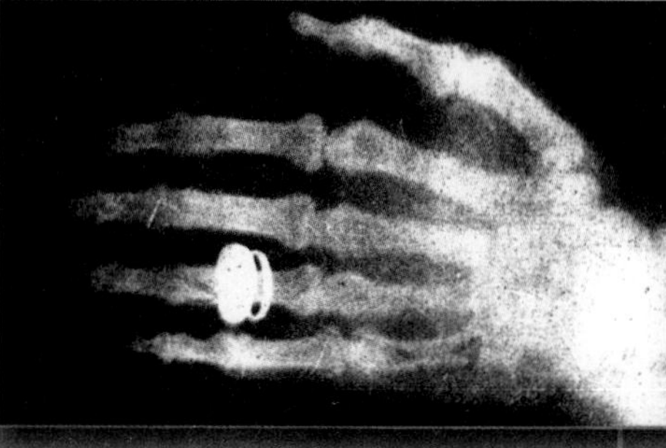

EVENTS

1880 | 1890

APPLICATIONS AND TECHNOLOGY

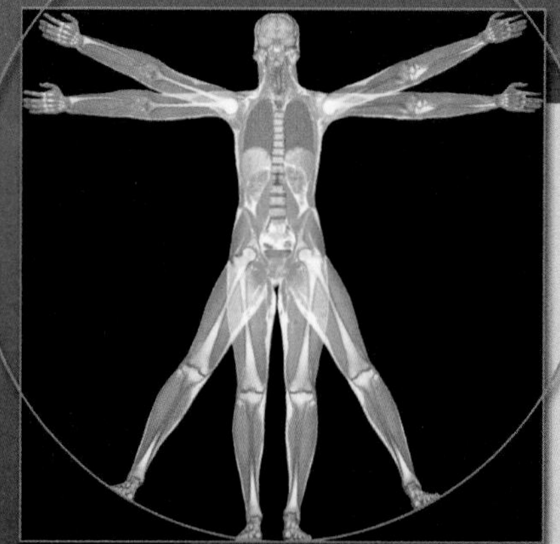

APPLICATION

Doctor Detectives

Within a year of Roentgen's discovery, X-rays were used in medicine for examining patients. By the 1920s, their use was wide-spread. Modern day X-ray tubes are based on the design of William Coolidge. Around 1913, Coolidge developed a new X-ray tube which, unlike the old gas tube, provides consistent exposure and quality. X-ray imaging changed the practice of medicine by allowing doctors to look inside the body without using surgery. Today, X-ray images, and other technologies, like the MRI used to produce the image at the left, show bones, organs, and tissues.

DIFFERENTIATE INSTRUCTION

Below Level Have students make a timeline of their lives with index cards. Each card can represent a year. They can tape the cards together and record special events in news, sports, weather, or their own lives.

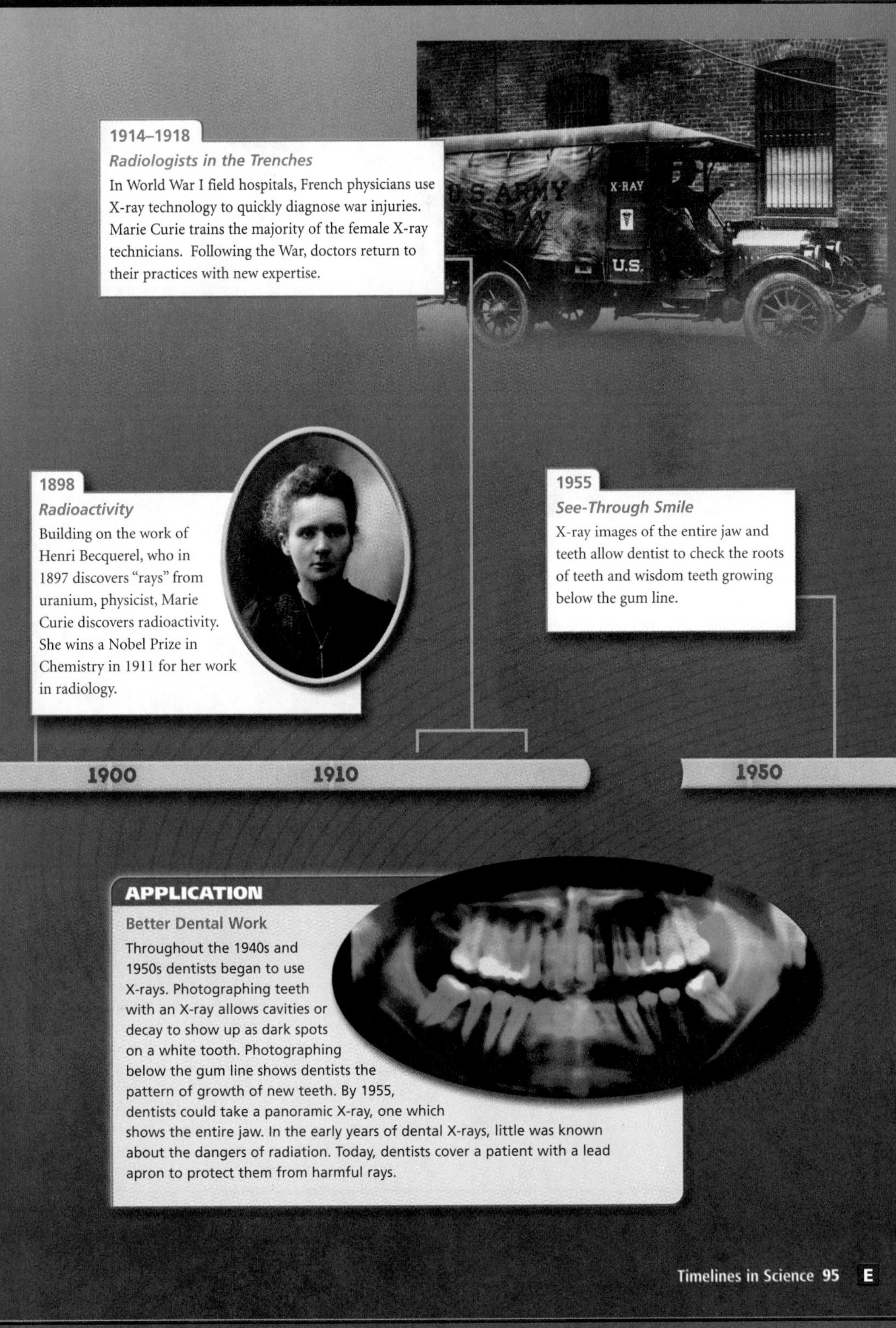

1914–1918
Radiologists in the Trenches
In World War I field hospitals, French physicians use X-ray technology to quickly diagnose war injuries. Marie Curie trains the majority of the female X-ray technicians. Following the War, doctors return to their practices with new expertise.

1898
Radioactivity
Building on the work of Henri Becquerel, who in 1897 discovers "rays" from uranium, physicist, Marie Curie discovers radioactivity. She wins a Nobel Prize in Chemistry in 1911 for her work in radiology.

1955
See-Through Smile
X-ray images of the entire jaw and teeth allow dentist to check the roots of teeth and wisdom teeth growing below the gum line.

1900 1910 1950

APPLICATION
Better Dental Work
Throughout the 1940s and 1950s dentists began to use X-rays. Photographing teeth with an X-ray allows cavities or decay to show up as dark spots on a white tooth. Photographing below the gum line shows dentists the pattern of growth of new teeth. By 1955, dentists could take a panoramic X-ray, one which shows the entire jaw. In the early years of dental X-rays, little was known about the dangers of radiation. Today, dentists cover a patient with a lead apron to protect them from harmful rays.

Timelines in Science 95 E

DIFFERENTIATE INSTRUCTION

Advanced One of Marie Curie's legacies is the use of radiation to treat cancer. Have students research this topic and give an oral presentation to the class.

Scientific Process

Have students name the skill that scientists are practicing when they use technology to see inside the body. *making observations*

Technology

1898 Tell students that radioactive elements such as uranium and radium, with which Marie Curie worked extensively, give off radiation that can make a person sick. Ask: What technological step was important in using radiation in medicine? *improvement in safe handling procedures*

Integrate the Sciences

The radioactivity of different elements can vary widely. For example, a gram of radium emits about one million times more radiation than a gram of uranium. Therefore, there are two common units for measuring radioactivity: the becquerel (Bq) and the curie (Ci). One Bq equals one particle of radiation released per second. One Ci is equal to the number of particles of radiation released per second by one gram of radium.

Scientific Process

Have students identify the part of the scientific process undertaken by Damadian, Minkoff, and Goldsmith when they worked on the first MRI, despite being told that it couldn't be done. *proving old theories false*

Application

1976 The use of computers in medicine has opened many doors for doctors today. An entire new field, medical informatics, is developing. For instance, the computer can develop models that take the patient's information, including scan results, and plan a course of treatment.

Technology

ULTRASOUND An echocardiogram uses ultrasound to show a heart beating. It helps doctors diagnose some types of heart disease, find tumors or blood clots in the heart, and monitor the heart after a heart attack.

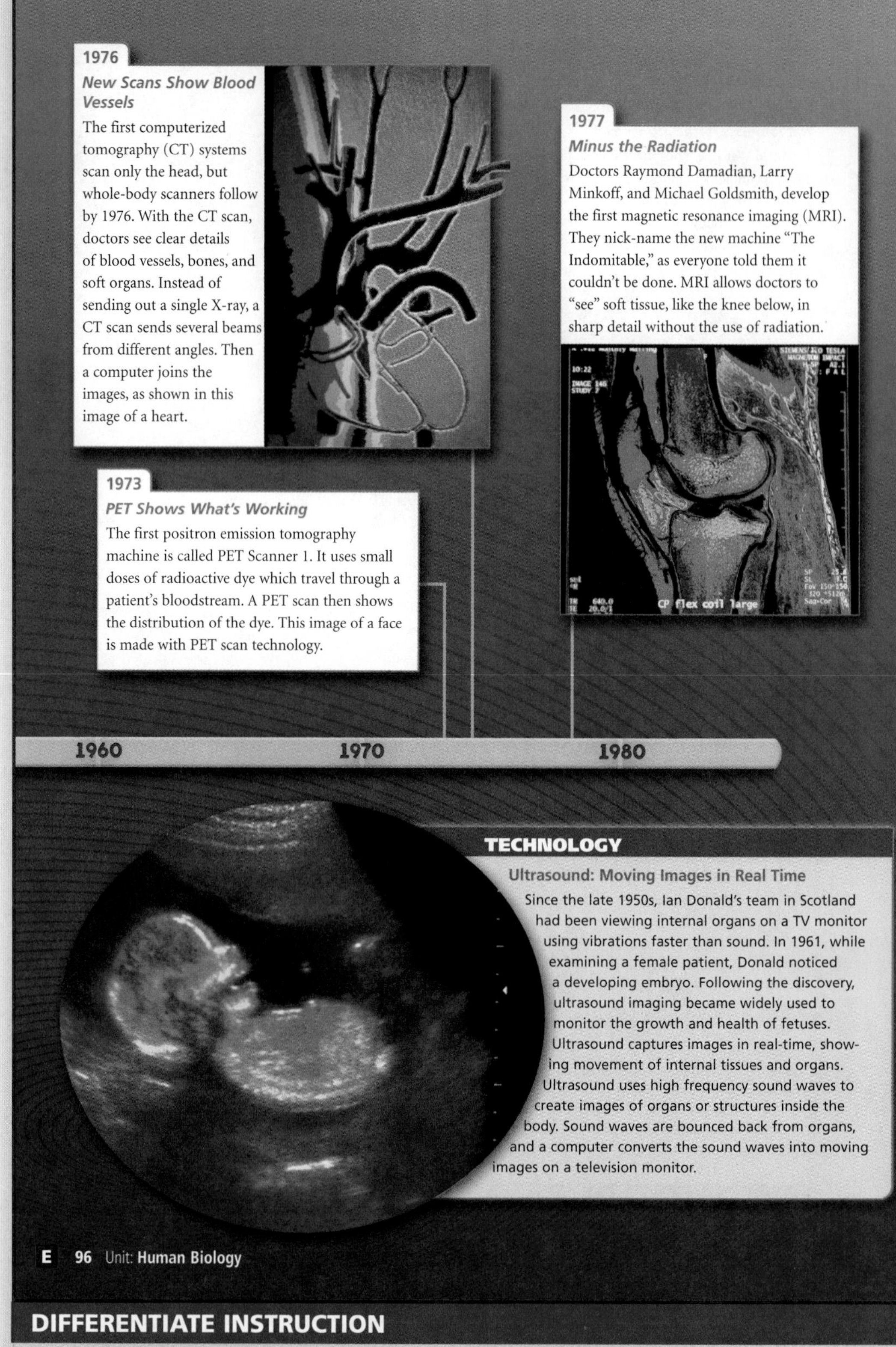

1976

New Scans Show Blood Vessels

The first computerized tomography (CT) systems scan only the head, but whole-body scanners follow by 1976. With the CT scan, doctors see clear details of blood vessels, bones, and soft organs. Instead of sending out a single X-ray, a CT scan sends several beams from different angles. Then a computer joins the images, as shown in this image of a heart.

1977

Minus the Radiation

Doctors Raymond Damadian, Larry Minkoff, and Michael Goldsmith, develop the first magnetic resonance imaging (MRI). They nick-name the new machine "The Indomitable," as everyone told them it couldn't be done. MRI allows doctors to "see" soft tissue, like the knee below, in sharp detail without the use of radiation.

1973

PET Shows What's Working

The first positron emission tomography machine is called PET Scanner 1. It uses small doses of radioactive dye which travel through a patient's bloodstream. A PET scan then shows the distribution of the dye. This image of a face is made with PET scan technology.

1960 1970 1980

TECHNOLOGY

Ultrasound: Moving Images in Real Time

Since the late 1950s, Ian Donald's team in Scotland had been viewing internal organs on a TV monitor using vibrations faster than sound. In 1961, while examining a female patient, Donald noticed a developing embryo. Following the discovery, ultrasound imaging became widely used to monitor the growth and health of fetuses. Ultrasound captures images in real-time, showing movement of internal tissues and organs. Ultrasound uses high frequency sound waves to create images of organs or structures inside the body. Sound waves are bounced back from organs, and a computer converts the sound waves into moving images on a television monitor.

E 96 Unit: Human Biology

DIFFERENTIATE INSTRUCTION

Advanced Have students research the costs of medical imaging, including the cost of buying the machines, training personnel, and performing the procedure. Have them calculate how many patients must use the technology before it starts to make money for the clinic or hospital.

1990s

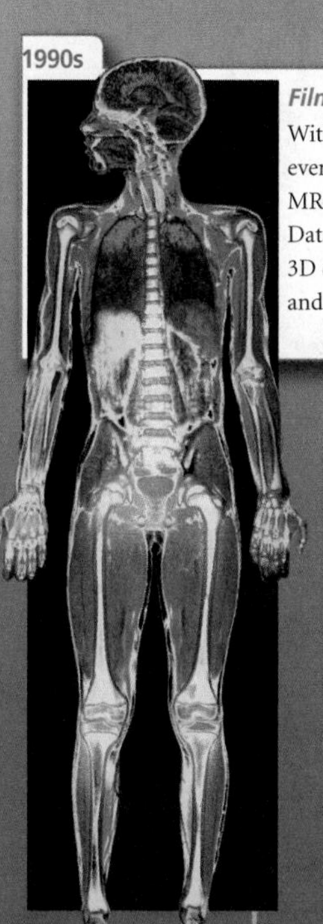

Filmless Images

With digital imaging, everything from X-rays to MRIs is now filmless. Data moves directly into 3D computer programs and shared databases.

2003

Multi-Slice CT

By 2003, 8- and 16-slice CT scanners offer detail and speed. A multi-slice scanner reduces exam time from 45 minutes to under 10 seconds.

RESOURCE CENTER
CLASSZONE.COM
Find more on advances in medical imaging.

1990 2000

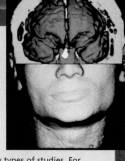

TECHNOLOGY

3-D Images and Brain Surgery

In operating rooms, surgeons are beginning to use another type of 3D ultrasound known as interventional MRI. They watch 3-D images in real time and observe details of tissues while they operate. These integrated technologies now allow scientists to conduct entirely new types of studies. For example, 3-D brain images of many patients with one disease—can now be integrated into a composite image of a "typical" brain of someone with that disease.

INTO THE FUTURE

Although discovered over 100 years ago X-rays are certain to remain a key tool of health workers for many years. What will be different in the future? Dentists have begun the trend to stop using film images, and rely on digital X-rays instead. In the future, all scans may be viewed and stored on computers. Going digital allows doctors across the globe to share images quickly by email.

Magnetic resonance imaging has only been in widespread use for about 20 years. Look for increased brain mapping—ability to scan the brain during a certain task. The greater the collective data on brain-mapping, the better scientists will understand how the brain works. To produce such an image requires thousands of patients and trillions of bytes of computer memory.

Also look for increased speed and mobile MRI scanners, which will be used in emergency rooms and doctor's offices to quickly assess internal damage after an accident or injury.

ACTIVITIES

Writing About Science: Brochure

Make a chart of the different types of medical imaging used to diagnose one body system. Include an explanation of how the technique works and list the pros and cons of using it.

Reliving History

X-rays use radioactivity which can be dangerous. You can use visible light to shine through thin materials that you don't normally see through. Try using a flashlight to illuminate a leaf. Discuss or draw what you see.

Technology

1990s Using digital imaging for dental X-rays reduces the radiation to which patients are exposed by about 90 percent. Instead of inserting film into a patient's mouth, dentists insert electronic sensors. These sensors develop images that are viewed on a computer screen. Ask: Why is this technology an improvement over the old X-rays? *It exposes patients to less radiation, which can be harmful in large doses; it reduces use of film.*

INTO THE FUTURE

Have students brainstorm some technological advances that might further improve our understanding of the inner body. Discuss the feasibility of the fictional *Star Trek* medical tricorder, which reveals a patient's condition after merely skimming the patient's body.

ACTIVITIES

Writing About Science: Brochure

Have students produce a decision tree that a physician could show patients to explain why one imaging technique is best for their circumstances.

Reliving History

Students can also aim a flashlight inside their mouths and see through their cheeks.

Technology Resources

Students can visit **ClassZone.com** for current news about medical imaging.

DIFFERENTIATE INSTRUCTION

Below Level Have a group of students work on the chart, each student tackling a different kind of imaging. They could find and include photographs of the procedure being done.

CHAPTER

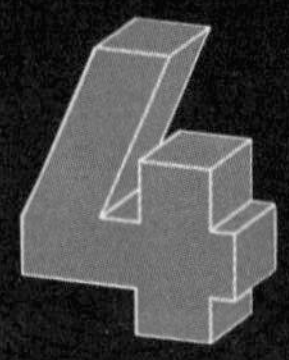

Control and Reproduction

Life Science
UNIFYING PRINCIPLES

PRINCIPLE 1
All living things share common characteristics.

PRINCIPLE 2
All living things share common needs.

PRINCIPLE 3
Living things meet their needs through interactions with the environment.

PRINCIPLE 4
The types and numbers of living things change over time.

Unit: Human Biology
BIG IDEAS

CHAPTER 1
Systems, Support, and Movement
The human body is made up of systems that work together to perform necessary functions.

CHAPTER 2
Absorption, Digestion, and Exchange
Systems in the body obtain and process materials and remove waste.

CHAPTER 3
Transport and Protection
Systems function to transport materials and to defend and protect the body.

CHAPTER 4
Control and Reproduction
The nervous and endocrine systems allow the body to respond to internal and external conditions.

CHAPTER 5
Growth, Development, and Health
The body develops and maintains itself over time.

CHAPTER 4
KEY CONCEPTS

SECTION 4.1

The nervous system responds and controls.

1. Senses connect the human body to its environment.
2. The central nervous system controls functions.
3. The peripheral nervous system is a network of nerves.

SECTION 4.2

The endocrine system helps regulate body conditions.

1. Hormones are the body's chemical messengers.
2. Glands produce and release hormones.
3. Control of the endocrine system includes feedback mechanisms.

SECTION 4.3

The reproductive system allows the production of offspring.

1. The reproductive system produces specialized cells.
2. The reproduction of offspring includes fertilization, pregnancy, and birth.

T The Big Idea Flow Chart is available on p. T25 in the **UNIT TRANSPARENCY BOOK.**

Previewing Content

SECTION

The nervous system responds and controls. pp. 101–109

1. **Senses connect the human body to its environment.**
   Senses monitor the environment to help maintain homeostasis.
   - The eye collects, bends, and focuses light into an image.
   - Sensory receptors in the skin detect pressure, texture, pain, and temperature.
   - Ears perceive sound waves, which are produced by vibrations.
   - Olfactory nerves in the nose perceive the presence of chemicals in the air.
   - Taste buds on the tongue perceive chemicals in food.

2. **The central nervous system controls functions.**
   The brain and spinal cord are the **central nervous system.** They communicate with the rest of the nervous system through electrical signals sent to and from neurons.
   - Every area of the brain has a specific function, controlling both voluntary and involuntary responses. Some of these areas appear in the figure below.

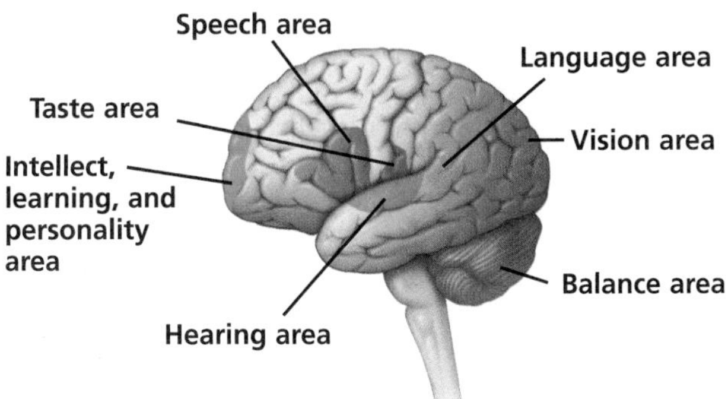
Speech area
Language area
Taste area
Vision area
Intellect, learning, and personality area
Balance area
Hearing area

   - The spinal cord is the main pathway for information. It is protected and supported by the vertebral column.

3. **The peripheral nervous system is a network of nerves.**
   The **autonomic nervous system** controls involuntary responses such as digestion and heartbeat. The voluntary nervous system monitors and controls conscious functions.

SECTION

The endocrine system helps regulate body conditions. pp. 110–117

1. **Hormones are the body's chemical messengers.**
   Hormones are chemicals made in one organ to trigger response in another organ. The **endocrine system** uses hormones to control conditions in the body.

2. **Glands produce and release hormones.**
   - The pituitary gland directs the endocrine system. Its hormones control growth, sexual development, and water absorption.
   - The hypothalamus is the part of the brain that controls the pituitary gland.
   - The pineal gland regulates sleep, body temperature, reproduction, and aging.
   - The thyroid gland produces hormones for growth and metabolism.
   - The thymus helps the body fight disease.
   - The adrenal glands produce about 30 hormones that regulate food metabolism, water and salt levels, immunity, and response to stress.
   - The pancreas produces insulin, which regulates glucose levels in the blood.

3. **Control of the endocrine system includes feedback mechanisms.**
   - Negative feedback mechanisms help maintain homeostasis.

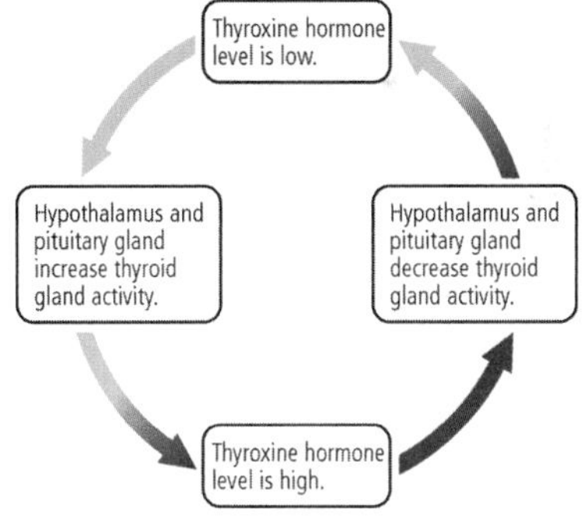
Thyroxine hormone level is low.
Hypothalamus and pituitary gland decrease thyroid gland activity.
Thyroxine hormone level is high.
Hypothalamus and pituitary gland increase thyroid gland activity.

   - Positive feedback mechanisms, like blood clotting, result in extreme responses.
   - Some hormone action is regulated by a hormone that produces an opposite action.
   - Hormone imbalance can cause serious disease.

Common Misconceptions

THE BRAIN Students sometimes think that the body performs some functions without needing the brain. In fact, the brain is in control of all functions of the body.

 This misconception is addressed on p. 106.

MISCONCEPTION DATABASE

CLASSZONE.COM Background on student misconceptions

Previewing Content

SECTION

The reproductive system allows the production of offspring.

pp. 118–125

1. **The reproductive system produces specialized cells.**
   The female produces eggs; the male produces sperm. These specialized cells contain genetic material. The female reproductive system has two functions: to produce eggs and to protect and nourish the offspring until birth.
   - The pituitary gland stimulates eggs in the ovary to grow.
   - An egg is released into the fallopian tube, where it may be fertilized by a sperm cell.
   - The egg moves to the uterus, where, if fertilized, it implants in the uterine wall.
   - If not fertilized, the egg and the uterine lining exit the body through the vagina, in a process called **menstruation.**

   The male reproductive system contains the testes, which produce sperm, and the urethra, which is a canal through the penis. Sperm move through the urethra and exit the penis.

2. **The reproduction of offspring includes fertilization, pregnancy, and birth.**
   - **Fertilization** is the joining of one egg cell and one sperm cell.
   - Once fertilized, the egg divides, producing an **embryo.** Cells in the embryo continue dividing.
   - The embryo implants in the spongy lining of the uterus.
   - During the nine months of pregnancy, the embryo grows and develops, fed through the placenta.
   - The first stage of birth begins with muscular contractions of the uterus.
   - In the second stage, the cervix is fully dilated and the baby leaves the mother.
   - In the third stage the umbilical cord is cut and the placenta exits the body.

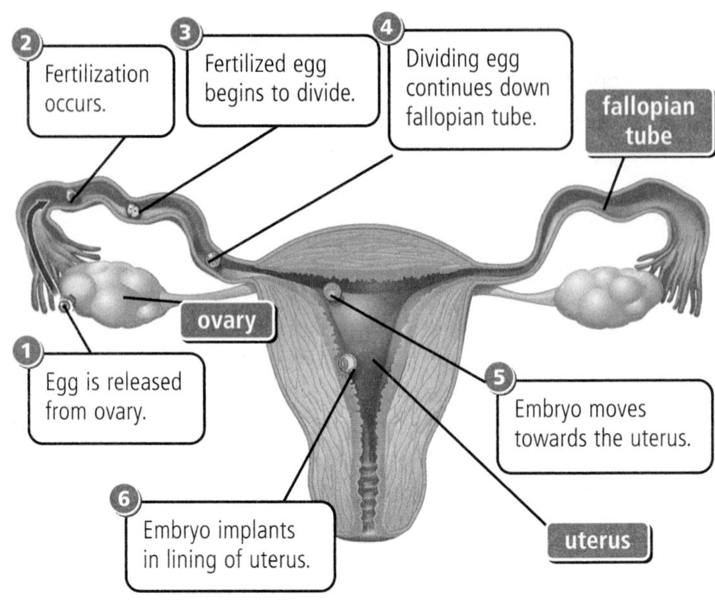

2 Fertilization occurs.
3 Fertilized egg begins to divide.
4 Dividing egg continues down fallopian tube.
fallopian tube
ovary
1 Egg is released from ovary.
5 Embryo moves towards the uterus.
6 Embryo implants in lining of uterus.
uterus

An embryo implants in the spongy lining of the uterus, where it grows and develops during the nine months of pregnancy.

Common Misconceptions

MISCONCEPTION DATABASE
CLASSZONE.COM Background on student misconceptions

THE UNION OF EGG AND SPERM Students sometimes think that the egg contains life, with the sperm needed to trigger growth, or that the sperm contains life, with the egg needed for protection and nourishment. Neither contain life as neither type of cell can divide or reproduce on its own. Life results when they join.

This misconception is addressed on p. 119.

Previewing Labs

Lab Generator CD-ROM
Edit these Pupil Edition labs and generate alternative labs.

EXPLORE the BIG idea

Color Confusion, p. 99 Students create conflicting visual stimuli with colors and words in order to observe sensory confusion.	**TIME** 10 minutes **MATERIALS** paper, 6 different-colored markers
Eggs, p. 99 Students observe a chicken egg to learn the parts and functions of an egg.	**TIME** 10 minutes **MATERIALS** egg, small bowl
Internet Activity: The Senses, p. 99 Students view how each sense sends specific information to the brain.	**TIME** 20 minutes **MATERIALS** computer with Internet access

SECTION  4.1

EXPLORE Smell, p. 101 Students explore their sense of smell, testing it with three unknowns.	**TIME** 10 minutes **MATERIALS** 3 bags, an orange, a pine branch, a cinnamon stick
CHAPTER INVESTIGATION **Are You a Supertaster?** pp. 108–109 Students take impressions of their fungiform papillae to infer possible connections between tastebuds and sensitivity to flavors.	**TIME** 40 minutes **MATERIALS** blue food coloring, paper cup, cotton swab, reinforcement circle, paper towel or napkin, white paper, notebook

SECTION  4.2

INVESTIGATE Response to Exercise, p. 114 Students record their body temperatures before and during exercise to observe a physical response.	**TIME** 30 minutes **MATERIALS** stopwatch, graph paper, mercury-free thermometer, rubbing alcohol or plastic thermometer covers

SECTION  4.3

EXPLORE Reproduction, p. 118 Students observe and contrast egg and sperm cells under a microscope.	**TIME** 20 minutes **MATERIALS** microscope, slides of egg and sperm cells, paper and pencil

 **Additional INVESTIGATION,** Taste That Smell!, A, B, & C, pp. 235–243; Teacher Instructions, pp. 305–306

Previewing Chapter Resources

CHAPTER 4
Control and Reproduction

INTEGRATED TECHNOLOGY

CLASSZONE.COM
- eEdition Plus
- EasyPlanner Plus
- Misconception Database
- Content Review
- Test Practice
- Visualization
- Resource Centers
- Internet Activity: The Senses
- Math Tutorial

SCILINKS.ORG

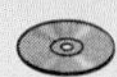
CD-ROMS
- eEdition
- EasyPlanner
- Power Presentations
- Content Review
- Lab Generator
- Test Generator

AUDIO CDS
- Audio Readings
- Audio Readings in Spanish

LABS AND ACTIVITIES

EXPLORE the Big Idea, p. 99
- Color Confusion
- Eggs
- Internet Activity: The Senses

UNIT RESOURCE BOOK
Unit Projects, pp. 5–10

Lab Generator CD-ROM
Generate customized labs.

The nervous system responds and controls.
pp. 101–109

Time: 3 periods (1.5 blocks)

 Lesson Plan, pp. 188–189

RESOURCE CENTER, Nervous System

UNIT TRANSPARENCY BOOK
- Big Idea Flow Chart, p. T25
- Daily Vocabulary Scaffolding, p. T26
- Note-Taking Model, p. T27
- 3-Minute Warm-Up, p. T28

- EXPLORE Smell, p. 101
- CHAPTER INVESTIGATION, Are You a Supertaster? pp. 108–109

UNIT RESOURCE BOOK
- CHAPTER INVESTIGATION, Are You a Supertaster? A, B, & C, pp. 226–234
- Additional INVESTIGATION, Taste That Smell!, A, B, & C, pp. 235–243

The endocrine system helps regulate body conditions.
pp. 110–117

Time: 2 periods (1 block)

 Lesson Plan, pp. 198–299

RESOURCE CENTER, Endocrine System

UNIT TRANSPARENCY BOOK
- Daily Vocabulary Scaffolding, p. T26
- 3-Minute Warm-Up, p. T28
- "Endocrine System" Visual, p. T30

- INVESTIGATE Response to Exercise, p. 114
- Connecting Sciences, p. 117

UNIT RESOURCE BOOK
Datasheet, Response to Exercise, p. 207

The reproductive system allows the production of offspring.
pp. 118–125

Time: 3 periods (1.5 blocks)

Lesson Plan, pp. 209–210

- **VISUALIZATION,** Fertilization
- **MATH TUTORIAL**

UNIT TRANSPARENCY BOOK
- Big Idea Flow Chart, p. T25
- Daily Vocabulary Scaffolding, p. T26
- 3-Minute Warm-Up, p. T29
- Chapter Outline, pp. T31–T32

- EXPLORE Reproduction, p. 118
- Math in Science, p. 125

UNIT RESOURCE BOOK
- Math Support, p. 224
- Math Practice, p. 225

KEY TO ICONS

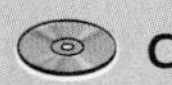 CD/CD-ROM
 Teacher Edition
 UNIT TRANSPARENCY BOOK
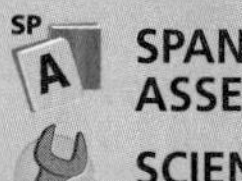 SPANISH ASSESSMENT BOOK
INTERNET
Pupil Edition
 UNIT RESOURCE BOOK
UNIT ASSESSMENT BOOK
SCIENCE TOOLKIT

READING AND REINFORCEMENT	ASSESSMENT	STANDARDS
 • Description Wheel, B20–21 • Choose Your Own Strategy, C35–44 • Daily Vocabulary Scaffolding, H1–8  **UNIT RESOURCE BOOK** • Vocabulary Practice, p. 221–222 • Decoding Support, p. 223 • Summarizing the Chapter, pp. 244–245  **Audio Readings CD** Listen to Pupil Edition.  **Audio Readings in Spanish CD** Listen to Pupil Edition in Spanish.	 • Chapter Review, pp. 127–128 • Standardized Test Practice, p. 129  **UNIT ASSESSMENT BOOK** • Diagnostic Test, pp. 58–59 • Chapter Test, A, B, & C, pp. 63–74 • Alternative Assessment, pp. 75–76  Spanish Chapter Test, pp. 93–96  **Test Generator CD-ROM** Generate customized tests.  **Lab Generator CD-ROM** Rubrics for Labs	National Standards A.1–8, A.9.a–c, A.9.e–g, C.1.d–e, C.2.a–c, G.1.b See p. 98 for the standards.
 **UNIT RESOURCE BOOK** • Reading Study Guide, A & B, pp. 190–193 • Spanish Reading Study Guide, pp. 194–195 • Challenge and Extension, p. 196 • Reinforcing Key Concepts, p. 197 • Challenge Reading, pp. 219–220	 Ongoing Assessment, pp. 102–105, 107  Section 4.1 Review, p. 107  **UNIT ASSESSMENT BOOK** Section 4.1 Quiz, p. 60	National Standards A.1–7, A.9.a–b, A.9.e–g, C.1.e, G.1.b
 **UNIT RESOURCE BOOK** • Reading Study Guide, A & B, pp. 200–203 • Spanish Reading Study Guide, pp. 204–205 • Challenge and Extension, p. 206 • Reinforcing Key Concepts, p. 208	 Ongoing Assessment, pp. 111–114, 116  Section 4.2 Review, p. 116  **UNIT ASSESSMENT BOOK** Section 4.2 Quiz, p. 61	National Standards A.2–7, A.9.a–b, A.9.e–f, C.1.e, G.1.b
 **UNIT RESOURCE BOOK** • Reading Study Guide, A & B, pp. 211–214 • Spanish Reading Study Guide, pp. 215–216 • Challenge and Extension, p. 217 • Reinforcing Key Concepts, p. 218	 Ongoing Assessment, pp. 119–124  Section 4.3 Review, p. 124  **UNIT ASSESSMENT BOOK** Section 4.3 Quiz, p. 62	National Standards A.2–8, A.9.a–c, A.9.e–f, C.1.d, C.2.a–c, G.1.b

Previewing Resources for Differentiated Instruction

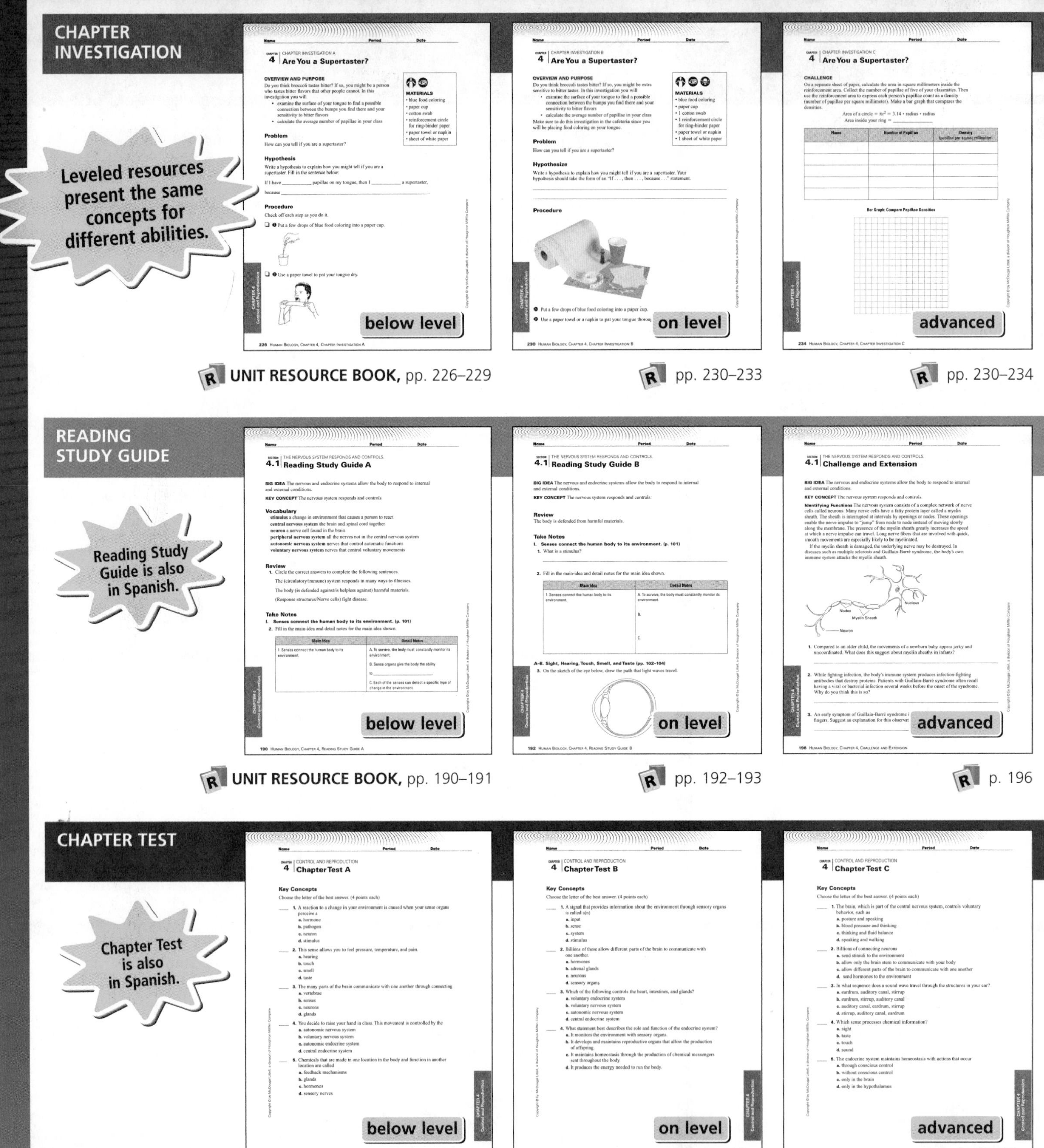
CHAPTER INVESTIGATION
Leveled resources present the same concepts for different abilities.
Chapter Investigation A
Are You a Supertaster?
below level
Chapter Investigation B
Are You a Supertaster?
on level
Chapter Investigation C
Are You a Supertaster?
advanced
UNIT RESOURCE BOOK, pp. 226–229
pp. 230–233
pp. 230–234
READING STUDY GUIDE
Reading Study Guide is also in Spanish.
4.1 Reading Study Guide A
below level
4.1 Reading Study Guide B
on level
4.1 Challenge and Extension
advanced
UNIT RESOURCE BOOK, pp. 190–191
pp. 192–193
p. 196
CHAPTER TEST
Chapter Test is also in Spanish.
4 Chapter Test A
below level
4 Chapter Test B
on level
4 Chapter Test C
advanced
UNIT ASSESSMENT BOOK, pp. 63–66
pp. 67–70
pp. 71–74

TECHNOLOGY

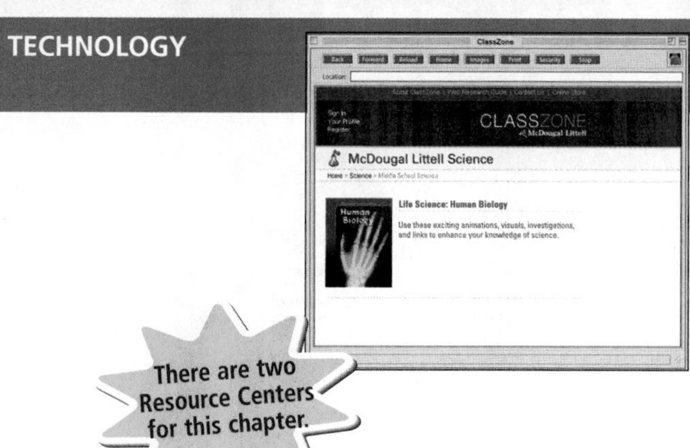

There are two Resource Centers for this chapter.

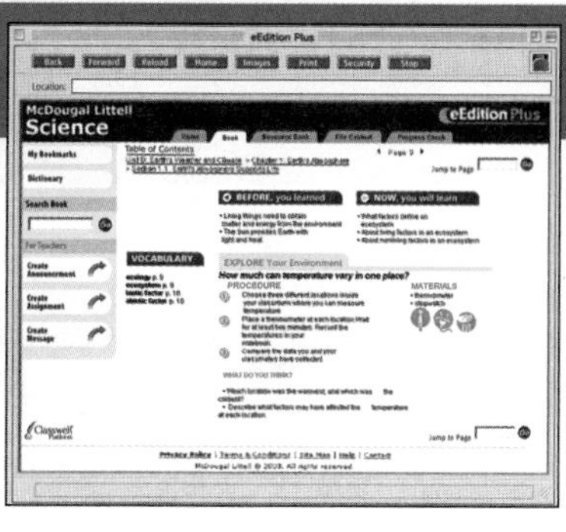

CLASSZONE.COM

CD/CD-ROMS

CLASSZONE.COM

VISUAL CONTENT

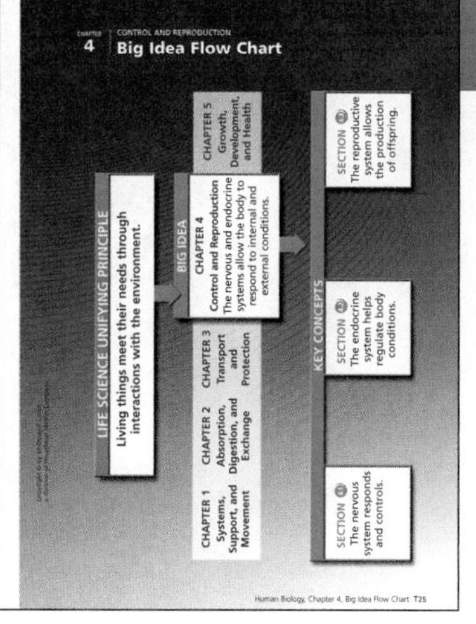

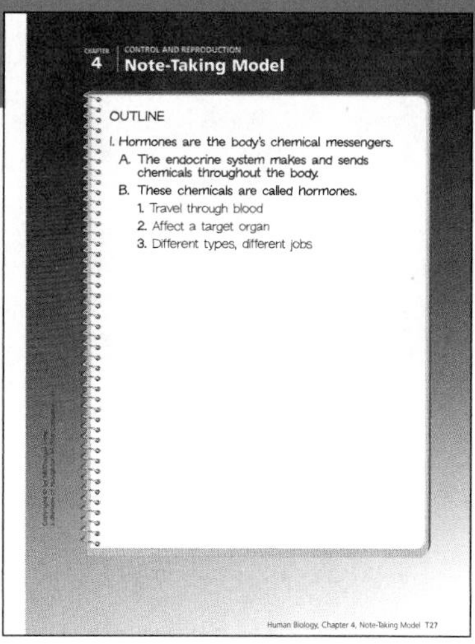

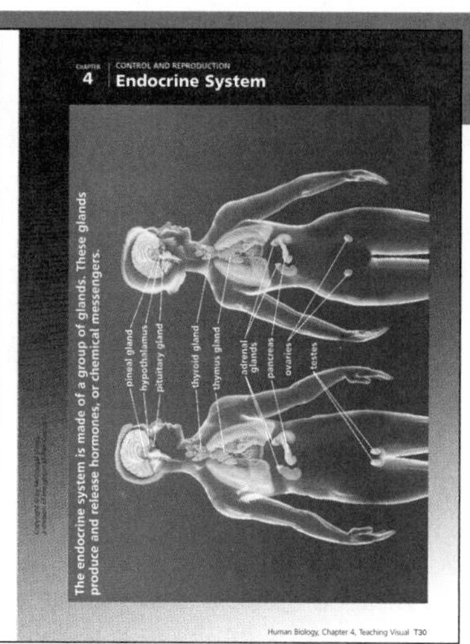

UNIT TRANSPARENCY BOOK, p. T25

p. T27

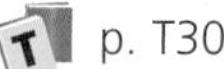
p. T30

MORE SUPPORT

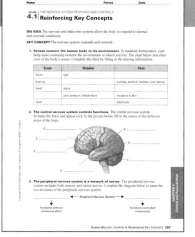

Reinforcing Key Concepts for each section

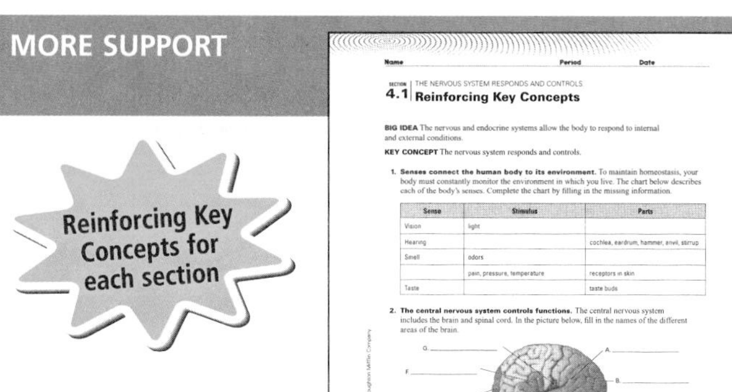

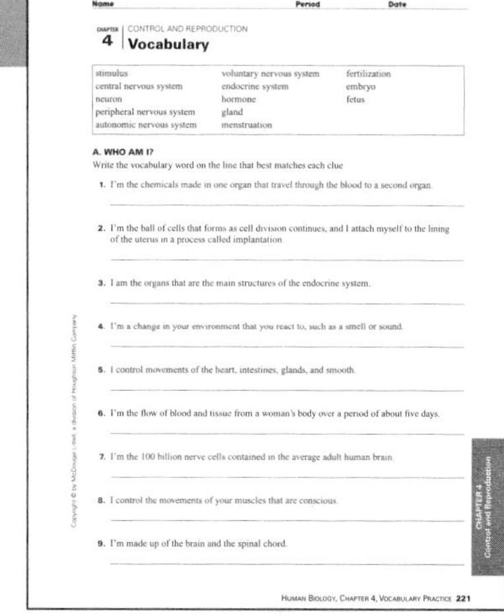

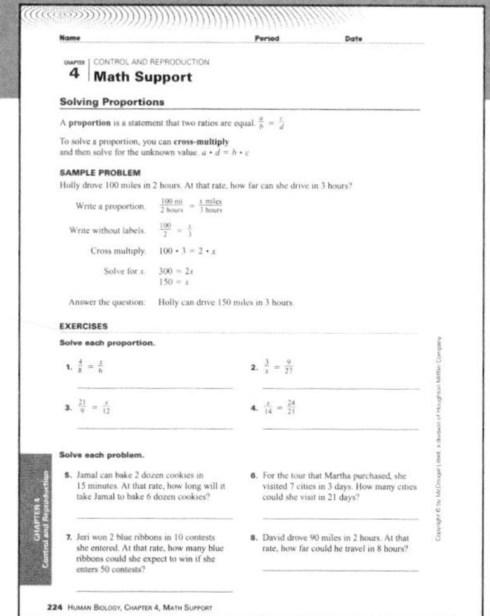

UNIT RESOURCE BOOK, p. 197

pp. 221–222

p. 224

CHAPTER 4 Control and Reproduction

INTRODUCE

the BIG idea

Have students look at the photograph of nerve cells and discuss how the question in the box relates to the Big Idea. Ask:

- How does your body get the information from words on a page?
- Why might it be necessary for different types of cells to be involved?

National Science Education Standards

Content

C.1.d Specialized cells perform specialized functions in multicellular organisms.

C.1.e The human organism has systems: digestion, respiration, reproduction, circulation, excretion, movement, control and coordination, and protection.

C.2.a Reproduction is essential to the continuation of every species.

C.2.b Females produce eggs, and males sperm, which unite in a new individual.

C.2.c Every organism requires a set of instructions for specifying its traits.

Process

A.1–8 Identify questions that can be answered through scientific investigations; design and conduct an investigation; use tools to gather and interpret data; use evidence to describe, predict, explain, model; think critically to make relationships between evidence and explanation; recognize different explanations and predictions; communicate scientific procedures and explanations; use mathematics.

A.9.a–c, A.9.e–g Understand scientific inquiry by using different investigations, methods, mathematics, explanations based on logic, evidence, and skepticism. Data often result in new investigations.

G.1.b Science requires different abilities.

CHAPTER 4 Control and Reproduction

the BIG idea

The nervous and endocrine systems allow the body to respond to internal and external conditions.

These are nerve cells. What do nerves in your body do?

Key Concepts

SECTION 4.1 **The nervous system responds and controls.** Learn how the senses help the body get information about the environment.

SECTION 4.2 **The endocrine system helps regulate conditions inside the body.** Learn the functions of different hormones.

SECTION 4.3 **The reproductive system allows the body to produce offspring.** Learn about the process of reproduction.

Internet Preview

CLASSZONE.COM

Chapter 4 online resources: Content Review, Visualization, three Resource Centers, Math Tutorial, Test Practice

INTERNET PREVIEW

CLASSZONE.COM For student use with the following pages:

Review and Practice

- Content Review, pp. 100, 126
- Math Tutorial: Solving Proportions, p. 125
- Test Practice, p. 129

Activities and Resources

- Internet Activity: The Senses, p. 99
- Resource Centers: Nervous System, p. 104; Endocrine System, p. 111
- Visualization: Fertilization, p. 120

Reproductive System

Code: MDL047

EXPLORE the BIG idea

Color Confusion

Make a list of six colors using a different color marker or colored pencil to write each one. Make sure not to write the color name with the same color marker or pencil. Read the list out loud as fast as you can. Now try quickly saying the color of each word out loud.

Observe and Think Did you notice a difference between reading the words in the list and saying the colors? If so, why do you think that is?

Eggs

Examine a raw chicken egg. Describe the appearance of the outside shell. Break it open into a small dish and note the different parts inside. Wash your hands when you have finished.

Observe and Think If this egg had been fertilized, which part do you think would have served as the food for the growing chicken embryo? Which part would protect the embryo from impact and serve to cushion it?

Internet Activity: The Senses

Go to **ClassZone.com** to learn how the senses allow the body to respond to external conditions. See how each sense sends specific information to the brain.

Observe and Think How do the different senses interact with one another?

Reproductive System **Code: MDL047**

TEACHING WITH TECHNOLOGY

Graphing Calculator Students can use a graphing calculator to collect and graph data on exercise on p. 114.

PC Microscope Students could use a PC microscope to view eggs and sperm, p. 118, and then create a presentation with screen captures.

EXPLORE the BIG idea

These inquiry-based activities are appropriate for use at home or as a supplement to classroom instruction.

Color Confusion

PURPOSE To look at conflicting stimuli (coloring and word meaning) and observe how the senses can be confused.

TIP *10 min.* To shorten the activity, print a master set of flash cards and show them to the class, or work with three of four colors instead of six.

Answer: It took some extra time to sort out what color was actually being referred to.

REVISIT after p. 102.

Eggs

PURPOSE To examine a chicken egg to infer the parts and functions of an egg.

TIPS *10 min.* To conserve eggs, students can work in groups as large as five or six, or the activity can be done as a demo. If possible, show students the membrane that surrounds the egg white. Point out that eggs and embryos in humans do not have shells because the body cushions them, but they have protective membranes.

Answer: the yolk; the white

REVISIT after p. 123.

Internet Activity: The Senses

PURPOSE To learn the function of each of the senses.

TIP *20 min.* Students might try answering the question before and after their observations.

Sample answer: Taste and smell interact; losing sense of smell can decrease ability to taste food.

REVISIT after p. 105.

PREPARE

CONCEPT REVIEW

Activate Prior Knowledge

- Ask student volunteers to name the functions of the circulatory, immune, and integumentary systems.
- You might create a matching game with three types of cards: system, organ, and function. Students can match functions and organs to systems. They can also arrange cards to show interactions and relationships.

TAKING NOTES

Choose Your Own Strategy

Remind students that different graphic organizers may appeal to different types of learners. An outline presents information in linear form, while a main idea web can organize more random thought processes. Main idea and detail notes are particularly helpful for those who draw and visualize what they learn.

Vocabulary Strategy

A description wheel can help students relate new or difficult concepts to familiar or concrete ones.

- Encourage students to create a wheel for each vocabulary term.
- They can set up their wheels in their Science Notebooks before reading the section.
- They can fill in the center of each wheel with a term, and then search or scan the text for details that explain or give context to the term.

Vocabulary and Note-Taking Resources

- Vocabulary Practice, pp. 222–223
- Decoding Support, p. 224

- Daily Vocabulary Scaffolding, p. T26
- Note-Taking Model, p. T27

- Description Wheel, B20–21
- Choose Your Own Strategy, C35–44
- Daily Vocabulary Scaffolding, H1–8

CHAPTER 4

Getting Ready to Learn

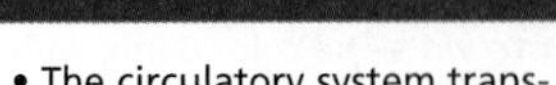

CONCEPT REVIEW

- The circulatory system transports materials.
- The immune system responds to foreign materials.
- The integumentary system protects the body.

VOCABULARY REVIEW

homeostasis p. 12
circulatory system p. 65
immune system p. 75
integumentary system p. 83

CONTENT REVIEW
CLASSZONE.COM
Review concepts and vocabulary.

TAKING NOTES

CHOOSE YOUR OWN STRATEGY

Take notes using one or more of the strategies from earlier chapters—**main idea webs, outlines,** or **main idea and detail notes.** You can also use other note-taking strategies that you might already know.

VOCABULARY STRATEGY

Place each vocabulary term at the center of a **description wheel** diagram. Write some words describing it on the spokes.

See the Note-Taking Handbook on pages R45–R51.

SCIENCE NOTEBOOK

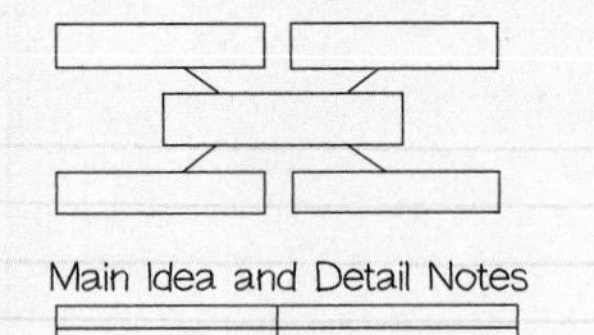

CHECK READINESS

Administer the Diagnostic Test to determine students' readiness for new science content and their mastery of requisite math skills.

 Diagnostic Test, pp. 58–59

Technology Resources

Students needing content and math skills should visit **ClassZone.com**.

- **CONTENT REVIEW**
- **MATH TUTORIAL**

 **CONTENT REVIEW CD-ROM**

KEY CONCEPT

The nervous system responds and controls.

BEFORE, you learned

- The body is defended from harmful materials
- Response structures fight disease
- The immune system responds to illness in many ways

NOW, you will learn

- How the body's senses help monitor the environment
- How the sensory organs respond to stimuli
- How the nervous system works with other body systems

VOCABULARY

stimulus p. 102
central nervous system p. 104
neuron p. 105
peripheral nervous system p. 106
autonomic nervous system p. 107
voluntary nervous system p. 107

EXPLORE Smell

Can you name the scent?

PROCEDURE

1. With a small group, take turns smelling the 3 mystery bags given to you by your teacher.
2. In your notebook, write down what you think is inside each bag without showing the people in your group.
3. Compare your answers with those in your group and then look inside the bags.

MATERIALS

three small paper bags

WHAT DO YOU THINK?

- Did you know what was in the bags before looking inside? If so, how did you know?
- What are some objects that would require more than a sense of smell to identify?

Senses connect the human body to its environment.

CHOOSE YOUR OWN STRATEGY
Use a strategy from an earlier chapter to take notes on the main idea. *Senses connect the human body to its environment.*

To maintain homeostasis and to survive, your body must constantly monitor the environment in which you live. This involves organs that interact so closely with the nervous system that they are often considered extensions of the nervous system. These are your sense organs. They give you the ability to see, smell, touch, hear, and taste.

Each of the senses can detect a specific type of change in the environment. For example, if you have begun to cross the street but suddenly hear a horn blowing, you may stop and step back onto the curb. Your sense of hearing allowed your brain to perceive that a car was coming and thus helped you to protect yourself.

RESOURCES FOR DIFFERENTIATED INSTRUCTION

Below Level

UNIT RESOURCE BOOK

- Reading Study Guide A, pp. 190–191
- Decoding Support, p. 223

AUDIO CDS

Additional INVESTIGATION, Taste That Smell!, A, B, & C, pp. 235–243; Teacher Instructions, pp. 305–306

Advanced

UNIT RESOURCE BOOK

- Challenge and Extension, p. 196
- Challenge Reading, pp. 219–220

English Learners

UNIT RESOURCE BOOK

Spanish Reading Study Guide, pp. 194–195

- Audio Readings in Spanish
- Audio Readings (English)

4.1 FOCUS

Set Learning Goals

Students will

- Describe how the body's senses help monitor the environment.
- Explain how the sensory organs respond to stimuli.
- Describe how the nervous system works with other body systems.
- Infer possible connections between taste buds and sensitivity to flavors.

3-Minute Warm-Up

Display Transparency 28 or copy this exercise on the board:

Imagine that for the past three days someone in your household has had a cold, yet you have not caught it. Detail what has been happening in your body to keep you healthy. *Every time the cold virus tried to enter your body, the immune system killed it or your skin and circulatory system repelled it.*

 3-Minute Warm-Up, p. T28

4.1 MOTIVATE

EXPLORE Smell

PURPOSE To explore the sense of smell

TIPS *10 min.* Have students close their eyes when smelling the contents of the bags. You may want to tape or staple bags shut to eliminate the chance of peeking. If possible, include mystery smells not pictured in the student book, such as mint, cloves, garlic, lemon, lime, coffee, cedar shavings, or onion. Remind students that they should never smell mystery items without a teacher's permission.

WHAT DO YOU THINK? *You could tell what the items were by their odor. Things such as paper or glass without distinctive odors would require more than the sense of smell to identify them.*

INSTRUCT

Teach from Visuals

To help students interpret the diagram of the eye, ask:

- Where are light rays bent? *cornea*
- Where are light rays focused? *lens*
- Where do rays form an image? *on the retina at the back of the eye*

Arts Connection

Students are usually fascinated by optical illusions. Show them a piece of M.C. Escher's work, or enter "optical illusions" into the search engine on your computer and download examples from the Internet. Also display samples of pointillism, a style of painting that relies on persistence of vision for viewers to see a field of colored dots as a seamless image.

Teacher Demo

Persistence of vision is an everyday optical illusion that can be easily demonstrated in two ways.

- Show a dollar bill, a page from the Sunday comics, or a printout from an inkjet printer photocopied at an enlarged (200% or greater) size. From a distance of a few inches away students see a seamless image, but their brains are really filling in the spaces between the dots.
- Create or find a flip book. Demonstrate by flipping two or more images in an action sequence how the brain fills in a fluid motion where there are really disjointed, still images.

EXPLORE the BIG idea

Revisit "Color Confusion" on p. 99. Have students explain their results.

Ongoing Assessment

Describe how the body's senses help monitor the environment.

Ask: What does your body do when there is a change in your environment? *The senses perceive the change and the brain makes sense of it, causing the body to respond to maintain homeostasis.*

The sound of the horn is a **stimulus.** A stimulus is a change in your environment that you react to, such as a smell, taste, sound, feeling, or sight. Your brain interprets any such change. If it did not, the information perceived by the senses would be meaningless.

Sight

If you have ever tried to find your way in the dark, you know how important light is for seeing. Light is a stimulus. You are able to detect it because your eyes, the sense organs of sight, capture light and help turn it into an image, which is processed by the brain.

Light enters the eye through the lens, a structure made of transparent tissue. Muscles surrounding the lens change its shape so the lens focuses light. Other muscles control the amount of light that enters the eye by altering the size of the pupil, a dark circle in the center of the eye. To reduce the amount of light, the area around the pupil, called the iris, contracts, making the pupil smaller thus allowing less light to enter. When the iris relaxes, more light can enter the eye.

At the back of the eye, the light strikes a layer called the retina. Among the many cells of the retina are two types of receptors, called rods and cones. Rods detect changes in brightness, while cones are sensitive to color.

Sight

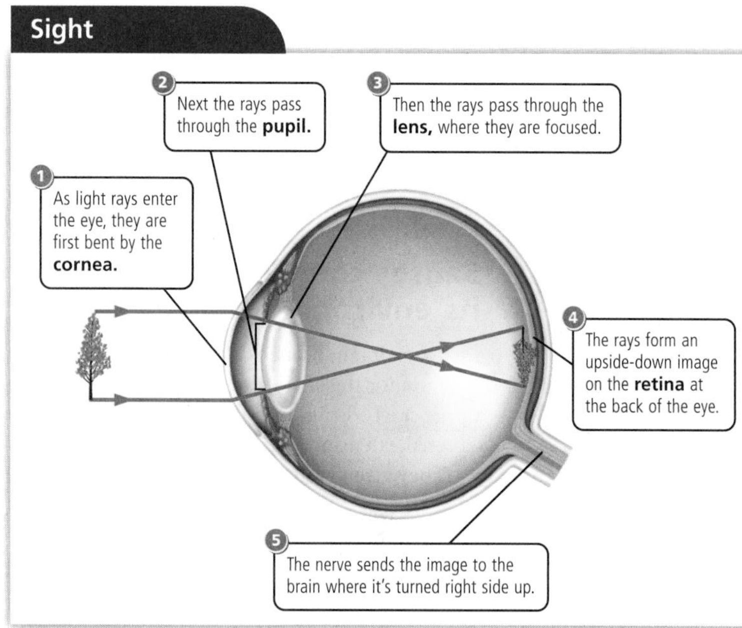

1 As light rays enter the eye, they are first bent by the **cornea.**
2 Next the rays pass through the **pupil.**
3 Then the rays pass through the **lens,** where they are focused.
4 The rays form an upside-down image on the **retina** at the back of the eye.
5 The nerve sends the image to the brain where it's turned right side up.

DIFFERENTIATE INSTRUCTION

More Reading Support

A What stimulus triggers sight? *light; variations in light*

B What is the lens made of? *transparent tissue*

English Learners Have English learners work in pairs or groups with advanced students to trace the diagram of sight in the human eye. They should leave the boxes blank and unnumbered except for the key terms: *cornea, pupil, lens, retina, nerve.* Then they should take turns explaining orally what happens when an image is seen. (They can continue with the Hearing diagram on p. 103). You may want students who are new to English to preview all the variants of sensory words used in this section, such as *vision, sight, seeing; sound, hearing; touch, feeling; perceive, sense,* and so on.

Hearing

Your eyes perceive light waves, but your ears perceive and interpret a different type of stimulus, sound waves. Sound waves are produced by vibrations. A reed on a clarinet vibrates, and so do your vocal cords. So does a bell after it has been hit by a mallet. The motion causes changes in the air that surrounds the bell. These changes can often be processed by the ear as sound, although many vibrations are too low or high to be heard by humans.

Sound waves enter the ear and are funneled into the auditory canal, a tube-shaped structure that ends at the eardrum. The eardrum vibrates when the sound waves strike it, and it transmits some of the vibrations to a tiny bone called the stirrup. Pressure caused by vibrations from the stirrup causes fluid in the ear to move. The movement of the fluid sends signals to the brain that are interpreted as sound.

 How are vibrations involved in hearing?

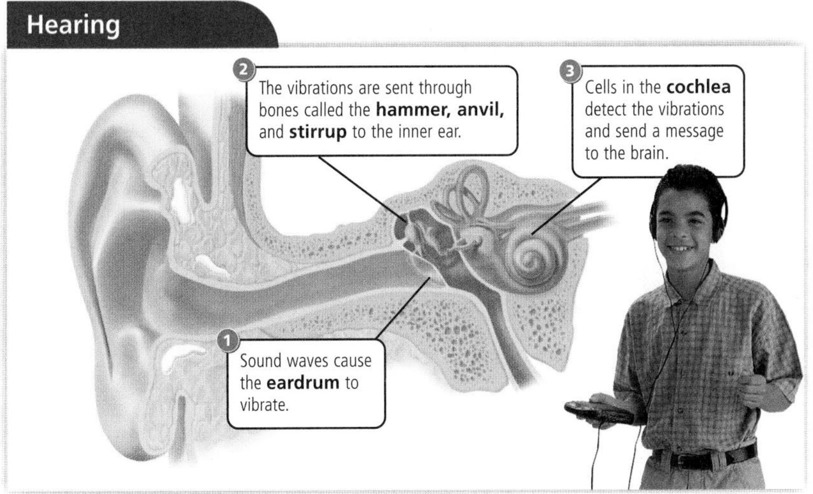

Hearing
1 Sound waves cause the eardrum to vibrate.
2 The vibrations are sent through bones called the hammer, anvil, and stirrup to the inner ear.
3 Cells in the cochlea detect the vibrations and send a message to the brain.

Touch

The sense of touch depends on tiny sensory receptors in the skin. Without these you wouldn't be able to feel pressure, temperature, or pain. Nerves in the outer layer of your skin, or epidermis, sense textures, like smooth glass or rough concrete. Nerves deeper in the skin, in the dermis, sense pressure. Receptors also sense how hot or cold an object is and can thus help protect you from burning yourself. The sense of touch is important in alerting your brain to danger. Though you might wish that you couldn't feel pain it serves a critical purpose. Without it, you could harm your body without realizing it.

Develop Critical Thinking

APPLY Have students describe incidents or situations where their senses have alerted them to an important change in their environment. For example, the sense of smell can alert you to burning food or smoke in the kitchen.

Integrate the Sciences

Many people who have difficulty hearing wear hearing aids. The microphone of a hearing aid converts sound into an electric current, and then the amplifier amplifies the current. An earphone converts the amplified electric current into a sound that is louder than the original one. Hearing aids were large, often handheld devices until transistors were invented in the 1950s. Now hearing aids can often fit inside the ear canal and are barely noticeable. A transistor is an electric device that controls electric current flow using a semiconductor such as silicon.

Teacher Demo

Have students clear their desks of everything but paper and pen. Then have everyone try to be completely quiet for two minutes. At the end of the time, have them write down every sound they heard while the classroom was "silent."

Arts Connection

Point out to students that many famous painters and fine artists have had visual impairments, and that some of the world's best-loved composers—Beethoven is the famous example—had hearing loss. As an exercise, students might try drawing or painting while squinting tightly or looking through a pinhole. Or they could try using earplugs and feeling music by touching a speaker or instrument while it is played.

DIFFERENTIATE INSTRUCTION

More Reading Support

C What produces sound waves? *vibrations*

D What structures give us the sense of touch? *sensory receptors in the skin*

Additional Investigation To reinforce Section 4.1 learning goals, use the following full-period investigation:

Additional INVESTIGATION, Taste That Smell!, A, B, & C, pp. 235–243, 305–306 (Advanced students should complete Levels B and C.)

Advanced Have students demonstrate sound waves traveling through a medium. They can use coils, water, rubber bands, or other materials to illustrate vibrations.

Ongoing Assessment

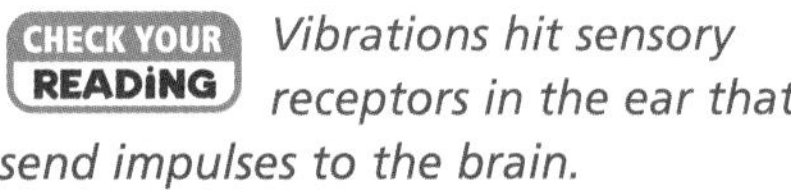

CHECK YOUR READING *Vibrations hit sensory receptors in the ear that send impulses to the brain.*

Real World Example

Anyone who has eaten the seasoning monosodium glutamate (MSG) on food has experienced what is sometimes called "the fifth taste, *umami*." Isolated from kombu seaweed by Kikunae Ikeda of Tokyo Imperial University, glutamate has no smell or texture of its own, but greatly enhances the flavor of other foods. MSG is now a longstanding flavor enhancer for meat, fish, and cheese dishes, and is widely used in processed foods such as canned soups and snack foods.

Teach from Visuals

To help students interpret the map of the taste buds, ask:

- If you bit into a lemon, where on your tongue would you taste it? *on the sides toward the back*
- Where would you taste ice cream? *on the tip of the tongue*

Health Connection

Like the other senses, taste can sometimes get skewed or scrambled. Some people have phantom taste perceptions, some have reduced ability to taste, and others perceive foul tastes from normally pleasant foods. Taste disorders can be congenital or can result from illness or injury. Injuries to the head, exposure to certain chemicals like pesticides, some surgery, and radiation therapy can cause taste disorders. A person with a taste disorder can face diminished quality of life, the loss of an important early warning system, and a reduced desire to eat.

Ongoing Assessment

Explain how the sensory organs respond to stimuli.

Ask: How do the different sense organs process a stimulus? *Eyes respond to light, touch is felt through sensory receptors in the skin, the ear responds to sound waves caused by vibrations, and taste receptors and olfactory nerves sense chemicals in food or the air.*

Smell

E Whereas sight, touch, and hearing involve processing physical information from the environment, the senses of smell and taste involve processing chemical information. Much as taste receptors sense chemicals in food, smelling receptors sense chemicals in the air. High in the back of your nose, a patch of tissue grows hairlike fibers covered in mucus. Scent molecules enter your nose, stick to the mucus, and then bind to receptors in the hairlike fibers. The receptors send an impulse to your brain, and you smell the scent.

Taste

Your tongue is covered with small sensory structures called taste buds, which are also found in the throat and on the roof of the mouth. Each taste bud includes about 100 sensory cells that are specialized to detect chemicals in foods. The thousands of tastes you experience are combinations of just four basic types of taste: sweet, sour, bitter, and salty.

Taste

Taste receptors on the tongue sense four types of taste: sweet, bitter, sour, and salty.

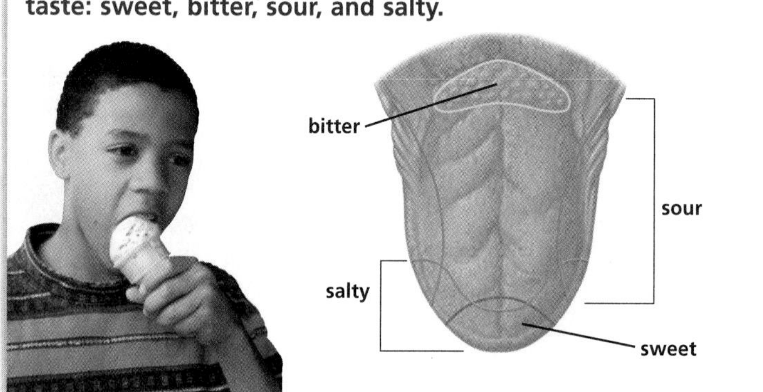

The central nervous system controls functions.

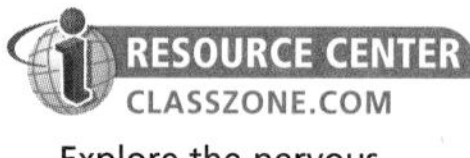

Explore the nervous system.

The **central nervous system** consists of the brain and spinal cord. The brain is located in and protected by the skull, and the spinal cord is located in and protected by the spine. The central nervous system communicates with the rest of the nervous system through electrical signals sent to and from neurons. Impulses travel very quickly, some as fast as 90 meters (295 ft) per second. That's like running almost the entire length of a soccer field in one second! F

DIFFERENTIATE INSTRUCTION

More Reading Support

E What kind of information do taste and smell perceive? *chemical*

F What body parts are in the central nervous system? *brain, spinal cord*

Advanced Have students research the structure of a taste bud and build a model to share with the class.

 Challenge and Extension, p. 196

Have students who are interested in the nervous system read the following article:

 Challenge Reading, pp. 219–220

The Brain

Different areas of the brain control different functions.

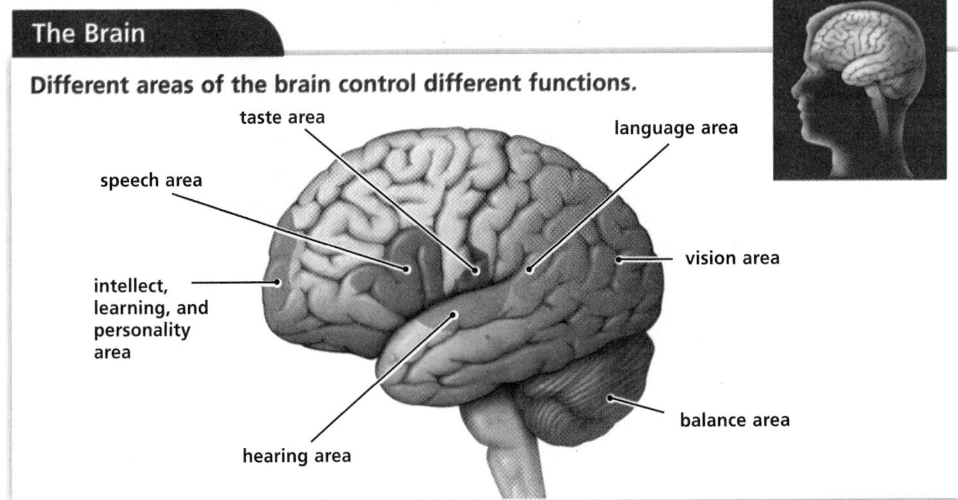

Brain

The average adult brain contains nearly 100 billion nerve cells, called **neurons.** The brain directly controls voluntary behavior, such as walking and thinking. It also allows the body to control most involuntary responses such as heartbeat, blood pressure, fluid balance, and posture.

VOCABULARY
Be sure to make a description wheel for the term *neuron.*

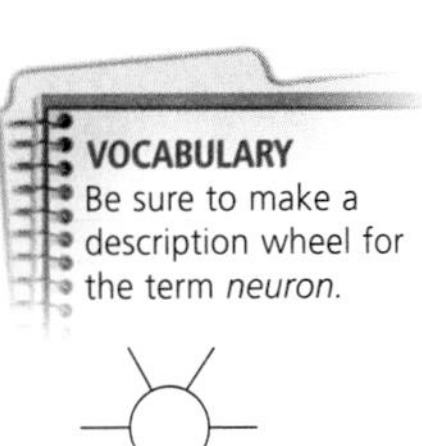

As you can see in the diagram, every area of the brain has a specific function, although many functions may involve more than one area. For example, certain areas in the brain control the senses, while other areas help you stand up straight. The lower part of the brain, called the brain stem, controls activities such as breathing and vomiting.

Spinal Cord

The spinal cord is about 44 centimeters (17 in.) long and weighs about 35–40 grams (1.25–1.4 oz). It is the main pathway for information, connecting the brain and the nerves throughout your body. The spinal cord is protected and supported by the vertebral column, which is made up of small bones called vertebrae. The spinal cord itself is a double-layered tube with an outer layer of nerve fibers wrapped in tissue, an inner layer of nerve cell bodies, and a central canal that runs the entire length of the cord. Connected to the spinal cord are 31 pairs of nerves, which send sensory impulses into the spinal cord, which in turn sends them to the brain. In a similar way, spinal nerves send impulses to muscles and glands.

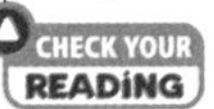

Describe the functions performed by the central nervous system.

DIFFERENTIATE INSTRUCTION

G What is the scientific name for a nerve cell? *neuron*

H What is the main pathway for information in the body? *the spinal cord*

Advanced Have students research neurons and make a model using beads and wire.

Below Level Students may easily confuse the meanings of *nerve, nerve cell, neuron,* and *nerve fiber*. Emphasize that *nerve cell* and *neuron* have the same meaning. They each refer to one cell of nerve tissue. *Nerve* and *nerve fiber* both refer to many cells working together.

EXPLORE the BIG idea

Revisit "Internet Activity: The Senses" on p. 99. Ask students how each sense sends specific information to the brain.

Teach from Visuals

To help students interpret the diagram of the brain, ask:

- What functions are handled by the back part of the brain? *vision, balance*
- What functions are handled by the front part of the brain? *intellect, learning, personality*

Real World Example

The first brain structure appeared in reptiles about 300 million years ago and dealt with breathing, heartbeat, and basic motor and foraging skills. This primitive hindbrain is now the human pons and the medulla, for our brains have evolved by adding on sections rather than replacing old ones with new. The hippocampus and cerebellum, or limbic system, developed in mammals 250 million years ago. It governs emotional behavior, balance, and coordination. The newest part of the brain is the cerebrum, which developed about 200 million years ago for functions such as language, thinking, and information processing.

History of Science

Wilder Penfield, the first person ever to perform surgery on the human brain, was also the first to map the brain according to function. In the 1950s, while trying to treat patients with epilepsy, Penfield stimulated parts of the patients' brains in an effort to discover the source of the seizures. He would then remove the disordered tissue. His treatment was often successful, and led to the discovery that memories are stored in particular areas of the temporal lobes.

Ongoing Assessment

Answer: communication with the rest of the nervous system, control of voluntary actions, and control of many involuntary activities

Teach from Visuals

To help students interpret the diagram of the nervous system, ask:

- What part of the nervous system contains sensory and motor nerves? *the peripheral nervous system*
- What part of the nervous system processes information from the peripheral nervous system? *the central nervous system*
- What part of the nervous system carries signals to and from muscles, glands, and other organs? *the peripheral nervous system*

Address Misconceptions

IDENTIFY Ask: What functions does your body carry out without using the brain? If students say anything other than "none" or "no functions," they hold the misconception that some behavior is not controlled by the brain.

CORRECT Use students' answers to the question as examples and take them through how the brain enters the process of the function they named.

REASSESS Ask: What organ is necessary to all human behavior and functions? *the brain*

Technology Resources

Visit **ClassZone.com** for background on common student misconceptions.

 **MISCONCEPTION DATABASE**

Teach Difficult Concepts

If students have trouble differentiating between the autonomic and voluntary nervous systems, have them look up the words *automatic* and *volunteer* in the dictionary, and then suggest actions or events in their lives that were examples of these words. *automatic transmission on a car; volunteering to help around the house.* Connect the meanings with the terms *autonomic* and *voluntary.*

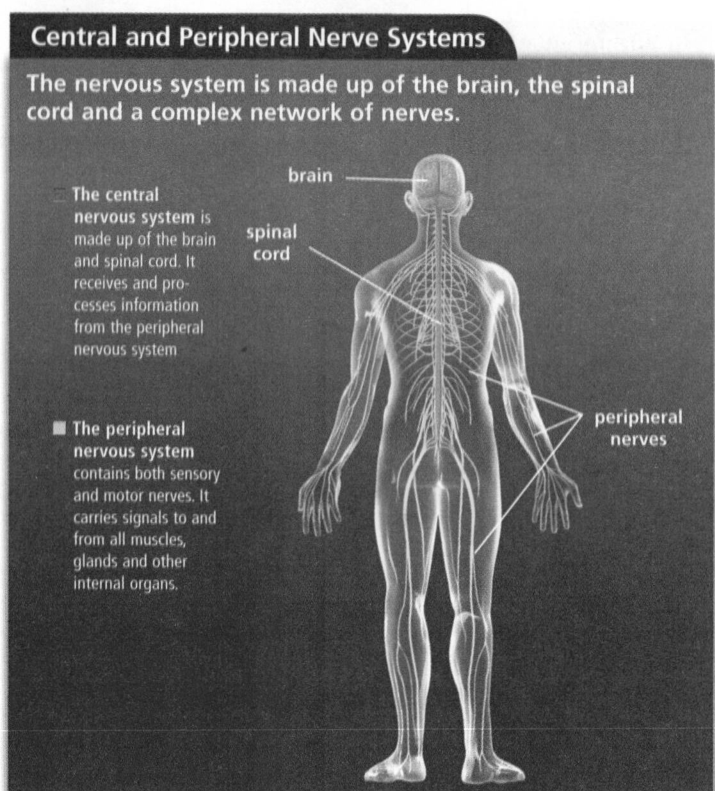

Central and Peripheral Nerve Systems
The nervous system is made up of the brain, the spinal cord and a complex network of nerves.
The central nervous system is made up of the brain and spinal cord. It receives and processes information from the peripheral nervous system
The peripheral nervous system contains both sensory and motor nerves. It carries signals to and from all muscles, glands and other internal organs.
brain
spinal cord
peripheral nerves

The peripheral nervous system is a network of nerves.

Nerves, which are found throughout the body, are often referred to altogether as the **peripheral nervous system.** Both sensory and motor nerves are parts of the peripheral nervous system. Sensory nerves receive information from the environment—such as heat or cold—and pass the information to the central nervous system. Motor nerves send signals to your muscles that allow you to move. The peripheral nervous system includes both voluntary motor nerves and involuntary responses.

Another type of motor nerves controls the involuntary responses of the body. In times of danger, there is no time to think. The body must respond immediately. In less stressful situations, the body maintains activities like breathing and digesting food. These functions go on without conscious thought. They are controlled by part of the peripheral nervous system called the the autonomic (AW-tuh-NAHM-ihk) nervous system.

DIFFERENTIATE INSTRUCTION

More Reading Support

I What is the system that runs throughout the body? *peripheral nervous system*

Below Level Help students organize information about the two parts of the nervous system into a table:

System	Functions	Example
autonomic	conserve and store energy, respond to change	digestion, breathing, heartbeat
voluntary	movement, conscious functions	locomotion, speech

The **autonomic nervous system** controls the movement of the heart, the smooth muscles in the stomach, the intestines, and the glands. The autonomic nervous system has two distinct functions: to conserve and store energy and to respond quickly to changes. You can think of the autonomic nervous system as having a division that performs each of these two main functions.

Each division is controlled by different locations on the spinal cord, or within the brain and the brain stem. The cerebellum, which is located at the rear of the brain, coordinates balance and related muscle activity. The brain stem, which lies between the spinal cord and the rest of the brain, controls heartbeat, respiration, and the smooth muscles in the blood vessels.

When you are under stress, one part of the autonomic nervous system causes what is called the "fight or flight response." Rapid changes in your body prepare you either to fight the danger or to take flight and run away from the danger. The response of your nervous system is the same, whether the stress is a real danger, like falling off a skateboard, or a perceived danger, like being worried or embarrassed.

The **voluntary nervous system** monitors movement and functions that can be controlled consciously. Every movement you think about is voluntary. The voluntary nervous system controls the skeletal muscles of the arms, the legs, and the rest of the body. It also controls the muscles that are responsible for speech and the senses.

What is the difference between the voluntary and the autonomic nervous systems?

The autonomic nervous system responds quickly to changes in balance.

4.1 Review

KEY CONCEPTS

1. Make a chart of the five senses that includes a definition and a stimulus for each sense.
2. Explain the process by which you hear a sound.
3. What are two body systems with which the nervous system interacts? How do these interactions take place?

CRITICAL THINKING

4. **Classify** Determine if the following actions involve the autonomic or the voluntary nervous system: chewing, eye blinking, jumping at a loud noise, and riding a bike.
5. **Apply** Describe what messages are sent by the nervous system when you go outside wearing a sweater on a hot day.

CHALLENGE

6. **Hypothesize** When people lose their sense of smell, their sense of taste is often affected as well. Why do you think the ability to taste would be decreased by the loss of the ability to smell?

ANSWERS

1. Charts should include definitions of senses and examples of stimuli for each.

2. Vibration causes sound waves. Waves enter ear through auditory canal and vibrate eardrum to stirrup, which moves liquid in ear, sensed by nerves in brain and interpreted as sound.

3. Sample answer: nervous/muscular: sends messages from brain to muscles; nervous/digestive: controls movements of smooth muscles in stomach

4. autonomic: jumping at a loud noise, eye blinking; voluntary: riding a bike, chewing

5. Autonomic signals to your sweat glands to begin sweating; voluntary signals to muscles to remove sweater.

6. Areas of the brain that process taste and smell area may be connected.

Ongoing Assessment

Describe how the nervous system works with other systems.

Ask: How do the various muscles and organs of the body know what to do? *The peripheral nervous system signals the central nervous system. The brain processes the signal and tells the autonomic and voluntary systems to respond.*

 *Answer: voluntary—can be controlled by conscious thought; autonomic—functions without the brain thinking thoughts about it*

Reinforce the BIG idea

Have students relate the section to the Big Idea.

 Reinforcing Key Concepts, p. 197

4.1 ASSESS & RETEACH

Assess

A Section 4.1 Quiz, p. 60

Reteach

Discuss how the human body maintains its well-being in an ever-changing environment. Have students sketch or discuss an example for each of the following concepts:

- Senses perceive stimuli in the environment.
- The central nervous system processes information collected by the peripheral nervous system.
- The autonomic nervous system controls functions that occur without conscious thought.
- The voluntary nervous system controls conscious actions.

Technology Resources

Have students visit **ClassZone.com** for reteaching of Key Concepts.

 **CONTENT REVIEW**

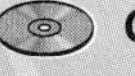 **CONTENT REVIEW CD-ROM**

CHAPTER INVESTIGATION

Focus

PURPOSE To test the ability to taste

OVERVIEW Students will count the number of taste buds in a sample area of their tongues to determine if they are extra-sensitive to tastes. Students will find the following:

- Different people have different numbers of taste buds.
- The number of taste buds on your tongue can affect your sensitivity to tastes.

Lab Preparation

Prior to the investigation, have students read through the investigation and prepare their data tables. Or you may wish to copy and distribute datasheets and rubrics.

UNIT RESOURCE BOOK, pp. 226–234

SCIENCE TOOLKIT, F14

Lab Management

- Caution students not to swallow too much food coloring and not to get food coloring on their clothing.
- Make sure students properly dispose of cotton swabs after painting their tongues.
- If students have their tongues in the air for long, they may start to drool. They should use paper towels and napkins to wipe away any saliva.
- It may be easier to count taste buds if students paint their tongues, put reinforcement circles directly on their tongues, and count the pink spots on their tongues.
- Note: For the "Calculate" section of "Observe and Analyze," p. 109, help students by keeping a list or chart on the board to provide whole-class data.

CHAPTER INVESTIGATION

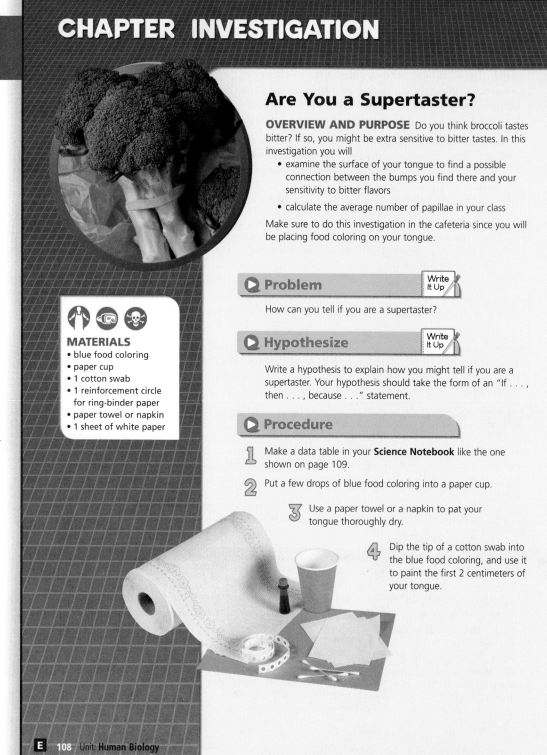

Are You a Supertaster?

OVERVIEW AND PURPOSE Do you think broccoli tastes bitter? If so, you might be extra sensitive to bitter tastes. In this investigation you will

- examine the surface of your tongue to find a possible connection between the bumps you find there and your sensitivity to bitter flavors
- calculate the average number of papillae in your class

Make sure to do this investigation in the cafeteria since you will be placing food coloring on your tongue.

MATERIALS
- blue food coloring
- paper cup
- 1 cotton swab
- 1 reinforcement circle for ring-binder paper
- paper towel or napkin
- 1 sheet of white paper

Problem

Write It Up

How can you tell if you are a supertaster?

Hypothesize

Write It Up

Write a hypothesis to explain how you might tell if you are a supertaster. Your hypothesis should take the form of an "If . . . , then . . . , because . . ." statement.

Procedure

1. Make a data table in your **Science Notebook** like the one shown on page 109.
2. Put a few drops of blue food coloring into a paper cup.
3. Use a paper towel or a napkin to pat your tongue thoroughly dry.
4. Dip the tip of a cotton swab into the blue food coloring, and use it to paint the first 2 centimeters of your tongue.

INVESTIGATION RESOURCES

CHAPTER INVESTIGATION, Are You a Supertaster?
- Level A, pp. 226–229
- Level B, pp. 230–233
- Level C, p. 234

Advanced students should complete Levels B & C.

Writing a Lab Report, D12–13

Technology Resources

Customize this student lab as needed or look for an alternative. Print rubrics to assess student lab reports.

Lab Generator CD-ROM

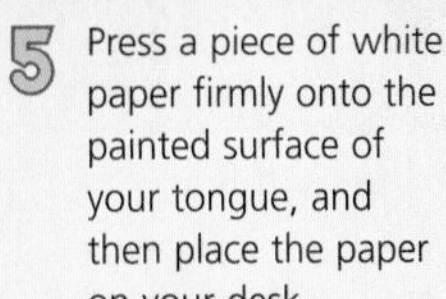

5 Press a piece of white paper firmly onto the painted surface of your tongue, and then place the paper on your desk.

step 5

6 Place a notebook reinforcement circle on the blue area.

7 You should see white circles in a field of blue. The white circles are the bumps on your tongue called fungiform papillae. Count the number of white circles inside the reinforcement circle. There may be many white circles crammed together that vary in size, or just a few. If there are just a few, they may be larger than the ones on someone who has many white circles close together. If there are too many to count, try to count the number in half of the circle and multiply this number by 2. Record your total count in your data table.

Observe and Analyze

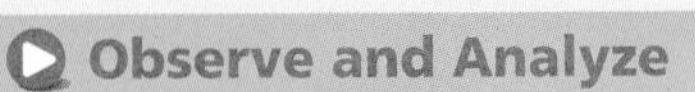

1. OBSERVE What did you observe while looking at the tongue print? Is the surface the same all over your tongue?

2. CALCULATE Record the number of papillae within the reinforcement circle of all the students in your class.

AVERAGE Calculate the average number of papillae counted in the class.

$$\text{average} = \frac{\text{sum of papillae in class}}{\text{number of students}}$$

Conclude

1. INTERPRET How do the number of fungiform papillae on your tongue compare with the number your partner counted?

2. INFER Do you think there is a relationship between the number of fungiform papillae and taste? If so, what is it?

3. IDENTIFY What foods might a supertaster not like?

4. APPLY Do you think that there are other taste perceptions besides bitterness that might be influenced by the number of fungiform papillae that an individual has? Why do you think so?

INVESTIGATE Further

CHALLENGE Calculate the area in square millimeters inside the reinforcement circle, and use this value to express each person's papillae count as a density (number of papillae per square millimeter).

Are You a Supertaster?

Table 1. Papillae

Name	Number of papillae

Observe and Analyze

1. Answers will vary. Students should observe different densities of white circles. The density of circles should vary over the surface of the tongue.

2. Answers will vary. Check student calculations of average number of papillae in the class.

Conclude

1. Answers will vary. Some students will have more papillae and some will have less than others.

2. Most fungiform papillae allow for more sensitivity in taste.

3. Supertasters might not like foods with a strong bitter taste.

4. Sour tastes might be influenced by the number of papillae, because strong sour taste can also cause dislike of foods.

INVESTIGATE Further

CHALLENGE Area within the reinforcement hole is equal to πr^2. For the typical reinforcement, $r = 3$ mm, giving an area of about 28 mm^2. To get density, divide the number of papillae by the area of the reinforcement hole.

Post-Lab Discussion

Ask: Do you think researchers could adapt this method to determine where on the tongue different types (for different tastes) of fungiform papillae occur? If so, how?

Explain to students that the sense of taste is an early warning system. It helps protect the body from such things as spoiled foods or poisons. Do you think there is an ideal number and size of fungiform papillae for humans? If so, what would it be?

4.2 FOCUS

Set Learning Goals

Students will

- Describe the role of hormones.
- Analyze the function of glands.
- Explain how the body uses feedback mechanisms.
- Determine how body temperature changes during an experiment.

3-Minute Warm-Up

Display Transparency 28 or copy this exercise on the board:

Decide if these statements are true. If not true, correct them.

1. The senses of sight and taste react to chemicals in the environment. *The senses of smell and taste react to chemicals in the environment.*
2. The central nervous system consists of the spinal cord. *The central nervous system consists of the spinal cord and brain.*
3. The peripheral nervous system is a network of sensory and motor nerves. *true*

 3-Minute Warm-Up, p. T28

4.2 MOTIVATE

THINK ABOUT

PURPOSE Determine how the body responds to surprise

DISCUSS Allow students to work in small groups, then share their ideas as a class. Point out that the body's reaction to surprise is an instinctive response to prepare a person to deal with danger. Increased heart rate gets more oxygen and nutrients to the muscles to prepare to fight or run from a source of danger.

KEY CONCEPT

4.2 The endocrine system helps regulate body conditions.

BEFORE, you learned

- Many body systems function without conscious control
- The body systems work automatically to maintain homeostasis
- Homeostasis is important to an organism's survival

NOW, you will learn

- About the role of hormones
- About the functions of glands
- How the body uses feedback mechanisms to help maintain homeostasis

VOCABULARY

endocrine system p. 110
hormone p. 111
gland p. 111

THINK ABOUT

How does your body react to surprise?

In a small group, determine how your body responds to a surprising situation. Have one student in the group pretend he or she is responding to a surprise. The other group members should determine how the body reacts physically to that event. How do your respiratory system, digestive system, circulatory system, muscle system, and skeletal system react?

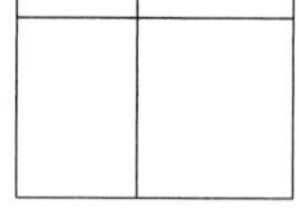

CHOOSE YOUR OWN STRATEGY
Begin taking notes on the main idea: *Hormones are the body's messengers.* Use a strategy from an earlier chapter or one of your own. Include a definition of *hormone* in your notes.

Hormones are the body's chemical messengers.

Imagine you're seated on a roller coaster climbing to the top of a steep incline. In a matter of moments your car drops hundreds of feet. You might notice that your heart starts beating faster. You grab the seat and notice that your palms are sweaty. These are normal physical responses to scary situations. The **endocrine system** controls the conditions in your body by making and sending chemicals from one part of the body to another. Most responses of the endocrine system are controlled by the nervous system.

RESOURCES FOR DIFFERENTIATED INSTRUCTION

Below Level

UNIT RESOURCE BOOK

- Reading Study Guide A, pp. 200–201
- Decoding Support, p. 223

Advanced

UNIT RESOURCE BOOK

Challenge and Extension, p. 206

English Learners

UNIT RESOURCE BOOK

Spanish Reading Study Guide, pp. 204–205

- Audio Readings in Spanish
- Audio Readings (English)

Hormones are chemicals that are made in one organ and travel through the blood to a second organ. The second organ, often called the target organ, responds to the chemical. Most hormones have more than one target organ. Many hormones, as you can see in the table below, affect all the cells in the body.

Because hormones are made at one location and function at another, they are often called chemical messengers. When the hormone reaches its target organ, it binds to receptors on the surface of or inside the organ's cells. There the hormone begins the chemical changes that cause the target organ to function in a specific way. All of the functions of the endocrine system work automatically, without your conscious control.

The individuals on this roller coaster are experiencing a burst of the hormone adrenaline.

Different types of hormones perform different jobs. Some of these jobs are to control the production of other hormones, to regulate the balance of chemicals such as glucose and salt in your blood, or to produce responses to changes in the environment. Some hormones are made only during specific times in a person's life. For example, hormones that control the development of sexual characteristics are not produced during childhood. When production begins in adolescence, these hormones cause major changes in a person's body.

How are hormones like messengers?

Hormones

Name	Where produced	Where travels to
Growth hormone	pituitary gland	all body cells
Antidiuretic hormone	pituitary gland	kidneys
Thyroxine	thyroid gland	all body cells
Cortisol	adrenal glands	all body cells
Adrenaline	adrenal glands	heart, lungs, stomach, intestines, glands
Insulin	pancreas	all body cells
Testosterone (males)	testes	all body cells
Estrogen (females)	ovaries	all body cells

Glands produce and release hormones.

The main structures of the endocrine system are organs called **glands.** Many glands in the body produce hormones and release them into your circulatory system. As you can see in the illustration on page 113, endocrine glands can be found in many parts of your body. However, all hormones move from the organ in which they are produced to target organs.

RESOURCE CENTER
CLASSZONE.COM

Learn more about the endocrine system.

4.2 INSTRUCT

Real World Example

The roller coaster shown in the photograph is a familiar example of people experiencing a "rush of adrenaline." Have students brainstorm aloud a quick list of other examples. They might give examples from their own life or from films, books, or television.

Ongoing Assessment

Describe the role of hormones.

Ask: What can your pituitary gland produce to communicate with another part of your body? *hormones*

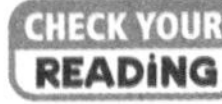

Answer: They travel from one location with signals for activity in another location.

DIFFERENTIATE INSTRUCTION

More Reading Support

A Summarize what a hormone is and does. *a chemical made in part of the body that causes reactions in another part*

B What are organs that produce hormones? *glands*

Below Level For students who have difficulty or who have limited attention or skills, you can teach the main concepts of hormones with only one example. Use the example of adrenaline, and refer students to the photograph and caption on p. 111, the paragraph on p. 112, and the illustration on p. 113. Then have them create a flow chart with drawings and/or captions to explain how the hormone adrenaline functions in the body.

Metacognitive Strategy

Have students discuss their own strategies for remembering the main ideas about glands and hormones. Some students will have ways of memorizing the details, while others may generalize. Point out that both are useful, as long as the main ideas are understood.

Health Connection

When the pancreas cannot make enough of the hormone insulin or cannot use insulin properly, diabetes results. The body cannot convert glucose into energy, and glucose builds up in the blood. Many people with diabetes develop it over the course of their lives. This type of diabetes is called adult-onset or Type II diabetes. Diabetes often begins with symptoms such as excessive thirst or hunger, fatigue, and nausea, and if left untreated, can progress to blindness, circulation problems, and kidney failure. It can be treated but not cured. Diabetics who exercise, are careful with diet, and monitor their blood glucose levels can often manage their diabetes and avoid having to take insulin.

History of Science

Until 1922, a diagnosis of diabetes was a death sentence. Canadian surgeon Frederick Banting and his assistant Charles Best were the first to isolate the hormone insulin, which is made in the islets of Langerhans in the pancreas. Once it was isolated, they administered insulin to a diabetic boy and found that it did, as hoped, lower blood glucose levels. It is estimated that today insulin helps 18 million people in the United States with diabetes.

Ongoing Assessment

Analyze the functions of glands.

Ask: What are the main structures of the endocrine system and how do they work? *the glands, which produce and secrete hormones that regulate such things as sleep, growth, calcium levels, and disease fighting*

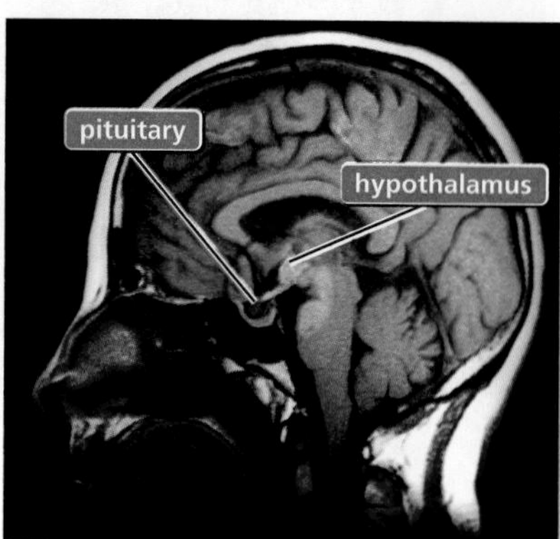

The hypothalamus and the pituitary are important endocrine glands.

Pituitary Gland The pituitary (pih-TOO-ih-TEHR-ee) gland can be thought of as the director of the endocrine system. The pituitary gland is the size of a pea and is located at the base of the brain—right above the roof of your mouth. Many important hormones are produced in the pituitary gland, including hormones that control growth, sexual development, and the absorption of water into the blood by the kidneys. C

Hypothalamus The hypothalamus (HY-poh-THAL-uh-muhs) is attached to the pituitary gland and is the primary connection between the nervous and endocrine systems. All of the secretions of the pituitary gland are controlled by the hypothalamus which produces releasing hormones.

Pineal Gland The pineal (PIHN-ee-uhl) gland is a tiny organ about the size of a pea. It is buried deep in the brain. The pineal gland is sensitive to different levels of light and is essential to rhythms such as sleep, body temperature, reproduction, and aging.

Thyroid Gland You can feel your thyroid gland if you place your hand on the part of your throat called the Adam's apple and swallow. What you feel is the cartilage surrounding your thyroid gland. The thyroid releases hormones necessary for growth and metabolism. The tissue of the thyroid is made of millions of tiny pouches, which store the thyroid hormone. The thyroid gland also produces the hormone calcitonin, which is involved in the regulation of calcium in the body.

Thymus The thymus is located in your chest. It is relatively large in the newborn baby and continues to grow until puberty. Following puberty, it gradually decreases in size. The thymus helps the body fight disease by controlling the production of white blood cells called T-cells.

D

Adrenal Glands The adrenal glands are located on top of your kidneys. The adrenal glands secrete about 30 different hormones that regulate carbohydrate, protein, and fat metabolism and water and salt levels in your body. Some other hormones produced by the adrenal glands help you fight allergies or infections. Roller coaster rides, loud noises, or stress can activate your adrenal glands to produce adrenaline, the hormone that makes your heart beat faster.

Pancreas The pancreas is part of both the digestive and the endocrine systems. The pancreas secretes two hormones, insulin and glucagon. These hormones regulate the level of glucose in your blood. The pancreas sits beneath the stomach and is connected to the small intestine.

Ovaries and Testes The ovaries and testes also secrete hormones that control sexual development.

DIFFERENTIATE INSTRUCTION

More Reading Support

C Which gland directs the endocrine system? *pituitary gland*

D Which glands secrete 30 different hormones? *adrenal glands*

Advanced Have students research diabetes and create an educational brochure about it for younger students.

Challenge and Extension, p. 206

E

Other Organs Some organs that are not considered part of the endocrine system do produce important hormones. The kidneys secrete a hormone that regulates the production of red blood cells. This hormone is secreted whenever the oxygen level in your blood decreases. By stimulating the red bone marrow to produce more red blood cells, the ability of the blood to carry oxygen increases. The heart produces two hormones that help regulate blood pressure. These hormones, secreted by one of the chambers of the heart, stimulate the kidneys to remove more salt.

CHECK YOUR READING Which organs are part of the endocrine system?

Endocrine System

The endocrine system is made of a group of glands. These glands produce and release hormones, or chemical messengers.

pineal gland
hypothalamus
pituitary gland
thyroid gland
thymus gland
adrenal glands
pancreas
ovaries
testes

DIFFERENTIATE INSTRUCTION

More Reading Support

E Which organs also function as glands? *the brain, heart, and kidney*

Below Level Refer students to the maps of the endocrine system. Have them create their own maps of either the male or female endocrine system by tracing the outline and all of the callout lines on p. 113. Then ask them to sketch or find pictures to illustrate the function of each gland's hormones.

Teach Difficult Concepts

Students may confuse glands with hormones. Point out that glands are a structure, while hormones are a chemical. You might also try the demonstration below.

Teacher Demo

Fill a small plastic bag with colored water. Poke a small hole in the bag with a pin. Hold the bag over a sponge and allow some colored water to drip on it. Ask:

- What correlates to a gland? *the bag*
- What correlates to a hormone? *the colored water*
- What is the sponge? *the target organ*

Develop Critical Thinking

PREDICT Have students choose one of the glands and study the functions performed by that gland's hormones. Ask them to predict what a human being's life would be like if that gland were missing or malfunctioning.

Teach from Visuals

What general area of the body are all the glands of the endocrine system in? *center* What areas do the hormones produced in the glands generally affect? *all parts*

Ask: How do the illustrations show the role of hormones as "chemical messengers"? *The glands are located throughout the body and are able to "deliver" hormones wherever the body needs them.*

This visual is also available as T30 in the Unit Transparency Book.

Ongoing Assessment

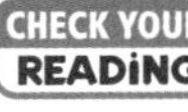

Answer: pineal gland, hypothalamus, pituitary gland, thyroid, thymus, adrenal glands, pancreas, ovaries, testes

INVESTIGATE Response to Exercise

PURPOSE To determine how body temperature changes during exercise

TIPS *30 min.* Encourage students to maintain an even level of intensity as they exercise. Use disposable thermometer covers or rubbing alcohol to maintain proper hygiene.

INCLUSION Sitting-down exercises are less strenuous and will allow inclusion of most students with physical or health limitations.

WHAT DO YOU THINK? *It increased. The amount of exercise, starting body temperature, weight, level of fitness, and gender could all be factors. Answers will vary. Students may say that the person looked more invigorated, rosier, had faster breathing, wider eyes, better posture, and so on.*

CHALLENGE *Graphs should have x-and y-axes labeled with appropriate units of measures. The graph should show an increase in temperature with each round of exercise.*

Datasheet, Response to Exercise, p. 207

Technology Resources

Customize this student lab as needed or look for an alternative. Print rubrics to assess student lab reports.

Lab Generator CD-ROM

Teaching with Technology

Students can use a graphing calculator to collect and graph exercise data.

Ongoing Assessment

CHECK YOUR READING *Answer: to help cells function at their best and for survival*

INVESTIGATE Response to Exercise

How does your body temperature change when you exercise?

SKILL FOCUS
Observing

MATERIALS
- stopwatch or timing device
- notebook
- graph paper
- mercury-free thermometer
- rubbing alcohol or plastic thermometer covers

TIME
30 minutes

PROCEDURE

1. Working in groups of two, read all the instructions in this activity first. Appoint one person to be the subject and one person to be the timer and note taker. Using a mercury-free thermometer, have the subject take his or her temperature. Record the temperature in your notebook.
2. While staying seated the subject begins to do sitting-down jumping jacks. The subjects does the jumping jacks for 1 minute and then immediately takes his or her temperature. Continue this procedure for a total of 3 times, measuring the temperature after each minute of exercise.

WHAT DO YOU THINK?
- How did the subject's temperature change while exercising?
- What factors may contribute to the rate at which the temperature changed in each person?
- How did the subject's physical appearance change from the beginning of the activity to the end?

CHALLENGE Graph the results on a line graph, with temperature on the *x*-axis and time on the *y*-axis.

Control of the endocrine system includes feedback mechanisms.

F

As you might recall, the cells in the human body function best within a specific set of conditions. Homeostasis (HOH-mee-oh-STAY-sihs) is the process by which the body maintains these internal conditions, even though conditions outside the body may change. The endocrine system is very important in maintaining homeostasis.

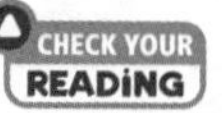
Why is homeostasis important?

Because hormones are powerful chemicals capable of producing dramatic changes, their levels in the body must be carefully regulated. The endocrine system has several levels of control. Most glands are regulated by the pituitary gland, which in turn is controlled by the hypothalamus, part of the brain. The endocrine system helps maintain homeostasis through the action of negative feedback mechanisms.

DIFFERENTIATE INSTRUCTION

More Reading Support

F What is the process of maintaining livable conditions in the body even if the environment outside changes? *homeostasis*

English Learners Demonstrate expected responses to frequently used prompts, such as "What do you think," "How do you know," "What would you predict," etc. For example, the Challenge portion of the Investigate (p. 114) asks the student to create a line graph. Help English learners by providing a model for this answer.

E 114 Unit: Human Biology

Negative Feedback

Most feedback mechanisms in the body are called negative mechanisms, because the final effect of the response is to turn off the response. An increase in the amount of a hormone in the body feeds back to inhibit the further production of that hormone.

? G The production of the hormone thyroxine by the thyroid gland is an example of a negative feedback mechanism. Thyroxine controls the body's metabolism, or the rate at which the cells in the body release energy by cellular respiration. When the body needs energy, the thyroid gland releases thyroxine into the blood to increase cellular respiration. However, the thyroid gland is controlled by the pituitary gland, which in turn is controlled by the hypothalamus. Increased levels of thyroxine in the blood inhibit the signals from the hypothalamus and the pituitary gland to the thyroid gland. Production of thyroxine in the thyroid gland decreases.

Negative and Positive Feedback

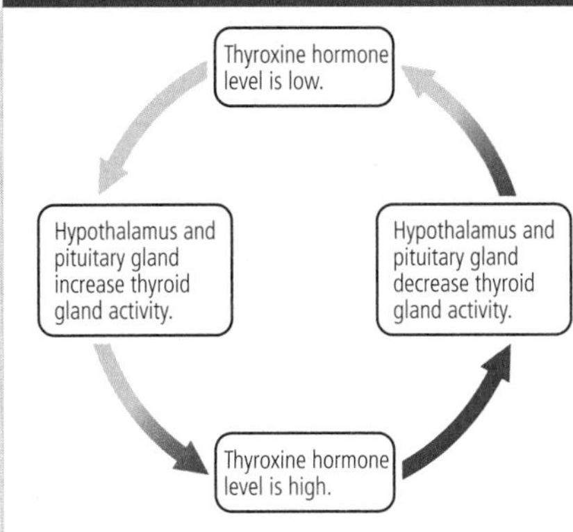
Thyroxine hormone level is low.
Hypothalamus and pituitary gland increase thyroid gland activity.
Hypothalamus and pituitary gland decrease thyroid gland activity.
Thyroxine hormone level is high.

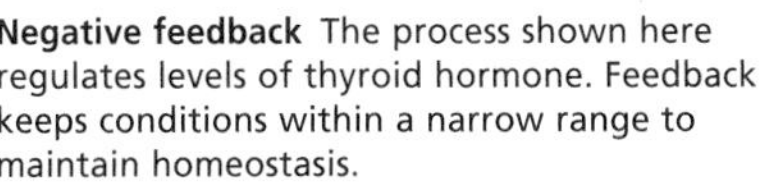

Negative feedback The process shown here regulates levels of thyroid hormone. Feedback keeps conditions within a narrow range to maintain homeostasis.

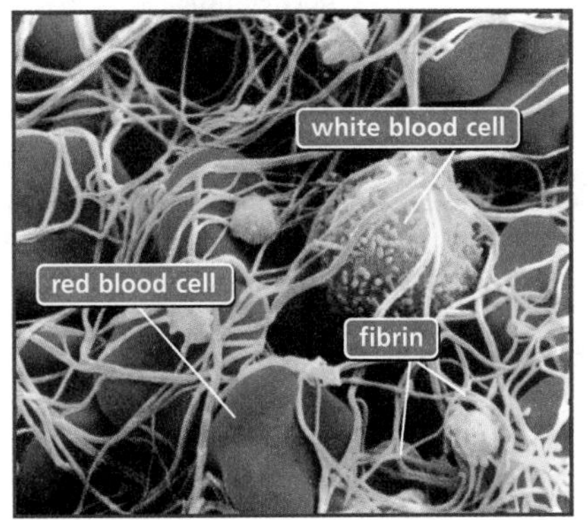
white blood cell
red blood cell
fibrin

Positive feedback These red blood cells are surrounded by fibrin, a protein that allows them to clot.

Positive Feedback

Some responses of the endocrine system, as well as other body systems, are controlled by positive feedback. The outcome of positive feedback mechanism is not to maintain homeostasis, but to produce a response that continues to increase. Most positive feedback mechanisms result in extreme responses that are necessary under extreme conditions.

? H For example, when you cut yourself, chemicals are released from the damaged tissue that signal the blood to clot. The process of clotting

Teach Difficult Concepts

If students are having trouble understanding homeostasis, call their attention to the *-stasis* part of the word, and relate that part to stability. Point out that homeostasis is a tendency to maintain stability or balance. When students attempt to keep their balance on a bicycle or balance beam, their process correlates to maintaining homeostasis.

DIFFERENTIATE INSTRUCTION

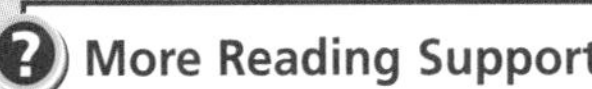

More Reading Support

G Give an example of a negative feedback mechanism. *thyroid hormone production*

H Give an example of a positive feedback mechanism. *blood clotting*

Below Level Students may need help to relate the negative feedback loop described in the text to the cycle diagram below it. Have pairs of students work to describe to each other in their own words what happens at each stage of the loop. They may want to sketch a simple metaphor as a reminder, such as a fuel pump, a gas gauge, a driver, and a gas tank.

Ongoing Assessment

Explain how the body uses feedback mechanisms.

Ask: How do feedback mechanisms involve cells? What is their general function in the body? *Feedback mechanisms signal cells to produce a response to a change in the environment. Their general function is to maintain healthy conditions.*

 *Answer: Negative feedback maintains homeostasis; positive feedback starts an increasing response after an extreme change occurs.*

Reinforce the BIG idea

Have students relate the section to the Big Idea.

 Reinforcing Key Concepts, p. 208

4.2 ASSESS & RETEACH

Assess

 Section 4.2 Quiz, p. 61

Reteach

Create a table on the board called Feedback Loops in the Body. Make four columns:

- Stimulus (starts activity)
- Response (active change)
- Body Structures (organs involved)
- Positive/Negative/Balance (+, −, =)

Complete the table for some examples in the text, such as cut skin, viral infection, puberty, or labor.

Have students search the text for information to fill in each column. Lead a discussion about how each "loop"—or mechanism—relates to homeostasis.

Technology Resources

Have students visit **ClassZone.com** for reteaching of Key Concepts.

 CONTENT REVIEW

 CONTENT REVIEW CD-ROM

causes more chemicals to form that increase the clotting action of the blood in the area of the injury. The upward spiral increases until a clot is formed that fills the injured area. Other examples of positive feedback include fever, the immune response, puberty, and labor.

 What is the difference between negative and positive feedback?

Balanced Hormone Action

In the body, the action of one hormone is often balanced by the action of another. When you ride a bicycle, you are able to ride in a straight line, despite bumps and dips in the road, by making constant steering adjustments. If the bicycle is pulled to the right, you adjust the handlebars by turning a tiny bit to the left.

Some hormones maintain homeostasis in the same way that you steer your bicycle. The pancreas, for example, produces two hormones. One hormone, insulin, decreases the level of sugar in the blood. The other hormone, glucagon, increases sugar levels in the blood. The balance of the levels of these hormones maintains stable blood sugar between meals.

Hormone Imbalance

Because hormones regulate critical functions in the body, too little or too much of any hormone can cause serious disease. When the pancreas produces too little insulin, sugar levels in the blood can rise to dangerous levels. Very high levels of blood sugar can damage the circulatory system and the kidneys. This condition, known as diabetes mellitus, is often treated by injecting synthetic insulin into the body to replace the insulin not being made by the pancreas.

4.2 Review

KEY CONCEPTS

1. List three different jobs that hormones perform.
2. Draw an outline of the human body. Add the locations and functions of the pituitary, thyroid, adrenal, and pineal glands to your drawing.
3. What is the function of a negative feedback mechanism?

CRITICAL THINKING

4. **Analyze** Explain why hormones are called chemical messengers.
5. **Analyze** List two sets of hormones that have opposing actions. How do the actions of these hormones help maintain homeostasis?

CHALLENGE

6. **Connect** Copy the diagram below and add three more stimuli and the resulting feedback mechanisms.

ANSWERS

1. Sample answer: regulating growth, sexual development, the balance of chemicals in the blood

2. Pituitary: base of brain; growth, sexual development, water absorption. Thyroid: neck; growth and metabolism. Adrenal: above kidneys; carbohydrate, protein, and fat metabolism and water and salt levels. Pineal: brain; body rhythms

3. to help the body maintain homeostasis

4. made in one organ and travel to another, which responds to the chemical

5. insulin and glucagon from the pancreas, insulin decreases blood sugar, glucagon increases blood sugar; hormones controlling salt removal by the kidney, one increases urine, one decreases urine

6. Sample answer: cut skin, clotting; infection, fever; developed baby, labor

CONNECTING SCIENCES

CONNECTING LIFE AND PHYSICAL SCIENCE

Heating and Cooling

The cells in our bodies can survive only within a limited temperature range. The body must maintain a constant core temperature at about 37°C (98.6°F). Body temperature is a measure of the average thermal energy in the body. To keep a constant temperature range, our bodies either lose or gain thermal energy.

Energy cannot be created or destroyed, but it can be transferred from one form or place to another. The major source of thermal energy in our bodies is food. When our bodies break down nutrients, some of the chemical energy is released as thermal energy that heats our bodies. Also, some of the kinetic energy from muscle movement is converted into thermal energy.

Body temperature is controlled by the hypothalamus region of the brain. The hypothalamus controls the rate of nutrient use. The hypothalamus also controls shivering and sweating.

Heat is the flow of energy from a warmer to a cooler object. Heat transfer between the body and its surroundings occurs in four ways.

1. **Evaporation:** When water evaporates, or changes from liquid to gas, energy is required. When perspiration evaporates from the surface of our skin, we lose thermal energy as heat.
2. **Radiation:** Heat transfer also occurs through waves that radiate out from a warm object or area. Sitting in the sunshine warms us because we gain thermal energy from the Sun's radiation. Our warm bodies can also radiate energy into cooler air.
3. **Conduction:** When two objects are in direct contact, heat flows by conduction from the warmer to the cooler object. If you stand barefoot on hot sand, heat quickly flows into your feet by conduction.
4. **Convection:** In convection, heat transfer occurs through the movement of particles in a gas or liquid. Your body loses some thermal energy because of convection in the air around you.

EXPLORE

1. CONNECT What are some behaviors that help you lose or gain thermal energy?
2. CHALLENGE Choose a behavior that either warms or cools your body. Draw a diagram and label it with the types of heat transfer that are occurring.

ansport and Regulation 117 E

EXPLORE

1. *CONNECT Sample answer: put on or take off clothes, sit in the shade, build houses, take cold or hot baths, use air conditioning or heating*
2. *CHALLENGE Diagrams should show that some activities promote or prevent radiation and some promote conduction.*

CONNECTING SCIENCES
Integration of Sciences

Set Learning Goal

To apply the physical science concepts of heat transfer and thermal energy to the study of responses in the human body

Present the Science

Energy is the ability to do work or cause change. Heat is a term that describes the felt or measured results of the rapid movement of tiny particles in matter, kinetic energy. Other types of energy can cause kinetic energy and kinetic energy can generate other forms of energy. Students experience kinetic energy and its connection to heat everyday. When they run around a lot, they feel warmer, for example. Hot cocoa and hot tea are liquids in which the tiny particles are set in rapid motion. The same liquids with the particles in slow motion might be cold chocolate milk and iced tea.

Heat transfer happens in all matter (think of ice cubes melting in the tea or steam rising from it). In a living system like an organism, however, heat transfer is controlled. Active parts of the system change in order to keep internal temperatures within a range necessary to survive.

Discussion Questions

Ask: How does shivering warm the body? *movement conduction from muscle*

Ask: Why do we lose heat when sweat evaporates from our skin? ***It takes energy to change the molecules of liquid into molecules of gas.***

Close

Ask: Which is more important to our bodies, heating or cooling? *Both are necessary to maintain normal cell temperature. Students in hot areas may say "cooling," and in cold areas may say "heat."*

4.3 FOCUS

Set Learning Goals

Students will

- Describe specialized cells and organs in the male and female reproductive systems.
- Describe the stages of fertilization.
- Describe the development of the embryo and fetus during pregnancy.
- Observe and contrast egg and sperm cells.

3-Minute Warm-Up

Display Transparency 29 or copy this exercise on the board:

Suppose that something occurs that frightens you. What will happen hormonally in your body? What will your physical responses be? *The adrenal glands produce adrenaline, which elevates heart rate. After a while, its effects wear off and you return to normal.*

 3-Minute Warm-Up, p. T29

4.3 MOTIVATE

EXPLORE Reproduction

PURPOSE To observe differences between egg and sperm cells

TIPS *20 min.* Help students focus the microscope and practice moving the slide slightly and slowly so cells can be clearly viewed. Provide images of sperm cells and egg cells or use a microscope projector, or digital images on a computer to assist students with vision limitations.

WHAT DO YOU THINK? *The shapes and sizes are different. The structures of the sperm and egg allow each cell to perform its function.*

Teaching with Technology

Students could use a PC microscope to view eggs and sperm, and then create a presentation with screen captures.

KEY CONCEPT

4.3 The reproductive system allows the production of offspring.

BEFORE, you learned

- Some hormones regulate sexual development
- Glands release hormones

NOW, you will learn

- About specialized cells and organs in male and female reproductive systems
- About fertilization
- About the development of the embryo and fetus during pregnancy

VOCABULARY

menstruation p. 119
fertilization p. 121
embryo p. 121
fetus p. 122

EXPLORE Reproduction

How are sperm and egg cells different?

PROCEDURE

1. From your teacher, gather slides of egg cells and sperm cells.
2. Put each slide under a microscope.
3. Draw a sketch of each cell.
4. With a partner, discuss the differences that you observed.

WHAT DO YOU THINK?

- What were the differences that you observed?
- What are the benefits of the different characteristics for each cell?

MATERIALS

- slides of egg and sperm cells
- microscope
- paper
- pencil

The reproductive system produces specialized cells.

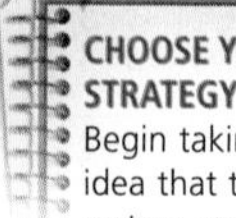

CHOOSE YOUR OWN STRATEGY
Begin taking notes on the idea that the reproductive system produces specialized cells. You might use an outline or another strategy of your choice.

Like all living organisms, humans reproduce. The reproductive system allows adults to produce offspring. Although males and females have different reproductive systems, both systems share an important characteristic. They both make specialized cells. In any organism or any system, a specialized cell is a cell that takes on a special job.

In the female these specialized cells are called egg cells. In the male they are called sperm cells. In the reproductive system, each specialized cell provides genetic material. Genetic material contains the information that an organism needs to form, develop, and grow.

RESOURCES FOR DIFFERENTIATED INSTRUCTION

Below Level

UNIT RESOURCE BOOK

- Reading Study Guide A, pp. 211–212
- Decoding Support, p. 223

Advanced

UNIT RESOURCE BOOK

Challenge and Extension, p. 217

English Learners

UNIT RESOURCE BOOK

Spanish Reading Study Guide, pp. 215–216

- Audio Readings in Spanish
- Audio Readings (English)

Both the male and female reproductive systems rely on hormones from the endocrine system. The hormones act as chemical messengers that signal the process of sexual development. Sexual development includes the growth of reproductive organs and the development of sexual characteristics. Once mature, the reproductive organs produce hormones to maintain secondary sexual characteristics.

The Female Reproductive System

The female reproductive system has two functions. The first is to produce egg cells, and the second is to protect and nourish the offspring until birth. The female has two reproductive organs called ovaries. Each ovary contains on average hundreds of egg cells. Every 28 days, the pituitary gland releases a hormone that stimulates some of the eggs to develop and grow.

Female Reproductive Organs

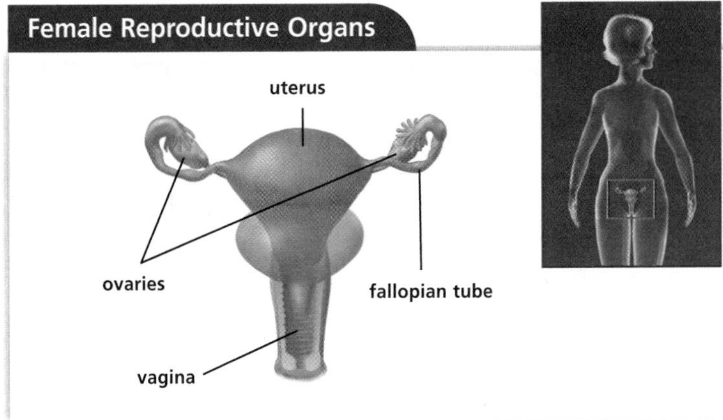

Menstruation

After an egg cell develops fully, another hormone signals the ovary to release the egg. The egg moves from the ovary into a fallopian tube. Within ten to twelve hours, the egg cell is fertilized by a sperm cell and moves to the uterus. Once inside the thick lining of the uterus, the fertilized egg cell rapidly grows and divides.

However, if fertilization does not occur within 24 hours after the egg cell leaves the ovary, both the egg and the lining of the uterus begin to break down. The muscles in the uterus contract in a process called **menstruation.** Menstruation is the flow of blood and tissue from the body through a canal called the vagina over a period of about five days.

CHECK YOUR READING Where does the egg travel to after it leaves the ovary?

DIFFERENTIATE INSTRUCTION

More Reading Support

A Where is an egg cell fertilized? *in the fallopian tube*

B What happens if an egg is not fertilized? *The egg and the lining of the uterus break down.*

English Learners Place the terms *fertilization, menstruation, embryo, sperm, egg,* and *reproduction* on a classroom Science Word Wall with brief definitions for each. Use the chapter summary page to help students focus on key concepts and vocabulary.

4.3 INSTRUCT

Address Misconceptions

IDENTIFY Ask students if the material to develop an embryo and then a new human baby is contained in the egg or in the sperm. If they answer either egg or sperm, they hold the misconception either that the developing baby exists in the egg, with the sperm triggering growth, or that the new life exists in the sperm, with the egg required for food and protection.

CORRECT Ask students what traits of each of their parents they have. Point out that they have traits from both parents, and that traits were determined by genetic material in both the egg and the sperm. Explain that neither the egg nor the sperm alone has life, and that what creates life in the process of reproduction is the fusion of egg and sperm.

REASSESS Ask: What makes sperm cells and egg cells different from almost all other types of cells? *They must join one another in order to reproduce.*

Technology Resources

Visit **ClassZone.com** for background on common student misconceptions.

MISCONCEPTION DATABASE

Develop Critical Thinking

COMPARE Have students compare the female and male reproductive systems, focusing on function, structure, and the specialized cells each produces.

Ongoing Assessment

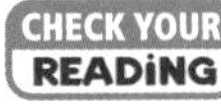

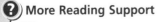

CHECK YOUR READING *Answer: the fallopian tube*

Teach from Visuals

Use the illustrations of the female and male reproductive systems (pp. 119–120) to trace the production of eggs and sperm and the process by which they join. Have students individually or in groups find the path of an egg cell in the female organs, and of a sperm cell in the male, and then the female organs.

Ongoing Assessment

Describe specialized cells and organs in the male and female reproductive systems.

Ask: What are the two main functions of the female reproductive system and where do they take place? *to produce egg cells (in the ovaries); to nourish offspring until birth (in the uterus)*

The Male Reproductive System

Testes The organs that produce sperm are called the testes (TEHS-teez). Inside the testes are tiny, coiled tubes hundreds of feet long. Sperm are produced inside these coiled tubes. The testes release a hormone that controls the development of sperm. This hormone is also responsible for the development of physical characteristics in men such as facial hair and a deep voice.

Sperm Sperm cells are the specialized cells of the male reproductive system. Males start producing sperm cells sometime during adolescence. The sperm is a single cell with a head and a tail. The sperm's head is filled with chromosomes, and the tail functions as a whip, making the sperm highly mobile. The sperm travel from the site of production, the testes, through several different structures of the reproductive system. While they travel, the sperm mix with fluids. This fluid is called semen and contains nutrients for the sperm cells. One drop of semen contains up to several million sperm cells.

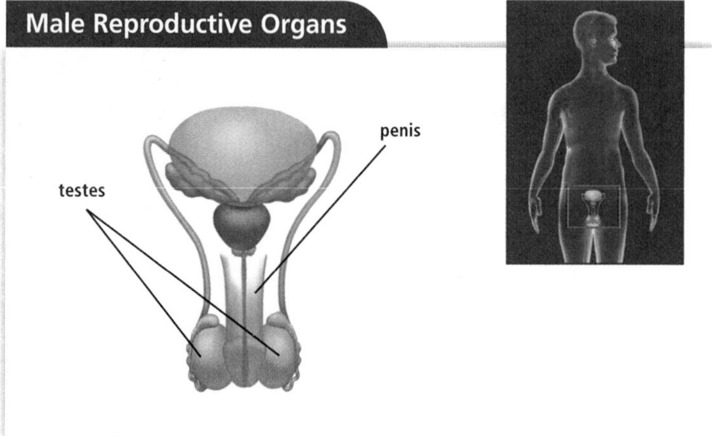

The reproduction of offspring includes fertilization, pregnancy, and birth.

VISUALIZATION CLASSZONE.COM

Follow an egg from fertilization to implantation.

Each sperm cell, like each egg cell, has half of the genetic material needed for a human being to grow and develop. During sexual intercourse, millions of sperm cells leave the testes. The sperm cells exit the male's body through the urethra, a tube that leads out of the penis. The sperm cells enter the female's body through the vagina. Next they travel into the uterus and continue on to the fallopian tube.

DIFFERENTIATE INSTRUCTION

More Reading Support

C Where are sperm produced? *in the testes*

D What is the function of the sperm's tail? *makes sperm mobile*

Below Level Help students organize information on the reproductive systems into a chart like the one below:

Gender	Specialized Cell	Organ That Produces Cell	Other Organs
female	egg	ovary	fallopian tube, uterus, vagina
male	sperm	testes	urethra, penis

Fertilization

Fertilization occurs when one sperm cell joins the egg cell. The fallopian tube is the site of fertilization. Immediately, chemical changes in the egg's surface prevent any more sperm from entering. Once inside the egg, the genetic material from the sperm combines with the genetic material of the egg cell. Fertilization is complete.

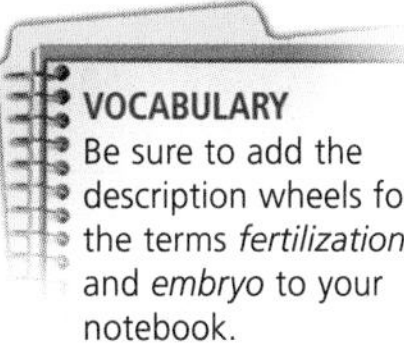

VOCABULARY
Be sure to add the description wheels for the terms *fertilization* and *embryo* to your notebook.

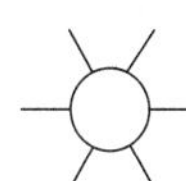

The fertilized egg cell then moves down the fallopian tube toward the uterus. You can trace the path of the egg cell in the diagram on this page. It divides into two cells. Each of those cells divides again, to form a total of four cells. Cell division continues, and a ball of cells forms, called an **embryo.** Within a few days, the embryo attaches itself to the thickened, spongy lining of the uterus in a process called implantation.

Fertilization

The egg cell moves down the fallopian tube following fertilization. Its final destination is the uterus.

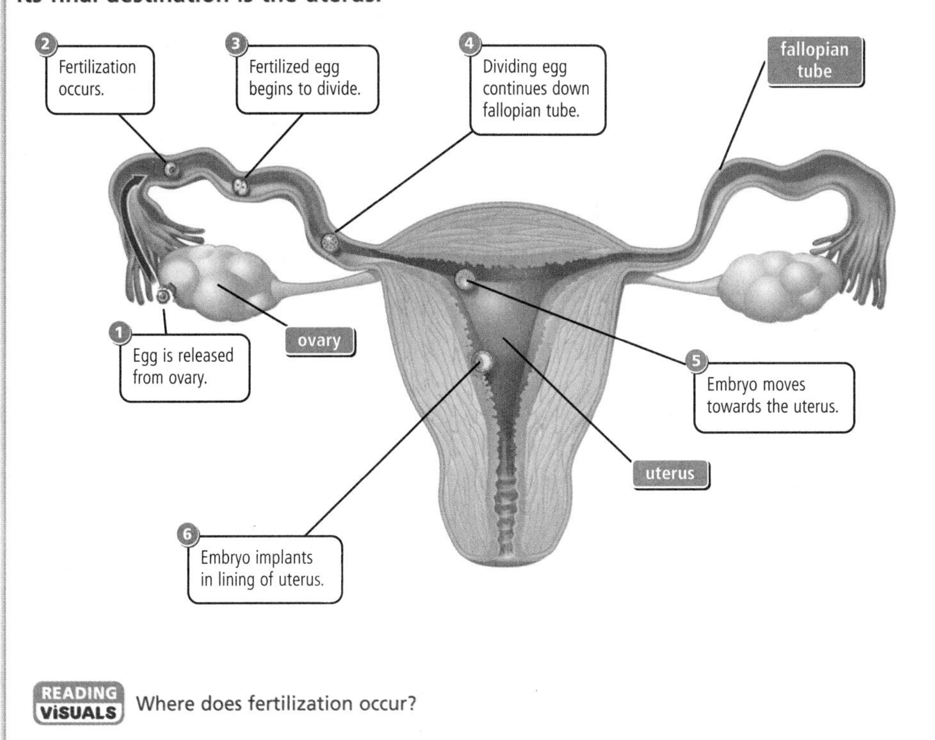

READING VISUALS Where does fertilization occur?

Teach from Visuals

Have students point to the ovary and fallopian tube on the right side of the diagram. Explain that both ovaries and both fallopian tubes are able to produce eggs and carry a fertilized egg to the uterus. To help students interpret the diagram of fertilization, ask:

- Where does the fertilized egg begin to divide? *in the fallopian tubes*
- Where does the embryo implant? *in the uterus*

Ongoing Assessment

Describe the stages of fertilization.

Ask: How does the egg progress to implantation? *A sperm fertilizes the egg in the fallopian tube. The fertilized egg begins to divide as it passes down the tube. Within a few days, a ball of cells is formed and this embryo attaches itself to the wall of the uterus.*

READING VISUALS *Answer: in fallopian tubes*

DIFFERENTIATE INSTRUCTION

More Reading Support

E What is fertilization? *when one sperm cell joins one egg cell*

F What is the ball of cells that grows from a fertilized egg? *an embryo*

Below Level Direct students to view the diagram (p. 121) one stage at a time, following the numbers. Students can render a simple sketch or tracing of the diagram, adding one word for each stage: 1. egg, 2. fertilization, 3. divides, 4. continues, 5. embryo, 6. implants.

Advanced

Challenge and Extension, p. 217

Health Connection

Drugs, including alcohol and nicotine, interfere with development of an embryo or fetus, killing cells as the body is forming. A classic example is the drug thalidomide, removed from the market in the 1960s, after causing severe birth defects, such as limb abnormalities, to babies of mothers who had taken the drug while pregnant. In the first trimester of a pregnancy a mother may not know she is pregnant and may be taking prescriptions, smoking, or drinking without realizing the impact on her child. Ask students:

- What are some of the reasons why drugs may be especially dangerous to a developing embryo during the first eight weeks of a pregnancy? *During the first trimester, especially when all major organ systems develop, the effects of chemical substances are not only multiplied due to body mass ratios, but also affect growth and health for life.*
- If a woman knows that there is a chance she may become pregnant, what actions could she take to promote health? *She should cease any recreational chemicals, and consult a doctor about prescription or over-the-counter medication.*

Teach Difficult Concepts

Students may have trouble understanding that an entire baby can grow from just two cells. Remind them that the cells contain the DNA instructions for how to grow, and that the mother's body furnishes the raw materials through the placenta.

Ongoing Assessment

CHECK YOUR READING *Answer: 2 weeks—embryo grows rapidly, placenta develops; 8 weeks—embryo becomes a fetus with developing facial features, organ systems, sexual traits, and skeleton; 12 weeks—bones of fetus develop further*

CHECK YOUR READING *Answer: First stage: labor: muscular contractions of the uterus; second stage: delivery: fetus is pushed out; third stage: delivery of the placenta*

Pregnancy

The nine months of pregnancy can be divided into three periods of about the same length. Each period marks specific stages of development. In the first week following implantation, the embryo continues to grow rapidly. Both the embryo and the uterus contribute cells to a new, shared organ called the placenta. The placenta has blood vessels that lead from the mother's circulatory system to the embryo through a large tube called the umbilical cord. Oxygen and nutrients from the mother's body will move through the placenta and umbilical cord to the growing embryo.

Uterus and Placenta

By the fifth week of pregnancy, the placenta is fully developed.

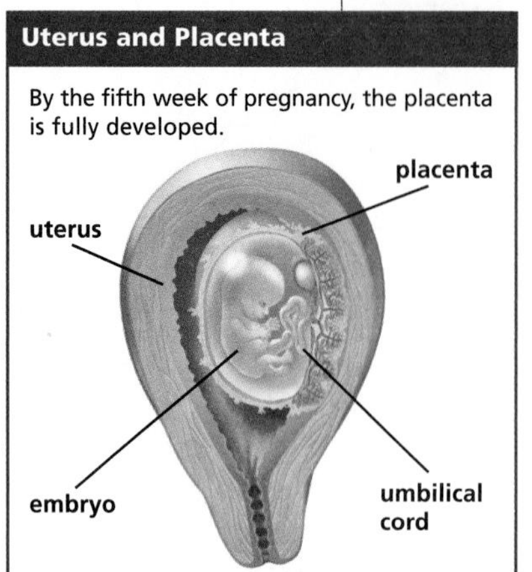

Around the eighth week of pregnancy, the developing embryo is called a **fetus.** The fetus begins to have facial features, major organ systems, and the beginnings of a skeleton. The fetus develops the sexual traits that are either male or female. In the twelfth week, the fetus continues to grow and its bones develop further. In the last twelve weeks the fetus and all its organ systems develop fully.

Describe the development of an embryo and fetus at two weeks, eight weeks, and twelve weeks.

Labor and Delivery

At the end of pregnancy, the fetus is fully developed and is ready to be born. The birth of a fetus is divided into three stages; labor, delivery, and birth of the placenta.

G

The first stage of birth begins with muscular contractions of the uterus. These contractions occur at intervals of 10 to 30 minutes and last about 40 seconds. They happen continually until the muscular contractions are occurring about every 2 minutes.

The second stage of birth is delivery. With each contraction the cervix dilates until it becomes wide enough for the mother's muscles to push the fetus out. During delivery the fetus is pushed out of the uterus, through the vagina, and out of the body. The fetus is still connected to the mother by the umbilical cord.

The umbilical cord is cut shortly after the fetus is delivered. Within minutes after birth, the placenta separates from the uterine wall and the mother pushes it out with more muscular contractions.

What happens during each of the three stages of birth?

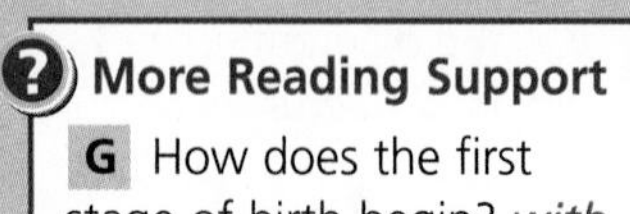

DIFFERENTIATE INSTRUCTION

More Reading Support

G How does the first stage of birth begin? *with muscular contractions of the uterus*

English Learners Help students with limited English create a sequence diagram to organize the text into stages of time. They should look for phrases that mention time in the text and diagram and use these phrases in headings in the diagram, as shown.

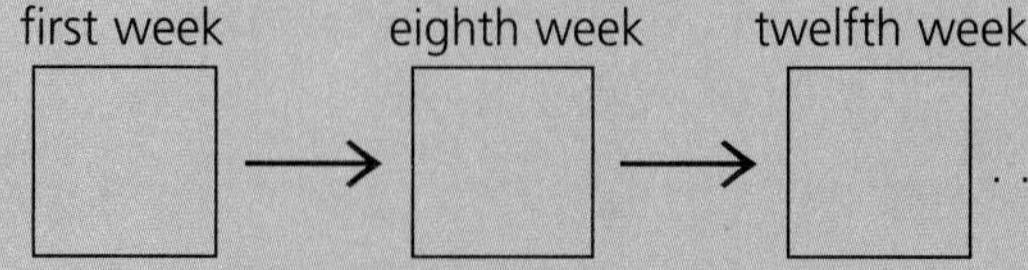

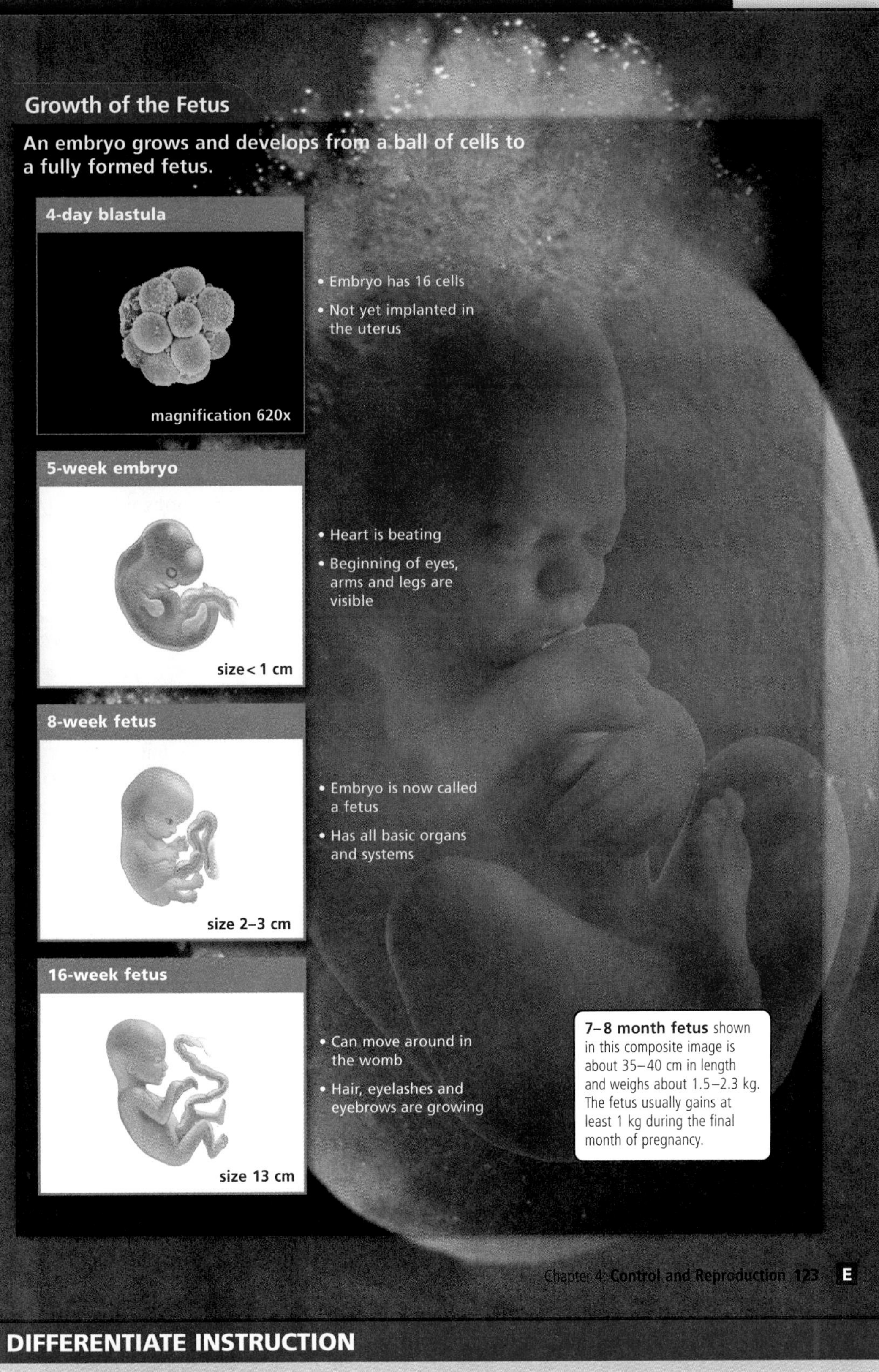
Growth of the Fetus
An embryo grows and develops from a ball of cells to a fully formed fetus.
4-day blastula
• Embryo has 16 cells
• Not yet implanted in the uterus
magnification 620x
5-week embryo
• Heart is beating
• Beginning of eyes, arms and legs are visible
size< 1 cm
8-week fetus
• Embryo is now called a fetus
• Has all basic organs and systems
size 2–3 cm
16-week fetus
• Can move around in the womb
• Hair, eyelashes and eyebrows are growing
size 13 cm
7–8 month fetus shown in this composite image is about 35–40 cm in length and weighs about 1.5–2.3 kg. The fetus usually gains at least 1 kg during the final month of pregnancy.
Chapter 4: Control and Reproduction 123 E

DIFFERENTIATE INSTRUCTION

Below Level Have students create sequence cards for the development of the fetus, drawing pictures of what happens at each stage.

Teach from Visuals

To help students understand the growth and development of a human fetus, ask:

- When does the heart start to beat? *by 5 weeks*
- How big is a 16-week fetus? *about 13 cm long*

Metacognitive Strategy

Have students suggest reasons why it is important to know about sexual reproduction.

EXPLORE the BIG idea

Revisit "Eggs" on p. 99. Have students compare chicken egg structures to human eggs.

Ongoing Assessment

Describe the development of the embryo and fetus during pregnancy.

Ask: How does the embryo change as pregnancy progresses? *By the eighth week the embryo has become a fetus. It begins to have facial features, major organ systems, some of its skeleton, and its sexual traits. In the second trimester the bones continue to grow, and during the third trimester the fetus and all of its systems develop fully.*

Ongoing Assessment

CHECK YOUR READING *Answer: Identical twins come from a single fertilized egg that divides in two. Fraternal twins come from two separate fertilized eggs.*

Reinforce the BIG idea

Have students relate the section to the Big Idea.

Reinforcing Key Concepts, p. 218

4.3 ASSESS & RETEACH

Assess

Section 4.3 Quiz, p. 62

Reteach

Draw a sequence or chain of events graphic organizer titled Reproductive System on the board. In the first box, write "male" and "female." Have students help you fill in the rest of the sequence, through fertilization, pregnancy, and birth.

Technology Resources

Have students visit **ClassZone.com** for reteaching of Key Concepts.

CONTENT REVIEW

CONTENT REVIEW CD-ROM

These twins provide an example of offspring born in a multiple birth.

Multiple Births

Do you have any friends who are twins or triplets? Perhaps you and your brothers or sisters are twins or triplets. The birth of more than one offspring is called a multiple birth. Multiple births are relatively uncommon in humans.

Identical twins are produced when a single fertilized egg divides in half. Each half then forms two complete organisms, or twins. Such twins are always of the same sex, look alike, and have identical blood types. Identical twins form early in pregnancy. Approximately 1 in 29 births is a set of identical twins.

Twins that are not identical are called fraternal twins. Fraternal twins are produced when two eggs are released at the same time and are fertilized by two different sperm. Consequently, fraternal twins may be very different from each other. Fraternal twins can be the same sex or different sexes.

CHECK YOUR READING Why are some twins identical and some are not?

4.3 Review

KEY CONCEPTS

1. Describe the function of the male reproductive system and the two main functions of the female reproductive system.
2. Explain how an egg travels from the ovary to the uterus.
3. How is an embryo different from a fetus?

CRITICAL THINKING

4. **Sequence** Describe the sequence of events that occurs between fertilization and the stage called implantation.
5. **Analyze** Detail two examples of hormones interacting with the reproductive system, one involving the male system and one involving the female system.

CHALLENGE

6. **Synthesize** Describe the interaction between the endocrine system and the reproductive system.

ANSWERS

1. male: makes sperm cells; female: makes egg cells; nourishes offspring until birth

2. Hormone stimulates eggs in ovary to develop. Hormone signals egg to move through fallopian tube to uterus.

3. Embryo is ball of cells; fetus has beginnings of features, organs, and skeleton.

4. Fertilized egg moves down fallopian tube; it divides and continues as a ball of cells; forms an embryo; embryo implants in uterus.

5. signal cells in testes to develop into sperm; signal ovaries to release eggs

6. Hormones from endocrine system signal sexual development. In females, hormones make egg develop and make ovary release it.

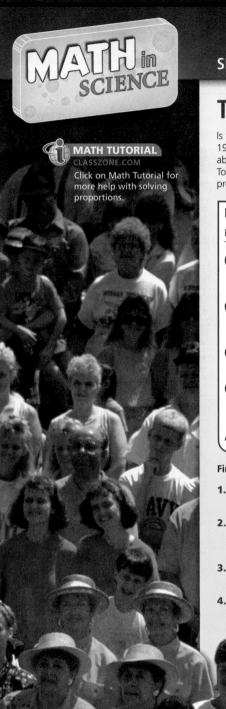

MATH TUTORIAL
CLASSZONE.COM
Click on Math Tutorial for more help with solving proportions.

SKILL: SOLVING PROPORTIONS

Twins and Triplets

Is the number of twins and triplets on the rise? Between 1980 and 1990, twin births in The United States rose from roughly 68,000 to about 105,000. In 1980, there were about 3,600,000 births total. To convert the data to birth rates, you can use proportions. A proportion is an equation. It shows two ratios that are equivalent.

Example

Find the birth rate of twins born in The United States for 1980. The rate is the number of twin births per 1000 births.

(1) Write the ratio of twin births to total births for that year.

$$\frac{68{,}000 \text{ twin births}}{3{,}600{,}000 \text{ total births}}$$

(2) Write a proportion, using x for the number you need to find.

$$\frac{68{,}000}{3{,}600{,}000} = \frac{x}{1000}$$

(3) In a proportion, the cross products are equal, so

$$68{,}000 \cdot 1000 = x \cdot 3{,}600{,}000$$

(4) Solve for x:

$$\frac{68{,}000{,}000}{3{,}600{,}000} = 18.9$$

ANSWER There were 18.9 twin births for every 1000 births in 1980.

Find the following birth rates.

1. In 1990, there were about 105,000 twin births and about 3,900,000 total births. What was the birth rate of twins?
2. In 1980, about 1,350 sets of triplets were born, and by 1990, this number had risen to about 6,750. What were the birth rates of triplets in 1987 and in 1990?
3. How much did the birth rate increase for twins between 1980 and 1990? for triplets?
4. Find the overall birth rate of twins and triplets in 1980.
5. Find the overall rate of twin and triplet births in 1990. How much did it increase between 1980 and 1990?

CHALLENGE In 1989, there were about 4 million total births, and the rate of triplets born per million births was about 700. How many triplets were born?

Chapter 4: **Control and Reproduction 125** **E**

ANSWERS

1. 27 per 1000 births

2. 1980: 0.38 per 1000 births; 1990: 1.7 per 1000 births

3. 42%; 347%

4. 19.3 per 1000 births

5. 28.7 per 1000 births; 49%

CHALLENGE 2800 sets of triplets, or 8400 triplets

MATH IN SCIENCE
Math Skills Practice for Science

Set Learning Goal

To solve proportions regarding birth rates for twins and triplets

Present the Science

A major reason for the rise in multiple pregnancies in recent decades is the use of fertility drugs and assisted reproductive technology. These advances have helped couples who were unable to have children, but they are not without risks. Multiple fetuses also increase complications of pregnancy and birth. Preterm births occur five times more often for twins than for single fetuses, and 92% of mothers of triplets give birth early. Incidences of gestational diabetes, preeclampsia, and placenta problems are also significantly higher. Anemia is also more common.

Develop Algebra Skills

- It may help students to read the proportions aloud as "68,000 is to 3,600,000 as x is to 1000."
- Describe how to calculate percentage of increase: Subtract the lower number from the greater, and divide the result by the lower number. Multiply by 100 to get a percentage.

Close

Ask: How does using rates help you adjust the data for population growth? *Rates are like percentages. They allow you to compare different-sized populations.*

- Math Support, p. 224
- Math Practice, p. 225

Technology Resources

Students can visit **ClassZone.com** for practice in solving proportions.

MATH TUTORIAL

BACK TO

the BIG idea

Ask students to explain the basic functions of the nervous and endocrine systems. *The nervous system regulates voluntary and involuntary movement of the body and allows it to respond to the environment through the senses. The endocrine system helps the body maintain homeostasis through regulation by hormones.*

KEY CONCEPTS SUMMARY

SECTION 4.1
Have students name an example of something that is controlled by the autonomic nervous system. *The autonomic nervous system controls involuntary movement such as the pumping of the heart and movement of smooth muscles during digestion.*

SECTION 4.2
Ask students why the diagram of the endocrine system is an example of a feedback system in the body. *Feedback systems allow the body to control the levels of hormones that are produced by certain glands, as illustrated in the diagram.*

SECTION 4.3
Ask students to describe the reproductive system shown in the illustration. *The illustration shows the female reproductive system and includes ovaries, fallopian tubes, and uterus.*

Review Concepts

- Big Idea Flow Chart, p. T25
- Chapter Outline, pp. T31–T32

4 Chapter Review

the BIG idea

The nervous and endocrine systems allow the body to respond to internal and external conditions.

CONTENT REVIEW
CLASSZONE.COM

KEY CONCEPTS SUMMARY

4.1 The nervous system responds and controls.

- The nervous system connects the body with its environment using five senses: sight, touch, hearing, smell, and taste. Central nervous system includes the brain, the control center, and the spinal cord.
- The peripheral nervous system includes the autonomic and voluntary systems

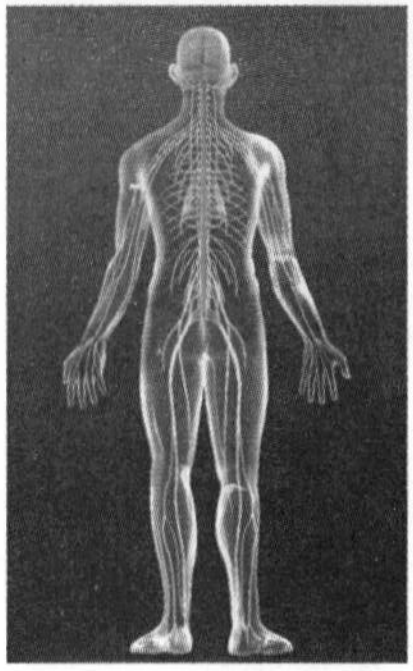

VOCABULARY
stimulus p. 102
central nervous system p. 104
neuron p.105
peripheral nervous system p. 106
autonomic nervous system p. 107
voluntary nervous system p. 107

4.2 The endocrine system helps regulate conditions inside the body.

The body has chemical messengers called **hormones** that are regulated by the **endocrine system. Glands** produce and release hormones. The endocrine system includes feedback systems that maintain homeostasis.

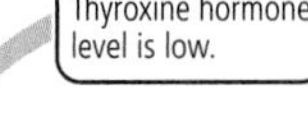
Thyroxine hormone level is low.
Hypothalamus and pituitary gland decrease thyroid gland activity.
Thyroxine hormone level is high.
Hypothalamus and pituitary gland increase thyroid gland activity.

VOCABULARY
endocrine system p. 110
hormone p. 111
gland p. 111

4.3 The reproductive system allows the body to produce offspring.

The female produces eggs, and the male produces sperm. Following **fertilization** the egg develops over a period of about nine months.

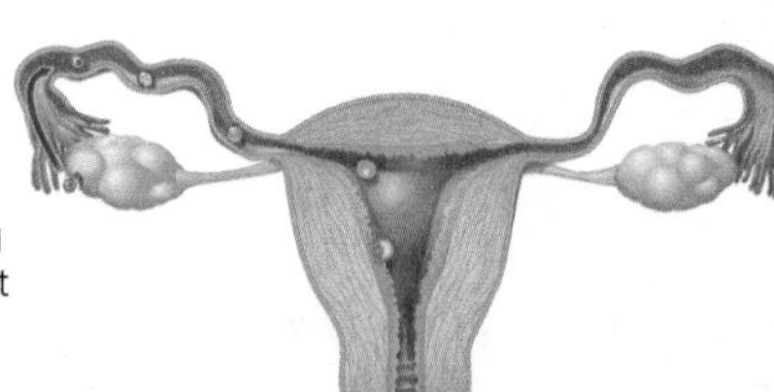

VOCABULARY
menstruation p. 119
fertilization p. 121
embryo p. 121
fetus p. 122

E 126 Unit: Human Biology

Technology Resources

Have students visit **ClassZone.com** or use the CD-ROM for a cumulative review of concepts.

 **CONTENT REVIEW**

 **CONTENT REVIEW CD-ROM**

Engage students in a whole-class interactive review of Key Concepts. Edit content as you wish.

 **POWER PRESENTATIONS**

Reviewing Vocabulary

Make a frame for each of the vocabulary words listed. Write the word in the center. Decide what information to frame it with. Use definitions, examples, descriptions, parts, or pictures.

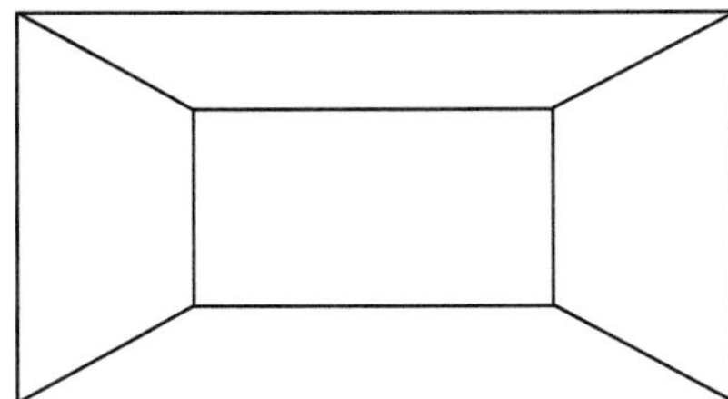

1. stimulus
2. neurons
3. hormones
4. fertilization
5. embryo

Reviewing Key Concepts

Multiple Choice *Choose the letter of the best answer.*

6. Which is a stimulus?
   a. a car horn blowing
   b. jumping at a loud noise
   c. taste buds on the tongue
   d. turning on a lamp
7. Light enters the eye through
   a. the lens
   b. the auditory canal
   c. the olfactory epithelium
   d. the taste buds
8. Which senses allow you to process chemical information?
   a. sight and smell
   b. taste and smell
   c. touch and hearing
   d. hearing and taste
9. What conserves energy and responds quickly to change?
   a. central nervous system
   b. peripheral nervous system
   c. autonomic nervous system
   d. voluntary nervous system
10. Which is not regulated by hormones?
    a. production of red blood cells
    b. physical growth
    c. blood pressure
    d. sexual development
11. Which gland releases hormones that are necessary for growth and metabolism?
    a. thyroid gland c. adrenal gland
    b. pituitary gland d. pineal gland
12. Eggs develop in the female reproductive organ called
    a. an ovary c. a uterus
    b. a fallopian tube d. a vagina
13. The joining of one sperm cell and one egg cell is an event called
    a. menstruation c. implantation
    b. fertilization d. birth
14. A ball of cells that is formed by fertilization is called the
    a. testes c. ovary
    b. urethra d. embryo
15. The period in which a fetus and all of its systems develop fully is the
    a. first three months
    b. second three months
    c. third three months
    d. pregnancy

Short Answer *Write a short answer to each question.*

16. List the parts of the body that are controlled by the autonomic nervous system.
17. What is a negative feedback mechanism? Give an example.
18. How are fertilization and menstruation related?

ASSESSMENT RESOURCES

UNIT ASSESSMENT BOOK
- Chapter Test, Level A, pp. 63–66
- Chapter Test, Level B, pp. 67–70
- Chapter Test, Level C, pp. 71–74
- Alternative Assessment, pp. 75–76

SPANISH ASSESSMENT BOOK
Spanish Chapter Test, pp. 93–96

Technology Resources

Edit test items and answer choices.

Test Generator CD-ROM

Visit **ClassZone.com** to extend test practice.

Test Practice

Reviewing Vocabulary

Sample answers:

1. sounds, scents, tastes, touch, light; change in environment that people react to; detected by senses; processed by brain

2. electrical signals travel through them between body, brain, and spinal cord; make up brain and spinal cord

3. made and sent around body by endocrine system; the body's chemical messengers; chemicals made in one structure that are sent to a second structure; growth, antidiuretic, thyroxine, cortisol, insulin, testosterone, estrogen, adrenaline; produced and released by glands

4. may lead to pregnancy; results when egg and sperm unite; happens in fallopian tube; may happen as a result of sexual intercourse

5. ball of cells that forms from a fertilized egg; implants in lining of uterus; contributes to formation of placenta; attached to placenta by umbilical cord; after eight weeks develops into fetus

Reviewing Key Concepts

6. a	*11. a*
7. a	*12. a*
8. b	*13. b*
9. c	*14. d*
10. c	*15. c*

16. The heart and the organs of the digestive system are controlled by the autonomic nervous system.

17. A negative feedback system controls the amount of a hormone in the body by signaling the body to stop producing it. Increased levels of thyroxine, for example, inhibit signals from the hypothalamus and pituitary glands that trigger production of thyroxine, and thyroxine production decreases.

18. If fertilization does not occur, the egg and the lining of the uterus break down, triggering muscle contractions that expel blood and tissue from the uterus in a process called menstruation.

Thinking Critically

19. *Light enters through lens; pupil gets smaller or larger to control the amount of light that enters; light focused by lens and hits retina at back of eye.*
20. *Image is like the object but inverted; is corrected by the brain after it receives information from optic nerve.*
21. *Central and peripheral nervous systems are involved; sensory nerves of peripheral nervous system feel pain and send impulses to central nervous system; motor nerves send signals to muscles to move foot.*
22. *Response to the stimulus increases stimulus, which then increases response; example: when endocrine system releases hormone that triggers blood clotting.*
23. *central nervous system: brain, spinal cord; peripheral nervous system: autonomic*
24. *thyroid gland because it controls metabolism; slow metabolism would likely cause lack of energy*
25. *Pituitary gland releases a hormone to stimulate egg development; another hormone signals ovary to release egg; egg travels down fallopian tube; not fertilized, egg and uterus lining break down and menstruation begins.*
26. *Both produce reproductive cells and are regulated by hormones; ovaries produce eggs; testes produce sperm.*

the BIG idea

27. *Answers will depend on students' original answers.*
28. *Nervous system controls endocrine and reproductive systems. Sample examples: when frightened, hormones cause heartbeat to increase; hormones control all aspects of male and female reproductive systems.*

UNIT PROJECTS

Check to make sure students are working on their projects. Check schedules and work in progress.

 Unit Projects, pp. 5–10

Thinking Critically

Use the diagram to answer the following two questions.

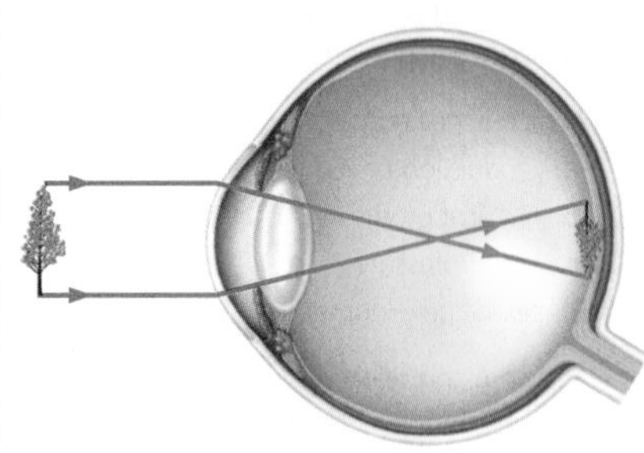

19. **SUMMARIZE** Use the diagram of the eye to describe how images are formed on the retina.
20. **COMPARE AND CONTRAST** How is the image that forms on the retina like the object? How is it different? Explain how the viewer interprets the image that forms on the retina.
21. **APPLY** A person steps on a sharp object with a bare foot and quickly pulls the foot back in pain. Describe the parts of the nervous system that are involved in this action.
22. **ANALYZE** Explain why positive feedback mechanisms do not help the body maintain homeostasis. Give an example.
23. **CONNECT** Copy the concept map and add the following terms to the correct box: brain, spinal cord, autonomic.

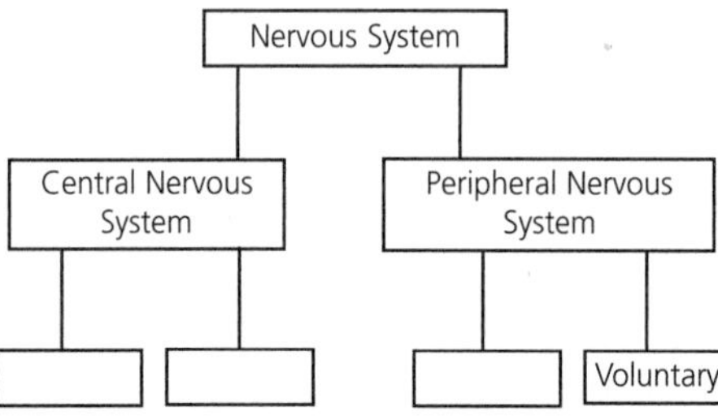

24. **DRAW CONCLUSIONS** A person who is normally very active begins to notice a significant decrease in energy level. After visiting a doctor, tests results show that one of the endocrine glands is not secreting enough of its hormone. Which gland could this be? Explain your answer.
25. **SUMMARIZE** Describe the events that occur during the female's 28-day menstrual cycle. Include in your answer how hormones are involved in the cycle.
26. **COMPARE AND CONTRAST** How are the functions of the ovaries and the testes alike? How are their functions different?

the BIG idea

27. **INFER** Look again at the picture on pages 98–99. Now that you have finished the chapter, how would you change or add details to your answer to the question on the photograph?
28. **SYNTHESIZE** How does the nervous system interact with the endocrine and reproductive systems? Give examples that support your answer.

UNIT PROJECTS

If you need to do an experiment for your unit project, gather the materials. Be sure to allow enough time to observe results before the project is due.

MONITOR AND RETEACH

If students are having trouble applying the concepts in items 21–23, suggest that they review the nervous system (pp. 104–107) and feedback mechanisms (pp. 114–116). Have them sketch the situation described in item 21. They can label the activities of various organs involved.

Students may benefit from summarizing one or more sections of the chapter.

 Summarizing the Chapter, pp. 244–245

Standardized Test Practice

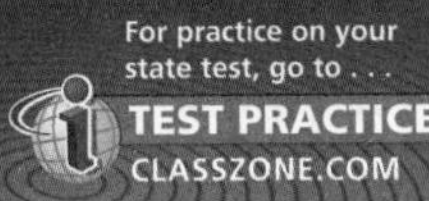
For practice on your state test, go to . . .
TEST PRACTICE
CLASSZONE.COM

Analyzing Data

This chart shows some of the stages of development of a typical fetus.

Week of Pregnancy	Average Length of Fetus	Developmental Changes in the Fetus
6	0.5 cm	Primitive heartbeat
10	2.5 cm	Face, fingers, and toes are formed
14	7.5 cm	Muscle and bone tissue have developed
18	12.5 cm	Fetus makes active movements
24	28 cm	Fingerprints and footprints forming
28	37 cm	Rapid brain development
36	45 cm	Increase in body fat
38	50 cm	Fetus is considered full term

Use the chart to answer the questions below.

1. What is the average length the fetus at 10 weeks?
   - **a.** 0.5 cm
   - **b.** 0.5 in.
   - **c.** 2.5 cm
   - **d.** 2.5 in.

2. At about which week of development does the fetus begin to make active movements?
   - **a.** week 10
   - **b.** week 14
   - **c.** week 18
   - **d.** week 24

3. At about which week of development does the fetus reach a length of about 28 cm?
   - **a.** week 18
   - **b.** week 24
   - **c.** week 36
   - **d.** week 38

4. Which statement is true?
   - **a.** The fetus grows about 5 cm in length during the last 2 weeks of development.
   - **b.** The fetus begins to develop fingerprints at about week 28.
   - **c.** During week 10, the average length of the fetus is about 7.5 cm
   - **d.** The fetus is about 12.5 cm long when muscle and bone tissue develop.

5. Between which two weeks of development does the greatest increase in length usually take place?
   - **a.** weeks 6 and 10
   - **b.** weeks 10 and 14
   - **c.** weeks 14 and 18
   - **d.** weeks 24 and 28

Extended Response

6. Describe the changes in length and development that occur during each trimester of pregnancy.
7. The endocrine system and the nervous system have similar functions. Compare and contrast the two systems including the terms in the box. Underline each term in your answer.

homeostasis	autonomic system	hormones
feedback	smooth muscles	

METACOGNITIVE ACTIVITY

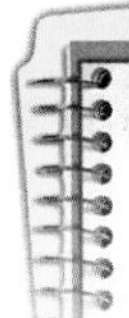

Have students answer the following questions in their **Science Notebook:**

1. What is the most pressing question that you still have about the nervous system?
2. Did you learn anything in this chapter about endocrine, nervous, or reproductive systems that surprised you? Explain.
3. What learning strategies did you use to master a difficult concept in this chapter?

Analyzing Data

1. c *3. b* *5. d*
2. c *4. a*

Extended Response

6. RUBRIC

4 points for a response that correctly answers the question and describes changes in length and development for all three trimesters.

Sample answer: First trimester: fetus grows up to 8.5 cm. Heartbeat, fingers, face, toes, and muscle and bone tissue formed. Second trimester: fetus reaches up to 30 cm. Fetus makes active movements, and fingerprints and footprints are forming. Third trimester: fully developed fetus reaches about 50 cm. Rapid brain development and increase in body fat occurs.

3 points correctly answers question and describes changes in length of all three trimesters of pregnancy and development for at least two trimesters

2 points correctly answers question and describes changes in length of all three trimesters of pregnancy and development for at least one trimester

1 point describes changes in length during all three trimesters or change in development for one trimester

7. RUBRIC

4 points for a response that includes at least four of the following terms:

- homeostasis
- feedback
- autonomic system
- smooth muscles
- hormones

Sample answer: Endocrine and nervous systems control body functions. Autonomic nervous system works with endocrine system to regulate heartbeat, blood pressure, operation of smooth muscles, and secretion of hormones by glands. Endocrine system controls homeostasis, or maintenance of correct body conditions, through negative feedback mechanisms in which increase in amount of a hormone in body triggers decrease in secretion of that hormone.

3 points includes at least three of the terms

2 points includes at least two of the terms

1 point includes at least one of the terms

CHAPTER 5 Growth, Development, and Health

Life Science
UNIFYING PRINCIPLES

PRINCIPLE 1
All living things share common characteristics.

PRINCIPLE 2
All living things share common needs.

PRINCIPLE 3
Living things meet their needs through interactions with the environment.

PRINCIPLE 4
The types and numbers of living things change over time.

Unit: Human Biology
BIG IDEAS

CHAPTER 1
Systems, Support, and Movement
The human body is made up of systems that work together to perform necessary functions.

CHAPTER 2
Absorption, Digestion, and Exchange
Systems in the body obtain and process materials and remove waste.

CHAPTER 3
Transport and Protection
Systems function to transport materials and to defend and protect the body.

CHAPTER 4
Control and Reproduction
The nervous and endocrine systems allow the body to respond to internal and external conditions.

CHAPTER 5
Growth, Development, and Health
The body develops and maintains itself over time.

CHAPTER 5
KEY CONCEPTS

SECTION 5.1

The human body changes over time.
1. The human body develops and grows.
2. Systems interact to maintain the human body.

SECTION 5.2

Systems in the body function to maintain health.
1. Diet affects the body's health.
2. Exercise is part of a healthy lifestyle.
3. Drug abuse, addiction, and eating disorders cause serious health problems.

SECTION 5.3

Science helps people prevent and treat disease.
1. Scientific understanding helps people fight disease.
2. The germ theory describes some causes of disease.
3. Infectious diseases spread in many ways.
4. Noninfectious diseases are not contagious.
5. Scientists continue efforts to prevent and treat illness.

The Big Idea Flow Chart is available on p. T33 in the **UNIT TRANSPARENCY BOOK.**

Previewing Content

SECTION

5.1 The human body changes over time. pp. 133–139

1. The human body develops and grows.

- There are several stages of development in a human life. **Infancy** begins at birth and continues as a child learns basic physical skills as well as mental and social skills.

Apgar Score	
Quality	**2 points**
Appearance	Completely pink or good color
Pulse	>100 beats per minute
Grimace	Grimace, cough/ sneeze with suction
Activity	Active motion
Respiration	Good, strong cry

Good health begins in infancy, as the qualities measured in the Apgar test show.

- **Childhood** is the period after infancy and before puberty. Children become more able to care for themselves and learn skills such as cooperation and sharing.
- **Adolescence** begins with puberty, the process of sexual maturation. Hormones signal the reproductive system to mature.
- **Adulthood** occurs when the human body completes its growth and reaches sexual maturity. Mental and social development continue into later adulthood and throughout life.

2. Systems interact to maintain the human body.
All systems in the body constantly interact with other systems to keep the body functioning. Body systems also interact with the external environment to maintain homeostasis.

SECTION

5.2 Systems in the body function to maintain health. pp. 140–147

1. Diet affects the body's health.
Nutrition is the study of the materials that affect the body. Six classes of nutrients include carbohydrates, proteins, fats, water, vitamins, and minerals. Food labels use terms such as *low-fat* and *organic* to describe the make-up of food.
More importantly, food labels give nutrition information based on recommended daily amounts, as shown in the chart below.

Nutrition Information		
Nutrient	**Function**	**Daily Amount**
Proteins	Build tissues used for growth and repair	20 percent of the diet
Carbohydrates	Body's most important source of energy	40 to 50 percent of the diet
Fats	Essential for energy	10 to 15 percent of the diet

2. Exercise is part of a healthy lifestyle.
Regular exercise improves respiratory and heart function. It keeps bones and muscles strong and helps to prevent health conditions later in life such as osteoporosis, excess body fat, and heart disease.

3. Drug abuse, addiction, and eating disorders cause serious health problems.
Unhealthy choices may lead to drug abuse, **addiction,** or eating disorders. A drug is a chemical substance that causes a change in the body's function. Tobacco and alcohol are drugs. Nicotine, which is the drug found in tobacco, is addictive. Drugs such as stimulants and narcotics are also addictive and can cause heart failure and death.

Common Misconceptions

FACTORS THAT AFFECT HEALTH Students of all ages focus on the physical dimensions of health and pay less attention to the mental and social dimensions. Students associate health primarily with food and fitness. Many students tend to think that factors important to health and life span are beyond their personal control.

This misconception is addressed on p. 143.

MISCONCEPTION DATABASE
CLASSZONE.COM Background on student misconceptions

Previewing Content

SECTION

Science helps people prevent and treat disease. pp. 148–153

1. Scientific understanding helps people fight disease.
Disease is a change that disturbs the normal functioning of the body. Infectious diseases are caused by **microorganisms.**

2. The germ theory describes some causes of disease.
Louis Pasteur's germ theory states that some infectious diseases are caused by germs. Bacteria and viruses are examples of pathogens, which are disease-causing agents. Bacterial diseases can be treated with antibiotics, first discovered by Alexander Fleming.

3. Infectious diseases spread in many ways.
- Pathogens spread in many ways and can be found in food, water, in the air, and on surfaces of objects.
- Animals and insects carry disease-causing organisms. Lyme disease is carried by deer ticks. Rabies is carried by bats, raccoons, and opossums.
- Pathogens can also be transmitted by contact with another person, or through the mouth, nose, eyes, and cuts.

4. Noninfectious diseases are not contagious.
Noninfectious diseases such as hemophilia may be present at birth. Other diseases, such as cancer and diabetes, may occur later in life.

5. Scientists continue efforts to prevent and treat illness.
Vaccinations have made some diseases nearly extinct. Overuse of antibiotics may cause **resistance** to certain strains of pathogens.

Washing hands and foods and boiling water can help prevent the spread of pathogens, such as this *E. coli* bacteria.

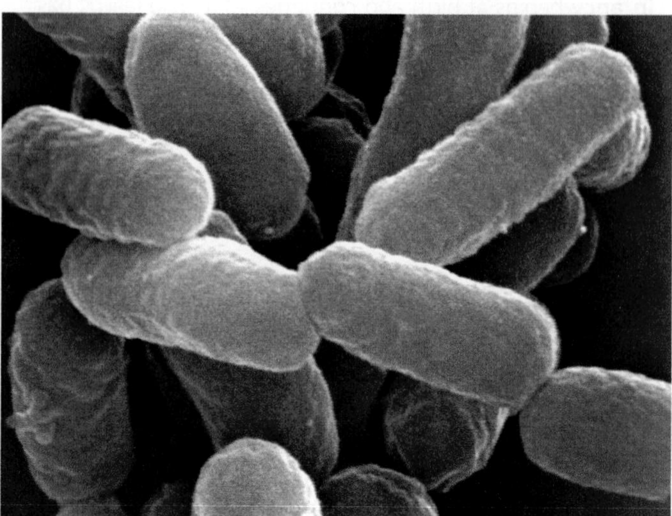

Common Misconceptions

MISCONCEPTION DATABASE
CLASSZONE.COM Background on student misconceptions

CAUSES OF ILLNESS Many students hold the misconception that all illnesses are caused by germs and that germs enter through the mouth by eating. Students may also think that every illness is contagious. As students grow older, their understanding of the causes of illness begins to include malfunctioning of internal organs and systems, poor health habits, and genetics. Upper elementary students can understand that a change in internal body state or the experience of symptoms is the consequence of illness.

This misconception is addressed on p. 150.

Previewing Labs

Lab Generator CD-ROM
Edit these Pupil Edition labs and generate alternative labs.

EXPLORE the BIG idea

Lab	Time and Materials
How Much Do You Exercise? p. 131 Students keep track of how much they exercise in a week in order to understand the connection between exercise and resting heart rate.	**TIME** 10 minutes (per day) **MATERIALS** clock with second hand
How Safe Is Your Food? p. 131 Students analyze freshness dates to learn about food safety.	**TIME** 15 minutes **MATERIALS** various food labels or containers
Internet Activity: Human Development, p. 131 Students observe changes to a person's face during aging.	**TIME** 20 minutes **MATERIALS** computer with Internet access

SECTION 5.1

Lab	Time and Materials
EXPLORE Growth, p. 133 Students compile data on body measurements to analyze patterns of growth.	**TIME** 15 minutes **MATERIALS** flexible tape measure, graph paper
INVESTIGATE Life Expectancy, p. 137 Students create a graph to analyze data about life expectancy from 1900–2000 to infer why average life expectancy has changed.	**TIME** 30 minutes **MATERIALS** graph paper, computer graphing program (optional), notebook, pen

SECTION 5.2

Lab	Time and Materials
INVESTIGATE Food Labels, p. 143 Students evaluate food labels to determine the nutritional values of a variety of foods.	**TIME** 30 minutes **MATERIALS** nutrition information labels from carbonated soft drink, bag of fresh carrots, canned spaghetti in sauce, potato chips, plain popcorn kernels, unsweetened applesauce, and fruit juice

SECTION 5.3

Lab	Time and Materials
EXPLORE The Immune System, p. 148 Students model how easily germs are spread.	**TIME** 10 minutes **MATERIALS** glitter
CHAPTER INVESTIGATION **Cleaning Your Hands,** pp. 154–155 Students sample bacteria on their hands and test the effectiveness of washing their hands with just water compared with soap and water.	**TIME** 40 minutes **MATERIALS** 3 covered petri dishes with sterile nutrient agar, tape, marker, hand soap, hand lens

Additional INVESTIGATION, Where's the Starch?, A, B, & C, pp. 294–302; Teacher Instructions, pp. 305–306

Previewing Chapter Resources

INTEGRATED TECHNOLOGY | **LABS AND ACTIVITIES**

CHAPTER 5
Growth, Development, and Health

Integrated Technology

CLASSZONE.COM
- eEdition Plus
- EasyPlanner Plus
- Misconception Database
- Content Review
- Test Practice
- Resource Centers
- Internet Activity: Human Development
- Math Tutorial

SCILINKS.ORG

CD-ROMS
- eEdition
- EasyPlanner Plus
- Power Presentations
- Content Review
- Lab Generator
- Test Generator

AUDIO CDS
- Audio Readings
- Audio Readings in Spanish

Labs and Activities

EXPLORE the Big Idea, p. 131
- How Much Do You Exercise?
- How Safe Is Your Food?
- Internet Activity: Human Development

UNIT RESOURCE BOOK
Unit Projects, pp. 5–10

Lab Generator CD-ROM
Generate customized labs.

SECTION

The human body changes over time.
pp. 133–139

Time: 2 periods (1 block)

 Lesson Plan, pp. 246–247

MATH TUTORIAL

UNIT TRANSPARENCY BOOK
- Big Idea Flow Chart, p. T33
- Daily Vocabulary Scaffolding, p. T34
- Note-Taking Model, p. T35
- 3-Minute Warm-Up, p. T36

- EXPLORE Growth, p. 133
- INVESTIGATE Life Expectancy, p. 137
- Science on the Job, p. 139

UNIT RESOURCE BOOK
Datasheet, Life Expectancy, p. 255

SECTION

Systems in the body function to maintain health.
pp. 140–147

Time: 2 periods (1 block)

R Lesson Plan, pp. 257–258

- **RESOURCE CENTERS,** Human Health, Principles of Nutrition
- **MATH TUTORIAL**

UNIT TRANSPARENCY BOOK
- Daily Vocabulary Scaffolding, p. T34
- 3-Minute Warm-Up, p. T36

- INVESTIGATE Food Labels, p. 143
- Math in Science, p. 147

UNIT RESOURCE BOOK
- Datasheet, Food Labels, p. 266
- Additional INVESTIGATION, Where's the Starch?, Levels A, B, & C, pp. 294–302
- Math Support, p. 283
- Math Practice, p. 284

SECTION

Science helps people prevent and treat disease.
pp. 148–153

Time: 4 periods (2 blocks)

 Lesson Plan, pp. 268–269

RESOURCE CENTER, Ways to Fight Disease

UNIT TRANSPARENCY BOOK
- Big Idea Flow Chart, p. T33
- Daily Vocabulary Scaffolding, p. T34
- 3-Minute Warm-Up, p. T36
- "Pathogens and Disease" Visual, p. T38
- Chapter Outline, pp. T39–T40

- EXPLORE The Immune System, p. 148
- CHAPTER INVESTIGATION, Cleaning Your Hands, pp. 154–155

UNIT RESOURCE BOOK
- CHAPTER INVESTIGATION, Cleaning Your Hands, Levels A, B, & C, pp. 285–293

KEY TO ICONS 

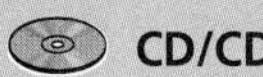 CD/CD-ROM | INTERNET | Pupil Edition |  Teacher Edition | UNIT RESOURCE BOOK |  UNIT TRANSPARENCY BOOK | UNIT ASSESSMENT BOOK |  SPANISH ASSESSMENT BOOK | SCIENCE TOOLKIT

READING AND REINFORCEMENT	ASSESSMENT	STANDARDS
 • Choose Your Own Strategy, B20–27 • Content Frame, C35 • Daily Vocabulary Scaffolding, H1–8  **UNIT RESOURCE BOOK** • Vocabulary Practice, pp. 280–281 • Decoding Support, p. 282 • Summarizing the Chapter, pp. 303–304  **Audio Reading CD** Listen to Pupil Edition.  **Audio Readings in Spanish CD** Listen to Pupil Edition in Spanish.	 • Chapter Review, pp. 157–158 • Standardized Test Practice, p. 159  **UNIT ASSESSMENT BOOK** • Diagnostic Test, pp. 77–78 • Chapter Test, A, B, & C, pp. 82–93 • Alternative Assessment, pp. 94–95 • Unit Test, A, B, C, pp. 96–107  • Spanish Chapter Test, pp. 97–100 • Spanish Unit Test, pp. 101–104  **Test Generator CD-ROM** Generate customized tests.  **Lab Generator CD-ROM** Rubrics for Labs	National Standards A.2–8, A.9.a–c, A.9.e–f, C.1.e–f, C.3.a–b, F.1.c–e, G.1.b See p. 130 for the standards.
 **UNIT RESOURCE BOOK** • Reading Study Guide, A & B, pp. 248–251 • Spanish Reading Study Guide, pp. 252–253 • Challenge and Extension, p. 254 • Reinforcing Key Concepts, p. 256 • Challenge Reading, pp. 278–279	 Ongoing Assessment, pp. 135–138  Section 5.1 Review, p. 138  **UNIT ASSESSMENT BOOK** Section 5.1 Quiz, p. 79	National Standards A.2–7, A.9.a–b, A.9.e–f, C.1.e–f, C.3.a–b, G.1.b
 **UNIT RESOURCE BOOK** • Reading Study Guide, A & B, pp. 259–262 • Spanish Reading Study Guide, pp. 263–264 • Challenge and Extension, p. 265 • Reinforcing Key Concepts, p. 267	Ongoing Assessment, pp. 140, 141, 144, 146  Section 5.2 Review, p. 146  **UNIT ASSESSMENT BOOK** Section 5.2 Quiz, p. 80	National Standards A.2–8, A.9.a–c, A.9.e–f, C.1.e, C.3.a, F.1.a, F.1.c–e, G.1.b
 **UNIT RESOURCE BOOK** • Reading Study Guide, A & B, pp. 270–273 • Spanish Reading Study Guide, pp. 274–275 • Challenge and Extension, p. 276 • Reinforcing Key Concepts, p. 277	Ongoing Assessment, pp. 150–153  Section 5.3 Review, p. 153  **UNIT ASSESSMENT BOOK** Section 5.3 Quiz, p. 81	National Standards A.2–7, A.9.a–b, A.9.e–f, C.1.e–f, G.1.b

Previewing Resources for Differentiated Instruction

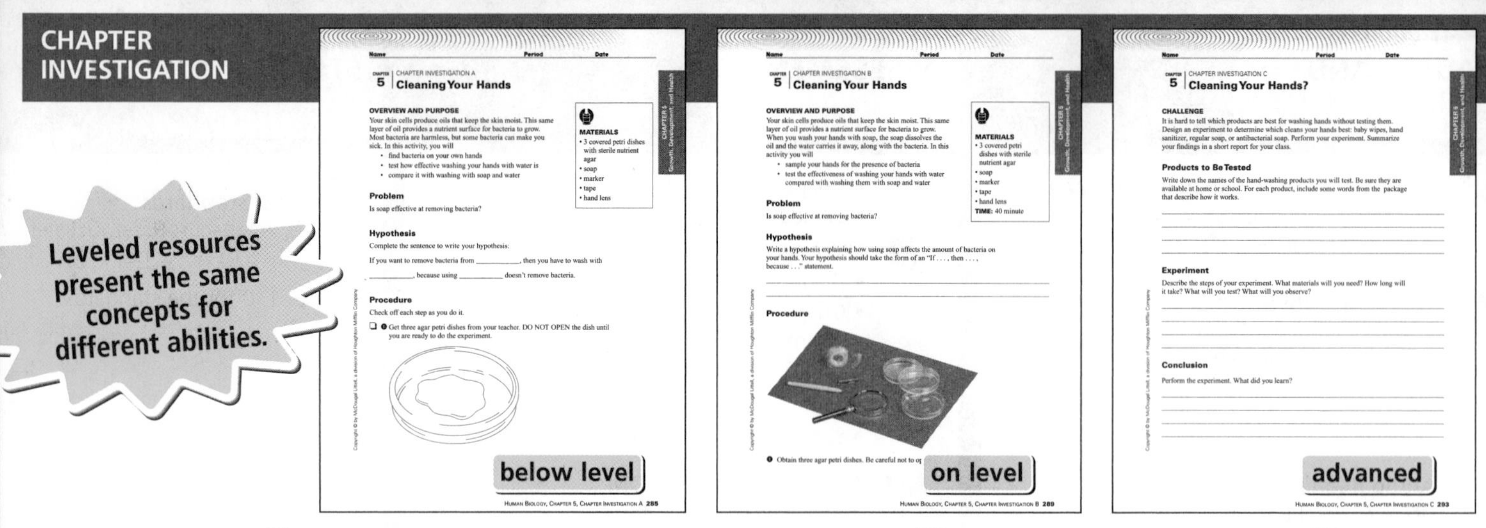

UNIT RESOURCE BOOK, pp. 285–288 pp. 289–292 pp. 289–293

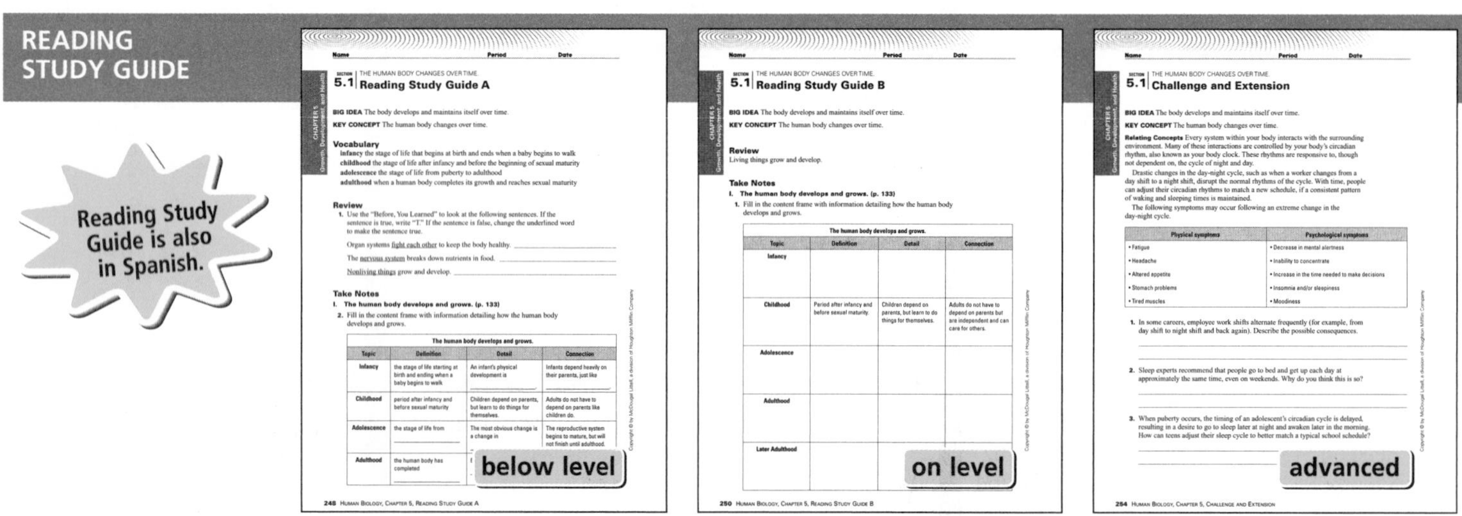

UNIT RESOURCE BOOK, pp. 248–249 pp. 250–251 p. 254

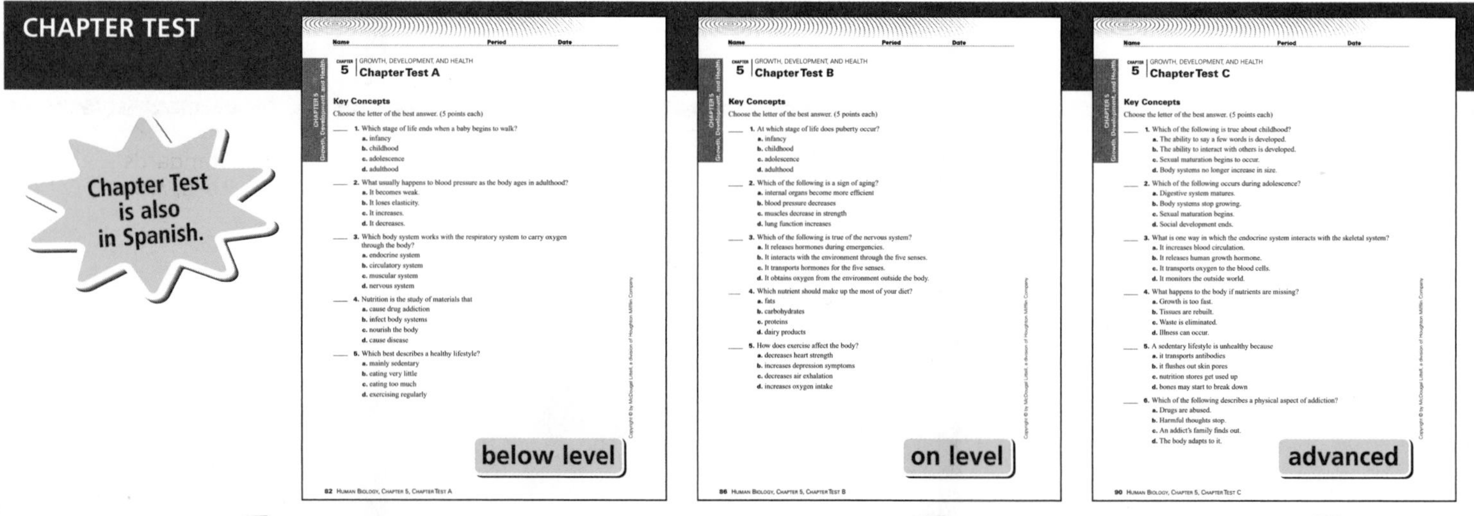

UNIT ASSESSMENT BOOK, pp. 82–85 pp. 86–89 pp. 90–93

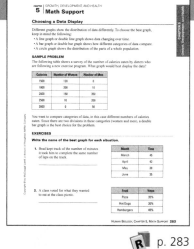

TECHNOLOGY

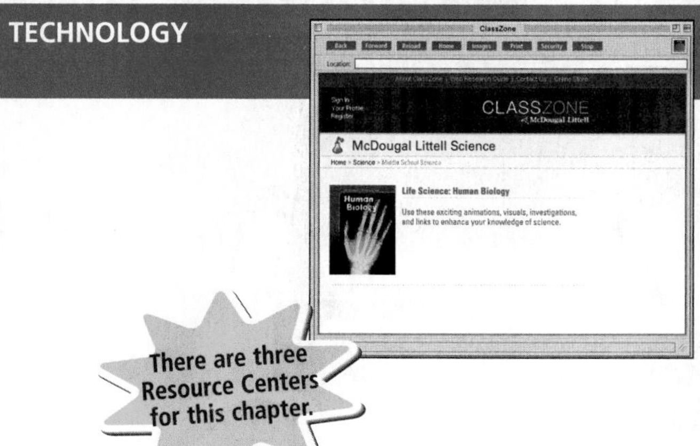

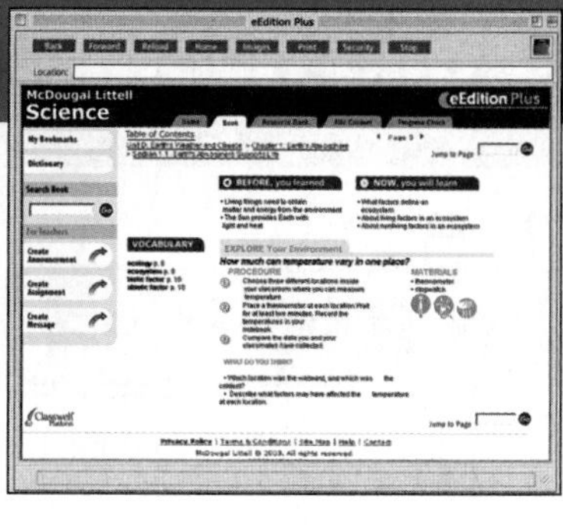

VISUAL CONTENT

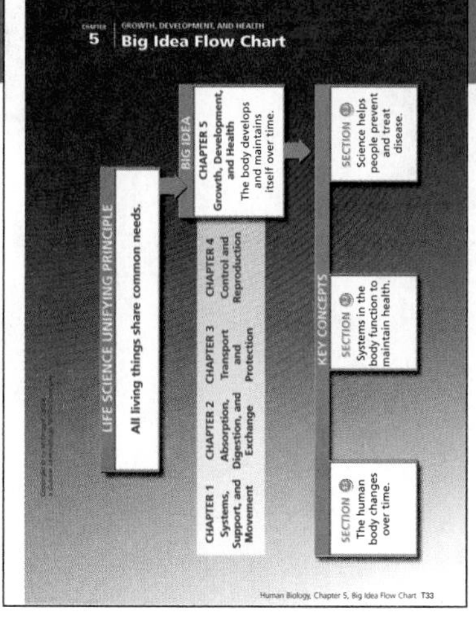

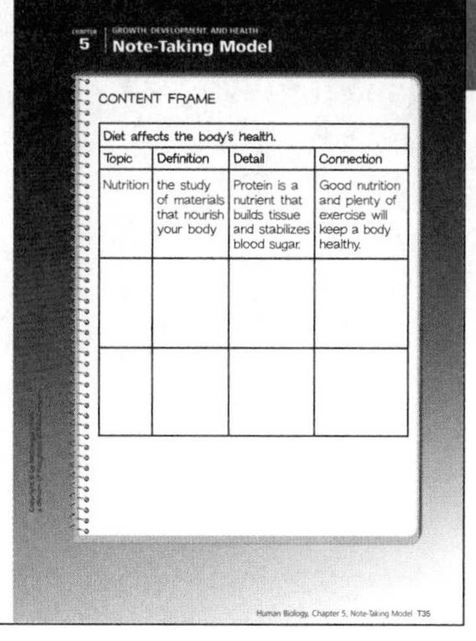

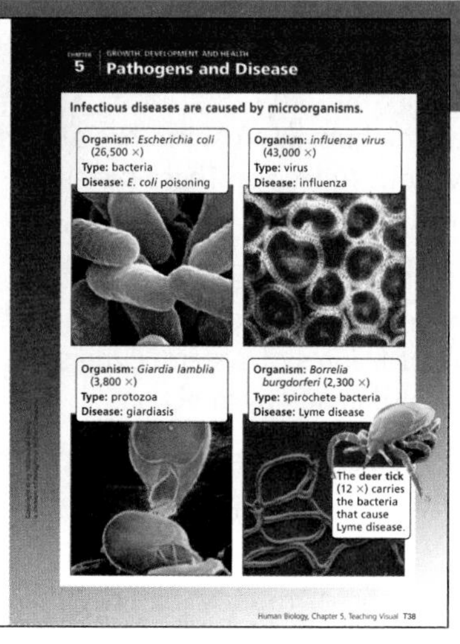

UNIT TRANSPARENCY BOOK, p. T33 p. T35 p. T38

MORE SUPPORT

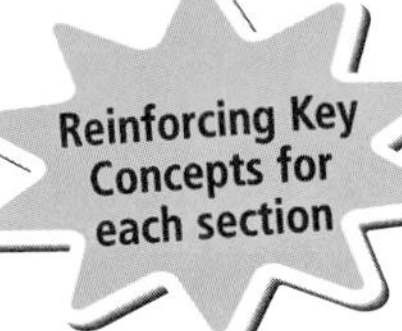

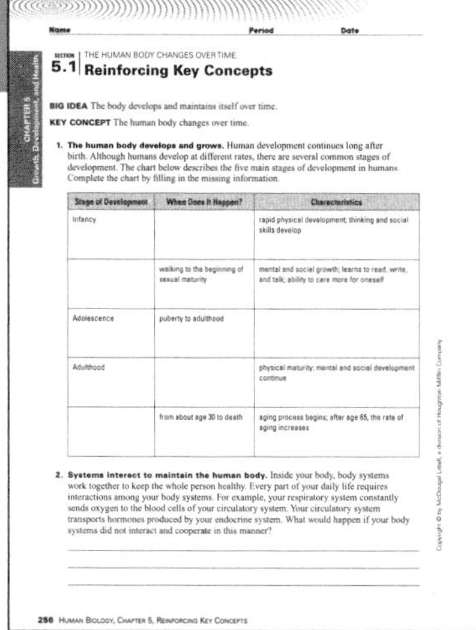

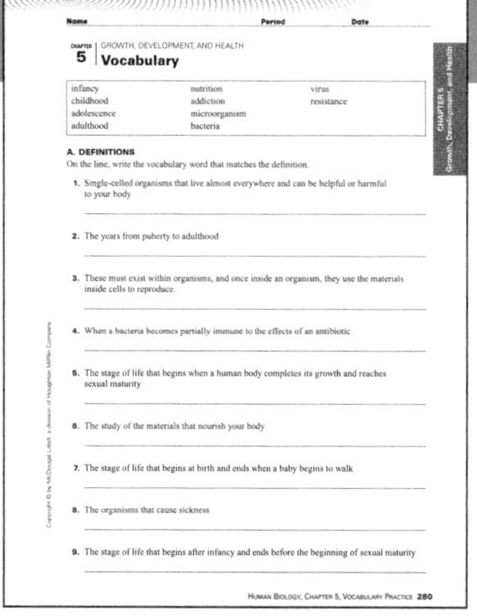

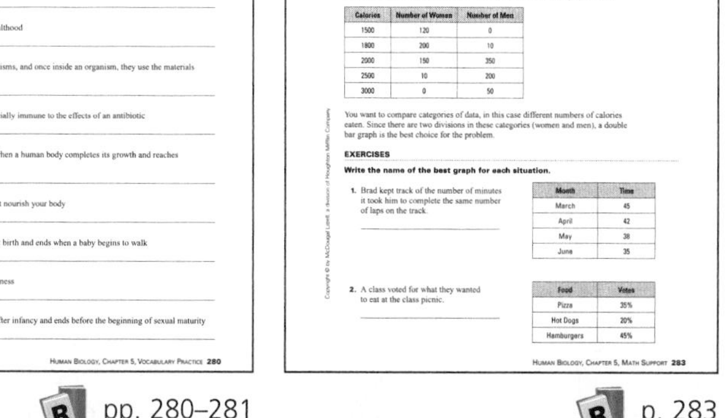

UNIT RESOURCE BOOK, p. 256 pp. 280–281 p. 283

CHAPTER

Growth, Development, and Health

INTRODUCE

the BIG idea

Have students look at the photograph of a family flying a kite and discuss how the question in the box links to the Big Idea:

- What differences can you notice between the three people?
- How does the photograph relate to being healthy?

National Science Education Standards

Content

C.1.e The human organism has systems for various functions. These systems interact.

C.1.f Disease is a breakdown in structures or functions. Some diseases are the result of failures of system. Others are a result of damage by infection.

F.1.a Regular exercise is important to the maintenance and improvement of health.

F.1.c The use of tobacco increases risk of illness. Short-term social and psychological factors lead to tobacco use.

F.1.d Alcohol and drugs change body function and can lead to addiction.

F.1.e Food provides energy and nutrients for growth and development.

Process

A.2–8 Design and conduct an investigation; use tools to gather and interpret data; use evidence to describe, predict, explain, model; think critically to make relationships between evidence and explanation; recognize different explanations and predictions; communicate scientific procedures and explanations; use mathematics.

A.9.a–c, A.9.e–f Understand scientific inquiry by using different investigations, methods, mathematics, and explanations based on logic, evidence, and skepticism.

CHAPTER

Growth, Development, and Health

the BIG idea

The body develops and maintains itself over time.

Key Concepts

SECTION

The human body changes over time.
Learn about the different stages of human development.

SECTION

Diet, exercise, and behaviors affect health.
Learn about what a body needs to be healthy.

SECTION

Science helps people prevent and treat disease.
Learn how to help prevent the spread of disease.

Internet Preview

CLASSZONE.COM

Chapter 5 online resources: Content Review, Visualization, three Resource Centers, Math Tutorial, Test Practice

E 130 Unit: Human Biology

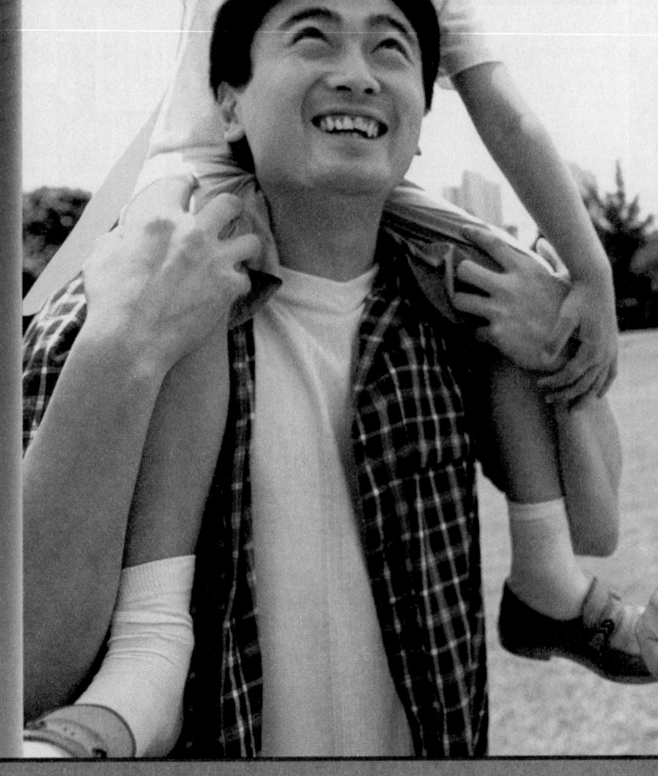

How do people change as they grow?

INTERNET PREVIEW

CLASSZONE.COM For student use with the following pages:

Review and Practice

- Content Review, pp. 132, 156
- Math Tutorial: Choosing a Data Display, p. 147
- Test Practice, p. 159

Activities and Resources

- Internet Activity: Human Development, p. 131
- Resource Centers: Human Health, p. 141; Principles of Nutrition, p. 142; Ways to Fight Disease, p. 149

Human Development
Code: MDL048

EXPLORE the BIG idea

How Much Do You Exercise?

In your notebook, create a chart to keep track of your exercise for a week. Each time you exercise, write down the type of activity and the amount of time you spend. If possible, measure your heart rate during the activity.

Observe and Think How does the exercise affect your heart rate? If you exercised regularly, what would be the effect on your heart rate while you were resting?

How Safe Is Your Food?

Almost all food that you buy in a store is dated for freshness. Look at the labels of various foods including cereal, juice, milk, eggs, cheese, and meats.

Observe and Think Why do you think some foods have a longer freshness period than others? What types of problems could you have from eating food that is past date?

Internet Activity: Human Development

Go to **ClassZone.com** to watch a movie of a person aging.

Observe and Think In what ways does a person's face change as he or she ages?

NSTA SCILINKS scilinks.org
Human Development **Code: MDL048**

TEACHING WITH TECHNOLOGY

Graphing Software Students can use graphing software while investigating how life expectancy has changed over time on p. 137.

Spreadsheet and Slide-Show Software Students could use spreadsheet software and a slide-show program for recording data and outlining the dangers of drugs on p. 145.

EXPLORE the BIG idea

These inquiry-based activities are appropriate for use at home or as a supplement to classroom instruction.

How Much Do You Exercise?

PURPOSE To keep a record of exercise for a week.

TIP *10 min.* Tell students to measure their heart rates by placing their first two fingers at the base of the jaw line to find their pulse.

Answer: Exercising longer and harder increases the heart rate more than exercising for less time or at less intensity. Resting heart rate should decrease with regular exercise. A lower resting heart rate decreases the amount of work the heart does.

REVISIT after p. 144.

How Safe Is Your Food?

PURPOSE To examine food labels and packaging for freshness dates.

TIP *15 min.* Have students look at food packaging at home for "sell by" dates.

Answer: Foods that are perishable have shorter freshness periods than foods that are dried or preserved in some way. Food that is past its freshness date could be contaminated with bacteria or other microorganisms that can make a person sick.

REVISIT after p. 150.

Internet Activity: Human Development

PURPOSE To observe the changes brought about by aging.

TIP *20 min.*

Answer: The skin loosens and wrinkles. Some parts of the face sag, or may fill out with fat deposits. The skin can also change color or develop "age spots."

REVISIT after p. 136.

PREPARE

CONCEPT REVIEW

Activate Prior Knowledge

- Ask students to describe how the skin helps to keep foreign materials from entering the body. *The epidermis is tough and protective. The cells form a thick, waterproof layer.*
- Discuss how white blood cells and antibodies work together to fight foreign materials that enter the body. *White blood cells travel through the circulatory and lymphatic systems to an infected part of the body. These cells produce antibodies, which help fight the foreign materials.*

TAKING NOTES

Content Frame

Students can use the vocabulary lists at the beginning of each section as a guide for topics to use in their content frames. They can add further topics by scanning the red headings in the text. Definitions and details can be found under each red heading, or following each bold-face word in their text. Encourage students to think about how each topic relates to another topic on their list, and have them write a sentence about the connection.

Choose Your Own Strategy

Encourage students to think about their own learning styles and choose a strategy that best suits them. Remind students that different strategies may be appropriate for different vocabulary words.

Vocabulary and Note-Taking Resources

- Vocabulary Practice, pp. 280–281
- Decoding Support, p. 282

- Daily Vocabulary Scaffolding, p. T34
- Note-Taking Model, p. T35

- Choose Your Own Strategy, B20–27
- Content Frame, C35
- Daily Vocabulary Scaffolding, H1–8

CHAPTER 5

Getting Ready to Learn

CONCEPT REVIEW

- The integumentary system protects the body.
- The immune system fights disease.
- A microscope is an instrument used to observe very small objects.

VOCABULARY REVIEW

nutrient p. 45
pathogen p. 74
antibiotic p. 81
hormone p. 111

Review concepts and vocabulary.

TAKING NOTES

CONTENT FRAME

Make a content frame for each main idea. Include the following columns: *Topic, Definition, Detail,* and *Connection.* In the first column, list topics about the title. In the second column, define the topic. In the third column, include a detail about the topic. In the fourth column, add a sentence that connects that topic to another topic in the chart.

CHOOSE YOUR OWN STRATEGY

For each new vocabulary term, take notes by choosing a strategy from earlier chapters—**four square, magnet word, frame game,** or **description wheel.** Or, use a strategy of your own.

See the Note-Taking Handbook on pages R45–R51.

SCIENCE NOTEBOOK

The human body develops and grows.			
Topic	Definition	Detail	Connection
Childhood	Period after infancy and before sexual maturity.	Children depend on parents, but learn to do things for themselves, such as get dressed.	Adults do not have to depend on parents; they are independent and can care for others.

Four Square

Frame Game

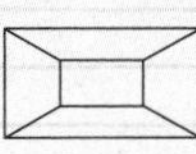

Magnet Word

Description Wheel

CHECK READINESS

Administer the Diagnostic Test to determine students' readiness for new science content and their mastery of requisite math skills.

 Diagnostic Test, pp. 77–78

Technology Resources

Students needing content and math skills should visit **ClassZone.com.**

- **CONTENT REVIEW**
- **MATH TUTORIAL**

 **CONTENT REVIEW CD-ROM**

KEY CONCEPT

The human body changes over time.

BEFORE, you learned
- Living things grow and develop
- The digestive system breaks down nutrients in food
- Organ systems interact to keep the body healthy

NOW, you will learn
- About four stages of human development
- About the changes that occur as the body develops
- How every body system interacts constantly with other systems to keep the body healthy

VOCABULARY

infancy p. 134
childhood p. 134
adolescence p. 135
adulthood p. 136

EXPLORE Growth

Are there patterns of growth?

PROCEDURE

1. Measure the circumference of your wrist by using the measuring tape as shown. Record the length. Now measure the length from your elbow to the tip of your middle finger. Record the length.
2. Create a table and enter all the data from each person in the class.

MATERIALS
- flexible tape measure
- graph paper

WHAT DO YOU THINK?
- How does the distance between the elbow and the fingertip compare with wrist circumference?
- Do you see a pattern between the size of one's wrist and the length of one's forearm?

CONTENT FRAME
Make a content frame for the first main idea: *The human body develops and grows.* List the red headings in this section in the topics column.

The human body develops and grows.

Have you noticed how rapidly your body has changed over the past few years? Only five years ago you were a young child in grade school. Today you are in middle school. How has your body changed? Growth is both physical and emotional. You are becoming more responsible and socially mature. What are some emotional changes that you have noticed?

Human development continues long after birth. Although humans develop at different rates, there are several stages of development common to human life. In this section we will describe some of the stages, including infancy, childhood, adolescence, and later adulthood

5.1 FOCUS

Set Learning Goals

Students will
- Describe the four stages of human development.
- Discuss the changes that happen as the body develops.
- Analyze how body systems interact.
- Graph data about life expectancy.

3-Minute Warm-Up

Display Transparency 36 or copy this exercise on the board:

Decide if these statements are true. If not true, correct them.

1. White blood cells transport oxygen throughout the body. *Red blood cells transport oxygen throughout the body.*
2. Cilia are tiny hairlike projections that trap foreign particles in the nose and lungs. *true*
3. Some white blood cells produce antigens, which attack specific foreign materials. *White blood cells produce antibodies.*

 3-Minute Warm-Up, p. T36

5.1 MOTIVATE

EXPLORE Growth

PURPOSE To compile data on body measurements and analyze patterns of growth

TIPS *15 min.* Have students work in pairs to help one another take the measurements. Demonstrate accurate use of the tape measure. If you do not have a soft tape measure, students can use string and a ruler.

WHAT DO YOU THINK? *There is a difference in length. Generally, students with wider wrists have longer lengths from the elbow to the tip of the finger. The fact that some people have bigger bones can account for this pattern of growth.*

RESOURCES FOR DIFFERENTIATED INSTRUCTION

Below Level

UNIT RESOURCE BOOK
- Reading Study Guide A, pp. 248–249
- Decoding Support, p. 282

Advanced

UNIT RESOURCE BOOK
- Challenge and Extension, p. 254
- Challenge Reading, pp. 278–279

English Learners

UNIT RESOURCE BOOK
Spanish Reading Study Guide, pp. 252–253

- Audio Readings in Spanish
- Audio Readings (English)

5.1 INSTRUCT

Teach from Visuals

To help students interpret the chart of the Apgar test, ask:

- Which body systems are assessed with the Apgar test? *circulatory, respiratory, muscular, and nervous systems*
- Which five items are tested? *Appearance (color), Pulse (heart rate), Grimace (responsiveness), Activity (muscle tone), and Respiration (breathing)*

History of Science

The Apgar score is named for Dr. Virginia Apgar (1909–1974), a pediatrician who specialized in newborn (neonatal) care. In 1952, she developed a simple way to evaluate a newborn's transition to life outside the uterus. Points are given for each sign at one minute and five minutes after the birth. A score of 7–10 is considered normal. A baby with an Apgar score of 4–7 may need medical attention, and a score of 3 and below requires immediate resuscitation.

Develop Critical Thinking

COMPARE Students can analyze the text by comparing infancy and childhood. Ask: In what ways does a child change between infancy and childhood? What are some things that remain the same? *The child learns to talk, walk, read, write, and communicate. The child is dependent on the parent for food and other needs.*

Infancy

A

The stage of life that begins at birth and ends when a baby begins to walk is called **infancy.** An infant's physical development is rapid. As the infant's body grows larger and stronger, it also learns physical skills. When you were first born, you could not lift your head. But as your muscles developed, you learned to lift your head, to roll over, to sit, to crawl, to stand, and finally to walk. You also learned to use your hands to grasp and hold objects.

Infants also develop thinking skills and social skills. At first, they simply cry when they are uncomfortable. Over time, they learn that people respond to those cries. They begin to expect help when they cry. They learn to recognize the people who care for them. Smiling, cooing, and eventually saying a few words are all part of an infant's social development.

Nearly every body system changes and grows during infancy. For example, as the digestive system matures, an infant becomes able to process solid foods. Changes in the nervous system, including the brain, allow an infant to see more clearly and to control parts of her or his body.

The Apgar score is used to evaluate the newborn's condition after delivery.

Apgar Score

Quality	0 points	1 point	2 points
Appearance	Completely blue or pale	Good color in body, blue hands or feet	Completely pink or good color
Pulse	No heart rate	<100 beats per minute	>100 beats per minute
Grimace	No response to airway suction	Grimace during suctioning	Grimace, cough/ sneeze with suction
Activity	Limp	Some arm and leg movement	Active motion
Respiration	Not breathing	Weak cry	Good, strong cry

Childhood

B

The stage called **childhood** lasts for several years. Childhood is the period after infancy and before the beginning of sexual maturity. During childhood children still depend very much upon their parents. As their bodies and body systems grow, children become more able to care for themselves. Although parents still provide food and other needs, children perform tasks such as eating and getting dressed. In addition, children are able to do more complex physical tasks such as running, jumping, and riding a bicycle.

Childhood is also a time of mental and social growth. During childhood a human being learns to talk, read, write, and communicate in other ways. Along with the ability to communicate come social skills such as cooperation and sharing. A child learns to interact with others.

DIFFERENTIATE INSTRUCTION

More Reading Support

A What are some skills that an infant learns? *lift head, roll over, sit, crawl, stand, walk, and grasp*

B When does childhood end? *at the beginning of sexual maturity*

English Learners Students new to English may find the following science words challenging: *infancy, childhood, adolescence, puberty, maturity, adulthood.* Preview vocabulary by adding these words and their definitions to a Science Word Wall.

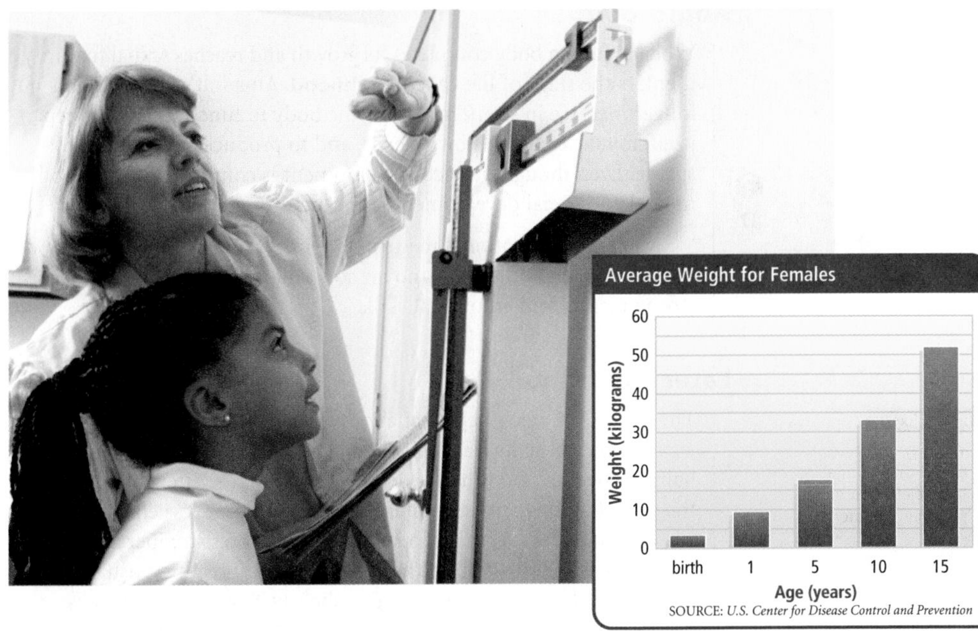

Adolescence

The years from puberty to adulthood are called **adolescence** (AD-uhl-EHS-uhns). You and your classmates are adolescents. Childhood ends when the body begins to mature sexually. This process of physical change is called puberty. Not all people reach puberty at the same age. For girls, the changes usually start between ages eight and fourteen; for boys, puberty often begins between ages ten and sixteen.

The human body changes greatly during adolescence. As you learned in Chapter 4, the endocrine system produces chemicals called hormones. During adolescence, hormones signal parts of the reproductive system to mature. At this stage a person's sexual organs become ready for reproduction. These changes are accompanied by other changes. Adolescents develop secondary sexual characteristics. Boys may notice their voices changing. Girls begin developing breasts. Boys and girls both begin growing more body hair.

Probably the change that is the most obvious is a change in height. Boys and girls grow taller by as much as 10 centimeters (3.9 in.) during adolescence. Most adolescents eat more as they grow. Food provides materials necessary for growth.

VOCABULARY
Choose a strategy from earlier chapters or one of your own to take notes on the term *adolescence*.

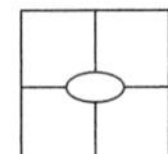

CHECK YOUR READING What are some of the ways the body changes during adolescence?

Teach from Visuals

To help students interpret the bar graph of Average Weight for Females, ask:

- Which period shows the greatest weight gain in kilograms? *between ages 10 and 15*
- Which period shows the greatest gain in terms of percent of body weight? Explain. *between birth and 5 years; The bar more than doubles in height.*

Point out to students that the vertical axis shows even intervals, but the horizontal axis shows a jump of one year between the first two bars and five years between the others.

Teach Difficult Concepts

Many students confuse the terms *adolescence* and *puberty* and often interchange the two. Ask students to define each term in their own words. Students may also think there is a "normal" timeline for specific body changes to occur during adolescence. Point out that individuals develop at certain rates depending on their family histories and their own biological clocks. Ask:

- What is the age range when girls reach puberty? *between 8 and 14 years*
- What is the age range when boys reach puberty? *between 10 and 16 years*

Ongoing Assessment

CHECK YOUR READING *Answer: During adolescence, a person grows taller, and sexual organs become ready for reproduction. Boys' voices change and girls develop breasts. There is more body hair.*

DIFFERENTIATE INSTRUCTION

More Reading Support

C What term describes the process of physical change during sexual maturation? *puberty*

Alternative Assessment Have students create a timeline of important events in their lives starting with their birthdays to the present. Students should note in their timelines the stage of life and each important physical, social, and mental change that took place at that stage.

Develop Critical Thinking

PREDICT Have students predict how they see themselves as adults and how their lives will change from adolescence to adulthood. Ask them at what age they think adulthood begins and why. *Sample answer: age 18, because the person will not get any taller and has completed sexual maturity*

Real World Example

Ask students to name some television shows that portray people in various stages of life. Ask students if the portrayals are realistic or not. Ask:

- How do the shows you have seen portray teens? Do any of them look and act like typical teens? *Be sure answers mention the variety of individuals as well as some of the traits that scientifically typify each stage.*
- How do the shows portray adults? Do you think the portrayal of adults is realistic or not? Explain. *Student answers should include comparison and contrast with adults they know as well as information in the text.*

EXPLORE the BIG idea

Revisit "Internet Activity: Human Development" on p. 131. Ask students to explain what observations are brought about by aging.

Ongoing Assessment

Describe the changes that happen as the body develops.

Ask: What changes take place in the body as a person reaches later adulthood? *skin begins to wrinkle, eyesight becomes poorer, hair loss, muscle strength loss, internal organs become less efficient, blood vessels are less elastic, blood pressure rises, lung function decreases*

Describe the four stages of human development.

Ask: What are the four stages of human development? *infancy, childhood, adolescence, adulthood*

Adulthood

When a human body completes its growth and reaches sexual maturity, it enters the stage of life called **adulthood.** An adult's body systems no longer increase in size. They allow the body to function fully, to repair itself, to take care of its own needs, and to produce and care for offspring. Even though physical development is complete at adulthood, mental and social development continue throughout life. D

Mental and emotional maturity are important parts of adulthood. To maintain an adult body and an adult lifestyle, an individual needs strong mental and emotional skills.

Later Adulthood

READING TIP You may find it helpful to review the information on the skeletal and muscular system in Chapter 1.

The process of aging begins at about the age of 30. Skin begins to wrinkle and loose its elasticity. Eyesight becomes increasingly poor, hair loss begins, and muscles decrease in strength. After the age of 65, the rate of aging increases. Internal organs become less efficient. Blood vessels become less elastic. The average blood pressure increases and may remain slightly high. Although the rate of breathing usually does not change, lung function decreases slightly. Body temperature is harder to regulate. However, one can slow the process of aging by a lifestyle of exercise and healthy diet.

Systems interact to maintain the human body.

It's easy to observe the external changes to the body during growth and development. Inside the body, every system interacts constantly with other systems to keep the whole person healthy throughout his or her lifetime. For example, the respiratory system constantly sends oxygen to the blood cells of the circulatory system. The circulatory system transports hormones produced by the endocrine system.

Your body systems also interact with the environment outside your body. Your nervous system monitors the outside world through your senses of taste, smell, hearing, vision, and touch. It allows you to respond to your environment. For example, your nervous system allows you to squint if the sun is too bright or to move indoors if the weather is cold. Your endocrine system releases hormones that allow you to have an increased heart rate and send more blood to your muscles if you have to respond to an emergency.  E

E 136

DIFFERENTIATE INSTRUCTION

More Reading Support

D What development continues in adulthood? *mental, social*

E How does the body know to respond to the outside? *nervous system and senses*

Below Level Have students imagine they are going to teach younger children about what to expect when they reach adolescence. Have students prepare a brochure describing the physical, social, and emotional changes that a young child will go through.

INVESTIGATE Life Expectancy

How has life expectancy changed over time?

In this activity, you will look for trends in the changes in average life expectancy over the past 100 years.

SKILL FOCUS
Graphing

MATERIALS
- graph paper
- computer graphing program

TIME
30 minutes

PROCEDURE

1. Using the following data, create a bar graph to chart changes in life expectancy over the last 100 years.

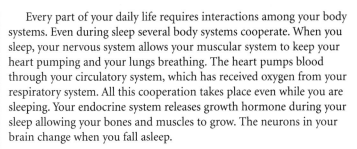

Life Expectancy 1900–2000

Year	1900	1910	1920	1930	1940	1950	1960	1970	1980	1990	2000
Average Life Expectancy (years)	47.3	50.0	54.1	59.7	62.9	68.2	69.7	70.8	73.7	75.4	76.9

SOURCE: National Center for Health Statistics

2. Study the graph. Observe any trends that you see. Record them in your notebook.

WHAT DO YOU THINK?

- In general, what do these data demonstrate about life expectancy?
- Between which decades did average life expectancy increase the most?

CHALLENGE Using a computer program, create a table and bar graph to chart the data shown above.

Every part of your daily life requires interactions among your body systems. Even during sleep several body systems cooperate. When you sleep, your nervous system allows your muscular system to keep your heart pumping and your lungs breathing. The heart pumps blood through your circulatory system, which has received oxygen from your respiratory system. All this cooperation takes place even while you are sleeping. Your endocrine system releases growth hormone during your sleep allowing your bones and muscles to grow. The neurons in your brain change when you fall asleep.

Keeping the body healthy is complex. The digestive and urinary systems eliminate solid and liquid wastes from the body. The circulatory and respiratory systems remove carbon dioxide gas. As you will learn in the next section, a healthy diet and regular exercise help the body to stay strong and function properly.

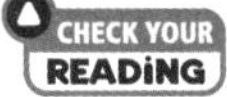

Name three systems that interact as your body grows and maintains itself.

DIFFERENTIATE INSTRUCTION

F Which two systems eliminate solid and liquid wastes? *digestive and urinary systems*

Advanced Write the words *peer pressure* on the board and have students give the definition in their own words. Ask students to brainstorm a list of ways that adolescents may experience both negative and positive peer pressure.

 Challenge and Extension, p. 254

Have students who are interested in the effects of stress on the body read the following article:

 Challenge Reading, pp. 278–279

INVESTIGATE Life Expectancy

PURPOSE To compare the average life expectancy over the last 100 years and to infer why it has changed

TIPS *30 min.* Ask students to think about scale and range for the *x*- and *y*-axes of their graphs before making the graph. Prior to the investigation, prepare tables for visually impaired students to use. Highlight the main heads in the table with colored markers.

WHAT DO YOU THINK? *Life expectancy over the last 100 years has increased. The greatest increases occurred between 1920–1930, and between 1940–1950.*

CHALLENGE *Computer-generated graphs should match data shown.*

 Datasheet, Life Expectancy, p. 255

Technology Resources

Customize this student lab as needed or look for an alternative. Print rubrics to assess student lab reports.

 **Lab Generator CD-ROM**

Teaching with Technology

If graphing software is available, have students use it to record the data for the investigation. They can also use a graphing calculator.

Ongoing Assessment

Analyze how body systems interact.

Ask: How do body systems keep you functioning while you sleep? *Sample answer: Circulatory and respiratory systems keep bringing oxygen to cells. The nervous system keeps signaling the other systems to function. The muscular system keeps the heart pumping. The digestive system brings water and nutrients to cells, while the excretory system removes waste.*

CHECK YOUR READING *Answer: The circulatory, respiratory, and endocrine systems interact when you grow.*

Ongoing Assessment

Answer: When you are active, all body systems are working to provide energy to fuel your muscles. When you rest, your body systems cooperate to recharge your body, and to help it grow and maintain itself.

Reinforce

Have students relate the section to the Big Idea.

Reinforcing Key Concepts, p. 256

5.1 ASSESS & RETEACH

Assess

Section 5.1 Quiz, p. 79

Reteach

List the four stages of human development on the board and draw a sketch next to each. Tell students that as a group you are going to create a "Progress Report" for each character you drew. Have them search the text or work cooperatively to find ways the person has progressed, or developed, at each stage.

For each of the four stages of development, ask students to name some of the physical, mental, and social changes that take place.

1. infancy—*Sample answers: learns to sit, to cry to show needs, to coo*
2. childhood—*Sample answers: grows taller, learns to read, learns to cooperate with others*
3. adolescence—*Sample answers: sexual organs mature, grows taller, boys' voices change*
4. adulthood—*Sample answers: body systems stop growing, emotional and mental maturity*

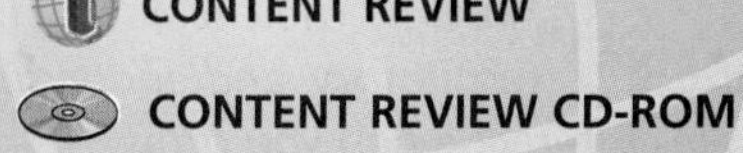

Technology Resources

Have students visit **ClassZone.com** for reteaching of Key Concepts.

CONTENT REVIEW

CONTENT REVIEW CD-ROM

READING VISUALS COMPARE AND CONTRAST How do the interactions of your body system change when you are active and when you rest?

When body systems fail to work together, the body can become ill. Stress, for example, can affect all the body systems. Some types of stress, such as fear, can be a healthy response to danger. However, if the body experiences stress over long periods of time, serious health problems such as heart disease, ulcers, headaches, muscle tension, and depression can arise.

All stages of life include different types of stress. Infants and children face stresses as they learn to become more independent and gain better control over their bodies. Adolescents can be challenged by school, by the changes of puberty, or by being socially accepted by their peers. Adults may encounter stress in their jobs or with their families. The stress of aging can be very difficult for some older adults.

5.1 Review

KEY CONCEPTS

1. Make a development timeline with four sections. Write the names of the stages in order under each section. Include a definition and two details.
2. List a physical characteristic of each stage of development.
3. Give an example of an activity that involves two or more body systems.

CRITICAL THINKING

4. **Compare and Contrast** Make a chart to compare and contrast the infancy and childhood stages of development.
5. **Identify Cause and Effect** How is the endocrine system involved in adolescence?

CHALLENGE

6. **Synthesis** How does each of the body systems described change as a human being develops from infancy to older adulthood?

ANSWERS

1. Timelines show stages with definitions and details.

2. Sample answer: infancy—unable to walk; childhood—learns to talk; adolescence—grows more body hair; adulthood—no new physical growth

3. Sample answer: eating: digestive, muscular, and circulatory systems interact.

4. Charts include ability to walk, increase of independence.

5. Hormones secreted by the endocrine system signal the reproductive system to mature in adolescence.

6. Answers should include details of development and aging of body systems: respiratory/circulatory, digestive, endocrine, integumentary, muscular, nervous, and reproductive.

STAGE MAKEUP ARTIST

Aging the Face

In a movie, characters may go through development stages of a whole lifetime in just over an hour. An actor playing such a role will need to look both older and younger than he or she really is. Stage makeup artists have a toolbox full of techniques to make the actor look the part.

Makeup Guide for Aging

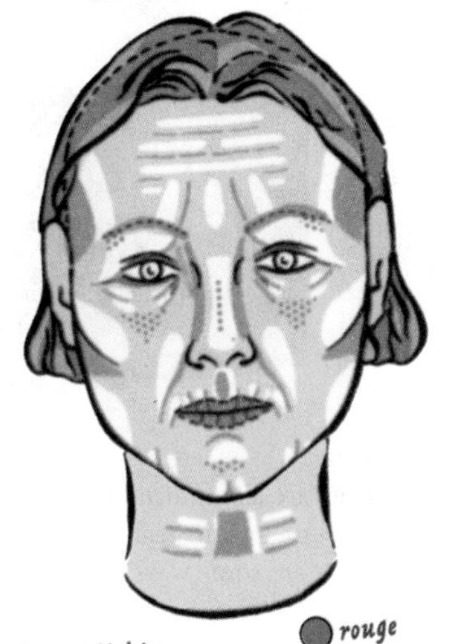

highlights

shadows

rouge

foundation

Hair

As humans go through adulthood, their hair may lose the pigments that make it dark. Makeup artists color hair with dyes or even talcum powder. Wigs and bald caps, made of latex rubber, cover an actor's real hair. Eyebrows can be colored or aged by rubbing them with makeup.

Features

For a bigger-looking nose or extra skin around the neck, makeup artists use foam rubber, or layers of liquid rubber, and sometimes, wads of paper tissue to build up facial features. For example, building up the cheekbones with layers of latex makes the cheeks appear sharper, less rounded, and more hollow.

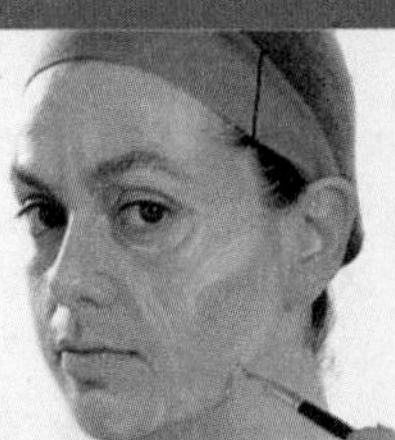

Skin

To make wrinkles or scars, makeup artists use light-colored makeup to for the raised highlights and dark-colored makeup for lower shadows and spots.

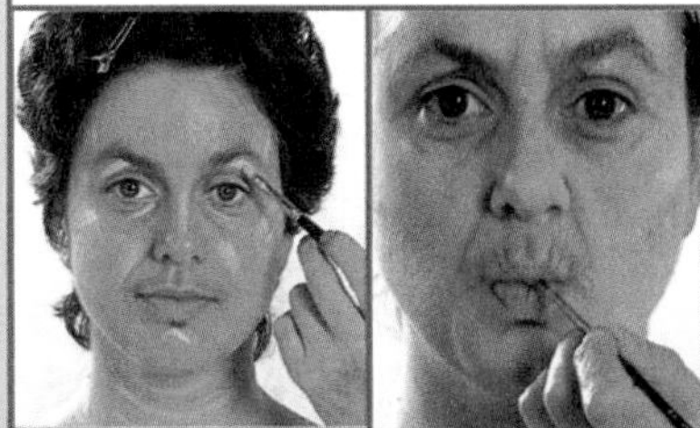

EXPLORE

1. **COMPARE** Look at photos of an older relative at three different stages of life, at about ten years apart. Describe how you might apply makeup to your own face if you were to portray this person's life in three movie scenes. What changes do you need to show?
2. **CHALLENGE** Research to find an image of a character portrayed in a movie, who needed to look very different than real life. From the picture, describe how the effect was achieved.

SCIENCE ON THE JOB

Relevance of Science to Non-science Jobs

Set Learning Goal

To understand why makeup artists need knowledge of the aging process

Present the Science

Remind students that the integumentary system includes the skin, hair, and nails. The skin helps regulate body temperature as well as protect the body from foreign materials. As we age, skin loses its elasticity and may begin to sag or wrinkle. Hair becomes drier and may begin to turn gray. Hands and other skin that is exposed also show age.

FEATURES Every fold or crease in the skin catches light and creates a different shadow. To begin the aging process with makeup, makeup artists begin by applying a foundation on the face and neck.

SKIN Highlights to make skin age are applied over the muscles of the face. Applying the makeup under an overhead light source can help the makeup artist find the highlights and shadows in the face. Surgical adhesive and liquid latex can create excessive wrinkles and a sagging neck. Age spots can be added to the hands.

HAIR Men can be aged by applying gray at the temples and sideburns as well as on beards and moustaches. Wigs can show thinning hair or a receding hairline.

Discussion Question

Ask: How might creating an aging face with stage makeup assist an actor in developing his or her character? *A realistic look can help the actor portray an age he has not yet experienced. It helps him or her to "become" the character.*

Close

Ask: Why do makeup artists need knowledge of the human body and how it ages? *This knowledge helps them create a realistic accurate look.*

EXPLORE

1. ***COMPARE*** *Students might say that as the person ages, they might want to apply makeup to show wrinkles. They might also need to apply latex or other rubber layers to show various facial features. Students might also suggest using wigs or hair coloring to make the hair look as close to the relative as possible.*
2. ***CHALLENGE*** *Accept all complete responses. Students' answers will vary depending on the movie character that they chose.*

5.2 FOCUS

Set Learning Goals

Students will

- Discuss the role of nutrients in health.
- Explain why exercise is needed to keep body systems healthy.
- Describe how drug abuse, eating disorders, and addiction can negatively affect the body.
- Analyze food labels for their nutritional values.

3-Minute Warm-Up

Display Transparency 36 or copy this exercise on the board:

Match the changes to the stages of development.

Changes

1. internal organs stop growing *d*
2. a person reaches puberty *c*
3. a person learns to cooperate with friends *b*

Stages of Development

a. infancy
b. childhood
c. adolescence
d. adulthood
e. later adulthood

 3-Minute Warm-up, p. T36

5.2 MOTIVATE

THINK ABOUT

PURPOSE To write a definition of health and describe ways to protect health

DISCUSS Allow students to work in small groups, then share their ideas as a class. Point out that good health includes factors that affect a person's physical, emotional, and social well-being. Write all ideas on the board and leave them for students to add to or revise as they read the chapter.

Ongoing Assessment

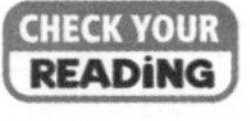
Answer: Your body needs proper nutrition to stay healthy and to function well.

KEY CONCEPT

5.2 Systems in the body function to maintain health.

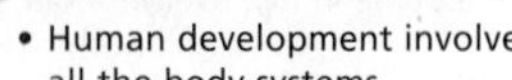

BEFORE, you learned

- Human development involves all the body systems
- The human body continues to develop until adulthood
- Every body system interacts constantly with other systems to keep the body healthy

NOW, you will learn

- About the role of nutrients in health
- Why exercise is needed to keep body systems healthy
- How drug abuse, eating disorders, and addiction can affect the body

VOCABULARY

nutrition p. 140
addiction p. 145

THINK ABOUT

What is health?

If you went online and searched under the word *health*, you would find millions of links. Clearly, health is important to most people. You may be most aware of your health when you aren't feeling well. But you know that clean water, food, exercise, and sleep are all important for health. Preventing illness is also part of staying healthy. How would you define health? What are some ways that you protect your health?

Diet affects the body's health.

What makes a meal healthy? The choices you make about what you eat are important. Nutrients from food are distributed to every cell in your body. You use those nutrients for energy and to maintain and build new body tissues. **Nutrition** is the study of the materials that nourish your body. It also refers to the process in which the different parts of food are used for maintenance, growth, and reproduction. When a vitamin or other nutrient is missing from your diet, illness can occur. Your body's systems can function only when they get the nutrients they need.

VOCABULARY
Choose a strategy from an earlier chapter, such as a magnet word diagram, for taking notes on the term *nutrition*. Or use any strategy that you think works well.

CHECK YOUR READING How is nutrition important to health?

RESOURCES FOR DIFFERENTIATED INSTRUCTION

Below Level

UNIT RESOURCE BOOK

- Reading Study Guide A, pp. 259–260
- Decoding Support, p. 282

AUDIO CDS

Advanced

UNIT RESOURCE BOOK

Challenge and Extension, p. 265

English Learners

UNIT RESOURCE BOOK

Spanish Reading Study Guide, pp. 263–264

- Audio Readings in Spanish
- Audio Readings (English)

 **Additional INVESTIGATION,** Where's the Starch?, A, B, & C, pp. 294–302; Teacher Instructions, pp. 305–306

This family is enjoying a healthy meal of proteins, carbohydrates, and fats.

Getting Nutrients

RESOURCE CENTER
CLASSZONE.COM

Discover more about human health.

In order to eat a healthy diet, you must first understand what good nutrition is. There are six classes of nutrients: carbohydrates, proteins, fats, vitamins, minerals, and water. All of these nutrients are necessary as sources of energy for your body. Also, they each contribute to the chemical reactions that must take place within your cells.

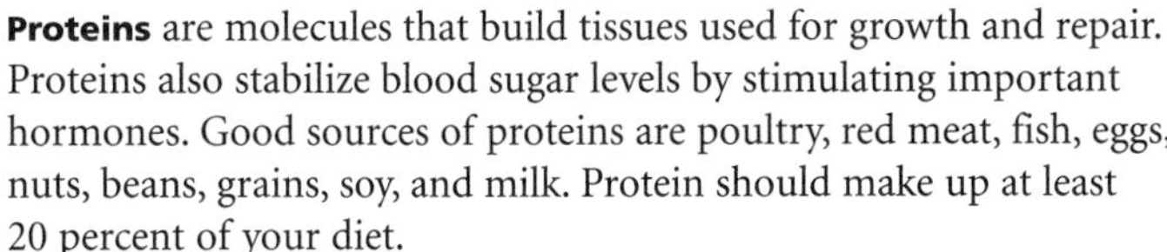

 A

Proteins are molecules that build tissues used for growth and repair. Proteins also stabilize blood sugar levels by stimulating important hormones. Good sources of proteins are poultry, red meat, fish, eggs, nuts, beans, grains, soy, and milk. Protein should make up at least 20 percent of your diet.

 B

Carbohydrates are the body's most important energy source and are found in starch, sugar, and fiber. Fiber provides little energy, but is important for regular elimination. Natural sugars such as those found in fruits and vegetables are the best kinds of sugars for your body. Carbohydrates are found in bread and pasta, fruits, and vegetables. Carbohydrates should make up about 40 to 50 percent of your diet.

Fats are essential for energy and should account for about 10 to 15 percent of your diet. Many people eliminate fats from their diets in order to lose weight. But a certain amount of fat is necessary. Fats made from plants have the greatest health benefits. For example, olive oil is better for you than the oil found in butter.

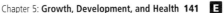

5.2 INSTRUCT

Real World Example

The Food Guide Pyramid was released by the Food and Drug Administration in 1992. Periodically, the FDA reviews and makes changes to the Pyramid. An updated Pyramid is expected to be released in 2005. Changes to the Pyramid include revised daily amounts of food from each of the food groups, suggested energy levels for food intake patterns for different age groups, and target goals for vitamins and minerals. For more information, go to www.cnpp.usda.gov/index.html

Develop Critical Thinking

ANALYZE Ask students to give examples of foods that contain proteins, carbohydrates, and fats. Ask:

- Why does a body need proper nutrition? What happens when nutrition is lacking? *Without the proper nutrients, the body may become ill or may not be able to function efficiently.*
- What are some signs of poor nutrition? *A person might feel tired or sluggish, or be underweight; bones may be weak; skin may look sallow; hair may look dull.*

Ongoing Assessment

Discuss the role of nutrients in health.

Ask: What are the essential nutrients? *carbohydrates, proteins, fats, vitamins, minerals, and water*

DIFFERENTIATE INSTRUCTION

More Reading Support

A What foods are a source of protein? *poultry, red meat, fish, eggs, nuts, grains, beans, milk, soy*

B Name a main food energy source. *carbohydrates*

English Learners On a classroom Science Word Wall, place the terms *protein, carbohydrate,* and *fat.* Ask students to list foods they eat regularly and classify them as one of the above. Have students read their lists aloud. Discuss foods that may be unfamiliar. Then, have students write the food names with a marker on index cards and tape them to the wall under the correct heading.

Teacher Demo

Ask students to list what they ate for breakfast on slips of paper. Read their responses aloud. As you read, have students classify whether the breakfast is a healthful one or if it needs improvement. Write the responses on the board in two columns. In the "needs improvement" category, have students suggest ways to improve the healthfulness of the meal.

Teach from Visuals

To help students interpret the table on vitamins and minerals, ask:

- Why do you think there is a range for the recommended daily allowance? *Vitamin and mineral requirements change during the different stages of life.*
- Which vitamin or mineral does your body need in greatest amounts? in the smallest amounts? *potassium; vitamin A*

Vitamins and minerals are needed by your body in small amounts. Vitamins are small molecules that regulate body growth and development. Minerals help build body tissues. While some vitamins can be made by your body, most of them are supplied to the body in food.

C

D

Water is necessary to for life. A human being could live for less than a month without food, but only about one week without water. Water has several functions. Water helps regulate your body temperature through evaporation when you sweat and breathe. Without water, important materials such as vitamins and other nutrients, could not be transported around the body. Water helps your body get rid of the waste products that move through the kidneys and pass out of the body in urine. Urine is composed mostly of water.

To make sure your body can function and maintain itself, you need to drink about two and one half liters or about eight glasses of water every day. You also get water when you eat foods with water in them, such as fresh fruit and vegetables.

Vitamins and Minerals

Vitamin or Mineral	Recommended Daily Allowance
Vitamin A	0.3 to 1.3 mg
Niacin	6–18 mg
Vitamin B_2	0.5–1.6 mg
Vitamin B_6	0.5–2.0 mg
Vitamin C	15–120 mg
Vitamin E	6–19 mg
Calcium	500–1300 mg
Phosphorus	460–1250 mg
Potassium	1600–2000 mg
Zinc	3–13 mg
Magnesium	80–420 mg
Iron	7–27 mg

Source: National Institutes of Health

Understanding Nutrition

RESOURCE CENTER
CLASSZONE.COM

Examine the basic principles of nutrition.

Ever wonder what the word *lite* really means? What do labels saying that food is fresh or natural or organic mean? Not all advertising about nutrition is reliable. It is important to know what the labels on food really mean. Groups within the government, such as the United States Department of Agriculture, have defined terms that are used to describe food products. For example, if a food label says the food is "all natural," that means it does not contain any artificial flavor, color, or preservative.

Another example is the term *low-fat*. That label means that the food provides no more than 3 grams of fat per serving. The word *organic* means that the produce has been grown using no human-made fertilizers or chemicals that kill pests or weeds. It also means that livestock has been raised on organic feed and has not been given antibiotics or growth hormones. It takes some effort and a lot of reading to stay informed, but the more you know, the better the choices you can make.

DIFFERENTIATE INSTRUCTION

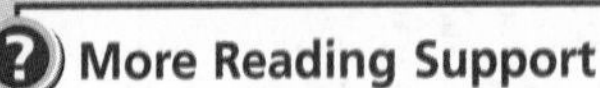
More Reading Support

C About how long can a human being live without water? *about one week*

D How does water help regulate body temperature? *through evaporation of sweat and breathing*

Additional Investigation To reinforce Section 5.2 learning goals, use the following full-period investigation:

R **Additional INVESTIGATION,** Where's the Starch?, A, B, & C, pp. 294–302, 305–306

Advanced Have students create a chart of essential vitamins A, C, D, and E, and the B complex vitamins. For each vitamin, students will list food sources in which the vitamins are found and how the vitamins help the body function.

Challenge and Extension, p. 265

INVESTIGATE Food Labels

What are you eating?

PROCEDURE

1. Gather nutrition labels from the following products: a carbonated soft drink, a bag of fresh carrots, canned spaghetti in sauce, potato chips, plain popcorn kernels, unsweetened applesauce, and fruit juice. Look at the percent of daily values of the major nutrients, as shown on the label for each food.
2. Make a list of ways to evaluate a food for high nutritional value. Include such criteria as nutrient levels and calories per serving.
3. Examine the nutrition labels and compare them with your list. Decide which of these foods would make a healthy snack.

WHAT DO YOU THINK?

- How does serving size affect the way you evaluate a nutritional label?
- What are some ways to snack and get nutrients at the same time?

CHALLENGE Design a full day's food menu that will give you all the nutrients you need. Use snacks as some of the foods that contribute these nutrients.

SKILL FOCUS
Analyzing

MATERIALS
nutrition labels

TIME
30 minutes

Spaghetti
IN TOMATO SAUCE WITH CHEESE

Nutrition Facts
Serving Size: 1 cup (252g)
Servings Per Container: about 2

Amount Per Serving	
Calories 210	Calories from Fat 20
	% Daily Value*
Total Fat 2g	3%
Saturated Fat 1g	5%
Cholesterol 5mg	2%
Sodium 1,020 mg	43%
Total Carbohydrate 41g	14%
Dietary Fiber 3g	12%
Sugars 14g	
Protein 7g	
Vitamin A 10%	Vitamin C 0%
Calcium 4%	Iron 10%

* Percent Daily Values are based on a 2,000 calorie diet. Your daily values may be higher or lower depending on your calorie needs.

Exercise is part of a healthy lifestyle.

Regular exercise allows all your body systems to stay strong and healthy. You learned that your lymphatic system doesn't include a structure like the heart to pump its fluid through the body. Instead, it relies on body movement and strong muscles to help it move antibodies and white blood cells. Exercise is good for the lymphatic system.

Exercise

When you exercise, you breathe harder and more quickly. You inhale and exhale more air, which exercises the muscles of your respiratory system and makes them stronger. Exercise also brings in extra oxygen. Oxygen is necessary for cellular respiration, which provides energy to other body systems. The circulatory system is strengthened by exercise. Your heart becomes stronger the more it is used. The skeletal system grows stronger with exercise as well. Studies show that older adults who lift weights have stronger bones than those who do not. In addition, physical activity can flush out skin pores by making you sweat, and it reduces the symptoms of depression.

DIFFERENTIATE INSTRUCTION

More Reading Support

E What happens to breathing rate during exercise? *It increases.*

Advanced Invite students to create a week's worth of healthy school lunch menus. Menus should have all of the necessary nutritional requirements for proteins, carbohydrates, and fats. For additional guidance, provide a copy of the USDA's most recent Food Guide Pyramid (www.fda.gov) so students can designate the number of servings for each of the food groups per meal.

INVESTIGATE Food Labels

PURPOSE To analyze food labels to determine their nutritional values

TIP *15 min.* Prior to the activity, ask students to bring in the nutrition label from a food they have eaten.

WHAT DO YOU THINK? *If a product has more than one serving and you eat the whole product, multiply the nutritional facts by the servings eaten to get the correct nutritional information; sample answer: applesauce, carrots, or fruit juice.*

CHALLENGE *Menus should include food from proteins, carbohydrates, and fats in percentages recommended on p. 141.*

Datasheet, Food Labels, p. 266

Technology Resources

Customize this student lab as needed or look for an alternative. Print rubrics to assess student lab reports.

Lab Generator CD-ROM

Address Misconceptions

IDENTIFY Ask students to describe what factors contribute to a healthful lifestyle. If students answer "Eating the right foods and exercising," prompt them to list all factors. They may hold the misconception that good health is associated only with food and physical fitness. Many students focus on the physical dimensions of health and pay no attention to mental and social aspects.

CORRECT Have students write *Physical, Mental,* and *Social* in three corners of a triangle. In the center of the triangle, have students write *Good Health.*

REASSESS Ask students to name one way that good health can be practiced for each corner. *Sample answer: P: exercise; M: relax; S: participate in community activities*

Technology Resources

Visit **ClassZone.com** for background on common student misconceptions.

MISCONCEPTION DATABASE

Teach Difficult Concepts

Some students may think that many factors they consider important to their health and life span are beyond their personal control. Remind students that all people make choices every day that affect the rest of their lives. Ask: What choices can you make today that will help you avoid health risks later on in life? *Sample answers: Not using tobacco can help prevent a person from developing lung cancer later in life. Daily exercise can help people avoid osteoporosis when they are older. Avoiding fatty foods can help a person avoid heart disease and diabetes.*

Teach from Visuals

To help students interpret the photograph showing exercise in a basketball game, ask: Which body systems and related body organs are the students exercising as they play basketball? *respiratory—lungs; circulatory—heart; muscular—muscles; skeletal—bones*

EXPLORE the BIG idea

Revisit "How Much Do You Exercise?" on p. 131. Ask students to explain how exercise affects heart rate.

Ongoing Assessment

Explain why exercise is needed to keep body systems healthy.

Ask: What happens when a person's body does not get an adequate amount of exercise? *The heart and lungs are not strengthened; muscles and bones may begin to break down; body fat is stored and excess fat may cause heart disease and diabetes.*

CHECK YOUR READING *Answer: An active lifestyle can promote good health; a sedentary lifestyle can promote excess body fat and disease.*

By eating healthy meals and exercising, you help your body to grow and develop.

Lifestyle

F

The lifestyles of many people involve regular exercise. Some lifestyles, however, include more sitting still than moving. A lifestyle that is sedentary, associated mostly with sitting down, can harm a person's health. Muscles and bones that are not exercised regularly can begin to break down. Your body stores unused energy from food as fat. The extra weight of body fat can make it harder for you to exercise. Therefore, it is harder to use up energy or to strengthen your skeletal, muscular, and immune systems. Researchers have also made connections between excess body fat and heart disease and diabetes.

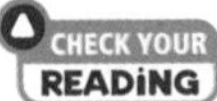
CHECK YOUR READING How does lifestyle affect health?

Drug abuse, addiction, and eating disorders cause serious health problems.

CONTENT FRAME
Make a content frame for the main idea: *Drug abuse, addiction, and eating disorders cause serious health problems.*

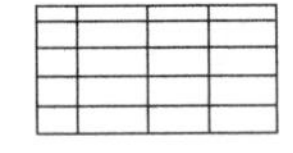

Every day, you make choices that influence your health. Some choices can have more serious health risks, or possibilities for harm, than others. You have the option to make healthy choices for yourself. Making unhealthy decisions about what you put into your body can lead to drug abuse, addiction, or eating disorders.

E 144 Unit: Human Biology

DIFFERENTIATE INSTRUCTION

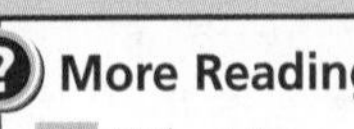

More Reading Support

F What term describes a lifestyle that is mostly inactive? *sedentary*

Below Level Invite students to work in small groups to create a public service announcement about one aspect of living healthy lifestyles. Record students' announcements and play them for the class. If video equipment is available, have students film and present their public service announcements to the class. You may want to suggest or assign topics and have students focus on the advantages and/or consequences of making healthy or unhealthy choices.

Drug Abuse

A drug is any chemical substance that you take in to change your body's functions. Doctors use drugs to treat and prevent disease and illness. The use of a drug for any other reason is an abuse of that drug. Abuse can also include using too much of a substance that is not harmful in small amounts. People abuse different drugs for different reasons. Drugs often do allow an individual to feel better for the moment. But they can also cause serious harm to an individual's health.

Tobacco Nicotine, the drug in tobacco, increases heart rate and blood pressure and makes it seem as if the user has more energy. Nicotine is also a poison; in fact, some farmers use it to kill insects. Tobacco smoke contains thousands of chemicals. Tar and carbon monoxide are two harmful chemicals in smoke. Tar is a sticky substance that is commonly used to pave roads. Carbon monoxide is one of the gases that cars release in their exhaust. People who smoke or chew tobacco have a high risk of cancer, and smokers are also at risk for heart disease.

Compounds Found in Unfiltered Tobacco Smoke

Compound	Amount in First-Hand Smoke (per cigarette)	Amount in Second-Hand Smoke (per cigarette)
Nicotine	1–3 mg	2.1–46 mg
Tar	15–40 mg	14–30 mg
Carbon monoxide	14–23 mg	27–61 mg
Benzene	0.012–0.05 mg	0.4 mg
Formaldehyde	0.07–0.1 mg	1.5 mg
Hydrogen cyanide	0.4–0.5 mg	0.014–11 mg
Phenol	0.08–0.16 mg	0.07–0.25 mg

Source: U.S. Department of Health and Human Services

Alcohol Even a small amount of alcohol can affect a person's ability to think and reason. Alcohol can affect behavior and the ability to make decisions. Many people are killed every year, especially in automobile accidents, because of choices they made while drinking alcohol. Alcohol abuse damages the heart, the liver, the nervous system, and the digestive system.

Other Drugs Some drugs, such as cocaine and amphetamines, can make people feel more energetic and even powerful because they stimulate the nervous system and speed up the heart. These drugs are very dangerous. They can cause nervous disorders and heart attacks.

Drugs called narcotics also affect the nervous system. Instead of stimulating it, however, they decrease its activity. Narcotics are prescribed by doctors to relieve pain and to help people sleep. Abuse of narcotics can lead to addiction. More and more narcotics are then needed to gain the same effect. Because narcotics work by decreasing nerve function, large amounts of these drugs can cause the heart and lungs to stop.

DIFFERENTIATE INSTRUCTION

More Reading Support

G What term names the drug found in tobacco? *nicotine*

H What are two harmful substances in tobacco smoke? *tar and carbon monoxide*

Below Level Ask students to find out the cost of smoking one pack of cigarettes each day for one year. Then have them make a list of how they could use the money instead of smoking.

History of Science

The tobacco plant is native to the Americas. The practice of smoking tobacco dates back more than 2,000 years ago to the Mayan culture, from which it spread north to Native American tribes. By the mid-16th century, the practice of smoking was introduced in Europe by explorers. Controversy surrounding the ill effects of smoking has existed since then, when James I forbade cultivation of tobacco in England. By the 1960s, clinical studies on smoking and disease concluded that most lung cancer deaths were caused by the use of tobacco. It was also found to be responsible for respiratory illnesses including chronic bronchitis, emphysema, and cardiovascular disease. By 1965, health warnings became mandatory on all cigarette packages. In 1984, the law was expanded to include health warnings on all tobacco advertisements.

Teach from Visuals

To help students interpret the chart of compounds found in tobacco, ask:

- What compounds are found in tobacco? *nicotine, tar, carbon monoxide, benzene, formaldehyde, hydrogen cyanide, phenol*
- Why does second-hand smoke pose a health risk? *Second-hand smoke contains all the dangerous compounds found in first-hand smoke.*

Teach Difficult Concepts

Many students think that tobacco and alcohol are not drugs. Remind students that a drug is any substance that changes the way the body functions. Have students make a list of the effects of both tobacco and alcohol on the body. Point out the the effects of drugs on a user are multiplied when taken in by a growing embryo or fetus. All drugs can cause birth defects if taken by a pregnant woman.

Teaching with Technology

If spreadsheet software is available, have students use it to record and outline the dangers of tobacco, alcohol, and other drugs.

Ongoing Assessment

Describe how drug abuse, eating disorders, and addiction can negatively affect the body.

Ask: How do stimulants and narcotics affect the body? *Stimulants speed up the nervous system and the heart; narcotics decrease the activity of the nervous system.*

Reinforce (the BIG idea)

Have students relate the section to the Big Idea.

 Reinforcing Key Concepts, p. 267

5.2 ASSESS & RETEACH

Assess

 Section 5.2 Quiz, p. 80

Reteach

Have students complete the following graphic organizer. Students should write one detail per category as shown.

Good Health Practices Help Maintain a Healthy Body	
Nutrition	*Eating the proper amount of proteins, carbohydrates, and fats helps your body function properly.*
Exercise	*Regular exercise strengthens the heart and keeps muscles and bones strong.*
Lifestyle	*Avoiding drugs keeps your body from becoming addicted to harmful substances that could lead to disease or death.*

Technology Resources

Have students visit **ClassZone.com** for reteaching of Key Concepts.

 CONTENT REVIEW

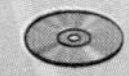 CONTENT REVIEW CD-ROM

Students can be active in protesting drug abuse.

Addiction

Drug abuse can often lead to addiction. **Addiction** is an illness in which a person becomes dependent on a substance or behavior. Repeated use of drugs such as alcohol, tobacco, and narcotics can cause the body to become physically dependent. When a person is dependent on a drug, taking away that drug can cause withdrawal. If affected by withdrawal, a person may become physically ill, sometimes within a very short period of time. Symptoms of withdrawal can include fever, muscle cramps, vomiting, and hallucinations.

Another type of addiction can result from the effect a drug, or even a behavior, produces. Although physical dependency may not occur, a person can become emotionally dependent. Gambling, overeating, and risk-taking are some examples of addictive behaviors. With both physical and emotional addictions, increasing amounts of a drug or behavior are necessary to achieve the effects. Someone who suffers from an addiction can be treated and can work to live a healthy life, but most addictions never go away completely.

Eating Disorders

An eating disorder is a condition in which people continually eat too much or too little food. One example of an eating disorder is anorexia nervosa. People with this disorder eat so little and exercise so hard that they become unhealthy. No matter how thin they are, they believe they need to be thinner. People with anorexia do not receive necessary nutrients because they don't eat. When the energy used by the body exceeds the energy taken in from food, tissues in the body are broken down to provide fuel. Bones and muscles, including the heart, can be damaged, and the person can die.

5.2 Review

KEY CONCEPTS

1. How do nutrients affect health?
2. Explain the effects of exercise on the respiratory and circulatory systems.
3. Make a chart showing the effects of tobacco, alcohol, and other drugs on the body.

CRITICAL THINKING

4. **Explain** How would you define health? Write your own definition.
5. **Synthesize** Explain how water can be considered a nutrient. Include a definition of *nutrient* in your explanation.

CHALLENGE

6. **Apply** You have heard about a popular new diet. All of the foods in the diet are fat-free, and the diet promises fast weight loss. How might this diet affect health? Explain your answer.

ANSWERS

1. Nutrients keep the body functioning properly.

2. Exercise causes deeper inhales and exhales that make the heart and muscles of the respiratory system stronger.

3. tobacco—damages respiratory and circulatory systems, high risk of cancer and heart disease; alcohol—damages liver, heart, nervous, and digestive systems; drugs—damage the nervous system and can cause heart and lungs to stop

4. Answers should include a relaxing lifestyle, good nutrition, exercise, and avoiding harmful drugs.

5. Nutrients are sources of energy for the body. A person needs enough water each day to keep body systems functioning.

6. This diet would cause poor health. A diet should contain about 10 to 15 percent fats.

MATH TUTORIAL
CLASSZONE.COM
Click on Math Tutorial for more help with choosing a data display.

SKILL: CHOOSING A DATA DISPLAY

Pumping Up the Heart

Heart rates differ with age, level of activity, and fitness. To communicate the differences clearly, you need to display the data visually. Choosing the appropriate display is important.

Example

The fitness trainer at a gym wants to display the following data:

Maximum heart rate while exercising (beats per minute)		
Age 21	Men	197
	Women	194
Age 45	Men	178
	Women	177
Age 65	Men	162
	Women	164

Here are some different displays the trainer could use:

- A bar graph shows how different categories of data compare. Data can be broken into 2 or even 3 bars per category.
- A line graph shows how data changes over time.
- A circle graph represents data as parts of a whole.

ANSWER The fitness trainer wants to show heart rate according to both age and gender, so a double bar graph would be the clearest.

What would be an appropriate way to display data in the following situations?

1. A doctor wants to display how a child's average heart rate changes as the child grows.
2. A doctor wants to display data showing how a person's resting heart rate changes the more the person exercises.
3. A scientist is studying each type of diet that the people in an experiment follow. She will show what percentage of the people with each diet had heart disease.

CHALLENGE **Describe a situation in which a double line graph is the most appropriate data display.**

ANSWERS

1. *line graph, to show change over time*
2. *line graph, to show change over time*
3. *circle graph, to show parts of a whole (subsets of a set)*
4. ***CHALLENGE*** *Answers will vary but should express the idea that two sets of data need to be compared to use a double bar graph.*

MATH IN SCIENCE
Math Skills Practice for Science

Set Learning Goal

To determine the appropriate way to display data

Present the Science

Aerobic exercise contributes to cardiovascular health. As the body works to exercise, the aerobic system uses glycogen, fats, and proteins as fuel sources. As the heart rate increases, cardiac output increases. Increased blood flow increases the amount of oxygen sent to the muscles to help them work.

Develop Data Skills

Review the different types of graphs. Line graphs generally show change over a period of time. Circle graphs represent data as percentages of one hundred. Bar graphs compare sets of data using bars across the *x*- or *y*-axis. Remind students that the *x*-axis is the horizontal axis and the *y*-axis is the vertical axis. Both axes intersect at zero.

DIFFERENTIATION TIP Prepare examples of the various types of graphs copied from your school's mathematics texts. Enlarge the graphs for students to examine.

Close

Ask students why neither a line graph nor a circle graph is a good choice for displaying the data in the chart. *Neither graph is useful in comparing two sets of data.*

- Math Support, p. 283
- Math Practice, p. 284

Technology Resources

Students can visit **ClassZone.com** to practice graphing skills.

MATH TUTORIAL

5.3 FOCUS

Set Learning Goals

Students will

- Discuss some of the causes of disease.
- Describe how diseases can be treated.
- Identify ways to prevent the spread of disease.
- Model how germs are spread.

3-Minute Warm-Up

Display Transparency 36 or copy this exercise on the board:

Complete the chart describing how each nutrient benefits the body.

Benefits		Nutrients
main source of energy	←	*carbohydrates*
build tissues, used for growth and repair	←	*proteins*
essential for energy	←	*fats*
transports materials through the body	←	*water*

 3-Minute Warm-Up, p. T36

5.3 MOTIVATE

EXPLORE The Immune System

PURPOSE To model how easily germs are spread

TIP *10 min.* Have students make a list of all the places where they observed the glitter.

WHAT DO YOU THINK? ***The glitter was easily transferred to other people and places. Germs are also easily spread by people touching things.***

KEY CONCEPT

5.3 Science helps people prevent and treat disease.

BEFORE, you learned

- Good nutrition and exercise help keep the body healthy
- Drug abuse can endanger health
- Eating disorders can affect the body's health

NOW, you will learn

- About some of the causes of disease
- How diseases can be treated
- How to help prevent the spread of disease

VOCABULARY

microorganism p. 148
bacteria p. 149
virus p. 149
resistance p. 153

EXPLORE The Immune System

How easily do germs spread?

PROCEDURE

1. Early in the day, place a small amount of glitter in the palm of one hand. Rub your hands together to spread the glitter to both palms. Go about your day normally.
2. At the end of the day, examine your environment, including the people around you. Where does the glitter show up?

WHAT DO YOU THINK?

- How easily did the glitter transfer to other people and objects?
- What do you think this might mean about how diseases might spread?

MATERIALS
glitter

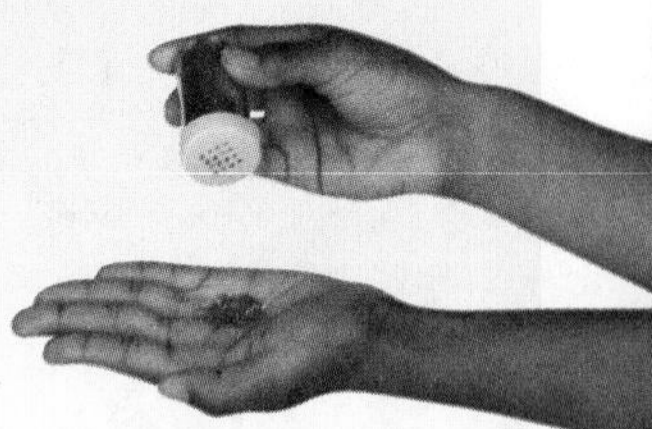

Scientific understanding helps people fight disease.

Disease is a change that disturbs the normal functioning of the body's systems. If you have ever had a cold, you have experienced a disease that affected your respiratory system. What are the causes of disease? Many diseases are classified as infectious diseases, or diseases that can be spread. Viruses, bacteria and other materials cause infectious disease. The organisms that cause sickness are called **microorganisms.**

Before the invention of the microscope, people didn't know about microorganisms that cause disease. They observed that people who lived near each other sometimes caught the same illness, but they didn't understand why. Understanding disease has helped people prevent and treat many illnesses.

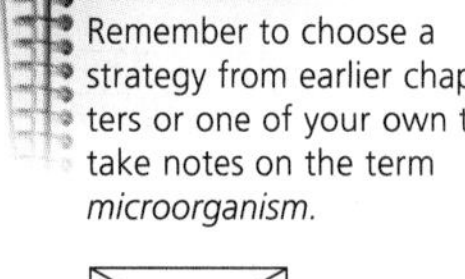

VOCABULARY
Remember to choose a strategy from earlier chapters or one of your own to take notes on the term *microorganism.*

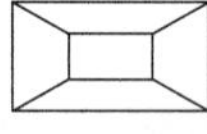

RESOURCES FOR DIFFERENTIATED INSTRUCTION

Below Level

UNIT RESOURCE BOOK

- Reading Study Guide A, pp. 270–271
- Decoding Support, p. 282

Advanced

UNIT RESOURCE BOOK
Challenge and Extension, p. 276

English Learners

UNIT RESOURCE BOOK
Spanish Reading Study Guide, pp. 274–275

- Audio Readings in Spanish
- Audio Readings (English)

The germ theory describes some causes of disease.

In the 1800s, questions about the causes of some diseases were answered. Scientists showed through experiments that diseases could be caused by very small living things. In 1857, French chemist Louis Pasteur did experiments that showed that microorganisms caused food to decay. Later, Pasteur's work and the work of Robert Koch and Robert Lister contributed to the germ theory. Pasteur's germ theory states that some diseases, called infectious diseases, are caused by germs.

Bacteria and Viruses

Germs are the general name given to organisms that cause disease. Germs include **bacteria** (bak-TEER-ee-uh), single-celled organisms that live almost everywhere. Within your intestines, bacteria function to digest food. Some bacteria, however, cause disease. Pneumonia (nu-MOHN-yuh), ear infections (ihn-FEHK-shuhnz), and strep throat can be caused by bacteria.

Explore ways to fight disease.

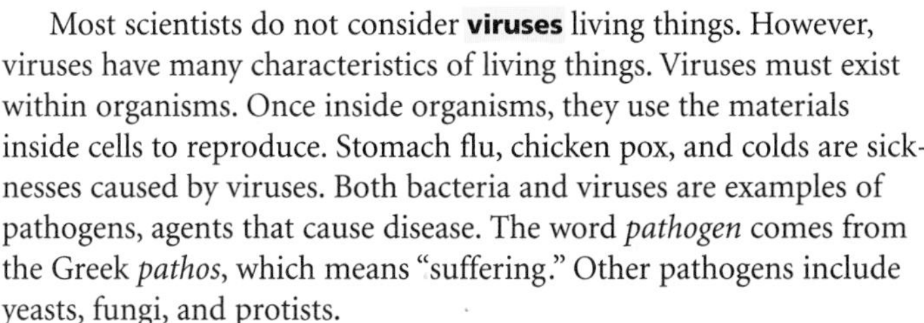

Most scientists do not consider **viruses** living things. However, viruses have many characteristics of living things. Viruses must exist within organisms. Once inside organisms, they use the materials inside cells to reproduce. Stomach flu, chicken pox, and colds are sicknesses caused by viruses. Both bacteria and viruses are examples of pathogens, agents that cause disease. The word *pathogen* comes from the Greek *pathos*, which means "suffering." Other pathogens include yeasts, fungi, and protists.

Treating Infectious Diseases

Diseases caused by bacteria can be treated with medicines that contain antibiotics. An antibiotic is a substance that can destroy bacteria. The first antibiotics were discovered in 1928 when a scientist named Alexander Fleming was performing experiments on bacteria. Fleming found mold growing on his bacteria samples. While most of the bacteria samples looked cloudy, the area around the mold was clear. From this observation, Fleming concluded that a substance in the mold had killed the bacteria.

Fleming had not intended to grow mold in his laboratory, but the accident led to the discovery of penicillin. Since the discovery of penicillin, many antibiotics have been developed. Antibiotics have saved the lives of millions of people.

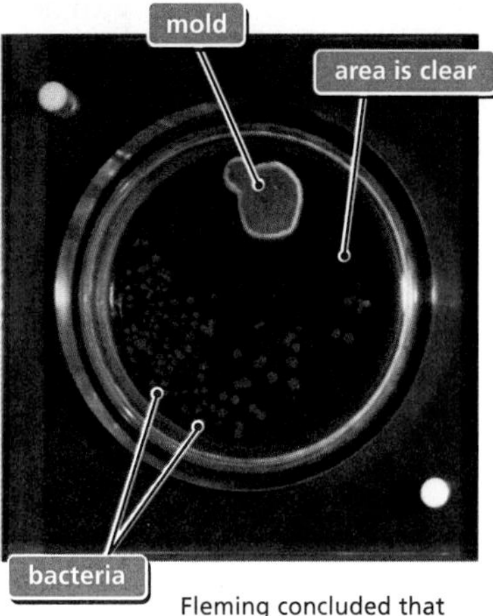

Fleming concluded that something in the mold had killed the bacteria.

5.3 INSTRUCT

History of Science

Prior to the mid 1860s, surgery was risky, with many patients dying of infection after the procedure. In 1865, British surgeon Joseph Lister began to apply Koch's and Pasteur's germ theories to his surgical practices. He hypothesized that microorganisms cause infections and began to use carbolic acid to wash his hands and surgical instruments prior to surgery. He used carbolic acid to cleanse the wounds and bandages. His practice greatly decreased the rate of death from surgical infections.

Real World Example

Bacteria have many industrial uses, including a role in the production of dairy products, vinegar, sauerkraut, and the preparation of certain antibiotics. Bacteria are also used in sewage disposal plants to clean up organic wastes. Certain bacteria decompose oil into small harmless molecules. These are useful when cleaning up oil spills that contaminate plant and animal life.

Teach Difficult Concepts

Some students may think that antibiotics are a cure for most illnesses. In fact, antibiotics are effective against most kinds of bacterial infections, but do not have much effect against viral infections such as influenza, chicken pox, and colds.

DIFFERENTIATE INSTRUCTION

More Reading Support

A What term describes agents of disease, such as bacteria and viruses? *pathogens*

B What substances found in medicines destroy bacteria? *antibiotics*

English Learners Have students write the definitions for the terms *germ, bacteria, virus, pathogen, antibiotic,* and *resistance* in their Science Word Dictionaries. Use the Key Concepts Summary, p. 156, to help students focus on key concepts and vocabulary.

Address Misconceptions

IDENTIFY Ask students to explain what illnesses are caused by germs and how germs cause illness. Many students hold the misconception that every illness is caused by germs, that all diseases are contagious and caused by the same kind of germ. Students may also hold the misconception that all germs enter the body through the mouth when eating.

CORRECT Have students create a chart with three columns: What are germs? What causes diseases? How do germs enter the body? Ask students to fill in the chart as they read pp. 149–155.

REASSESS Have students re-evaluate their charts after the chapter. Have them summarize the ways germs enter the body and the varied causes of illness.

Technology Resources

Visit **ClassZone.com** for background on common student misconceptions.

 **MISCONCEPTION DATABASE**

Social Studies Connection

Interested students may wish to find out about the bubonic plague that occurred in 14th century Europe. Also known as the Black Death, the plague killed over 25 million people. Ask students what caused it and how it changed the history and social structure of medieval Europe. Allow students to share their findings.

EXPLORE the BIG idea

Revisit "How Safe Is Your Food?" on p. 131. Ask students to reassess the dangers of eating food that is past its freshness date.

Ongoing Assessment

Discuss some of the causes of disease.

Ask: What are two types of pathogens that cause disease? *bacteria and viruses*

Identify ways to prevent the spread of disease.

Ask: How can harmful bacteria be killed in food and water? *Wash fruits and vegetables before eating; boil water for cooking.*

Infectious diseases spread in many ways.

READING TIP

As you read the text on this page, notice how each pathogen shown on p. 151 spreads.

One of the best ways to protect your health is by being informed and by avoiding pathogens. Pathogens can be found in many places, including air, water, and on the surfaces of objects. By knowing how pathogens travel, people are able to limit the spreading of disease.

Food, Air, and Water

Sometimes people get sick when they breathe in pathogens from the air. The viruses that cause colds can travel through air. If you cover your mouth when you sneeze or cough, you can avoid sending pathogens through the air. Pathogens also enter the body in food or water. Washing fruits and vegetables and cooking meats and eggs kills bacteria. Most cities in the United States add substances, such as chlorine, to the supply of public water. These substances kill pathogens. Boiling water also kills pathogens. People sometimes boil water if their community loses power or experiences a flood. Campers often need to boil or filter water before they use it.

Contact with Insects and Other Animals

Animals and insects can also carry organisms that cause disease. The animal itself does not cause the illness, but you can become sick if you take in the pathogen that the animal carries. Lyme disease, for example, is caused by bacteria that inhabit ticks. The ticks are not the illness, but if an infected tick bites you, you will get Lyme disease.

A deadly central nervous system infection called rabies can also come from animal contact. The virus that causes rabies is found in the saliva of an infected animal, such as a bat, raccoon, or opossum. If that animal bites you, you may get the disease. A veterinarian can give your pet an injection to prevent rabies. You can get other infections from pets. These infections include worms that enter through your mouth or nose and live in your intestines. You can also get a skin infection called ringworm, which is actually a fungus rather than a worm.

Person-to-Person Contact

Most of the illnesses you have had have probably been passed to you by another person. Even someone who does not feel sick can have pathogens on his or her skin. If you touch that person or if that person touches something and then you touch it, the pathogens will move to your skin. If the pathogens then enter your body through a cut or through your nose, mouth, or eyes, they can infect your body. The simplest way to avoid giving or receiving pathogens is to wash your hands often and well.

DIFFERENTIATE INSTRUCTION

More Reading Support

C What are two diseases spread by animals? *Lyme disease, rabies*

D How can pathogens enter the body? *nose, eyes, mouth, a cut*

Advanced Invite interested students to investigate some recent diseases or pathogens, their origin, symptoms, and what is being done to contain them. Topics for research might include Ebola virus, giardia, and SARS.

 Challenge and Extension, p. 276

Pathogens and Disease

Infectious diseases are caused by microorganisms.

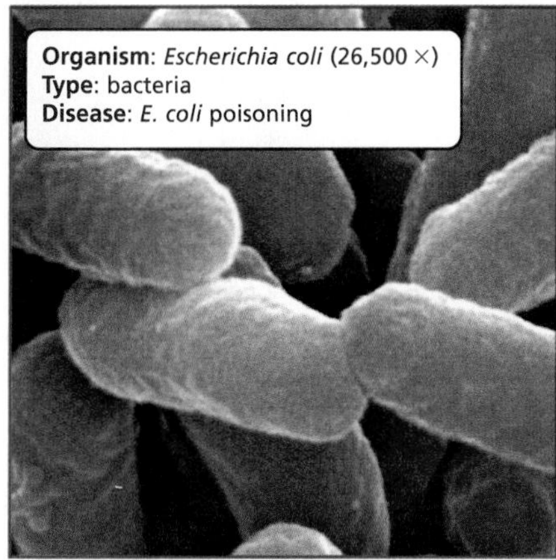

Organism: *Escherichia coli* (26,500 ×)
Type: bacteria
Disease: *E. coli* poisoning

Spread: contaminated food or water

Prevention: handwashing, thoroughly cooking meat, boiling contaminated water, washing fruits and vegetables, drinking only pasteurized milk, juice, or cider

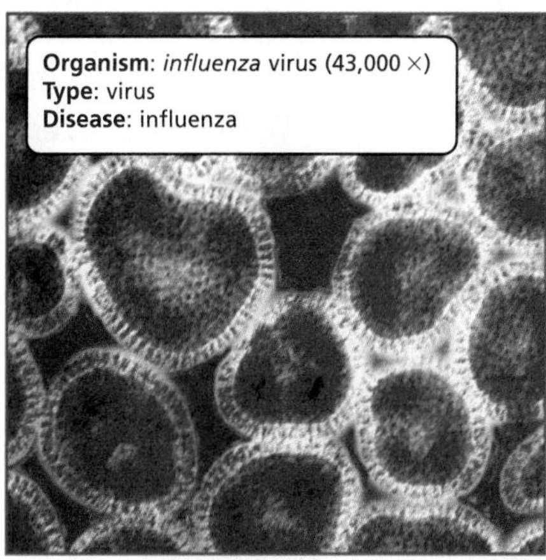

Organism: *influenza* virus (43,000 ×)
Type: virus
Disease: influenza

Spread: inhaling virus from sneezes or coughs of infected person

Prevention: vaccination

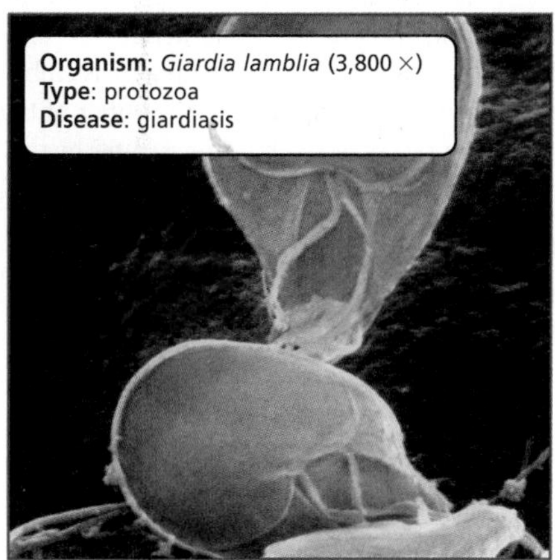

Organism: *Giardia lamblia* (3,800 ×)
Type: protozoa
Disease: giardiasis

Spread: contaminated food or water, close contact with infected person

Prevention: handwashing, thoroughly cooking meat, boiling contaminated water, washing fruits and vegetables, drinking only pasteurized milk, juice, or cider

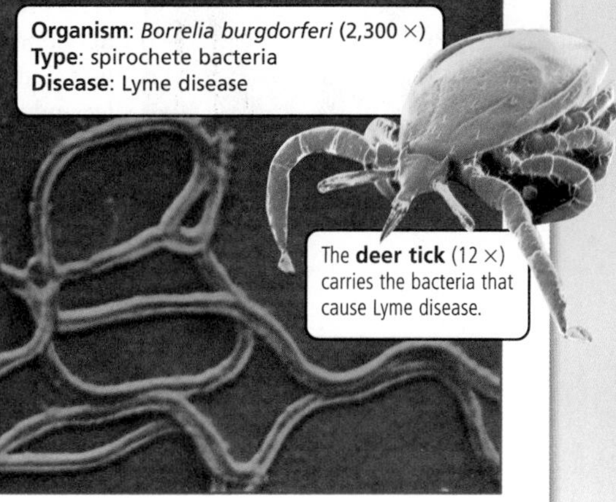

Organism: *Borrelia burgdorferi* (2,300 ×)
Type: spirochete bacteria
Disease: Lyme disease

The **deer tick** (12 ×) carries the bacteria that cause Lyme disease.

Spread: tick bite

Prevention: wear light-colored clothing, tuck pants into socks or shoes, check for ticks after outdoor activities, use repellents containing DEET

READING VISUALS How can people prevent each of these pathogens from spreading?

DIFFERENTIATE INSTRUCTION

Below Level Explain that the common cold can be caused by over 200 different viruses. Persons infected with cold viruses often experience similar symptoms even though the viruses differ. Ask students to develop an inference about why there is still no cure for the common cold. They should support the inference with evidence and try to convince their peers that the inference is valid. *Sample answer: Because there are so many viruses that cause the cold, there would have to be that many cures, and doctors would need to determine which virus is causing the symptoms.*

Teacher Demo

To model how disease is spread, try this demonstration. Prior to class, prepare a class set of envelopes. Place a green strip of paper inside three of the envelopes. Place white strips in the remaining envelopes. Give each student an envelope. Instruct students that the green strips are "infected." Tell students to move about the room for 60 seconds until you say *Stop.* They should whisper the color of the strip that they hold to the nearest person. The students who hear green will now whisper *green* at all future stops. At the end of two rounds, instruct students who have a green slip or who heard green at the first stop to sit down. After the third round, students who heard green at the second stop should sit down. At this point, many students should be "infected." Work with the class to trace the infections back to the original source of infection. Students should record all of their contacts to track how the virus spread.

Teach from Visuals

To help students interpret the photographs showing pathogens, ask: What are some different ways shown that diseases are spread? *through contaminated food or water, inhaling viruses, insect bites*

This visual is also available as T38 in the Unit Transparency Book.

Ongoing Assessment

READING VISUALS *Answer: E. coli and Giardia can be stopped by handwashing and avoiding contaminated food and water. Influenza can be prevented by vaccination. Lyme disease can be prevented by wearing proper protective clothing or repellent, and checking for ticks after being outdoors.*

Metacognitive Strategy

Have students make a list of diseases and choose methods of classifying the diseases by what causes them. Encourage students to write definitions for each category of disease classification. Examples include contagious and noncontagious; genetic or environmentally caused.

Health Connection

Point out that some noninfectious diseases can be prevented by practicing good health habits. For example, a person can cut down on the risks of getting lung cancer by not smoking cigarettes and avoiding second-hand smoke. Ask students to give examples of other ways that they can prevent the onset of noninfectious diseases.

Ongoing Assessment

CHECK YOUR READING *Answer: Hemophilia is an inherited disease that is present at birth. Cancer and diabetes are diseases that may come later in life.*

Noninfectious diseases are not contagious.

E Noninfectious diseases are diseases that cannot be spread by pathogens. They are not contagious. You are born with some of these, and others can develop during life.

Diseases Present at Birth

Some diseases present at birth are inherited. Genes, which act as instructions for your cells, are inherited from your parents. Some forms of a gene produce cells that do not function properly. Most genetic disorders are due to recessive forms of a gene, which means that while both parents carry the defective form, neither one has the disorder. Cystic fibrosis, sickle cell anemia, and hemophilia are diseases inherited this way.

Asthma is a noncontagious disease often present at birth.

The symptoms of some genetic diseases may not be present immediately at birth. Huntington's chorea, even though it is an inherited condition, does not begin to produce symptoms until a person reaches adulthood. Other genes can increase the chances of developing a disease later in life, such as cancer or diabetes, but the pattern of inheritance is not totally understood.

The process of human development is complex. Some diseases present at birth may involve both inherited factors and development. Talipes, a disorder commonly known as clubfoot, is due to the improper development of the bones of the leg and foot. Talipes can be corrected by surgery after birth.

Diseases in Later Life

Some diseases, including heart disease, certain forms of cancer, and many respiratory disorders, have much less to do with genetics and more to do with environment and lifestyle. You have learned about the ways in which you can lead a healthy lifestyle. Good nutrition, exercise, and avoiding substances that can damage the body systems not only increase the length of life, but also the quality of life.

While people with family histories of cancer are at higher risk of getting it, environmental factors can influence risk as well. Tar and other chemicals from cigarettes can damage the lungs, in addition to causing cancer. Much is still not known about the causes of cancer.

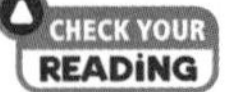

CHECK YOUR READING Name a noninfectious disease that is present at birth and one that may occur later in life.

DIFFERENTIATE INSTRUCTION

More Reading Support

E What are noninfectious diseases? *those not spread by pathogens*

Below Level Ask students to create a news conference about a disease outbreak such as SARS or Ebola. Students can role-play doctors and news reporters at the conference. Students should come prepared with questions and answers and be ready to explain how doctors differentiate the outbreak from other illnesses including those that are noninfectious.

Scientists continue efforts to prevent and treat illness.

In spite of all that scientists have learned, disease is still a problem all over the world. Illnesses such as AIDS and cancer are better understood than they used to be, but researchers must still find ways to cure them.

Even though progress is sometimes slow, it does occur. Better education has led to better nutrition. The use of vaccines has made some diseases nearly extinct. However, new types of illness sometimes appear. AIDS was first identified in the 1980s and spread quickly before it was identified. More recently, the West Nile virus appeared in the United States. This virus is transmitted by infected mosquitoes and can cause the brain to become inflamed. Efforts to control the disease continue.

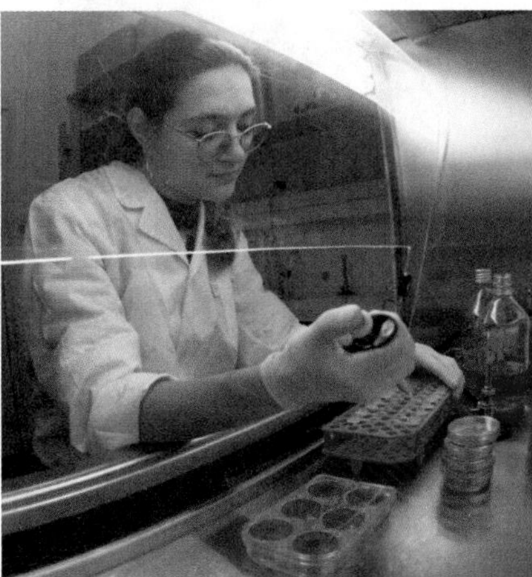

Scientists work hard to fight disease.

Antibiotics fight pathogens, but they can also lead to changes in them. When an antibiotic is used too often, bacteria can develop **resistance,** or become partially immune, to its effects. This means that the next time those bacteria invade, that particular antibiotic will not stop the infection. For this reason, it is best not to use pathogen-killing drugs or chemicals unless you really need them.

CHECK YOUR READING Describe the advantages and disadvantages of using an antibiotic when you are sick.

5.3 Review

KEY CONCEPTS

1. Define microorganism and explain how microorganisms can affect health.
2. What is an antibiotic?
3. Make a chart showing ways that infectious diseases spread and ways to keep them from spreading.

CRITICAL THINKING

4. **Connect** Make a list of things you can do to avoid getting Lyme disease or the West Nile virus.
5. **Apply** How does washing your hands before eating help protect your health?

CHALLENGE

6. **Synthesize** How can nutrition help in the prevention of disease? Use these terms in your answer: *nutrients, pathogens,* and *white blood cells.*

ANSWERS

1. a very small organism that can be seen only with a microscope; can bring disease, which disturbs the normal functioning of the body

2. An antibiotic is a substance that can destroy bacteria.

3. Charts should include food, air, water transmission, animal contact, and a prevention method for each transmission.

4. Answers may include wearing long sleeves and pants tucked into boots and using insect repellent.

5. Washing kills pathogens on your skin. This prevents pathogens from moving onto your food.

6. If the body is not healthy, it may not be able to produce enough white blood cells to kill p...

Ongoing Assessment

Describe how diseases can be treated.

Ask: What are the pathogen-killing drugs used to treat infectious disease? *antibiotics*

CHECK YOUR READING *Answer: Antibiotics fight diseases, but if they are used too often, bacteria can develop resistance to them and they are no longer effective in treating that bacteria.*

Reinforce the BIG idea

Have students relate the section to the Big Idea.

Reinforcing Key Concepts, p. 278

5.3 ASSESS & RETEACH

Assess

Section 5.3 Quiz, p. 81

Reteach

Discuss each concept. Have students name examples from the text or elsewhere.

1. Bacteria and viruses are pathogens that can cause disease. (pp. 150–151)
2. Antibiotics are used to treat infectious diseases. (p. 149)
3. Noninfectious diseases cannot be spread by pathogens. (p. 152)

Technology Resources

Have students visit **ClassZone.com** for reteaching of Key Concepts.

CONTENT REVIEW

CONTENT REVIEW CD-ROM

CHAPTER INVESTIGATION

Focus

PURPOSE To determine the effectiveness of cleaning hands with soap and water to remove bacteria

OVERVIEW Students will prepare petri dishes with samples of bacteria from their hands before and after washing. Students will do the following:

- sample their hands for the presence of bacteria
- test the effectiveness of washing with water compared with soap and water

Lab Preparation

- Sterile petri dishes may be obtained from biological supply companies.
- Prior to the investigation, have students read through the investigation and prepare their data tables. Or you may wish to copy and distribute datasheets and rubrics.

 UNIT RESOURCE BOOK, pp. 285–293

 SCIENCE TOOLKIT, F14

Lab Management

- Remind students to leave the petri dishes closed. Place the petri dishes in a cool place for several days.
- Dispose of petri dishes in a plastic container filled with dilute bleach. Do not allow students to open the dishes when they record their observations.
- You can prepare your own nutrient agar: boil 31 grams per liter of water for 10–15 minutes. Use a fume hood or leave windows open.

SAFETY Make sure students wash their hands thoroughly after the activity.

INCLUSION Provide a simplified checklist of the procedure steps for students to follow. Give each step a number and a name. Students can check off each step as they complete it. If you have students with limited attention, be sure to review the caution statements first.

CHAPTER INVESTIGATION

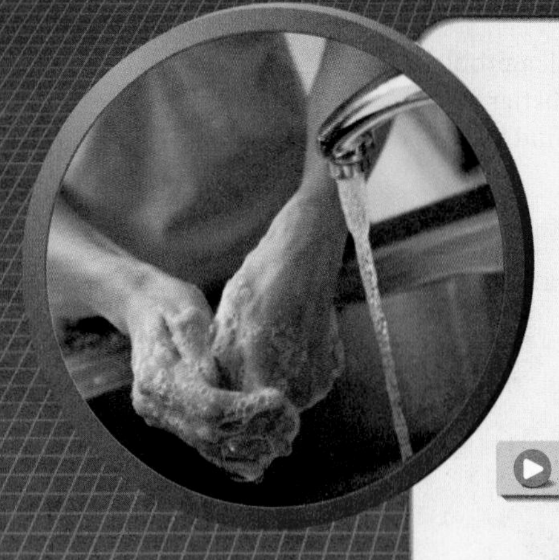

Cleaning Your Hands

OVERVIEW AND PURPOSE Your skin cells produce oils that keep the skin moist. This same layer of oil provides a nutrient surface for bacteria to grow. When you wash your hands with soap, the soap dissolves the oil and the water carries it away, along with the bacteria. In this activity you will

- sample your hands for the presence of bacteria
- test the effectiveness of washing your hands with water compared with washing them with soap and water

Problem

Is soap effective at removing bacteria?

Hypothesize

Write a hypothesis explaining how using soap affects the amount of bacteria on your hands. Your hypothesis should take the form of an "If . . . , then . . . , because . . ." statement.

MATERIALS
- 3 covered petri dishes with sterile nutrient agar
- soap
- marker
- tape
- hand lens

Procedure

1 Make a data table in your **Science Notebook** like the one shown on page 155.

2 Obtain three agar petri dishes. Be careful not to open the dishes.

3 Remove the lid from one dish and gently press two fingers from your right hand onto the surface of the agar. Close the lid immediately. Tape the dish closed. Mark the tape with the letter *A*. Include your initials and the date.

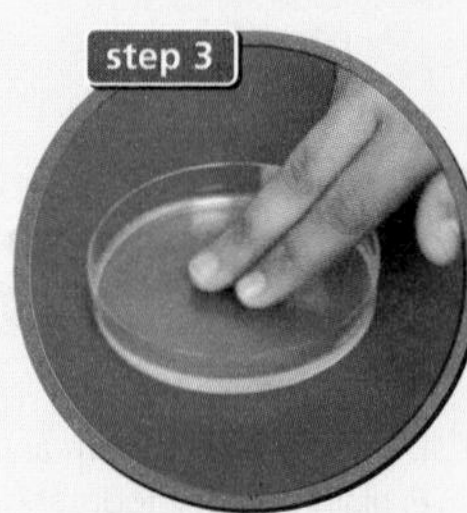

INVESTIGATION RESOURCES

 CHAPTER INVESTIGATION, Cleaning Your Hands
- Level A, pp. 285–288
- Level B, pp. 289–292
- Level C, p. 293

Advanced students should complete Levels B & C.

 Writing a Lab Report, D12–13

Technology Resources

Customize this student lab as needed or look for an alternative. Print rubrics to assess student lab reports.

 **Lab Generator CD-ROM**

4 Wash your hands in water and let them air-dry. Open the second dish with your right hand and press two fingers of your left hand into the agar. Close the lid immediately. Tape and mark the dish *B*, as in step 3.

5 Wash your hands in soap and water and let them air-dry. Open the third dish with one hand and press two fingers of the other hand into the agar. Close the lid immediately. Tape and mark the dish *C*, as in step 3.

6 Place the agar plates upside down in a dark, warm place for two to three days. **Caution:** Do not open the dishes. Wash your hands.

Observe and Analyze

Write It Up

1. **OBSERVE** Use a hand lens to observe the amounts of bacterial growth in each dish, and record your observations in Table 1. Which dish has the most bacterial growth? the least growth?
2. **OBSERVE** Is there anything you notice about the bacterial growth in each dish other than the amount of bacterial growth?
3. Return the petri dishes to your teacher for disposal. **Caution:** Do not open the dishes. Wash your hands thoroughly with warm water and soap when you have finished.

Conclude

Write It Up

1. **INFER** Why is it necessary to air-dry your hands instead of using a towel?
2. **INFER** Why is it important to use your right hand in step 3 and your left hand in step 4?
3. **INTERPRET** Compare your results with your hypothesis. Do your observations support your hypothesis?
4. **EVALUATE** Is there much value in washing your hands simply in water?
5. **EVALUATE** How might the temperature of the water you used when you washed your hands affect the results of your experiment?
6. **EVALUATE** Given the setup of your experiment, could you have prepared a fourth sample, for example to test the effectiveness of antibacterial soap?

INVESTIGATE Further

CHALLENGE It is hard to tell which products are best for washing hands without testing them. Design an experiment to determine which cleans your hands best: baby wipes, hand sanitizer, regular soap, or antibacterial soap.

Cleaning Your Hands

Table 1. Observations

Petri Dish	Source	Amount of Bacteria
A	hand	
B	hand washed with water	
C	hand washed with soap and water	

Observe and Analyze

1. Dish A should have the most bacterial growth. Dish C should have the least growth.

2. Observations may vary.

Conclude

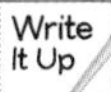

1. Towel drying would remove additional bacteria by rubbing them off, or might add bacteria to the hands from the towel.

2. Pressing the fingers of the right hand to the agar in step 3 removed the bacteria from the fingers and transferred it to the agar.

3. Students should provide evidence or explanation for describing whether their hypotheses are correct or incorrect.

4. Using water alone did remove some bacteria from the hands, but not as much as using water and soap.

5. Hot water would be more effective at killing bacteria than cold water.

6. It would be possible to test the effectiveness of antibacterial soap by using two different fingers from those used in steps 3, 4, or 5.

INVESTIGATE Further

CHALLENGE Students' experiments should include a hypothesis, steps for a procedure, and a method of gathering data. Students may use sterile petri dishes with agar to test the bacteria present on their hands before and after using soap and the antibacterial ingredients.

Post-Lab Discussion

- On the board, keep a running tally of results for each student or group. Have students think about conditions that might cause differences in results. Students can make inferences as to why one surface might contain more bacteria than another.
- Remind students that not all bacteria is harmful. Human beings have bacteria that line the digestive tract to help digest food and keep the digestive system functioning properly.

BACK TO

the BIG idea

Have students choose one of the life stages discussed in the chapter and describe the physical and developmental changes that occur during that stage. *Sample answer: In adulthood the body systems are not increasing in size. The systems continue to function and to repair tissue and to take care of the organism's needs. In adulthood the body can produce and care for off-spring. An adult continues to develop mental and social skills and knowledge.*

KEY CONCEPTS SUMMARY

SECTION 5.1
Ask students what is represented by the data in the graph. *average weight for females, birth to age 15*

SECTION 5.2
Have students describe why it is important to read and understand the nutrition labels that are given on food packaging. *Nutrition labels contain information about the nutrients and vitamins that are present in each serving of food. Labels help a person maintain and be knowledgeable about a healthy diet.*

SECTION 5.3
Have students describe different ways that they can prevent infectious disease. *Sample answers: wash hands, get vaccinations, wear insect repellent, avoid contact with persons who are ill, and maintain a healthy lifestyle by eating the right foods and getting exercise.*

Review Concepts

- Big Idea Flow Chart, p. T33
- Chapter Outline, pp. T39–T40

5 Chapter Review

the BIG idea

The body develops and maintains itself over time.

KEY CONCEPTS SUMMARY

5.1 The human body changes over time.

Your body develops and grows throughout your entire life. Some changes are physical and some are emotional. The stages of life are infancy, childhood, adolescence, adulthood, and later adulthood. All the different systems in the body interact to maintain your health.

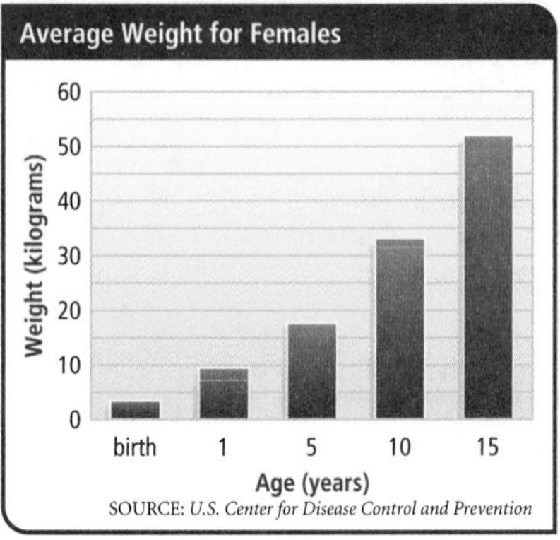

VOCABULARY
infancy p. 134
childhood p. 134
adolescence p. 135
adulthood p. 136

5.2 Diet, exercise, and behaviors affect health.

Your diet affects your health. Important nutrients include proteins, carbohydrates, fats vitamins, minerals, and water. Water is also essential to healthy living. Exercise is the final ingredient to a healthy life. Problems that can interfere with a healthy life are drug abuse, addiction, and eating disorders.

Spaghetti
IN TOMATO SAUCE WITH CHEESE
Nutrition Facts
Serving Size: 1 cup (252g)
Servings Per Container: about 2
Amount Per Serving
Calories 210
Calories from Fat 20
% Daily Value*
Total Fat 2g
3%
Saturated Fat 1g
Cholesterol 5mg

VOCABULARY
nutrition p. 140
addiction p. 146

5.3 Science helps people prevent and treat disease.

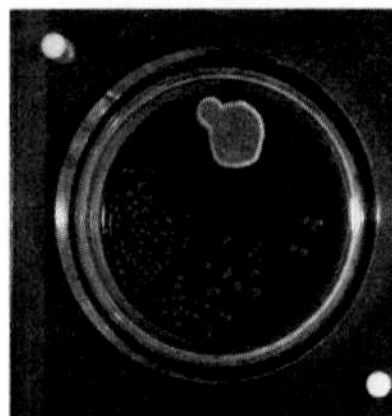

- Science helps people fight disease.
- Antibiotics are used to fight diseases caused by bacteria.
- Infectious disease can spread in many ways including food, air, water, insects, animals, and person-to-person contact.
- Noninfectious diseases are not contagious. Some of these noninfectious diseases are present at birth and others occur in later life.

VOCABULARY
microorganism p. 148
bacteria p. 149
virus p. 149
resistance p. 153

E 156 Unit: Human Biology

Technology Resources

Have students visit **ClassZone.com** or use the CD-ROM for a cumulative review of concepts.

CONTENT REVIEW

CONTENT REVIEW CD-ROM

Engage students in a whole-class interactive review of Key Concepts. Edit content as you wish.

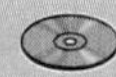

POWER PRESENTATIONS

Reviewing Vocabulary

Make a frame for each of the vocabulary words listed below. Write the word in the center. Decide what information to frame it with. Use definitions, examples, descriptions, parts, or pictures.

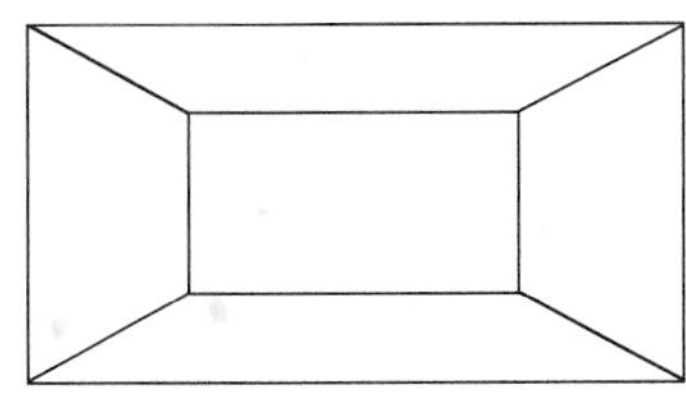

1. infancy
2. childhood
3. adolescence
4. adulthood

Reviewing Key Concepts

Multiple Choice *Choose the letter of the best answer.*

5. The stage of life known as infancy ends when an infant
- **a.** begins to cry
- **b.** learns to walk
- **c.** holds up his head
- **d.** sees more clearly

6. The process in which the body begins to mature sexually is called
- **a.** adolescence
- **b.** adulthood
- **c.** nutrition
- **d.** puberty

7. Which nutrients are sources of energy for the body?
- **a.** fats and carbohydrates
- **b.** water and protein
- **c.** fats and proteins
- **d.** water and carbohydrates

8. Which is *not* a benefit of regular exercise?
- **a.** flushed-out skin pores
- **b.** stronger skeletal system
- **c.** increased body fat
- **d.** strengthened heart

9. A sedentary life style is associated with
- **a.** a stronger immune system
- **b.** more sitting than moving
- **c.** regular exercise
- **d.** an eating disorder

10. The chemical found in tobacco that increases heart rate and blood pressure is
- **a.** cocaine
- **b.** carbon monoxide
- **c.** tar
- **d.** nicotine

11. Which term includes all of the others?
- **a.** bacteria
- **b.** germ
- **c.** virus
- **d.** pathogen

12. An example of a disease caused by bacteria is
- **a.** an ear infection
- **b.** stomach flu
- **c.** chicken pox
- **d.** a cold

13. Which statement about viruses is true?
- **a.** Viruses function to digest food.
- **b.** Viruses are one-celled organisms.
- **c.** Viruses are not living.
- **d.** Examples of viruses are fungi and yeasts.

14. A substance that can destroy bacteria is called
- **a.** a virus
- **b.** a pathogen
- **c.** an antibiotic
- **d.** a mold

15. Lyme disease is spread through
- **a.** drinking unfiltered water
- **b.** uncooked meats
- **c.** the bite of a dog
- **d.** the bite of a tick

Short Answer *Write a short answer to each question.*

16. In your own words, define *nutrition.*

17. What are pathogens? Give three examples.

18. Explain what happens if antibiotics are used too often.

ASSESSMENT RESOURCES

UNIT ASSESSMENT BOOK
- Chapter Test, Level A, pp. 82–85
- Chapter Test, Level B, pp. 86–89
- Chapter Test, Level C, pp. 90–93
- Alternative Assessment, pp. 94–95
- Unit Test, pp. 96–107

SPANISH ASSESSMENT BOOK
- Spanish Chapter Test, pp. 97–100
- Spanish Unit Test, pp. 101–104

Technology Resources

Edit test items and answer choices.

Test Generator CD-ROM

Visit **ClassZone.com** to extend test practice.

Test Practice

Reviewing Vocabulary

Sample answers:

1. infancy—rapid physical development; infants learn to roll over, sit, crawl, stand, and walk; development of thinking and social skills; much body system growth and change

2. childhood—period between infancy and adolescence; children learn to care for themselves; mental and social growth—learn to read, write, and communicate

3. adolescence—period between childhood and adulthood; puberty occurs; period of physical growth in boys and girls (as much as 10 cm)

4. adulthood—completes growth and reaches sexual maturity; mental and social development continue throughout life; mental and emotional maturity

Reviewing Key Concepts

5. b
6. d
7. a
8. c
9. b
10. d
11. d
12. a
13. c
14. c
15. d

16. Nutrition is the study of the nutrients, such as proteins and carbohydrates, that keep your body healthy and how these nutrients are used for maintenance, growth, and reproduction.

17. Pathogens are agents that cause disease. Examples may include bacteria, viruses, yeasts, fungi, and protists.

18. When an antibiotic is used too often, bacteria can develop resistance to it.

Thinking Critically

19. Infants learn that people respond to their crying when they are hungry or have other needs.

20. Sample answer: Physical: development of a deeper voice; Mental: learning to communicate better with others; Social: ability to cooperate with others

21. Fats provide energy that helps the body absorb vitamins and form parts of cells. A no-fat diet might result in not enough energy, being unable to absorb vitamins, and improper cell function.

22. Lack of activity causes muscles and bones to break down. Unused energy from food is stored as fat; excess fat can cause health problems.

23. Alcohol damages heart, liver, and body systems. Tobacco affects lungs, heart rate, and blood pressure. The fetus is connected to the mother's circulatory system; alcohol and tobacco could affect the fetus.

24. Both are eating disorders. Anorexia: eat so little they become unhealthy and can die; Bulimia: overeat, then force themselves to vomit

25. Pasteur showed that tiny organisms caused food decay. Pasteur's germ theory stated that some diseases are caused by germs.

26. The disease probably came from water contaminated by a pathogen. Boiling water would kill pathogens.

27. Eat nutritious foods, get exercise, stay out of the Sun, avoid tobacco and alcohol use.

the BIG idea

28. Answers will depend on students' original answers to the Big Idea question.

29. Answers should include concepts from the unit and terminology from the box.

UNIT PROJECTS

Have students present their projects. Use the appropriate rubrics from the URB to evaluate their work.

 Unit Projects, pp. 5–10

Thinking Critically

19. **ANALYZE** Why do you think crying is an example of a social skill that develops during infancy?

20. **ANALYZE** Describe one physical, one mental, and one social change that a ten-year-old boy might experience over the next five years.

21. **EVALUATE** Explain why a diet that doesn't contain any fat would be unhealthy for most people?

22. **APPLY** Explain why people who live sedentary lifestyles should get more exercise.

23. **SYNTHESIZE** Discuss why doctors recommend that women avoid alcohol and tobacco use during pregnancy.

24. **COMPARE AND CONTRAST** How are anorexia and bulimia alike? How do they differ?

25. **ANALYZE** Explain why the work of Louis Pasteur was important in the understanding of infectious disease.

26. **HYPOTHESIZE** In 1854 a disease called cholera spread through the city of London. Most of the people who contracted the disease lived near the city's various water pumps. What might you hypothesize about the cause of the disease? How could you prevent people from contracting the disease in the future?

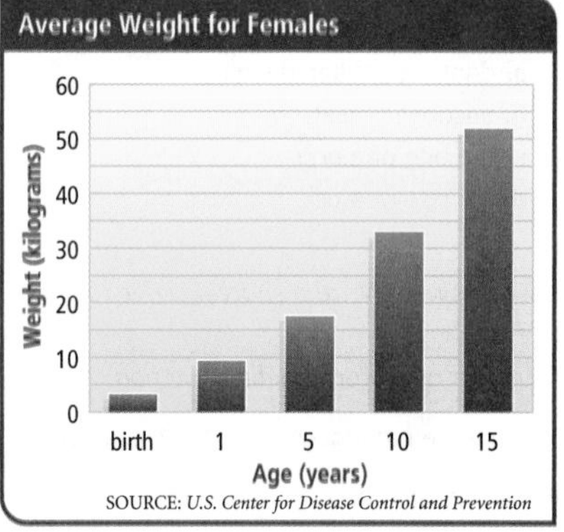

27. **PROVIDE EXAMPLES** What are some ways that a person can prevent noninfectious diseases such as cancer or diabetes?

the BIG idea

28. **INFER** Look again at the picture on pages 130–131. Now that you have finished the chapter, how would you change or add details to your answer to the question on the photograph?

29. **SUMMARIZE** Write one or more paragraphs explaining how lifestyle can lead to a healthy body and a longer life. Include these terms in your description.

nutrition	alcohol
exercise	infectious disease
germs	noninfectious disease
tobacco	

UNIT PROJECTS

Evaluate all the data, results, and information from your project folder. Prepare to present your project.

MONITOR AND RETEACH

If students are having trouble applying the concepts in Chapter Review items 19 and 20, suggest that they review the stages of life on pp. 134–135. Then have students create a concept map detailing the events of each stage.

Students may benefit from summarizing one or more sections of the chapter.

 Summarizing the Chapter, pp. 303–304

Standardized Test Practice

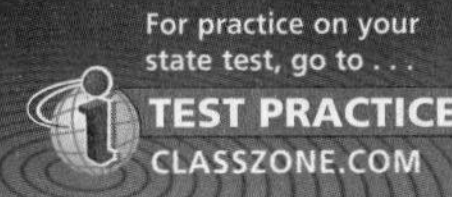
For practice on your state test, go to . . .
TEST PRACTICE
CLASSZONE.COM

Analyzing Data

The table below presents information about causes of death in the United States.

Leading Causes of Death in the United States

2000		1900	
Cause of Death	**Percent of Deaths**	**Cause of Death**	**Percent of Deaths**
heart disease	31%	pneumonia*	12%
cancer	23%	tuberculosis*	11%
stroke	9%	diarrhea*	11%
lung disease	5%	heart disease	6%
accident	4%	liver disease	5%
pneumonia*	4%	accident	4%
diabetes	3%	cancer	4%
kidney disease	1%	senility	2%
liver disease	1%	diphtheria*	2%

* infectious disease

Use the table to answer the questions below.

1. What was the leading cause of death in 1900?
   a. heart disease **c.** pneumonia
   b. cancer **d.** tuberculosis

2. Which infectious disease was a leading cause of death both in 1900 and 2000?
   a. tuberculosis **c.** stroke
   b. diphtheria **d.** pneumonia

3. Which was the leading noninfectious cause of death in both 1900 and 2000?
   a. pneumonia **c.** heart disease
   b. cancer **d.** accidents

4. Which cause of death showed the greatest increase between 1900 and 2000?
   a. heart disease **c.** liver disease
   b. cancer **d.** pneumonia

5. Which statement is true?
   a. The rate of infectious disease as a leading cause of death increased from 1900 to 2000.
   b. The rate of infectious disease as a leading cause of death decreased from 1900 to 2000.
   c. The rate of noninfectious disease as a leading cause of death decreased from 1900 to 2000.
   d. The rate of noninfectious disease as a leading cause of death remained the same.

6. How much did the rate of heart disease increase from 1900 to 2000?
   a. 37% **c.** 25%
   b. 31% **d.** 6%

Extended Response

bacterium	virus	pathogen
antibiotic	resistance	microorganism

7. Write a paragraph explaining the change in the number of deaths due to infectious diseases from 1900 to 2000. Use the information in the data and what you know about infectious disease in your description. Use the vocabulary words in the box in your answer.

8. The spread of infectious disease can be controlled in many different ways. Write a paragraph describing how the spread of infectious disease may be limited. Give at least two examples and describe how these diseases can be prevented or contained.

METACOGNITIVE ACTIVITY

Have students answer the following questions in their **Science Notebook:**

1. Are there any concepts in this chapter that you are still unclear about? Explain.
2. Is there anything else that you would like to know about how human beings grow and develop? Explain.
3. How does working in a group help you learn? Give an example from this chapter.

Analyzing Data

1. c *3. c* *5. b*
2. d *4. a* *6. c*

Extended Response

7. RUBRIC

4 points for a response that correctly answers the question and includes at least four of the following terms:

- bacterium
- antibiotic
- virus
- resistance
- pathogen
- microorganism

Sample answer: The discovery of pathogens such as bacteria and viruses led people to a greater understanding of disease and how people become ill. Antibiotics such as penicillin were developed to fight infectious diseases, so people have a better chance of surviving. Improvement in sanitation such as the treatment of water has limited the spread of disease. Vaccinations also help to prevent the spread of many infectious diseases.

3 points correctly answers the question and includes at least three of the terms shown
2 points correctly answers the question and includes at least two of the terms shown
1 point correctly answers the question and includes at least one term shown

8. RUBRIC

4 points for a response that includes two examples and describes how both diseases can be prevented.

Sample answer: Infectious diseases are caused by pathogens such as bacteria and viruses. Diseases caused by bacteria can enter the body through food and water. Washing fruits and vegetables and cooking meats thoroughly can help prevent the spread of disease. Boiling water that may be contaminated is another way to prevent disease. Some infectious diseases like rabies are transmitted by certain animals and are prevented by avoiding contact with these animals.

3 points includes two examples and describes how one disease can be prevented
2 points includes one example and describes how that disease can be prevented
1 point includes two examples, but no description of how the disease can be prevented

Student Resource Handbooks

SCIENTIFIC THINKING HANDBOOK

Making Observations

An **observation** is an act of noting and recording an event, characteristic, behavior, or anything else detected with an instrument or with the senses.

Observations allow you to make informed hypotheses and to gather data for experiments. Careful observations often lead to ideas for new experiments. There are two categories of observations:

- **Quantitative observations** can be expressed in numbers and include records of time, temperature, mass, distance, and volume.
- **Qualitative observations** include descriptions of sights, sounds, smells, and textures.

EXAMPLE

A student dissolved 30 grams of Epsom salts in water, poured the solution into a dish, and let the dish sit out uncovered overnight. The next day, she made the following observations of the Epsom salt crystals that grew in the dish.

To determine the mass, the student found the mass of the dish before and after growing the crystals and then used subtraction to find the difference.

The student measured several crystals and calculated the mean length. (To learn how to calculate the mean of a data set, see page R36.)

Table 1. Observations of Epsom Salt Crystals

Quantitative Observations	Qualitative Observations
• mass = 30 g • mean crystal length = 0.5 cm • longest crystal length = 2 cm	• Crystals are clear. • Crystals are long, thin, and rectangular. • White crust has formed around edge of dish.  Epsom salt crystals

Photographs or sketches are useful for recording qualitative observations.

MORE ABOUT OBSERVING

- Make quantitative observations whenever possible. That way, others will know exactly what you observed and be able to compare their results with yours.
- It is always a good idea to make qualitative observations too. You never know when you might observe something unexpected.

Predicting and Hypothesizing

A **prediction** is an expectation of what will be observed or what will happen. A **hypothesis** is a tentative explanation for an observation or scientific problem that can be tested by further investigation.

SCIENTIFIC THINKING HANDBOOK

EXAMPLE

Suppose you have made two paper airplanes and you wonder why one of them tends to glide farther than the other one.

1. Start by asking a question.
2. Make an educated guess. After examination, you notice that the wings of the airplane that flies farther are slightly larger than the wings of the other airplane.
3. Write a prediction based upon your educated guess, in the form of an "If . . . , then . . ." statement. Write the independent variable after the word *if,* and the dependent variable after the word *then.*
4. To make a hypothesis, explain why you think what you predicted will occur. Write the explanation after the word *because.*

1. Why does one of the paper airplanes glide farther than the other?

2. The size of an airplane's wings may affect how far the airplane will glide.

3. Prediction: If I make a paper airplane with larger wings, then the airplane will glide farther.

To read about independent and dependent variables, see page R30.

4. Hypothesis: If I make a paper airplane with larger wings, then the airplane will glide farther, because the additional surface area of the wing will produce more lift.

Notice that the part of the hypothesis after *because* adds an explanation of why the airplane will glide farther.

MORE ABOUT HYPOTHESES

- The results of an experiment cannot prove that a hypothesis is correct. Rather, the results either support or do not support the hypothesis.
- Valuable information is gained even when your hypothesis is not supported by your results. For example, it would be an important discovery to find that wing size is not related to how far an airplane glides.
- In science, a hypothesis is supported only after many scientists have conducted many experiments and produced consistent results.

Inferring

An **inference** is a logical conclusion drawn from the available evidence and prior knowledge. Inferences are often made from observations.

SCIENTIFIC THINKING HANDBOOK

EXAMPLE

A student observing a set of acorns noticed something unexpected about one of them. He noticed a white, soft-bodied insect eating its way out of the acorn.

The student recorded these observations.

Observations

- There is a hole in the acorn, about 0.5 cm in diameter, where the insect crawled out.
- There is a second hole, which is about the size of a pinhole, on the other side of the acorn.
- The inside of the acorn is hollow.

Here are some inferences that can be made on the basis of the observations.

Inferences

- The insect formed from the material inside the acorn, grew to its present size, and ate its way out of the acorn.
- The insect crawled through the smaller hole, ate the inside of the acorn, grew to its present size, and ate its way out of the acorn.
- An egg was laid in the acorn through the smaller hole. The egg hatched into a larva that ate the inside of the acorn, grew to its present size, and ate its way out of the acorn.

When you make inferences, be sure to look at all of the evidence available and combine it with what you already know.

MORE ABOUT INFERENCES

Inferences depend both on observations and on the knowledge of the people making the inferences. Ancient people who did not know that organisms are produced only by similar organisms might have made an inference like the first one. A student today might look at the same observations and make the second inference. A third student might have knowledge about this particular insect and know that it is never small enough to fit through the smaller hole, leading her to the third inference.

Identifying Cause and Effect

In a **cause-and-effect relationship,** one event or characteristic is the result of another. Usually an effect follows its cause in time.

There are many examples of cause-and-effect relationships in everyday life.

Cause	Effect
Turn off a light.	Room gets dark.
Drop a glass.	Glass breaks.
Blow a whistle.	Sound is heard.

Scientists must be careful not to infer a cause-and-effect relationship just because one event happens after another event. When one event occurs after another, you cannot infer a cause-and-effect relationship on the basis of that information alone. You also cannot conclude that one event caused another if there are alternative ways to explain the second event. A scientist must demonstrate through experimentation or continued observation that an event was truly caused by another event.

EXAMPLE

Make an Observation

Suppose you have a few plants growing outside. When the weather starts getting colder, you bring one of the plants indoors. You notice that the plant you brought indoors is growing faster than the others are growing. You cannot conclude from your observation that the change in temperature was the cause of the increased plant growth, because there are alternative explanations for the observation. Some possible explanations are given below.

- The humidity indoors caused the plant to grow faster.
- The level of sunlight indoors caused the plant to grow faster.
- The indoor plant's being noticed more often and watered more often than the outdoor plants caused it to grow faster.
- The plant that was brought indoors was healthier than the other plants to begin with.

To determine which of these factors, if any, caused the indoor plant to grow faster than the outdoor plants, you would need to design and conduct an experiment.

See pages R28–R35 for information about designing experiments.

Recognizing Bias

Television, newspapers, and the Internet are full of experts claiming to have scientific evidence to back up their claims. How do you know whether the claims are really backed up by good science?

Bias is a slanted point of view, or personal prejudice. The goal of scientists is to be as objective as possible and to base their findings on facts instead of opinions. However, bias often affects the conclusions of researchers, and it is important to learn to recognize bias.

When scientific results are reported, you should consider the source of the information as well as the information itself. It is important to critically analyze the information that you see and read.

SOURCES OF BIAS

There are several ways in which a report of scientific information may be biased. Here are some questions that you can ask yourself:

1. Who is sponsoring the research?

   Sometimes, the results of an investigation are biased because an organization paying for the research is looking for a specific answer. This type of bias can affect how data are gathered and interpreted.

2. Is the research sample large enough?

   Sometimes research does not include enough data. The larger the sample size, the more likely that the results are accurate, assuming a truly random sample.

3. In a survey, who is answering the questions?

   The results of a survey or poll can be biased. The people taking part in the survey may have been specifically chosen because of how they would answer. They may have the same ideas or lifestyles. A survey or poll should make use of a random sample of people.

4. Are the people who take part in a survey biased?

   People who take part in surveys sometimes try to answer the questions the way they think the researcher wants them to answer. Also, in surveys or polls that ask for personal information, people may be unwilling to answer questions truthfully.

SCIENTIFIC BIAS

It is also important to realize that scientists have their own biases because of the types of research they do and because of their scientific viewpoints. Two scientists may look at the same set of data and come to completely different conclusions because of these biases. However, such disagreements are not necessarily bad. In fact, a critical analysis of disagreements is often responsible for moving science forward.

Identifying Faulty Reasoning

Faulty reasoning is wrong or incorrect thinking. It leads to mistakes and to wrong conclusions. Scientists are careful not to draw unreasonable conclusions from experimental data. Without such caution, the results of scientific investigations may be misleading.

EXAMPLE

Scientists try to make generalizations based on their data to explain as much about nature as possible. If only a small sample of data is looked at, however, a conclusion may be faulty. Suppose a scientist has studied the effects of the El Niño and La Niña weather patterns on flood damage in California from 1989 to 1995. The scientist organized the data in the bar graph below.

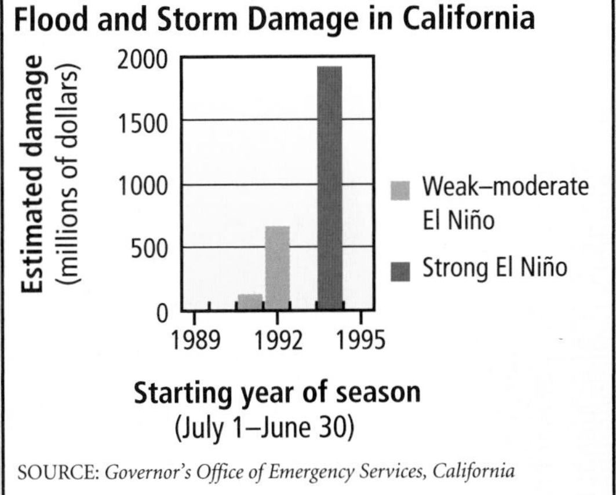

The scientist drew the following conclusions:

1. The La Niña weather pattern has no effect on flooding in California.
2. When neither weather pattern occurs, there is almost no flood damage.
3. A weak or moderate El Niño produces a small or moderate amount of flooding.
4. A strong El Niño produces a lot of flooding.

For the six-year period of the scientist's investigation, these conclusions may seem to be reasonable. However, a six-year study of weather patterns may be too small of a sample for the conclusions to be supported. Consider the following graph, which shows information that was gathered from 1949 to 1997.

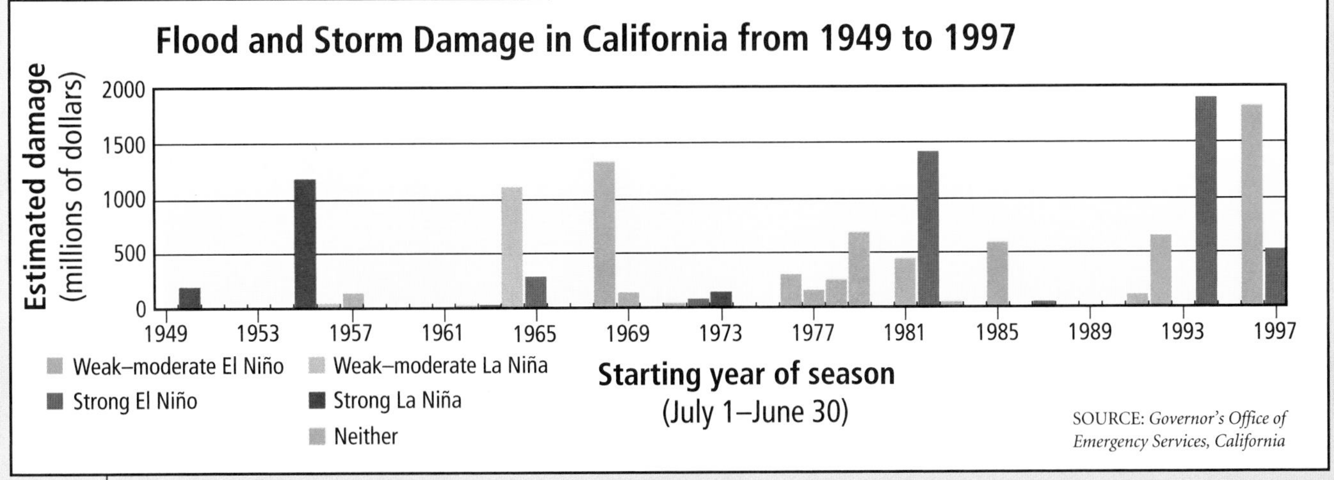

The only one of the conclusions that all of this information supports is number 3: a weak or moderate El Niño produces a small or moderate amount of flooding. By collecting more data, scientists can be more certain of their conclusions and can avoid faulty reasoning.

SCIENTIFIC THINKING HANDBOOK

Analyzing Statements

To **analyze** a statement is to examine its parts carefully. Scientific findings are often reported through media such as television or the Internet. A report that is made public often focuses on only a small part of research. As a result, it is important to question the sources of information.

Evaluate Media Claims

To **evaluate** a statement is to judge it on the basis of criteria you've established. Sometimes evaluating means deciding whether a statement is true.

Reports of scientific research and findings in the media may be misleading or incomplete. When you are exposed to this information, you should ask yourself some questions so that you can make informed judgments about the information.

1. **Does the information come from a credible source?**

   Suppose you learn about a new product and it is stated that scientific evidence proves that the product works. A report from a respected news source may be more believable than an advertisement paid for by the product's manufacturer.

2. **How much evidence supports the claim?**

   Often, it may seem that there is new evidence every day of something in the world that either causes or cures an illness. However, information that is the result of several years of work by several different scientists is more credible than an advertisement that does not even cite the subjects of the experiment.

3. **How much information is being presented?**

   Science cannot solve all questions, and scientific experiments often have flaws. A report that discusses problems in a scientific study may be more believable than a report that addresses only positive experimental findings.

4. **Is scientific evidence being presented by a specific source?**

   Sometimes scientific findings are reported by people who are called experts or leaders in a scientific field. But if their names are not given or their scientific credentials are not reported, their statements may be less credible than those of recognized experts.

Differentiate Between Fact and Opinion

Sometimes information is presented as a fact when it may be an opinion. When scientific conclusions are reported, it is important to recognize whether they are based on solid evidence. Again, you may find it helpful to ask yourself some questions.

1. **What is the difference between a fact and an opinion?**

   A **fact** is a piece of information that can be strictly defined and proved true. An **opinion** is a statement that expresses a belief, value, or feeling. An opinion cannot be proved true or false. For example, a person's age is a fact, but if someone is asked how old they feel, it is impossible to prove the person's answer to be true or false.

2. **Can opinions be measured?**

   Yes, opinions can be measured. In fact, surveys often ask for people's opinions on a topic. But there is no way to know whether or not an opinion is the truth.

HOW TO DIFFERENTIATE FACT FROM OPINION

Human Activities and the Environment

Unfortunately, human use of fossil fuels is one of the most significant developments of the past few centuries. Humans rely on fossil fuels, a non-renewable energy resource, for more than 90 percent of their energy needs.

This careless misuse of our planet's resources has resulted in pollution, global warming, and the destruction of fragile ecosystems. For example, oil pipelines carry more than one million barrels of oil each day across tundra regions. Transporting oil across such areas can only result in oil spills that poison the land for decades.

Opinions
Notice words or phrases that express beliefs or feelings. The words *unfortunately* and *careless* show that opinions are being expressed.

Opinion
Look for statements that speculate about events. These statements are opinions, because they cannot be proved.

Facts
Statements that contain statistics tend to be facts. Writers often use facts to support their opinions.

Lab Handbook

LAB HANDBOOK

Safety Rules

Before you work in the laboratory, read these safety rules twice. Ask your teacher to explain any rules that you do not completely understand. Refer to these rules later on if you have questions about safety in the science classroom.

Directions

- Read all directions and make sure that you understand them before starting an investigation or lab activity. If you do not understand how to do a procedure or how to use a piece of equipment, ask your teacher.
- Do not begin any investigation or touch any equipment until your teacher has told you to start.
- Never experiment on your own. If you want to try a procedure that the directions do not call for, ask your teacher for permission first.
- If you are hurt or injured in any way, tell your teacher immediately.

Dress Code

goggles

apron

gloves

- Wear goggles when
  - — using glassware, sharp objects, or chemicals
  - — heating an object
  - — working with anything that can easily fly up into the air and hurt someone's eye
- Tie back long hair or hair that hangs in front of your eyes.
- Remove any article of clothing—such as a loose sweater or a scarf—that hangs down and may touch a flame, chemical, or piece of equipment.
- Observe all safety icons calling for the wearing of eye protection, gloves, and aprons.

Heating and Fire Safety

fire safety

heating safety

- Keep your work area neat, clean, and free of extra materials.
- Never reach over a flame or heat source.
- Point objects being heated away from you and others.
- Never heat a substance or an object in a closed container.
- Never touch an object that has been heated. If you are unsure whether something is hot, treat it as though it is. Use oven mitts, clamps, tongs, or a test-tube holder.
- Know where the fire extinguisher and fire blanket are kept in your classroom.
- Do not throw hot substances into the trash. Wait for them to cool or use the container your teacher puts out for disposal.

Electrical Safety

electrical safety

- Never use lamps or other electrical equipment with frayed cords.
- Make sure no cord is lying on the floor where someone can trip over it.
- Do not let a cord hang over the side of a counter or table so that the equipment can easily be pulled or knocked to the floor.
- Never let cords hang into sinks or other places where water can be found.
- Never try to fix electrical problems. Inform your teacher of any problems immediately.
- Unplug an electrical cord by pulling on the plug, not the cord.

Chemical Safety

chemical safety

poison

fumes

- If you spill a chemical or get one on your skin or in your eyes, tell your teacher right away.
- Never touch, taste, or sniff any chemicals in the lab. If you need to determine odor, waft. Wafting consists of holding the chemical in its container 15 centimeters (6 in.) away from your nose, and using your fingers to bring fumes from the container to your nose.
- Keep lids on all chemicals you are not using.
- Never put unused chemicals back into the original containers. Throw away extra chemicals where your teacher tells you to.
- Pour chemicals over a sink or your work area, not over the floor.
- If you get a chemical in your eye, use the eyewash right away.
- Always wash your hands after handling chemicals, plants, or soil.

Wafting

Glassware and Sharp-Object Safety

sharp objects

- If you break glassware, tell your teacher right away.
- Do not use broken or chipped glassware. Give these to your teacher.
- Use knives and other cutting instruments carefully. Always wear eye protection and cut away from you.

Animal Safety

- Never hurt an animal.
- Touch animals only when necessary. Follow your teacher's instructions for handling animals.
- Always wash your hands after working with animals.

Cleanup

disposal

- Follow your teacher's instructions for throwing away or putting away supplies.
- Clean your work area and pick up anything that has dropped to the floor.
- Wash your hands.

Using Lab Equipment

Different experiments require different types of equipment. But even though experiments differ, the ways in which the equipment is used are the same.

Beakers

- Use beakers for holding and pouring liquids.
- Do not use a beaker to measure the volume of a liquid. Use a graduated cylinder instead. (See page R16.)
- Use a beaker that holds about twice as much liquid as you need. For example, if you need 100 milliliters of water, you should use a 200- or 250-milliliter beaker.

Test Tubes

- Use test tubes to hold small amounts of substances.
- Do not use a test tube to measure the volume of a liquid.
- Use a test tube when heating a substance over a flame. Aim the mouth of the tube away from yourself and other people.
- Liquids easily spill or splash from test tubes, so it is important to use only small amounts of liquids.

Test-Tube Holder

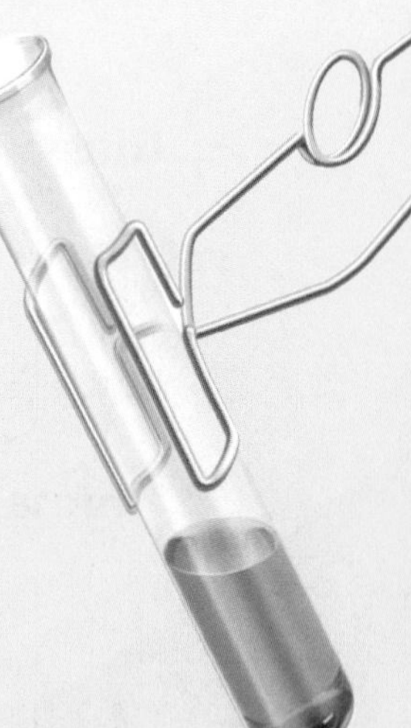

- Use a test-tube holder when heating a substance in a test tube.
- Use a test-tube holder if the substance in a test tube is dangerous to touch.
- Make sure the test-tube holder tightly grips the test tube so that the test tube will not slide out of the holder.
- Make sure that the test-tube holder is above the surface of the substance in the test tube so that you can observe the substance.

LAB HANDBOOK

R12

Test-Tube Rack

- Use a test-tube rack to organize test tubes before, during, and after an experiment.
- Use a test-tube rack to keep test tubes upright so that they do not fall over and spill their contents.
- Use a test-tube rack that is the correct size for the test tubes that you are using. If the rack is too small, a test tube may become stuck. If the rack is too large, a test tube may lean over, and some of its contents may spill or splash.

Forceps

- Use forceps when you need to pick up or hold a very small object that should not be touched with your hands.
- Do not use forceps to hold anything over a flame, because forceps are not long enough to keep your hand safely away from the flame. Plastic forceps will melt, and metal forceps will conduct heat and burn your hand.

Hot Plate

- Use a hot plate when a substance needs to be kept warmer than room temperature for a long period of time.
- Use a hot plate instead of a Bunsen burner or a candle when you need to carefully control temperature.
- Do not use a hot plate when a substance needs to be burned in an experiment.
- Always use "hot hands" safety mitts or oven mitts when handling anything that has been heated on a hot plate.

Microscope

Scientists use microscopes to see very small objects that cannot easily be seen with the eye alone. A microscope magnifies the image of an object so that small details may be observed. A microscope that you may use can magnify an object 400 times—the object will appear 400 times larger than its actual size.

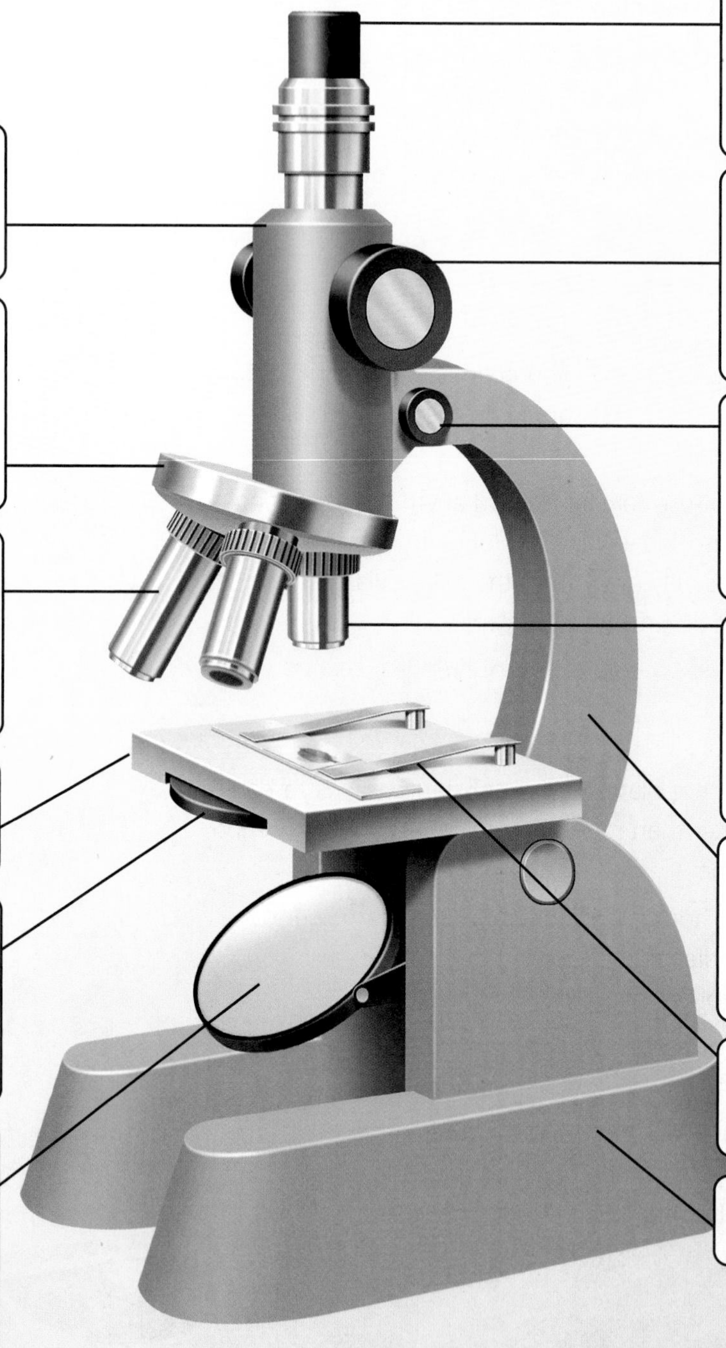

Eyepiece Objects are viewed through the eyepiece. The eyepiece contains a lens that commonly magnifies an image 10 times.

Body The body separates the lens in the eyepiece from the objective lenses below.

Coarse Adjustment This knob is used to focus the image of an object when it is viewed through the low-power lens.

Nosepiece The nosepiece holds the objective lenses above the stage and rotates so that all lenses may be used.

Fine Adjustment This knob is used to focus the image of an object when it is viewed through the high-power lens.

High-Power Objective Lens This is the largest lens on the nosepiece. It magnifies an image approximately 40 times.

Low-Power Objective Lens This is the smallest lens on the nosepiece. It magnifies an image approximately 10 times.

Stage The stage supports the object being viewed.

Arm The arm supports the body above the stage. Always carry a microscope by the arm and base.

Diaphragm The diaphragm is used to adjust the amount of light passing through the slide and into an objective lens.

Stage Clip The stage clip holds a slide in place on the stage.

Mirror or Light Source Some microscopes use light that is reflected through the stage by a mirror. Other microscopes have their own light sources.

Base The base supports the microscope.

LAB HANDBOOK

VIEWING AN OBJECT

1. Use the coarse adjustment knob to raise the body tube.
2. Adjust the diaphragm so that you can see a bright circle of light through the eyepiece.
3. Place the object or slide on the stage. Be sure that it is centered over the hole in the stage.
4. Turn the nosepiece to click the low-power lens into place.
5. Using the coarse adjustment knob, slowly lower the lens and focus on the specimen being viewed. Be sure not to touch the slide or object with the lens.
6. When switching from the low-power lens to the high-power lens, first raise the body tube with the coarse adjustment knob so that the high-power lens will not hit the slide.
7. Turn the nosepiece to click the high-power lens into place.
8. Use the fine adjustment knob to focus on the specimen being viewed. Again, be sure not to touch the slide or object with the lens.

MAKING A SLIDE, OR WET MOUNT

1 Place the specimen in the center of a clean slide.

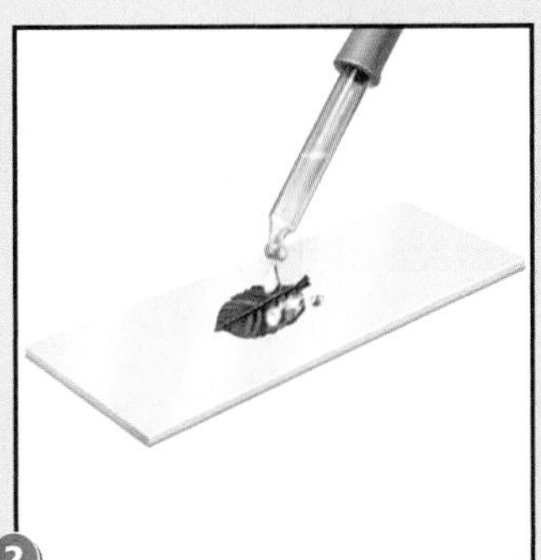

2 Place a drop of water on the specimen.

3 Place a cover slip on the slide. Put one edge of the cover slip into the drop of water and slowly lower it over the specimen.

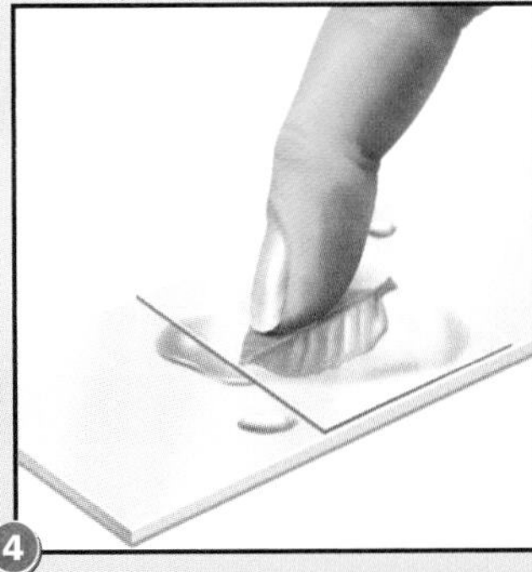

4 Remove any air bubbles from under the cover slip by gently tapping the cover slip.

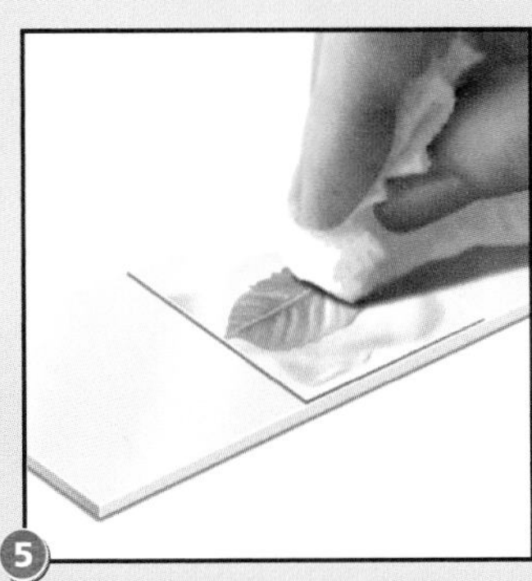

5 Dry any excess water before placing the slide on the microscope stage for viewing.

Spring Scale (Force Meter)

- Use a spring scale to measure a force pulling on the scale.
- Use a spring scale to measure the force of gravity exerted on an object by Earth.
- To measure a force accurately, a spring scale must be zeroed before it is used. The scale is zeroed when no weight is attached and the indicator is positioned at zero.
- Do not attach a weight that is either too heavy or too light to a spring scale. A weight that is too heavy could break the scale or exert too great a force for the scale to measure. A weight that is too light may not exert enough force to be measured accurately.

Graduated Cylinder

- Use a graduated cylinder to measure the volume of a liquid.
- Be sure that the graduated cylinder is on a flat surface so that your measurement will be accurate.
- When reading the scale on a graduated cylinder, be sure to have your eyes at the level of the surface of the liquid.
- The surface of the liquid will be curved in the graduated cylinder. Read the volume of the liquid at the bottom of the curve, or meniscus (muh-NIHS-kuhs).
- You can use a graduated cylinder to find the volume of a solid object by measuring the increase in a liquid's level after you add the object to the cylinder.

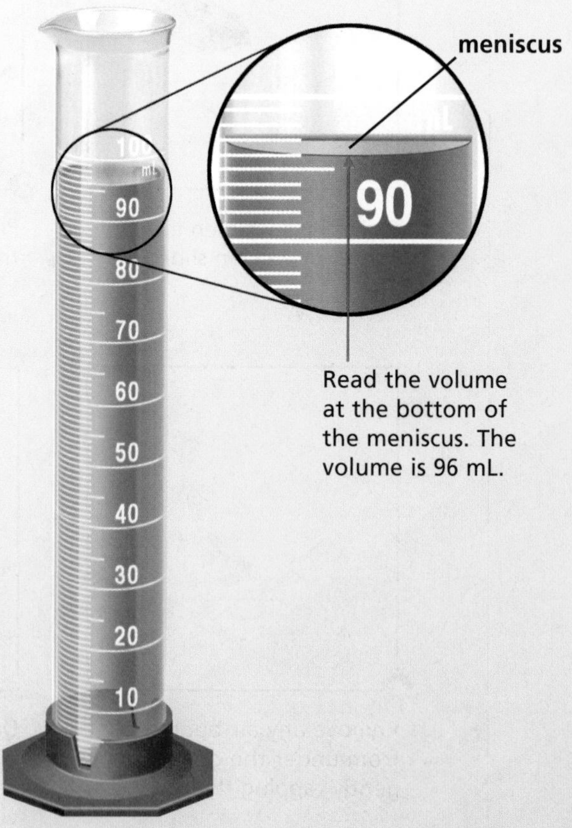

Read the volume at the bottom of the meniscus. The volume is 96 mL.

Metric Rulers

- Use metric rulers or meter sticks to measure objects' lengths.
- Do not measure an object from the end of a metric ruler or meter stick, because the end is often imperfect. Instead, measure from the 1-centimeter mark, but remember to subtract a centimeter from the apparent measurement.
- Estimate any lengths that extend between marked units. For example, if a meter stick shows centimeters but not millimeters, you can estimate the length that an object extends between centimeter marks to measure it to the nearest millimeter.
- **Controlling Variables** If you are taking repeated measurements, always measure from the same point each time. For example, if you're measuring how high two different balls bounce when dropped from the same height, measure both bounces at the same point on the balls—either the top or the bottom. Do not measure at the top of one ball and the bottom of the other.

EXAMPLE

How to Measure a Leaf

1. Lay a ruler flat on top of the leaf so that the 1-centimeter mark lines up with one end. Make sure the ruler and the leaf do not move between the time you line them up and the time you take the measurement.
2. Look straight down on the ruler so that you can see exactly how the marks line up with the other end of the leaf.
3. Estimate the length by which the leaf extends beyond a marking. For example, the leaf below extends about halfway between the 4.2-centimeter and 4.3-centimeter marks, so the apparent measurement is about 4.25 centimeters.
4. Remember to subtract 1 centimeter from your apparent measurement, since you started at the 1-centimeter mark on the ruler and not at the end. The leaf is about 3.25 centimeters long (4.25 cm – 1 cm = 3.25 cm).

LAB HANDBOOK

Triple-Beam Balance

This balance has a pan and three beams with sliding masses, called riders. At one end of the beams is a pointer that indicates whether the mass on the pan is equal to the masses shown on the beams.

1. Make sure the balance is zeroed before measuring the mass of an object. The balance is zeroed if the pointer is at zero when nothing is on the pan and the riders are at their zero points. Use the adjustment knob at the base of the balance to zero it.
2. Place the object to be measured on the pan.
3. Move the riders one notch at a time away from the pan. Begin with the largest rider. If moving the largest rider one notch brings the pointer below zero, begin measuring the mass of the object with the next smaller rider.
4. Change the positions of the riders until they balance the mass on the pan and the pointer is at zero. Then add the readings from the three beams to determine the mass of the object.

300 g	position of largest rider
90 g	position of middle rider
+ 3 g	position of smallest rider
393 g	mass of beaker

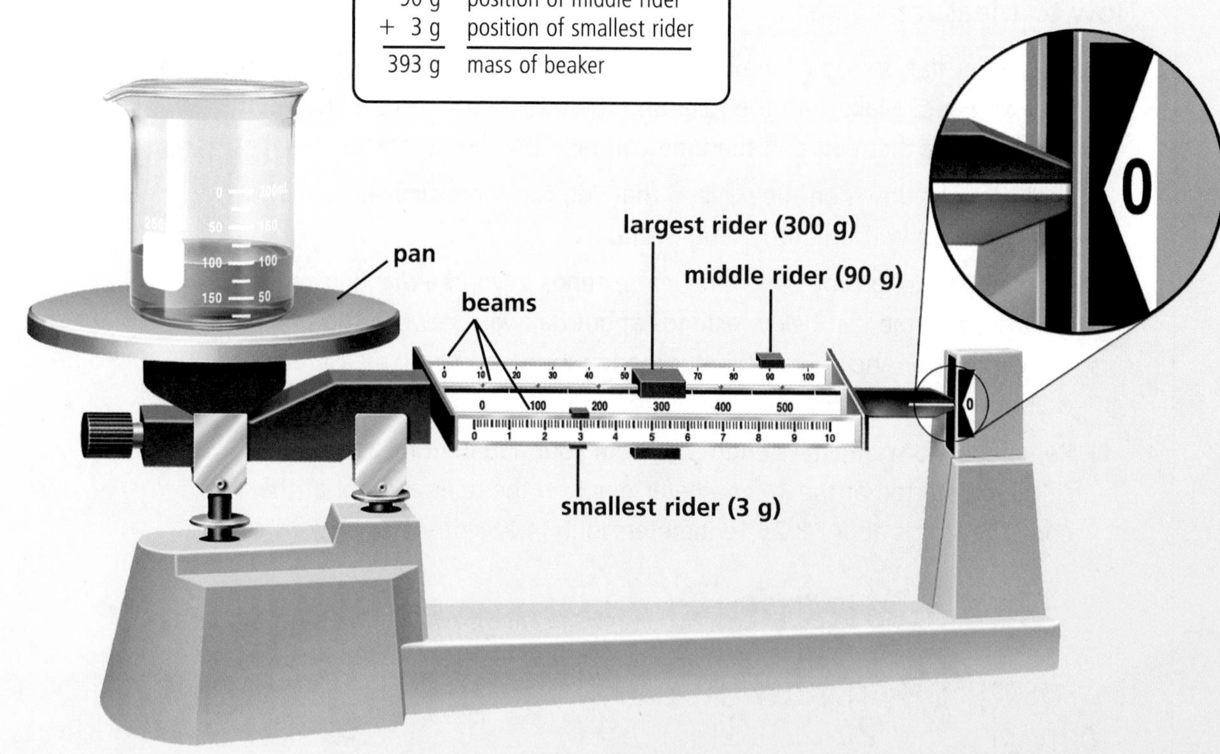

Double-Pan Balance

This type of balance has two pans. Between the pans is a pointer that indicates whether the masses on the pans are equal.

1. Make sure the balance is zeroed before measuring the mass of an object. The balance is zeroed if the pointer is at zero when there is nothing on either of the pans. Many double-pan balances have sliding knobs that can be used to zero them.
2. Place the object to be measured on one of the pans.
3. Begin adding standard masses to the other pan. Begin with the largest standard mass. If this adds too much mass to the balance, begin measuring the mass of the object with the next smaller standard mass.
4. Add standard masses until the masses on both pans are balanced and the pointer is at zero. Then add the standard masses together to determine the mass of the object being measured.

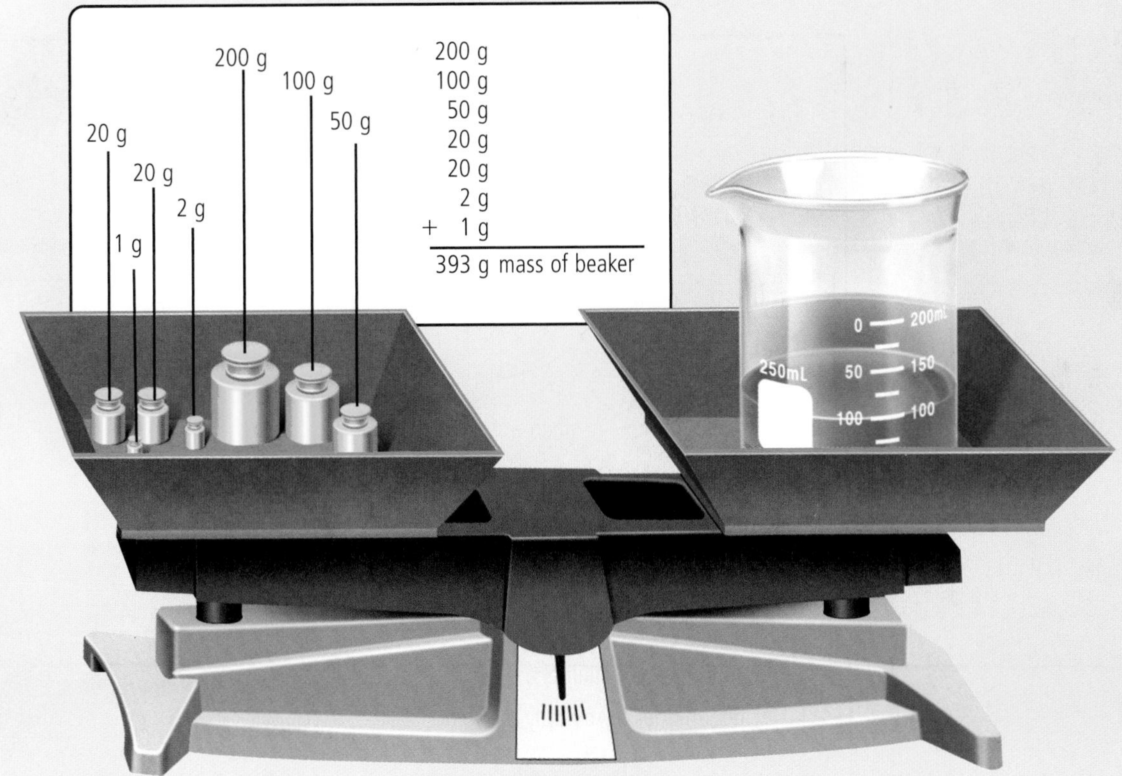

Never place chemicals or liquids directly on a pan. Instead, use the following procedure:

1. Determine the mass of an empty container, such as a beaker.
2. Pour the substance into the container, and measure the total mass of the substance and the container.
3. Subtract the mass of the empty container from the total mass to find the mass of the substance.

LAB HANDBOOK

The Metric System and SI Units

Scientists use International System (SI) units for measurements of distance, volume, mass, and temperature. The International System is based on multiples of ten and the metric system of measurement.

Basic SI Units		
Property	**Name**	**Symbol**
length	meter	m
volume	liter	L
mass	kilogram	kg
temperature	kelvin	K

SI Prefixes		
Prefix	**Symbol**	**Multiple of 10**
kilo-	k	1000
hecto-	h	100
deca-	da	10
deci-	d	0.1 $\left(\frac{1}{10}\right)$
centi-	c	0.01 $\left(\frac{1}{100}\right)$
milli-	m	0.001 $\left(\frac{1}{1000}\right)$

Changing Metric Units

You can change from one unit to another in the metric system by multiplying or dividing by a power of 10.

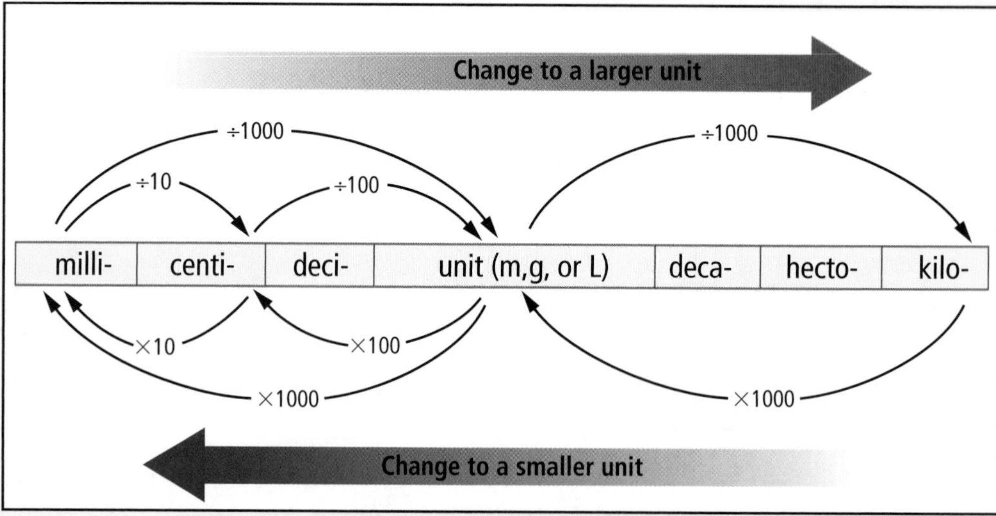

Example

Change 0.64 liters to milliliters.

(1) Decide whether to multiply or divide.

(2) Select the power of 10.

ANSWER 0.64 L = 640 mL

Change to a smaller unit by multiplying.

mL ← × 1000 — **L**

0.64 × 1000 = **640.**

Example

Change 23.6 grams to kilograms.

(1) Decide whether to multiply or divide.

(2) Select the power of 10.

ANSWER 23.6 g = 0.0236 kg

Change to a larger unit by dividing.

g — ÷ 1000 → **kg**

23.6 ÷ 1000 = **0.0236**

LAB HANDBOOK

Temperature Conversions

Even though the kelvin is the SI base unit of temperature, the degree Celsius will be the unit you use most often in your science studies. The formulas below show the relationships between temperatures in degrees Fahrenheit (°F), degrees Celsius (°C), and kelvins (K).

$$°C = \frac{5}{9}(°F - 32)$$

$$°F = \frac{9}{5}°C + 32$$

$$K = °C + 273$$

See page R42 for help with using formulas.

Examples of Temperature Conversions

Condition	Degrees Celsius	Degrees Fahrenheit
Freezing point of water	0	32
Cool day	10	50
Mild day	20	68
Warm day	30	86
Normal body temperature	37	98.6
Very hot day	40	104
Boiling point of water	100	212

Converting Between SI and U.S. Customary Units

Use the chart below when you need to convert between SI units and U.S. customary units.

SI Unit	From SI to U.S. Customary			From U.S. Customary to SI		
Length	**When you know**	**multiply by**	**to find**	**When you know**	**multiply by**	**to find**
kilometer (km) = 1000 m	kilometers	0.62	miles	miles	1.61	kilometers
meter (m) = 100 cm	meters	3.28	feet	feet	0.3048	meters
centimeter (cm) = 10 mm	centimeters	0.39	inches	inches	2.54	centimeters
millimeter (mm) = 0.1 cm	millimeters	0.04	inches	inches	25.4	millimeters
Area	**When you know**	**multiply by**	**to find**	**When you know**	**multiply by**	**to find**
square kilometer (km^2)	square kilometers	0.39	square miles	square miles	2.59	square kilometers
square meter (m^2)	square meters	1.2	square yards	square yards	0.84	square meters
square centimeter (cm^2)	square centimeters	0.155	square inches	square inches	6.45	square centimeters
Volume	**When you know**	**multiply by**	**to find**	**When you know**	**multiply by**	**to find**
liter (L) = 1000 mL	liters	1.06	quarts	quarts	0.95	liters
	liters	0.26	gallons	gallons	3.79	liters
	liters	4.23	cups	cups	0.24	liters
	liters	2.12	pints	pints	0.47	liters
milliliter (mL) = 0.001 L	milliliters	0.20	teaspoons	teaspoons	4.93	milliliters
	milliliters	0.07	tablespoons	tablespoons	14.79	milliliters
	milliliters	0.03	fluid ounces	fluid ounces	29.57	milliliters
Mass	**When you know**	**multiply by**	**to find**	**When you know**	**multiply by**	**to find**
kilogram (kg) = 1000 g	kilograms	2.2	pounds	pounds	0.45	kilograms
gram (g) = 1000 mg	grams	0.035	ounces	ounces	28.35	grams

LAB HANDBOOK

Precision and Accuracy

When you do an experiment, it is important that your methods, observations, and data be both precise and accurate.

low precision

precision, but not accuracy

precision and accuracy

Precision

In science, **precision** is the exactness and consistency of measurements. For example, measurements made with a ruler that has both centimeter and millimeter markings would be more precise than measurements made with a ruler that has only centimeter markings. Another indicator of precision is the care taken to make sure that methods and observations are as exact and consistent as possible. Every time a particular experiment is done, the same procedure should be used. Precision is necessary because experiments are repeated several times and if the procedure changes, the results will change.

EXAMPLE

Suppose you are measuring temperatures over a two-week period. Your precision will be greater if you measure each temperature at the same place, at the same time of day, and with the same thermometer than if you change any of these factors from one day to the next.

Accuracy

In science, it is possible to be precise but not accurate. **Accuracy** depends on the difference between a measurement and an actual value. The smaller the difference, the more accurate the measurement.

EXAMPLE

Suppose you look at a stream and estimate that it is about 1 meter wide at a particular place. You decide to check your estimate by measuring the stream with a meter stick, and you determine that the stream is 1.32 meters wide. However, because it is hard to measure the width of a stream with a meter stick, it turns out that you didn't do a very good job. The stream is actually 1.14 meters wide. Therefore, even though your estimate was less precise than your measurement, your estimate was actually more accurate.

Making Data Tables and Graphs

Data tables and graphs are useful tools for both recording and communicating scientific data.

Making Data Tables

You can use a **data table** to organize and record the measurements that you make. Some examples of information that might be recorded in data tables are frequencies, times, and amounts.

EXAMPLE

Suppose you are investigating photosynthesis in two elodea plants. One sits in direct sunlight, and the other sits in a dimly lit room. You measure the rate of photosynthesis by counting the number of bubbles in the jar every ten minutes.

1. Title and number your data table.
2. Decide how you will organize the table into columns and rows.
3. Any units, such as seconds or degrees, should be included in column headings, not in the individual cells.

Table 1. Number of Bubbles from Elodea

Always number and title data tables.

Time (min)	Sunlight	Dim Light
0	0	0
10	15	5
20	25	8
30	32	7
40	41	10
50	47	9
60	42	9

The data in the table above could also be organized in a different way.

Table 1. Number of Bubbles from Elodea

Put units in column heading.

Light Condition	Time (min)						
	0	10	20	30	40	50	60
Sunlight	0	15	25	32	41	47	42
Dim light	0	5	8	7	10	9	9

LAB HANDBOOK

Making Line Graphs

You can use a **line graph** to show a relationship between variables. Line graphs are particularly useful for showing changes in variables over time.

EXAMPLE

Suppose you are interested in graphing temperature data that you collected over the course of a day.

Table 1. Outside Temperature During the Day on March 7

	Time of Day						
	7:00 A.M.	9:00 A.M.	11:00 A.M.	1:00 P.M.	3:00 P.M.	5:00 P.M.	7:00 P.M.
Temp (°C)	8	9	11	14	12	10	6

1. Use the vertical axis of your line graph for the variable that you are measuring—temperature.
2. Choose scales for both the horizontal axis and the vertical axis of the graph. You should have two points more than you need on the vertical axis, and the horizontal axis should be long enough for all of the data points to fit.
3. Draw and label each axis.
4. Graph each value. First find the appropriate point on the scale of the horizontal axis. Imagine a line that rises vertically from that place on the scale. Then find the corresponding value on the vertical axis, and imagine a line that moves horizontally from that value. The point where these two imaginary lines intersect is where the value should be plotted.
5. Connect the points with straight lines.

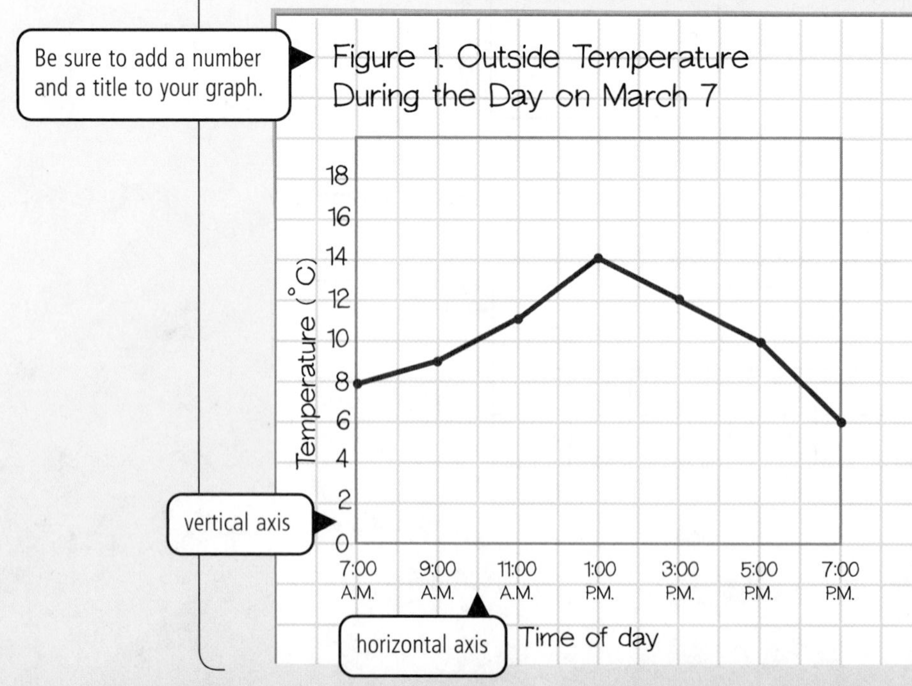

Making Circle Graphs

You can use a **circle graph,** sometimes called a pie chart, to represent data as parts of a circle. Circle graphs are used only when the data can be expressed as percentages of a whole. The entire circle shown in a circle graph is equal to 100 percent of the data.

EXAMPLE

Suppose you identified the species of each mature tree growing in a small wooded area. You organized your data in a table, but you also want to show the data in a circle graph.

1. To begin, find the total number of mature trees.

   $56 + 34 + 22 + 10 + 28 = 150$

2. To find the degree measure for each sector of the circle, write a fraction comparing the number of each tree species with the total number of trees. Then multiply the fraction by 360°.

   Oak: $\frac{56}{150} \times 360° = 134.4°$

3. Draw a circle. Use a protractor to draw the angle for each sector of the graph.
4. Color and label each sector of the graph.
5. Give the graph a number and title.

Table 1. Tree Species in Wooded Area

Species	Number of Specimens
Oak	56
Maple	34
Birch	22
Willow	10
Pine	28

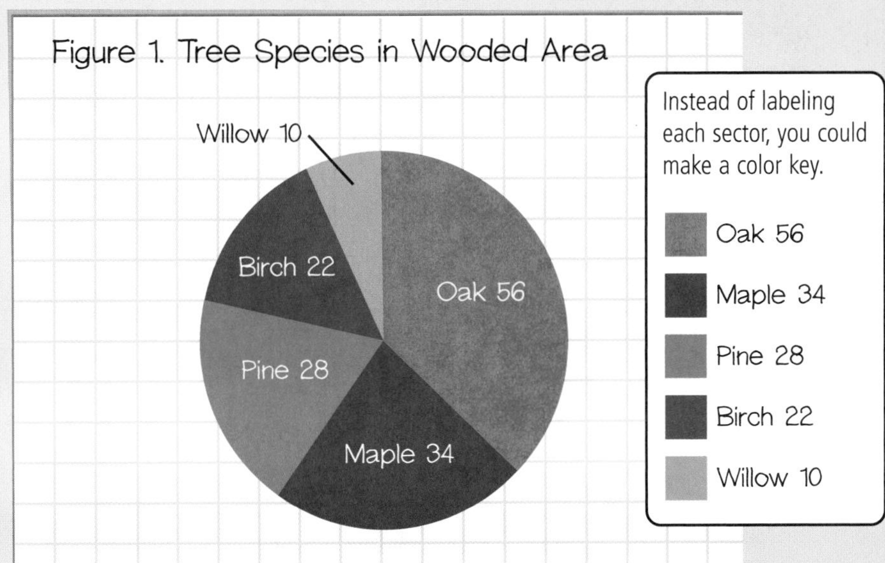

Figure 1. Tree Species in Wooded Area

LAB HANDBOOK

Bar Graph

A **bar graph** is a type of graph in which the lengths of the bars are used to represent and compare data. A numerical scale is used to determine the lengths of the bars.

EXAMPLE

To determine the effect of water on seed sprouting, three cups were filled with sand, and ten seeds were planted in each. Different amounts of water were added to each cup over a three-day period.

Table 1. Effect of Water on Seed Sprouting

Daily Amount of Water (mL)	Number of Seeds That Sprouted After 3 Days in Sand
0	1
10	4
20	8

1. Choose a numerical scale. The greatest value is 8, so the end of the scale should have a value greater than 8, such as 10. Use equal increments along the scale, such as increments of 2.
2. Draw and label the axes. Mark intervals on the vertical axis according to the scale you chose.
3. Draw a bar for each data value. Use the scale to decide how long to make each bar.

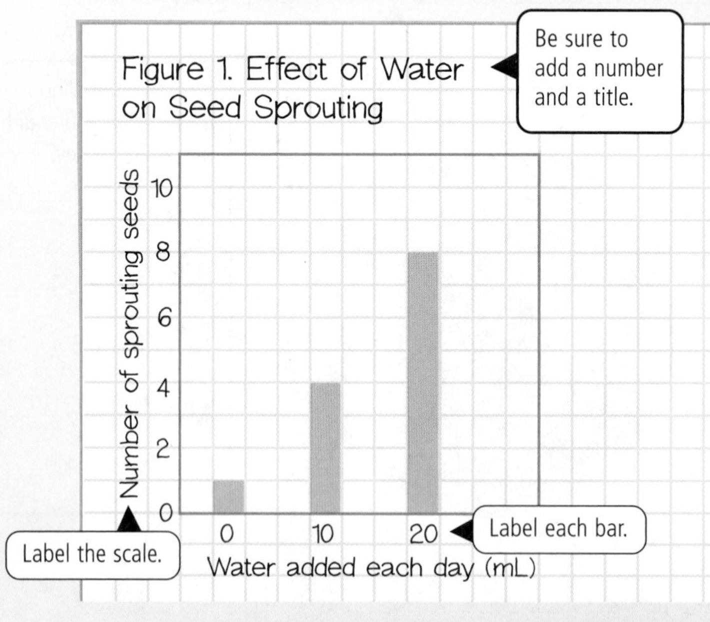

Double Bar Graph

A **double bar graph** is a bar graph that shows two sets of data. The two bars for each measurement are drawn next to each other.

EXAMPLE

The same seed-sprouting experiment was repeated with potting soil. The data for sand and potting soil can be plotted on one graph.

1. Draw one set of bars, using the data for sand, as shown below.
2. Draw bars for the potting-soil data next to the bars for the sand data. Shade them a different color. Add a key.

Table 2. Effect of Water and Soil on Seed Sprouting

Daily Amount of Water (mL)	Number of Seeds That Sprouted After 3 Days in Sand	Number of Seeds That Sprouted After 3 Days in Potting Soil
0	1	2
10	4	5
20	8	9

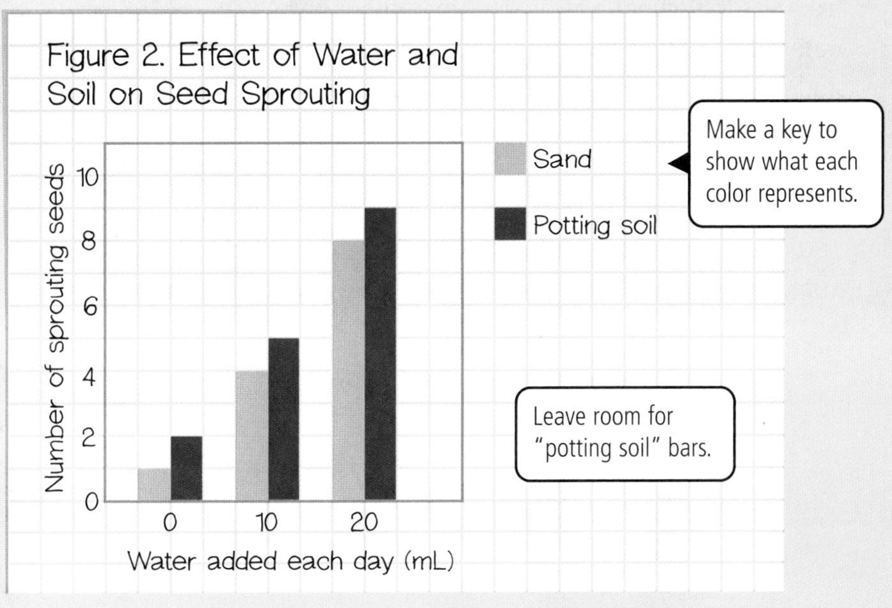

LAB HANDBOOK

Designing an Experiment

Use this section when designing or conducting an experiment.

Determining a Purpose

You can find a purpose for an experiment by doing research, by examining the results of a previous experiment, or by observing the world around you. An **experiment** is an organized procedure to study something under controlled conditions.

Don't forget to learn as much as possible about your topic before you begin.

1. Write the purpose of your experiment as a question or problem that you want to investigate.
2. Write down research questions and begin searching for information that will help you design an experiment. Consult the library, the Internet, and other people as you conduct your research.

EXAMPLE

Middle school students observed an odor near the lake by their school. They also noticed that the water on the side of the lake near the school was greener than the water on the other side of the lake. The students did some research to learn more about their observations. They discovered that the odor and green color in the lake came from algae. They also discovered that a new fertilizer was being used on a field nearby. The students inferred that the use of the fertilizer might be related to the presence of the algae and designed a controlled experiment to find out whether they were right.

Problem

How does fertilizer affect the presence of algae in a lake?

Research Questions

- Have other experiments been done on this problem? If so, what did those experiments show?
- What kind of fertilizer is used on the field? How much?
- How do algae grow?
- How do people measure algae?
- Can fertilizer and algae be used safely in a lab? How?

Research

As you research, you may find a topic that is more interesting to you than your original topic, or learn that a procedure you wanted to use is not practical or safe. It is OK to change your purpose as you research.

LAB HANDBOOK

Writing a Hypothesis

A **hypothesis** is a tentative explanation for an observation or scientific problem that can be tested by further investigation. You can write your hypothesis in the form of an "If . . . , then . . . , because . . ." statement.

Hypothesis

If the amount of fertilizer in lake water is increased, then the amount of algae will also increase, because fertilizers provide nutrients that algae need to grow.

Hypotheses
For help with hypotheses, refer to page R3.

Determining Materials

Make a list of all the materials you will need to do your experiment. Be specific, especially if someone else is helping you obtain the materials. Try to think of everything you will need.

Materials

- 1 large jar or container
- 4 identical smaller containers
- rubber gloves that also cover the arms
- sample of fertilizer-and-water solution
- eyedropper
- clear plastic wrap
- scissors
- masking tape
- marker
- ruler

Determining Variables and Constants

EXPERIMENTAL GROUP AND CONTROL GROUP

An experiment to determine how two factors are related always has two groups—a control group and an experimental group.

1. Design an experimental group. Include as many trials as possible in the experimental group in order to obtain reliable results.
2. Design a control group that is the same as the experimental group in every way possible, except for the factor you wish to test.

Experimental Group: two containers of lake water with one drop of fertilizer solution added to each

Control Group: two containers of lake water with no fertilizer solution added

Go back to your materials list and make sure you have enough items listed to cover both your experimental group and your control group.

VARIABLES AND CONSTANTS

Identify the variables and constants in your experiment. In a controlled experiment, a **variable** is any factor that can change. **Constants** are all of the factors that are the same in both the experimental group and the control group.

1. Read your hypothesis. The **independent variable** is the factor that you wish to test and that is manipulated or changed so that it can be tested. The independent variable is expressed in your hypothesis after the word *if*. Identify the independent variable in your laboratory report.
2. The **dependent variable** is the factor that you measure to gather results. It is expressed in your hypothesis after the word *then*. Identify the dependent variable in your laboratory report.

Hypothesis
If the amount of fertilizer in lake water is increased, then the amount of algae will also increase, because fertilizers provide nutrients that algae need to grow.

Table 1. Variables and Constants in Algae Experiment

Independent Variable	Dependent Variable	Constants
Amount of fertilizer in lake water	Amount of algae that grow	• Where the lake water is obtained • Type of container used • Light and temperature conditions where water will be stored

Set up your experiment so that you will test only one variable.

MEASURING THE DEPENDENT VARIABLE

Before starting your experiment, you need to define how you will measure the dependent variable. An **operational definition** is a description of the one particular way in which you will measure the dependent variable.

Your operational definition is important for several reasons. First, in any experiment there are several ways in which a dependent variable can be measured. Second, the procedure of the experiment depends on how you decide to measure the dependent variable. Third, your operational definition makes it possible for other people to evaluate and build on your experiment.

EXAMPLE 1

An operational definition of a dependent variable can be qualitative. That is, your measurement of the dependent variable can simply be an observation of whether a change occurs as a result of a change in the independent variable. This type of operational definition can be thought of as a "yes or no" measurement.

Table 2. Qualitative Operational Definition of Algae Growth

Independent Variable	Dependent Variable	Operational Definition
Amount of fertilizer in lake water	Amount of algae that grow	Algae grow in lake water

A qualitative measurement of a dependent variable is often easy to make and record. However, this type of information does not provide a great deal of detail in your experimental results.

EXAMPLE 2

An operational definition of a dependent variable can be quantitative. That is, your measurement of the dependent variable can be a number that shows how much change occurs as a result of a change in the independent variable.

Table 3. Quantitative Operational Definition of Algae Growth

Independent Variable	Dependent Variable	Operational Definition
Amount of fertilizer in lake water	Amount of algae that grow	Diameter of largest algal growth (in mm)

A quantitative measurement of a dependent variable can be more difficult to make and analyze than a qualitative measurement. However, this type of data provides much more information about your experiment and is often more useful.

Writing a Procedure

Write each step of your procedure. Start each step with a verb, or action word, and keep the steps short. Your procedure should be clear enough for someone else to use as instructions for repeating your experiment.

If necessary, go back to your materials list and add any materials that you left out.

Procedure

1. Put on your gloves. Use the large container to obtain a sample of lake water.
2. Divide the sample of lake water equally among the four smaller containers.
3. Use the eyedropper to add one drop of fertilizer solution to two of the containers.
4. Use the masking tape and the marker to label the containers with your initials, the date, and the identifiers "Jar 1 with Fertilizer," "Jar 2 with Fertilizer," "Jar 1 without Fertilizer," and "Jar 2 without Fertilizer."
5. Cover the containers with clear plastic wrap. Use the scissors to punch ten holes in each of the covers.
6. Place all four containers on a window ledge. Make sure that they all receive the same amount of light.
7. Observe the containers every day for one week.
8. Use the ruler to measure the diameter of the largest clump of algae in each container, and record your measurements daily.

Controlling Variables
The same amount of fertilizer solution must be added to two of the four containers.

Controlling Variables
All four containers must receive the same amount of light.

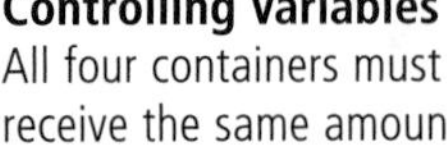

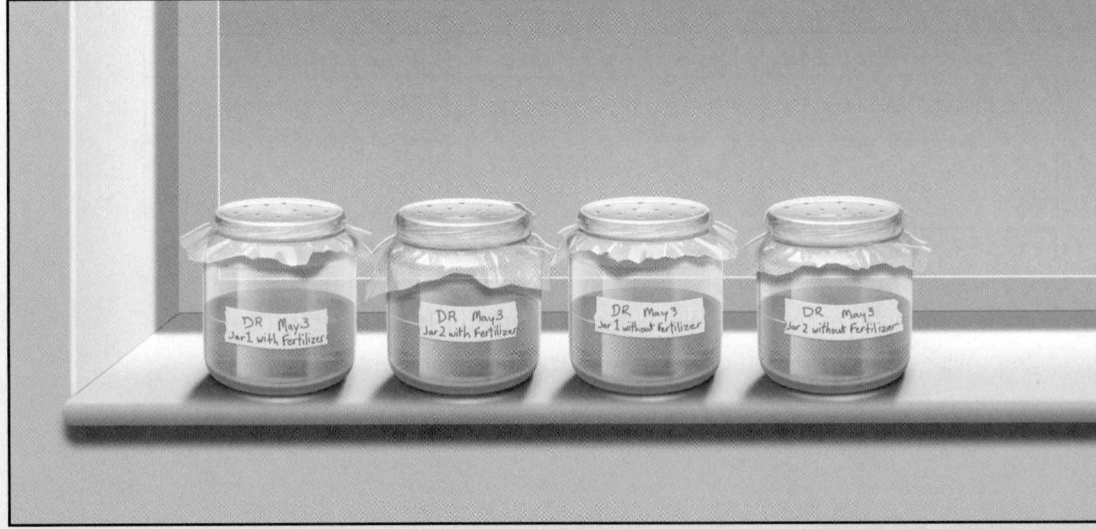

Recording Observations

Once you have obtained all of your materials and your procedure has been approved, you can begin making experimental observations. Gather both quantitative and qualitative data. If something goes wrong during your procedure, make sure you record that too.

Observations
For help with making qualitative and quantitative observations, refer to page R2.

For more examples of data tables, see page R23.

Table 4. Fertilizer and Algae Growth

	Experimental Group		Control Group		
Date and Time	Jar 1 with Fertilizer (diameter of algae in mm)	Jar 2 with Fertilizer (diameter of algae in mm)	Jar 1 without Fertilizer (diameter of algae in mm)	Jar 2 without Fertilizer (diameter of algae in mm)	Observations
5/3 4:00 P.M.	0	0	0	0	condensation in all containers
5/4 4:00 P.M.	0	3	0	0	tiny green blobs in jar 2 with fertilizer
5/5 4:15 P.M.	4	5	0	3	green blobs in jars 1 and 2 with fertilizer and jar 2 without fertilizer
5/6 4:00 P.M.	5	6	0	4	water light green in jar 2 with fertilizer
5/7 4:00 P.M.	8	10	0	6	water light green in jars 1 and 2 with fertilizer and in jar 2 without fertilizer
5/8 3:30 P.M.	10	18	0	6	cover off jar 2 with fertilizer
5/9 3:30 P.M.	14	23	0	8	drew sketches of each container

Notice that on the sixth day, the observer found that the cover was off one of the containers. It is important to record observations of unintended factors because they might affect the results of the experiment.

Drawings of Samples Viewed Under Microscope on 5/9 at 100x

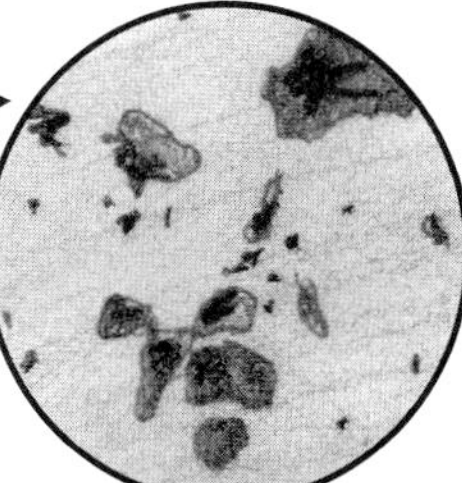
Jar 1 with Fertilizer

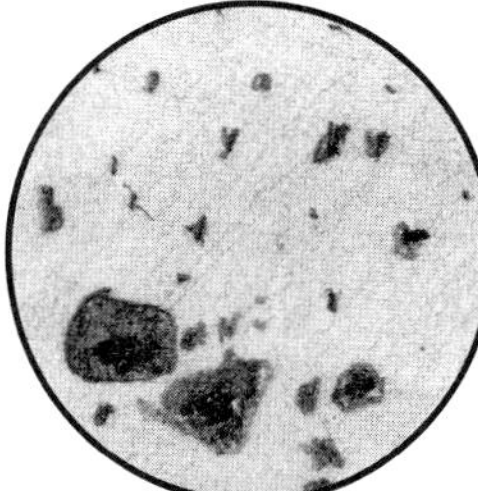
Jar 2 with Fertilizer

Jar 1 without Fertilizer

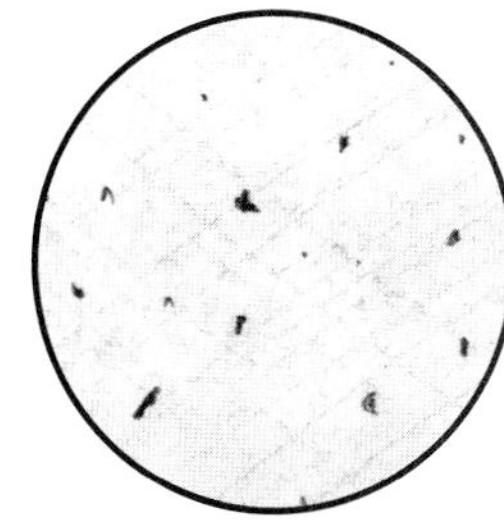
Jar 2 without Fertilizer

Use technology, such as a microscope, to help you make observations when possible.

LAB HANDBOOK

LAB HANDBOOK

Summarizing Results

To summarize your data, look at all of your observations together. Look for meaningful ways to present your observations. For example, you might average your data or make a graph to look for patterns. When possible, use spreadsheet software to help you analyze and present your data. The two graphs below show the same data.

EXAMPLE 1

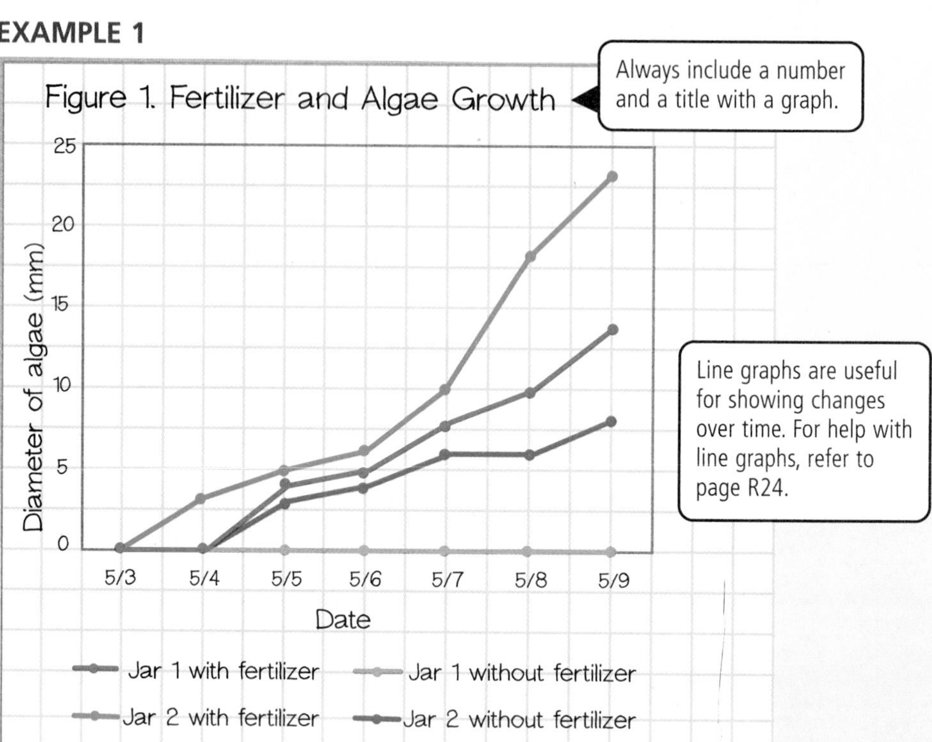

Always include a number and a title with a graph.

Line graphs are useful for showing changes over time. For help with line graphs, refer to page R24.

EXAMPLE 2

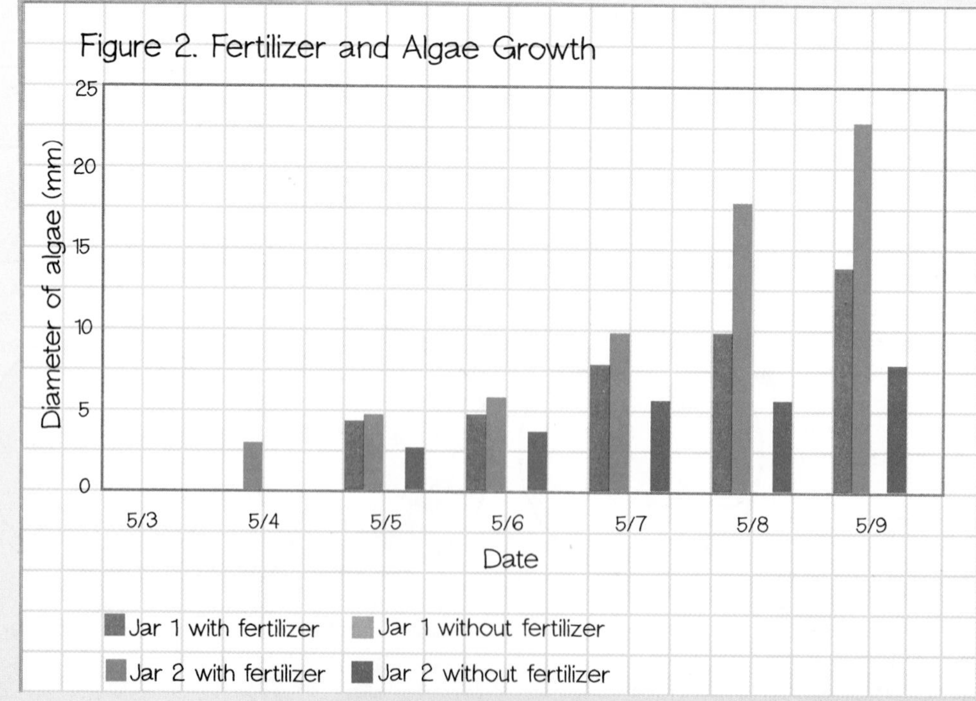

Bar graphs are useful for comparing different data sets. This bar graph has four bars for each day. Another way to present the data would be to calculate averages for the tests and the controls, and to show one test bar and one control bar for each day.

Drawing Conclusions

RESULTS AND INFERENCES

To draw conclusions from your experiment, first write your results. Then compare your results with your hypothesis. Do your results support your hypothesis? Be careful not to make inferences about factors that you did not test.

For help with making inferences, see page R4.

Results and Inferences

The results of my experiment show that more algae grew in lake water to which fertilizer had been added than in lake water to which no fertilizer had been added. My hypothesis was supported. I infer that it is possible that the growth of algae in the lake was caused by the fertilizer used on the field.

Notice that you cannot conclude from this experiment that the presence of algae in the lake was due only to the fertilizer.

QUESTIONS FOR FURTHER RESEARCH

Write a list of questions for further research and investigation. Your ideas may lead you to new experiments and discoveries.

Questions for Further Research

- What is the connection between the amount of fertilizer and algae growth?
- How do different brands of fertilizer affect algae growth?
- How would algae growth in the lake be affected if no fertilizer were used on the field?
- How do algae affect the lake and the other life in and around it?
- How does fertilizer affect the lake and the life in and around it?
- If fertilizer is getting into the lake, how is it getting there?

Describing a Set of Data

Means, medians, modes, and ranges are important math tools for describing data sets such as the following widths of fossilized clamshells.

13 mm 25 mm 14 mm 21 mm 16 mm 23 mm 14 mm

Mean

The **mean** of a data set is the sum of the values divided by the number of values.

Example

To find the mean of the clamshell data, add the values and then divide the sum by the number of values.

$$\frac{13\text{ mm} + 25\text{ mm} + 14\text{ mm} + 21\text{ mm} + 16\text{ mm} + 23\text{ mm} + 14\text{ mm}}{7} = \frac{126\text{ mm}}{7} = 18\text{ mm}$$

ANSWER The mean is 18 mm.

Median

The **median** of a data set is the middle value when the values are written in numerical order. If a data set has an even number of values, the median is the mean of the two middle values.

Example

To find the median of the clamshell data, arrange the values in order from least to greatest. The median is the middle value.

13 mm 14 mm 14 mm 16 mm 21 mm 23 mm 25 mm

ANSWER The median is 16 mm.

Mode

The **mode** of a data set is the value that occurs most often.

Example

To find the mode of the clamshell data, arrange the values in order from least to greatest and determine the value that occurs most often.

13 mm 14 mm 14 mm 16 mm 21 mm 23 mm 25 mm

ANSWER The mode is 14 mm.

A data set can have more than one mode or no mode. For example, the following data set has modes of 2 mm and 4 mm:

2 mm 2 mm 3 mm 4 mm 4 mm

The data set below has no mode, because no value occurs more often than any other.

2 mm 3 mm 4 mm 5 mm

Range

The **range** of a data set is the difference between the greatest value and the least value.

Example

To find the range of the clamshell data, arrange the values in order from least to greatest.

13 mm 14 mm 14 mm 16 mm 21 mm 23 mm 25 mm

Subtract the least value from the greatest value.

13 mm is the least value.
25 mm is the greatest value.

25 mm – 13 mm = 12 mm

ANSWER The range is 12 mm.

MATH HANDBOOK

Using Ratios, Rates, and Proportions

You can use ratios and rates to compare values in data sets. You can use proportions to find unknown values.

Ratios

A **ratio** uses division to compare two values. The ratio of a value a to a nonzero value b can be written as $\frac{a}{b}$.

Example

The height of one plant is 8 centimeters. The height of another plant is 6 centimeters. To find the ratio of the height of the first plant to the height of the second plant, write a fraction and simplify it.

$$\frac{8\text{ cm}}{6\text{ cm}} = \frac{4 \times \overset{1}{\cancel{2}}}{3 \times \underset{1}{\cancel{2}}} = \frac{4}{3}$$

ANSWER The ratio of the plant heights is $\frac{4}{3}$.

You can also write the ratio $\frac{a}{b}$ as "a to b" or as a:b. For example, you can write the ratio of the plant heights as "4 to 3" or as 4:3.

Rates

A **rate** is a ratio of two values expressed in different units. A unit rate is a rate with a denominator of 1 unit.

Example

A plant grew 6 centimeters in 2 days. The plant's rate of growth was $\frac{6\text{ cm}}{2\text{ days}}$. To describe the plant's growth in centimeters per day, write a unit rate.

Divide numerator and denominator by 2: $\frac{6\text{ cm}}{2\text{ days}} = \frac{6\text{ cm} \div 2}{2\text{ days} \div 2}$

Simplify: $= \frac{3\text{ cm}}{1\text{ day}}$

You divide 2 days by 2 to get 1 day, so divide 6 cm by 2 also.

ANSWER The plant's rate of growth is 3 centimeters per day.

MATH HANDBOOK

R38

Proportions

A **proportion** is an equation stating that two ratios are equivalent. To solve for an unknown value in a proportion, you can use cross products.

Example

If a plant grew 6 centimeters in 2 days, how many centimeters would it grow in 3 days (if its rate of growth is constant)?

Write a proportion: $\frac{6\text{ cm}}{2\text{ days}} = \frac{x\text{ cm}}{3\text{ days}}$

Set cross products: $6 \cdot 3 = 2x$

Multiply 6 and 3: $18 = 2x$

Divide each side by 2: $\frac{18}{2} = \frac{2x}{2}$

Simplify: $9 = x$

ANSWER The plant would grow 9 centimeters in 3 days.

Using Decimals, Fractions, and Percents

Decimals, fractions, and percentages are all ways of recording and representing data.

Decimals

A **decimal** is a number that is written in the base-ten place value system, in which a decimal point separates the ones and tenths digits. The values of each place is ten times that of the place to its right.

Example

A caterpillar traveled from point *A* to point *C* along the path shown.

A — 36.9 cm — B — 52.4 cm — C

ADDING DECIMALS To find the total distance traveled by the caterpillar, add the distance from *A* to *B* and the distance from *B* to *C*. Begin by lining up the decimal points. Then add the figures as you would whole numbers and bring down the decimal point.

$$\begin{array}{r} 36.9\text{ cm} \\ +\ 52.4\text{ cm} \\ \hline 89.3\text{ cm} \end{array}$$

ANSWER The caterpillar traveled a total distance of 89.3 centimeters.

MATH HANDBOOK

Example *continued*

SUBTRACTING DECIMALS To find how much farther the caterpillar traveled on the second leg of the journey, subtract the distance from *A* to *B* from the distance from *B* to *C*.

$$\begin{array}{r} 52.4 \text{ cm} \\ -\ 36.9 \text{ cm} \\ \hline 15.5 \text{ cm} \end{array}$$

ANSWER The caterpillar traveled 15.5 centimeters farther on the second leg of the journey.

Example

A caterpillar is traveling from point *D* to point *F* along the path shown. The caterpillar travels at a speed of 9.6 centimeters per minute.

D E 33.6 cm F

MULTIPLYING DECIMALS You can multiply decimals as you would whole numbers. The number of decimal places in the product is equal to the sum of the number of decimal places in the factors.

For instance, suppose it takes the caterpillar 1.5 minutes to go from *D* to *E*. To find the distance from *D* to *E*, multiply the caterpillar's speed by the time it took.

$$\begin{array}{r} 9.6 \\ \times\ 1.5 \\ \hline 480 \\ 96\ \\ \hline 14.40 \end{array} \qquad \begin{array}{rl} 1 & \text{decimal place} \\ +\ 1 & \text{decimal place} \\ \hline & \\ & \\ 2 & \text{decimal places} \end{array}$$

Align as shown.

ANSWER The distance from *D* to *E* is 14.4 centimeters.

DIVIDING DECIMALS When you divide by a decimal, move the decimal points the same number of places in the divisor and the dividend to make the divisor a whole number.

For instance, to find the time it will take the caterpillar to travel from *E* to *F*, divide the distance from *E* to *F* by the caterpillar's speed.

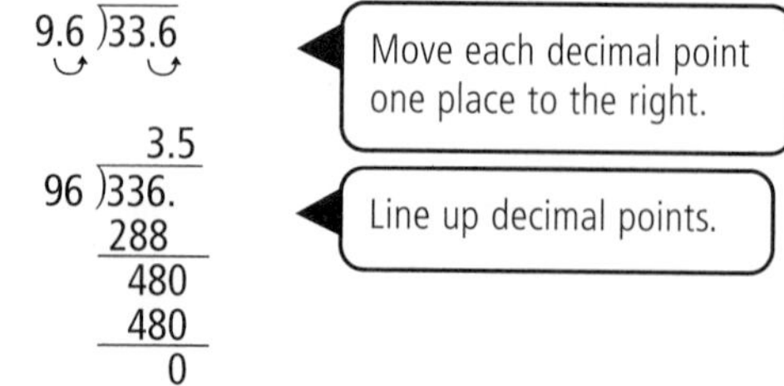

9.6)33.6
Move each decimal point one place to the right.
3.5
96)336.
288
480
480
0
Line up decimal points.

ANSWER The caterpillar will travel from *E* to *F* in 3.5 minutes.

MATH HANDBOOK

Fractions

A **fraction** is a number in the form $\frac{a}{b}$, where b is not equal to 0. A fraction is in **simplest form** if its numerator and denominator have a greatest common factor (GCF) of 1. To simplify a fraction, divide its numerator and denominator by their GCF.

Example

A caterpillar is 40 millimeters long. The head of the caterpillar is 6 millimeters long. To compare the length of the caterpillar's head with the caterpillar's total length, you can write and simplify a fraction that expresses the ratio of the two lengths.

Write the ratio of the two lengths: $\frac{\text{Length of head}}{\text{Total length}} = \frac{6 \text{ mm}}{40 \text{ mm}}$

Write numerator and denominator as products of numbers and the GCF: $= \frac{3 \times 2}{20 \times 2}$

Divide numerator and denominator by the GCF: $= \frac{3 \times \overset{1}{\cancel{2}}}{20 \times \underset{1}{\cancel{2}}}$

Simplify: $= \frac{3}{20}$

ANSWER In simplest form, the ratio of the lengths is $\frac{3}{20}$.

Percents

A **percent** is a ratio that compares a number to 100. The word *percent* means "per hundred" or "out of 100." The symbol for *percent* is %.

For instance, suppose 43 out of 100 caterpillars are female. You can represent this ratio as a percent, a decimal, or a fraction.

Percent	Decimal	Fraction
43%	0.43	$\frac{43}{100}$

Example

In the preceding example, the ratio of the length of the caterpillar's head to the caterpillar's total length is $\frac{3}{20}$. To write this ratio as a percent, write an equivalent fraction that has a denominator of 100.

Multiply numerator and denominator by 5: $\frac{3}{20} = \frac{3 \times 5}{20 \times 5}$

$= \frac{15}{100}$

Write as a percent: $= 15\%$

ANSWER The caterpillar's head represents 15 percent of its total length.

MATH HANDBOOK

Using Formulas

A mathematical **formula** is a statement of a fact, rule, or principle. It is usually expressed as an equation.

In science, a formula often has a word form and a symbolic form. The formula below expresses Ohm's law.

Word Form

$$\text{Current} = \frac{\text{voltage}}{\text{resistance}}$$

Symbolic Form

$$I = \frac{V}{R}$$

The term *variable* is also used in science to refer to a factor that can change during an experiment.

In this formula, *I*, *V*, and *R* are variables. A mathematical **variable** is a symbol or letter that is used to represent one or more numbers.

Example

Suppose that you measure a voltage of 1.5 volts and a resistance of 15 ohms. You can use the formula for Ohm's law to find the current in amperes.

Write the formula for Ohm's law: $I = \frac{V}{R}$

Substitute 1.5 volts for V *and 15 ohms for* R: $I = \frac{1.5 \text{ volts}}{15 \text{ ohms}}$

Simplify: $I = 0.1$ amp

ANSWER The current is 0.1 ampere.

If you know the values of all variables but one in a formula, you can solve for the value of the unknown variable. For instance, Ohm's law can be used to find a voltage if you know the current and the resistance.

Example

Suppose that you know that a current is 0.2 amperes and the resistance is 18 ohms. Use the formula for Ohm's law to find the voltage in volts.

Write the formula for Ohm's law: $I = \frac{V}{R}$

Substitute 0.2 amp for I *and 18 ohms for* R: $0.2 \text{ amp} = \frac{V}{18 \text{ ohms}}$

Multiply both sides by 18 ohms: $0.2 \text{ amp} \cdot 18 \text{ ohms} = V$

Simplify: $3.6 \text{ volts} = V$

ANSWER The voltage is 3.6 volts.

MATH HANDBOOK

Finding Areas

The area of a figure is the amount of surface the figure covers.

Area is measured in square units, such as square meters (m^2) or square centimeters (cm^2). Formulas for the areas of three common geometric figures are shown below.

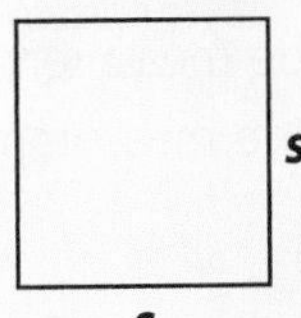

Area = (side length)^2
$A = s^2$

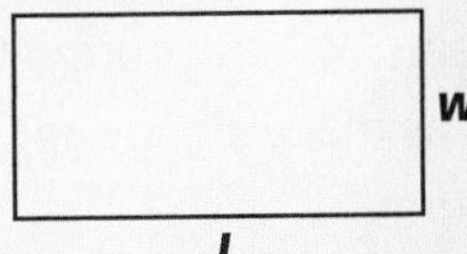

Area = length × width
$A = lw$

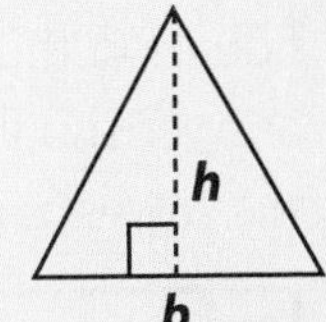

Area = $\frac{1}{2}$ × base × height
$A = \frac{1}{2}bh$

Example

Each face of a halite crystal is a square like the one shown. You can find the area of the square by using the steps below.

3 mm
3 mm

Write the formula for the area of a square: $A = s^2$

Substitute 3 mm for s: $= (3\text{ mm})^2$

Simplify: $= 9\text{ mm}^2$

ANSWER The area of the square is 9 square millimeters.

Finding Volumes

The volume of a solid is the amount of space contained by the solid.

Volume is measured in cubic units, such as cubic meters (m^3) or cubic centimeters (cm^3). The volume of a rectangular prism is given by the formula shown below.

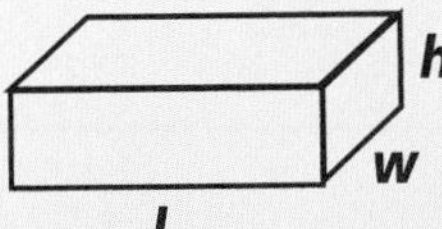

Volume = length × width × height
$V = lwh$

Example

A topaz crystal is a rectangular prism like the one shown. You can find the volume of the prism by using the steps below.

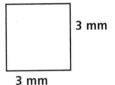

Write the formula for the volume of a rectangular prism: $V = lwh$

Substitute dimensions: $= 20\text{ mm} \times 12\text{ mm} \times 10\text{ mm}$

Simplify: $= 2400\text{ mm}^3$

ANSWER The volume of the rectangular prism is 2400 cubic millimeters.

Using Significant Figures

The **significant figures** in a decimal are the digits that are warranted by the accuracy of a measuring device.

When you perform a calculation with measurements, the number of significant figures to include in the result depends in part on the number of significant figures in the measurements. When you multiply or divide measurements, your answer should have only as many significant figures as the measurement with the fewest significant figures.

Example

Using a balance and a graduated cylinder filled with water, you determined that a marble has a mass of 8.0 grams and a volume of 3.5 cubic centimeters. To calculate the density of the marble, divide the mass by the volume.

Write the formula for density: $\text{Density} = \frac{\text{mass}}{\text{Volume}}$

Substitute measurements: $= \frac{8.0 \text{ g}}{3.5 \text{ cm}^3}$

Use a calculator to divide: $\approx 2.285714286 \text{ g/cm}^3$

ANSWER Because the mass and the volume have two significant figures each, give the density to two significant figures. The marble has a density of 2.3 grams per cubic centimeter.

Using Scientific Notation

Scientific notation is a shorthand way to write very large or very small numbers. For example, 73,500,000,000,000,000,000,000 kg is the mass of the Moon. In scientific notation, it is 7.35×10^{22} kg.

Example

You can convert from standard form to scientific notation.

Standard Form	**Scientific Notation**
720,000	7.2×10^5
5 decimal places left	Exponent is 5.
0.000291	2.91×10^{-4}
4 decimal places right	Exponent is –4.

You can convert from scientific notation to standard form.

Scientific Notation	**Standard Form**
4.63×10^7	46,300,000
Exponent is 7.	7 decimal places right
1.08×10^{-6}	0.00000108
Exponent is –6.	6 decimal places left

MATH HANDBOOK

Note-Taking Handbook

Note-Taking Strategies

Taking notes as you read helps you understand the information. The notes you take can also be used as a study guide for later review. This handbook presents several ways to organize your notes.

Content Frame

1. Make a chart in which each column represents a category.
2. Give each column a heading.
3. Write details under the headings.

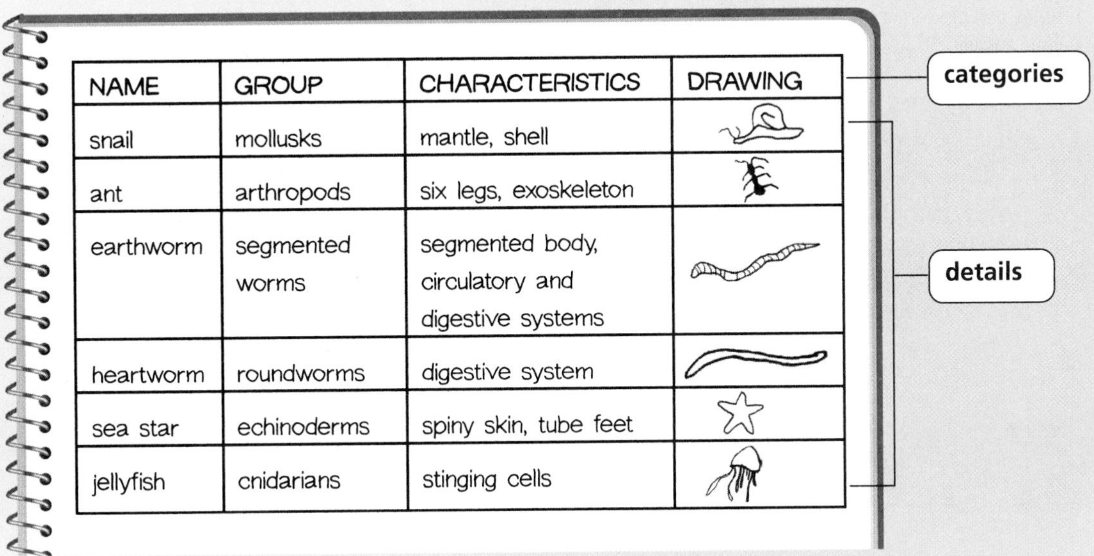

NAME	GROUP	CHARACTERISTICS	DRAWING
snail	mollusks	mantle, shell	
ant	arthropods	six legs, exoskeleton	
earthworm	segmented worms	segmented body, circulatory and digestive systems	
heartworm	roundworms	digestive system	
sea star	echinoderms	spiny skin, tube feet	
jellyfish	cnidarians	stinging cells	

Combination Notes

1. For each new idea or concept, write an informal outline of the information.
2. Make a sketch to illustrate the concept, and label it.

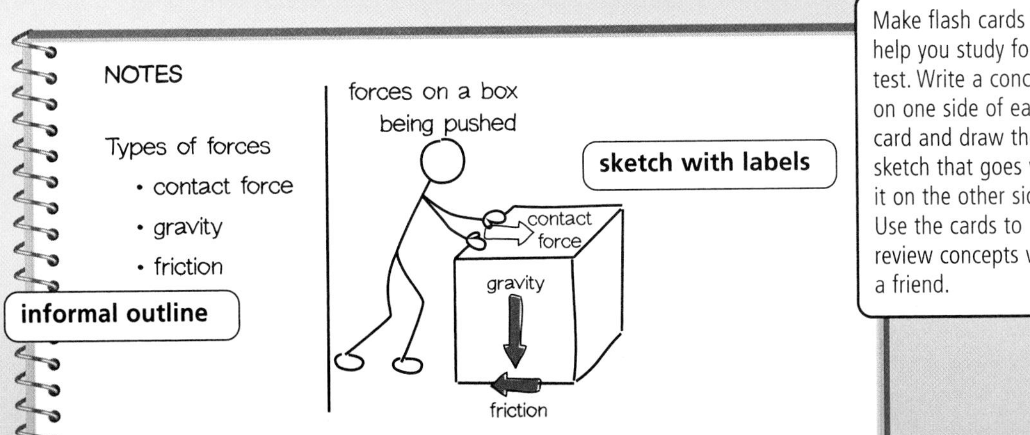

Make flash cards to help you study for a test. Write a concept on one side of each card and draw the sketch that goes with it on the other side. Use the cards to review concepts with a friend.

Main Idea and Detail Notes

1. In the left-hand column of a two-column chart, list main ideas. The blue headings express main ideas throughout this textbook.
2. In the right-hand column, write details that expand on each main idea.

You can shorten the headings in your chart. Be sure to use the most important words.

When studying for tests, cover up the detail notes column with a sheet of paper. Then use each main idea to form a question—such as "How does latitude affect climate?" Answer the question, and then uncover the detail notes column to check your answer.

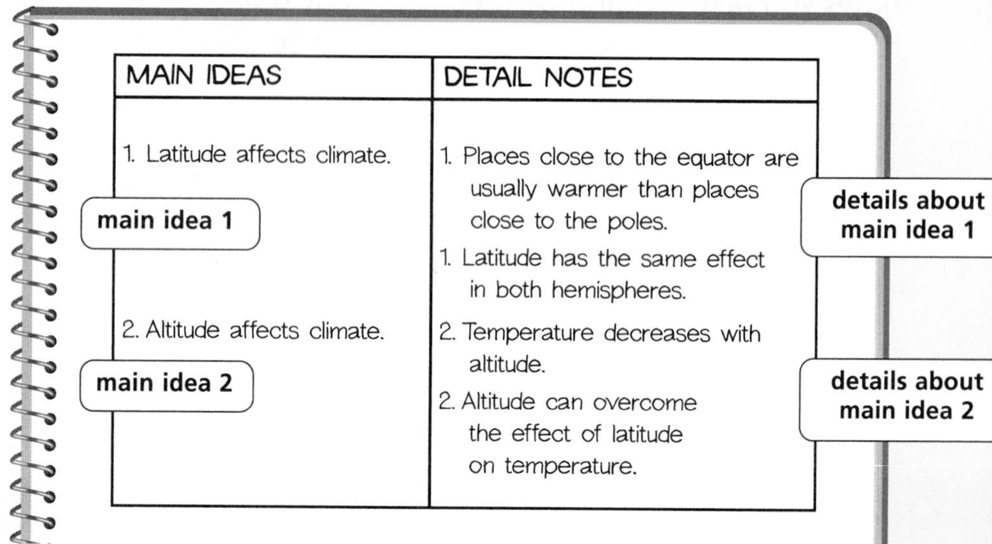

Main Idea Web

1. Write a main idea in a box.
2. Add boxes around it with related vocabulary terms and important details.

You can find definitions near highlighted terms.

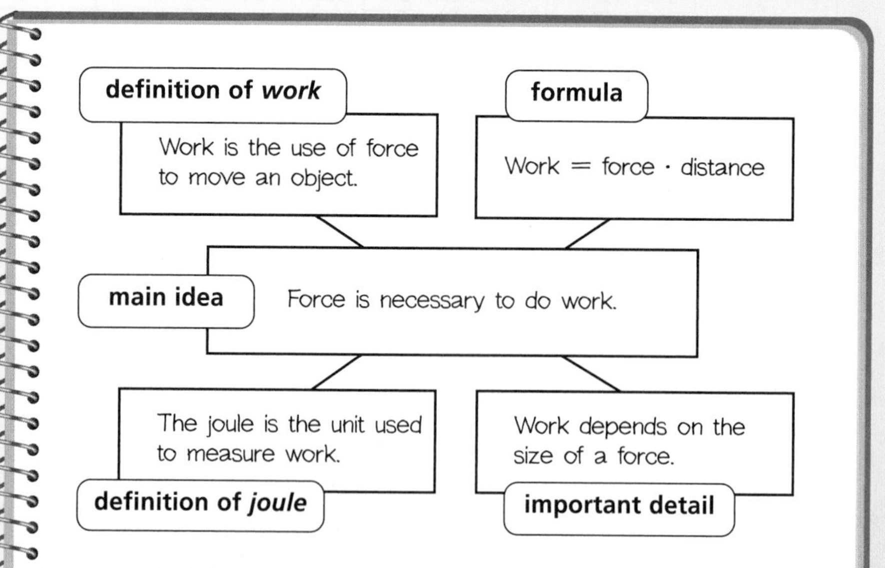

NOTE-TAKING HANDBOOK

Mind Map

1. Write a main idea in the center.
2. Add details that relate to one another and to the main idea.

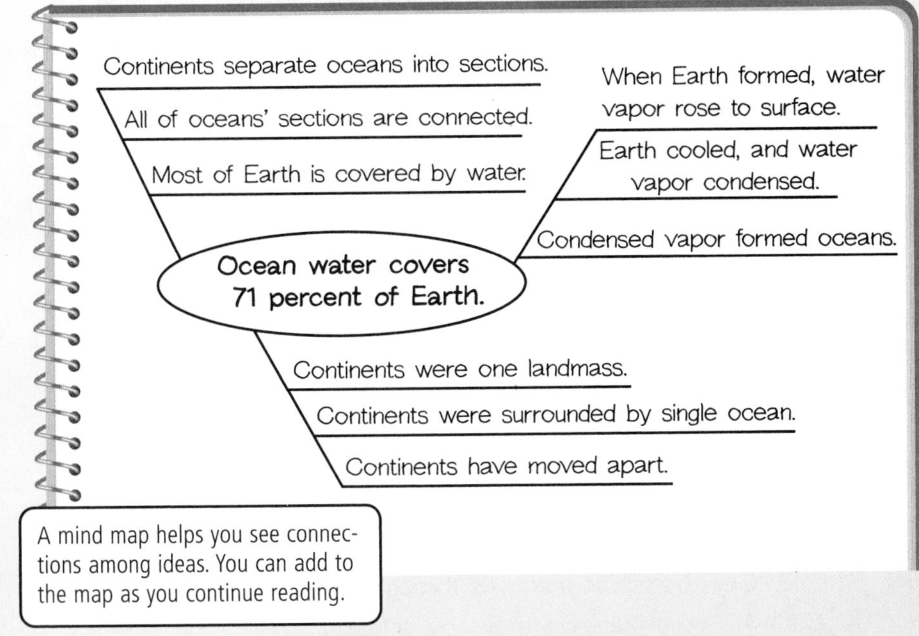

Supporting Main Ideas

1. Write a main idea in a box.
2. Add boxes underneath with information—such as reasons, explanations, and examples—that supports the main idea.

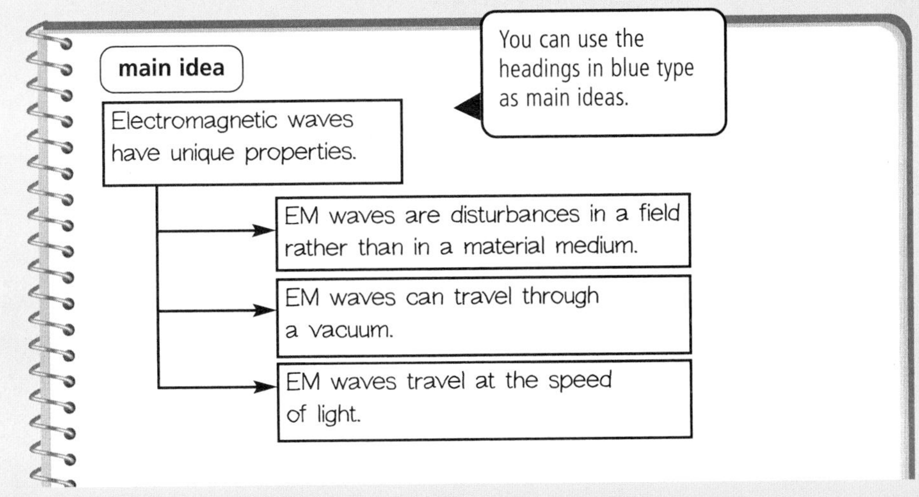

main idea
Electromagnetic waves have unique properties.
You can use the headings in blue type as main ideas.
EM waves are disturbances in a field rather than in a material medium.
EM waves can travel through a vacuum.
EM waves travel at the speed of light.

NOTE-TAKING HANDBOOK

Outline

1. Copy the chapter title and headings from the book in the form of an outline.
2. Add notes that summarize in your own words what you read.

Cell Processes

I. Cells capture and release energy. — 1st key idea
   A. All cells need energy. — 1st subpoint of I
   B. Some cells capture light energy. — 2nd subpoint of I
      1. Process of photosynthesis — 1st detail about B
      2. Chloroplasts (site of photosynthesis) — 2nd detail about B
      3. Carbon dioxide and water as raw materials
      4. Glucose and oxygen as products
   C. All cells release energy.
      1. Process of cellular respiration
      2. Fermentation of sugar to carbon dioxide
      3. Bacteria that carry out fermentation

II. Cells transport materials through membranes.
   A. Some materials move by diffusion.
      1. Particle movement from higher to lower concentrations
      2. Movement of water through membrane (osmosis)
   B. Some transport requires energy.
      1. Active transport
      2. Examples of active transport

Correct Outline Form
Include a title.

Arrange key ideas, subpoints, and details as shown.

Indent the divisions of the outline as shown.

Use the same grammatical form for items of the same rank. For example, if A is a sentence, B must also be a sentence.

You must have at least two main ideas or subpoints. That is, every A must be followed by a B, and every 1 must be followed by a 2.

Concept Map

1. Write an important concept in a large oval.
2. Add details related to the concept in smaller ovals.
3. Write linking words on arrows that connect the ovals.

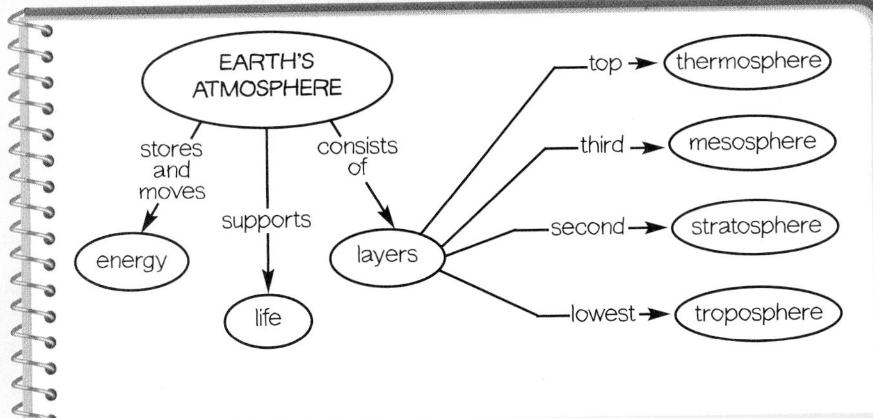

The main ideas or concepts can often be found in the blue headings. An example is "The atmosphere stores and moves energy." Use nouns from these concepts in the ovals, and use the verb or verbs on the lines.

Venn Diagram

1. Draw two overlapping circles, one for each item that you are comparing.
2. In the overlapping section, list the characteristics that are shared by both items.
3. In the outer sections, list the characteristics that are peculiar to each item.
4. Write a summary that describes the information in the Venn diagram.

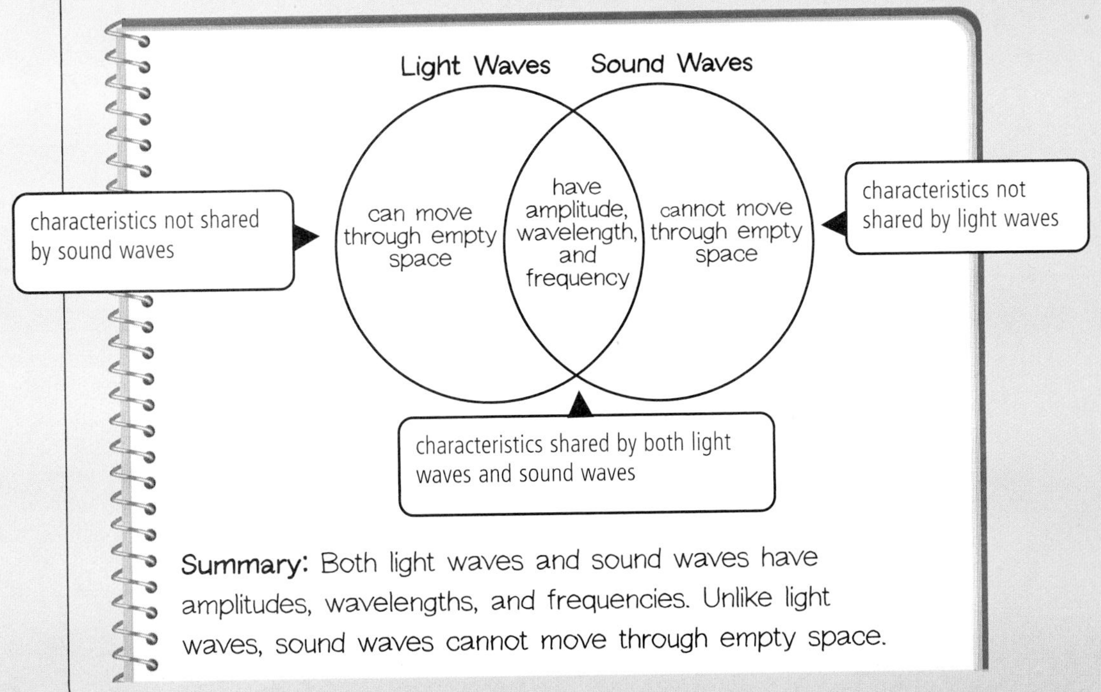

NOTE-TAKING HANDBOOK

Vocabulary Strategies

Important terms are highlighted in this book. A definition of each term can be found in the sentence or paragraph where the term appears. You can also find definitions in the Glossary. Taking notes about vocabulary terms helps you understand and remember what you read.

Description Wheel

1. Write a term inside a circle.
2. Write words that describe the term on "spokes" attached to the circle.

When studying for a test with a friend, read the phrases on the spokes one at a time until your friend identifies the correct term.

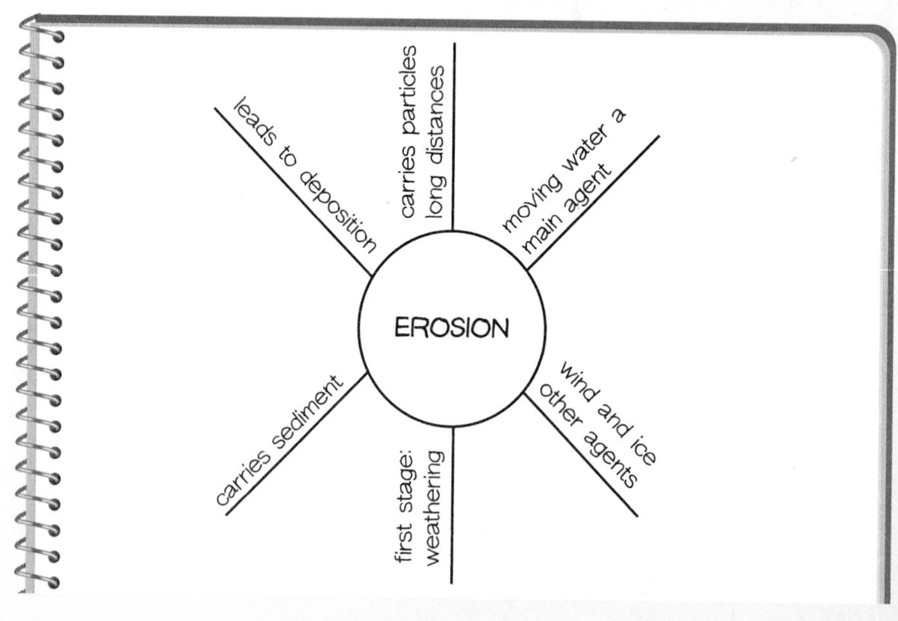

Four Square

1. Write a term in the center.
2. Write details in the four areas around the term.

Definition	Characteristics
any living thing	needs food, water, air; needs energy; grows, develops, reproduces
ORGANISM	
Examples	**Nonexamples**
dogs, cats, birds, insects, flowers, trees	rocks, water, dirt

Include a definition, some characteristics, and examples. You may want to add a formula, a sketch, or examples of things that the term does *not* name.

NOTE-TAKING HANDBOOK

Frame Game

1. Write a term in the center.
2. Frame the term with details.

Include examples, descriptions, sketches, or sentences that use the term in context. Change the frame to fit each new term.

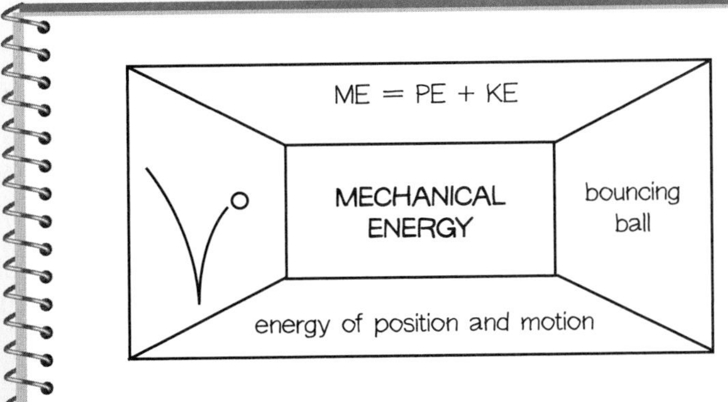

Magnet Word

1. Write a term on the magnet.
2. On the lines, add details related to the term.

You can also use phrases or sentences on the lines.

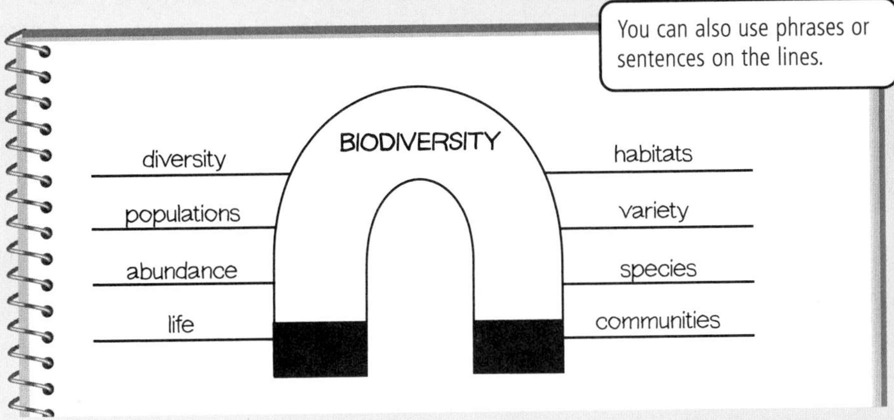

Word Triangle

1. Write a term and its definition in the bottom section.
2. In the middle section, write a sentence in which the term is used correctly.
3. In the top section, draw a small picture to illustrate the term.

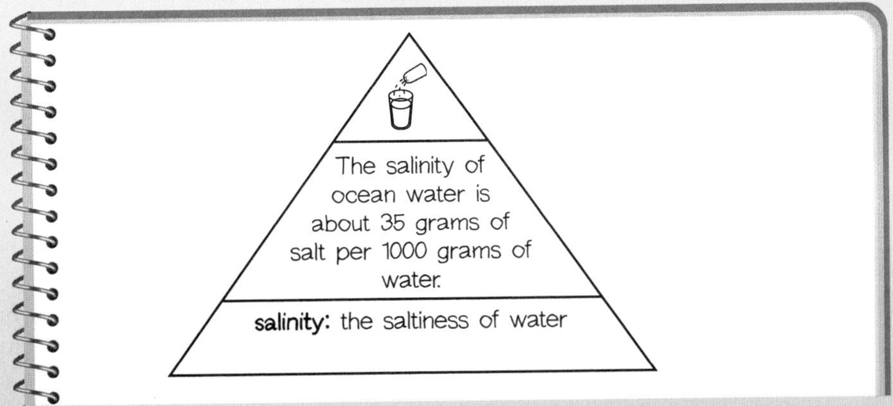

The salinity of ocean water is about 35 grams of salt per 1000 grams of water.
salinity: the saltiness of water

NOTE-TAKING HANDBOOK

Glossary

GLOSSARY

A

adaptation
A characteristic, a behavior, or any inherited trait that makes a species able to survive and reproduce in a particular environment. (p. xxi)

adaptación Una característica, un comportamiento o cualquier rasgo heredado que permite a una especie sobrevivir o reproducirse en un medio ambiente determinado.

addiction
A physical or psychological need for a habit-forming substance, such as alcohol or drugs. (p. 145)

adicción Una necesidad física o psicológica de una sustancia que forma hábito, como el alcohol o las drogas.

adolescence (AD-uhl-EHS-uhns)
The stage of life from the time a human body begins to mature sexually to adulthood. (p. 135)

adolescencia La etapa de la vida que va desde que el cuerpo humano empieza a madurar sexualmente hasta la edad adulta.

adulthood
The stage of life that begins once a human body completes its growth and reaches sexual maturity. (p. 136)

edad adulta La etapa de la vida que empieza una vez que el cuerpo humano completa su crecimiento y alcanza la madurez sexual.

antibiotic
A medicine that can block the growth and reproduction of bacteria. (p. 81)

antibiótico Una medicina que puede impedir el crecimiento y la reproducción de las bacterias.

antibody
A protein produced by some white blood cells to attack specific foreign materials. (p. 75)

anticuerpo Una proteína producida por algunos glóbulos blancos para atacar materiales extraños específicos.

antigen
A marker that a pathogen carries and that stimulates the production of antibodies. (p. 78)

antígeno Un marcador que lleva un patógeno y que estimula la producción de anticuerpos.

appendicular skeleton (AP-uhn-DIHK-yuh-luhr)
The bones of the skeleton that function to allow movement, such as arm and leg bones. (p. 16)

esqueleto apendicular Los huesos del esqueleto cuya función es permitir el movimiento, como los huesos del brazo y los huesos de la pierna.

artery
A blood vessel with strong walls that carries blood away from the heart. (p. 69)

arteria Un vaso sanguíneo con paredes fuertes que lleva la sangre del corazón hacia otras partes del cuerpo.

atom
The smallest particle of an element that has the chemical properties of that element.

átomo La partícula más pequeña de un elemento que tiene las propiedades químicas de ese elemento.

axial skeleton
The central part of the skeleton, which includes the cranium, the spinal column, and the ribs. (p. 16)

esqueleto axial La parte central del esqueleto que incluye al cráneo, a la columna vertebral y a las costillas.

B

bacteria (bak-TEER-ee-uh)
A large group of one-celled organisms that sometimes cause disease. *Bacteria* is a plural word; the singular is *bacterium.* (p. 149)

bacterias Un grupo grande de organismos unicelulares que algunas veces causan enfermedades.

biodiversity
The number and variety of living things found on Earth or within an ecosystem. (p. xxi)

biodiversidad La cantidad y variedad de organismos vivos que se encuentran en la Tierra o dentro de un ecosistema.

blood
A fluid in the body that delivers oxygen and other materials to cells and removes carbon dioxide and other wastes. (p. 65)

sangre Un fluido en el cuerpo que reparte oxígeno y otras sustancias a las células y elimina dióxido de carbono y otros desechos.

C

capillary
A narrow blood vessel that connects arteries with veins. (p. 69)

capilar Un vaso sanguíneo angosto que conecta a las arterias con las venas.

cardiac muscle
The muscle that makes up the heart. (p. 24)

músculo cardiaco El músculo del cual está compuesto el corazón.

cell
The smallest unit that is able to perform the basic functions of life. (p. xv)

célula La unidad más pequeña capaz de realizar las funciones básicas de la vida.

cellular respiration
A process in which cells use oxygen to release energy stored in sugars. (p. 38)

respiración celular Un proceso en el cual las células usan oxígeno para liberar energía almacenada en las azúcares.

central nervous system
The brain and spinal cord. The central nervous system communicates with the rest of the nervous system through electrical signals sent to and from neurons. (p. 104)

sistema nervioso central El cerebro y la médula espinal. El sistema nervioso central se comunica con el resto del sistema nervioso mediante señales eléctricas enviadas hacia y desde las neuronas.

childhood
The stage of life after infancy and before the beginning of sexual maturity. (p. 134)

niñez La etapa de la vida después de la infancia y antes del comienzo de la madurez sexual.

circulatory system
The group of organs, consisting of the heart and blood vessels, that circulates blood through the body. (p. 65)

sistema circulatorio El grupo de órganos, que consiste del corazón y los vasos sanguíneos, que hace circular la sangre por el cuerpo.

classification
The systematic grouping of different types of organisms by their shared characteristics.

clasificación La agrupación sistemática de diferentes tipos de organismos en base a las características que comparten.

compact bone
The tough, hard outer layer of a bone. (p. 15)

hueso compacto La capa exterior, resistente y dura de un hueso.

compound
A substance made up of two or more different types of atoms bonded together.

compuesto Una sustancia formada por dos o más diferentes tipos de átomos enlazados.

cycle
n. A series of events or actions that repeat themselves regularly; a physical and/or chemical process in which one material continually changes locations and/or forms. Examples include the water cycle, the carbon cycle, and the rock cycle.
v. To move through a repeating series of events or actions.

ciclo Una serie de eventos o acciones que se repiten regularmente; un proceso físico y/o químico en el cual un material cambia continuamente de lugar y/o forma. Ejemplos: el ciclo del agua, el ciclo del carbono y el ciclo de las rocas.

D

data
Information gathered by observation or experimentation that can be used in calculating or reasoning. *Data* is a plural word; the singular is *datum.*

datos Información reunida mediante observación o experimentación y que se puede usar para calcular o para razonar.

density
A property of matter representing the mass per unit volume.

densidad Una propiedad de la materia que representa la masa por unidad de volumen.

R53

GLOSSARY

dermis
The inner layer of the skin. (p. 84)

dermis La capa interior de la piel.

digestion
The process of breaking down food into usable materials. (p. 46)

digestión El proceso de descomponer el alimento en sustancias utilizables.

digestive system
The structures in the body that work together to transform the energy and materials in food into forms the body can use. (p. 46)

sistema digestivo Las estructuras en el cuerpo que trabajan juntas para transformar la energía y las sustancias en el alimento a formas que el cuerpo puede usar.

DNA
The genetic material found in all living cells that contains the information needed for an organism to grow, maintain itself, and reproduce. Deoxyribonucleic acid (dee-AHK-see-RY-boh-noo-KLEE-ihk).

ADN El material genético que se encuentra en todas las céulas vivas y que contiene la información necesaria para que un organismo crezca, se mantenga a sí mismo y se reproduzca. Ácido desoxiribunucleico.

E

element
A substance that cannot be broken down into a simpler substance by ordinary chemical changes. An element consists of atoms of only one type.

elemento Una sustancia que no puede descomponerse en otra sustancia más simple por medio de cambios químicos normales. Un elemento consta de átomos de un solo tipo.

embryo (EHM-bree-OH)
A multicellular organism, plant or animal, in its earliest stages of development. (p. 121)

embrión Una planta o un animal en su estadio mas temprano de desarrollo.

endocrine system
A group of organs called glands and the hormones they produce that help regulate conditions inside the body. (p. 110)

sistema endocrino Un grupo de órganos llamados glándulas y las hormonas que producen que ayudan a regular las condiciones dentro del cuerpo.

energy
The ability to do work or to cause a change. For example, the energy of a moving bowling ball knocks over pins; energy from food allows animals to move and to grow; and energy from the Sun heats Earth's surface and atmosphere, which causes air to move.

energía La capacidad para trabajar o causar un cambio. Por ejemplo, la energía de una bola de boliche en movimiento tumba los pinos; la energía proveniente de su alimento permite a los animales moverse y crecer; la energía del Sol calienta la superficie y la atmósfera de la Tierra, lo que ocasiona que el aire se mueva.

environment
Everything that surrounds a living thing. An environment is made up of both living and nonliving factors. (p. xix)

medio ambiente Todo lo que rodea a un organismo vivo. Un medio ambiente está compuesto de factores vivos y factores sin vida.

epidermis
The outer layer of the skin. (p. 84)

epidermis La capa exterior de la piel.

experiment
An organized procedure to study something under controlled conditions. (p. xxiv)

experimento Un procedimiento organizado para estudiar algo bajo condiciones controladas.

extinction
The permanent disappearance of a species. (p. xxi)

extinción La desaparición permanente de una especie.

F

fertilization
Part of the process of sexual reproduction in which a male reproductive cell and a female reproductive cell combine to form a new cell that can develop into a new organism. (p. 121)

fertilización El proceso mediante el cual una célula reproductiva masculina y una célula reproductiva femenina se combinan para formar una nueva célula que puede convertirse en un organismo nuevo.

G

genetic material
The nucleic acid DNA that is present in all living cells and contains the information needed for a cell's growth, maintenance, and reproduction.

material genético El ácido nucleico ADN, ue esta presente en todas las células vivas y que contiene la información necesaria para el crecimiento, el mantenimiento y la reproducción celular.

gland
An organ in the body that produces a specific substance, such as a hormone. (p. 112)

glándula Un órgano en el cuerpo que produce una sustancia específica, como una hormona.

H

homeostasis (HOH-mee-oh-STAY-sihs)
The process by which an organism or cell maintains the internal conditions needed for health and functioning, regardless of outside conditions. (p. 12)

homeostasis El proceso mediante el cual un organismo o una célula mantienen las condiciones internas necesarias para la salud y el funcionamiento, independientemente de las condiciones externas.

hormone
A chemical that is made in one organ and travels through the blood to another organ. (p. 111)

hormona Una sustancia química que se produce en un órgano y viaja por la sangre a otro órgano.

hypothesis
A tentative explanation for an observation or phenomenon. A hypothesis is used to make testable predictions. (p. xxiv)

hipótesis Una explicación provisional de una observación o de un fenómeno. Una hipótesis se usa para hacer predicciones que se pueden probar.

I, J, K

immune system
A group of organs that provides protection against disease-causing agents. (p. 75)

sistema immune o inmunológico Un grupo de órganos que provee protección contra agentes que causan enfermedades.

immunity
Resistance to a disease. Immunity can result from antibodies formed in the body during a previous attack of the same illness. (p. 80)

inmunidad La resistencia a una enfermedad. La inmunidad puede resultar de anticuerpos formados en el cuerpo durante un ataque previo de la misma enfermedad.

infancy
The stage of life that begins at birth and ends when a baby begins to walk. (p. 134)

infancia La etapa de la vida que inicia al nacer y termina cuando el bebe empieza a caminar.

integumentary system (ihn-TEHG-yu-MEHN-tuh-ree)
The body system that includes the skin and its associated structures. (p. 83)

sistema tegumentario El sistema corporal que incluye a la piel y a sus estructuras asociadas.

interaction
The condition of acting or having an influence upon something. Living things in an ecosystem interact with both the living and nonliving parts of their environment. (p. xix)

interacción La condición de actuar o influir sobre algo. Los organismos vivos en un ecosistema interactúan con las partes vivas y las partes sin vida de su medio ambiente.

involuntary muscle
A muscle that moves without conscious control. (p. 24)

músculo involuntario Un músculo que se mueve sin control consciente.

L

law
In science, a rule or principle describing a physical relationship that always works in the same way under the same conditions. The law of conservation of energy is an example.

ley En las ciencias, una regla o un principio que describe una relación física que siempre funciona de la misma manera bajo las mismas condiciones. La ley de la conservación de la energía es un ejemplo.

M

mass
A measure of how much matter an object is made of.

masa Una medida de la cantidad de materia de la que está compuesto un objeto.

matter
Anything that has mass and volume. Matter exists ordinarily as a solid, a liquid, or a gas.

materia Todo lo que tiene masa y volumen. Generalmente la materia existe como sólido, líquido o gas.

menstruation
A period of about five days during which blood and tissue exit the body through the vagina. (p. 119)

menstruación Un período de aproximadamente cinco días durante el cual salen del cuerpo sangre y tejido por la vagina.

microorganism
A very small organism that can be seen only with a microscope. Bacteria are examples of microorganisms. (p. 148)

microorganismo Un organismo muy pequeño que solamente puede verse con un microscopio. Las bacterias son ejemplos de microorganismos.

molecule
A group of atoms that are held together by covalent bonds so that they move as a single unit.

molécula Un grupo de átomos que están unidos mediante enlaces covalentes de tal manera que se mueven como una sola unidad.

muscular system
The muscles of the body that, together with the skeletal system, function to produce movement. (p. 23)

sistema muscular Los músculos del cuerpo que, junto con el sistema óseo, sirven para producir movimiento.

N

neuron
A nerve cell. (p. 105)

neurona Una célula nerviosa.

nutrient (NOO-tree-uhnt)
A substance that an organism needs to live. Examples include water, minerals, and materials that come from the breakdown of food particles. (p. 45)

nutriente Una sustancia que un organismo necesita para vivir. Ejemplos incluyen agua, minerales y sustancias que provienen de la descomposición de partículas de alimento.

nutrition
The study of the materials that nourish the body. (p. 140)

nutrición El estudio de las sustancias que dan sustento al cuerpo.

O

organ
A structure in a plant or animal that is made up of different tissues working together to perform a particular function. (p. 11)

órgano Una estructura en una planta o en un animal compuesta de diferentes tejidos que trabajan juntos para realizar una función determinada.

organism
An individual living thing, made up of one or many cells, that is capable of growing and reproducing.

organismo Un individuo vivo, compuesto de una o muchas células, que es capaz de crecer y reproducirse.

organ system
A group of organs that together perform a function that helps the body meet its needs for energy and materials. (p. 12)

sistema de órganos Un grupo de órganos que juntos realizan una función que ayuda al cuerpo a satisfacer sus necesidades energéticas y de materiales.

P, Q

pathogen
An agent that causes disease. (p. 74)

patógeno Un agente que causa una enfermedad.

peripheral nervous system
The part of the nervous system that lies outside the brain and spinal cord. (p. 106)

sistema nervioso periférico La parte del sistema nervioso que se encuentra fuera del cerebro y la médula espinal.

peristalsis (PEHR-ih-STAWL-sihs)
Wavelike contractions of smooth muscles in the organs of the digestive tract. The contractions move food through the digestive system. (p. 46)

peristalsis Contracciones ondulares de músculos lisos en los órganos del tracto digestivo. Las contracciones mueven el alimento por el sistema digestivo.

R

red blood cell
A type of blood cell that picks up oxygen in the lungs and delivers it to cells throughout the body. (p. 67)

glóbulos rojos Un tipo de célula sanguínea que toma oxígeno en los pulmones y lo transporta a células en todo el cuerpo.

resistance
The ability of an organism to protect itself from a disease or the effects of a substance. (p. 153)

resistencia La habilidad de un organismo para protegerse de una enfermedad o de los efectos de una sustancia.

respiratory system
A system that interacts with the environment and with other body systems to bring oxygen to the body and remove carbon dioxide. (p. 37)

sistema respiratorio Un sistema que interactúa con el medio ambiente y con otros sistemas corporales para traer oxígeno al cuerpo y eliminar dióxido de carbono.

S

skeletal muscle
A muscle that attaches to the skeleton. (p. 24)

músculo esquelético Un músculo que está sujeto al esqueleto.

skeletal system
The framework of bones that supports the body, protects internal organs, and anchors all the body's movement. (p. 14)

sistema óseo El armazón de huesos que sostiene al cuerpo, protege a los órganos internos y sirve de ancla para todo el movimiento del cuerpo.

smooth muscle
Muscle that performs involuntary movement and is found inside certain organs, such as the stomach. (p. 24)

músculo liso Músculos que realizan movimiento involuntario y se encuentran dentro de ciertos órganos, como el estómago.

species
A group of living things that are so closely related that they can breed with one another and produce offspring that can breed as well. (p. xxi)

especie Un grupo de organismos que están tan estrechamente relacionados que pueden aparearse entre sí y producir crías que también pueden aparearse.

spongy bone
Strong, lightweight tissue inside a bone. (p. 15)

hueso esponjoso Tejido fuerte y de peso ligero dentro de un hueso.

stimulus
Something that causes a response in an organism or a part of the body. (p. 102)

estímulo Algo que causa una respuesta en un organismo o en una parte del cuerpo.

system
A group of objects or phenomena that interact. A system can be as simple as a rope, a pulley, and a mass. It also can be as complex as the interaction of energy and matter in the four parts of the Earth system.

sistema Un grupo de objetos o fenómenos que interactúan. Un sistema puede ser algo tan sencillo como una cuerda, una polea y una masa. También puede ser algo tan complejo como la interacción de la energía y la materia en las cuatro partes del sistema de la Tierra.

T

technology
The use of scientific knowledge to solve problems or engineer new products, tools, or processes.

tecnología El uso de conocimientos científicos para resolver problemas o para diseñar nuevos productos, herramientas o procesos.

GLOSSARY

theory
In science, a set of widely accepted explanations of observations and phenomena. A theory is a well-tested explanation that is consistent with all available evidence.

teoría En las ciencias, un conjunto de explicaciones de observaciones y fenómenos que es ampliamente aceptado. Una teoría es una explicación bien probada que es consecuente con la evidencia disponible.

tissue
A group of similar cells that are organized to do a specific job. (p.10)

tejido Un grupo de células parecidas que juntas realizan una función específica en un organismo.

U

urinary system
A group of organs that filter waste from an organism's blood and excrete it in a liquid called urine. (p. 53)

sistema urinario Un grupo de órganos que filtran desechos de la sangre de un organismo y los excretan en un líquido llamado orina.

urine
Liquid waste that is secreted by the kidneys. (p. 53)

orina El desecho líquido que secretan los riñones.

V, W, X, Y, Z

vaccine
A small amount of a weakened pathogen that is introduced into the body to stimulate the production of antibodies. (p. 78)

vacuna Una pequeña cantidad de un patógeno debilitado que se introduce al cuerpo para estimular la producción de anticuerpos.

variable
Any factor that can change in a controlled experiment, observation, or model. (p. R30)

variable Cualquier factor que puede cambiar en un experimento controlado, en una observación o en un modelo.

vein
A blood vessel that carries blood back to the heart. (p. 69)

vena Un vaso sanguíneo que lleva la sangre de regreso al corazón.

virus
A nonliving, disease-causing particle that uses the materials inside cells to reproduce. A virus consists of genetic material enclosed in a protein coat. (p. 149)

virus Una particular sin vida, que causa enfermedad y que usa los materiales dentro de las células para reproducirse. Un virus consiste de material genético encerrado en una cubierta proteica.

volume
An amount of three-dimensional space, often used to describe the space that an object takes up.

volumen Una cantidad de espacio tridimensional; a menudo se usa este término para describir el espacio que ocupa un objeto.

voluntary muscle
A muscle that can be moved at will. (p. 24)

músculo voluntario Un músculo que puede moverse a voluntad.

Index

Page numbers for definitions are printed in **boldface** type.
Page numbers for illustrations, maps, and charts are printed in *italics.*

A

B

C

D

INDEX

E

F

G

H

I, J, K

INDEX

L

M

N

INDEX

P, Q

R

S

T

U

V

W, X, Y, Z

INDEX

Acknowledgements

Photography

Cover RNHRD NHS Trust; **iii** *left (top to bottom)* Photograph of James Trefil by Evan Cantwell; Photograph of Rita Ann Calvo by Joseph Calvo; Photograph of Linda Carnine by Amilcar Cifuentes; Photograph of Sam Miller by Samuel Miller; *right (top to bottom)* Photograph of Kenneth Cutler by Kenneth A. Cutler; Photograph of Donald Steely by Marni Stamm; Photograph of Vicky Vachon by Redfern Photographics; **vi** © Larry Dale Gordon/Getty Images; **vii** © Professors P.M. Motta & S. Correr/Photo Researchers, Inc.; **ix** *top* © Gunter Marx Photography/Corbis, *bottom left* Ken O'Donoghue; *bottom right* Frank Siteman; **xiv-xv** © Mark Hambin/Age Fotostock; **xvi-xvii** © Georgette Duowma/Taxi/Getty Images; **xviii-xix** © Ron Sanford/Corbis; **xx-xxi** © Nick Vedros & Assoc./Stone/Getty Images; **xxii** *left* © Michael Gadomski/Animals Animals, *right* © Shin Yoshino/ Minden Pictures; **xxiii** © Laif Elleringmann/Aurora Photos; **xxiv** © Pascal Goetgheluck/Photo Researchers, Inc.; **xxv** *left* © David Parker/Photo Researchers, Inc., *right* James King-Holmes/Photo Researchers, Inc., *bottom* Courtesy, Sinsheimer Labs/University of California, Santa Cruz; **xxvi-xxvii** *background* © Maximillian Stock/Photo Researchers, Inc.; **xxvi** *bottom* Courtesy, John Lair, Jewish Hospital, University of Louisville; **xxvii** *top* © Brand X Pictures/Alamy, *center* Courtesy, AbioMed; **xxxii** *bottom left* Chedd-Angier Production Company; **2–3** © Peter Byron/PhotoEdit; **3** *top right* © ISM/Phototake; **4** *top* © Wellcome Department of Cognitive Neurology/Photo Researchers, Inc., *bottom* Chedd-Angier Production Company; **5** © Myrleen Ferguson Cate/PhotoEdit; **6–7** © Chris Hamilton/Corbis; **7** *top* Frank Siteman, *bottom* Ken O'Donoghue; **9** © SuperStock; **10** Frank Siteman; **11** *left* © Martin Rotker/Phototake; **12** © SW Production/Index Stock Imagery/PictureQuest; **13** *background* © Hulton-Deutsch Collection/Corbis, *center* © Underwood & Underwood/Corbis; **14** Frank Siteman; **15** © Prof. P. Motta/Dept. of Anatomy/University "La Sapienza," Rome/Photo Researchers, Inc.; **16** © Photodisc/Getty Images; **18** *bottom* © Science Photo Library/Photo Researchers, Inc., *bottom left* © Zephyr/Photo Researchers, Inc.; **19** *top* © Zephyr/Photo Researchers, Inc., bottom Frank Siteman; **20** *top left* © Stock Image/SuperStock, top right © Science Photo Library/Photo Researchers, Inc.; **21** © Dennis Kunkel/Phototake; **22** Frank Siteman; **23** © Kevin R. Morris/Corbis; **25** *background* © Mary Kate Denny/PhotoEdit, *top* © Martin Rotker/Phototake, *left* © Triarch/Visuals Unlimited, *bottom* © Eric Grave/Phototake; **26** © Ron Frehm/AP Wide World Photos; **27** © Jeff Greenberg/PhotoEdit; **28** *top* © Gunter Marx Photography/Corbis, *bottom, all* Frank Siteman; **30** © Martin Rotker/Phototake; **31** *top* © Stock Image/SuperStock; **34–35** © Larry Dale Gordon/Getty Images; **35** *top* Frank Siteman, *bottom* Ken O'Donoghue; **37** Frank Siteman; **38** © Amos Nachoum/Corbis; **39** Ken O'Donoghue; **41** *bottom left* © Michael Newman/PhotoEdit, *bottom right* © Science Photo Library/Photo Researchers, Inc.; **43** © Kennan Harvey/Getty Images; **44** *background* © Jim Cummins/Getty Images, *center* © Steve Casimiro/Getty Images; **45** Ken O'Donoghue; **47** Ken O'Donoghue; **48** © Professors P. Motta & A. Familiari/University "La Sapienza," Rome/Photo Researchers, Inc.; **49** © David Young-Wolff/PhotoEdit; **50** © Dr. Gladden Willis/Visuals Unlimited; **51** © David Gifford/SPL/Custom Medical Stock Photo; **52** Frank Siteman; **55** © LWA-Dann Tardif/Corbis; **56** *top* © Myrleen Ferguson Cate/PhotoEdit, *bottom left* Ken O'Donoghue, *bottom right* Frank Siteman; **57** Frank Siteman; **59** © Professors P. Motta & A. Familiari/University "La Sapienza," Rome/Photo Researchers, Inc.; **62–63** © Professors P.M. Motta & S. Correr/Photo Researchers, Inc.; **63** *both* Frank Siteman; **65** Frank Siteman; **67** © Science Photo Library/Photo Researchers, Inc.; **68** © Myrleen Ferguson Cate/PhotoEdit; **69** © Susumu Nishinaga/Photo Researchers, Inc.; **71** © Journal-Courier/The Image Works; **72** *top left* © Michael Newman/PhotoEdit, *bottom left* Ken O'Donoghue, *center right, bottom right* Frank Siteman; **74** Frank Siteman; **75** *top* © Eddy Gray/Photo Researchers, Inc., *bottom* © Mary Kate Denny/PhotoEdit; **76** © Science Photo Library/Photo Researchers, Inc.; **77** *top* © Dr. P. Marazzi/Photo Researchers, Inc., *top right* © Dr. Jeremy Burgess/Photo Researchers, Inc.; **78** © Science Photo Library/Photo Researchers, Inc.; **79** Frank Siteman; **80** © Bob Daemmrich/The Image Works; **81** © Richard Lord/The Image Works; **82** *background* © SCIMAT/Photo Researchers, Inc., *inset* © Vision/Photo Researchers; 83 Ken O'Donoghue; **84** RMIP/Richard Haynes; **85** Frank Siteman; **86** *top inset* © Dennis Kunkel/Phototake, *bottom inset* © Andrew Syred/Photo Researchers, Inc., *center* © Photodisc/Getty Images; **87** *all* © Eric Schrempp/Photo Researchers, Inc.; **88** © The Image Bank/Getty Images; **89** *background* © James King-Holmes/Photo Researchers, Inc., *top right* © Sygma/Corbis, *bottom right* © David Hanson; **94** *top* © Hulton Archive/Getty Images, *bottom* © Simon Fraser/Photo Researchers, Inc.; **95** *top* © Bettmann/Corbis, *center* © Underwood & Underwood/Corbis, *bottom* © George Bernard/Photo Researchers, Inc.; **96** *top left* © Collection CNRI/Phototake, *top right* © Geoff Tompkinson/ Photo Researchers, Inc., *bottom* © Josh Sher/Photo Researchers, Inc.; **97** *top* © Simon Fraser/Photo Researchers, Inc., *bottom* © GJLP/Photo Researchers, Inc.; **98–99** © Photo Researchers, Inc.; **99** *top* Ken O'Donoghue, *center* © Photospin; **101** Ken O'Donoghue; **103** RMIP/Richard Haynes; **104** © David Young-Wolff/PhotoEdit; **107** © Royalty-Free/Corbis; **108** *top* © Ed Young/Corbis, *bottom* Ken O'Donoghue; **109** *top* Frank Siteman, *bottom* Ken O'Donoghue; **110** © David Young-Wolff/PhotoEdit; **111** © Kwame Zikomo/SuperStock; **112** © ISM/Phototake; **114** Frank Siteman; **115** © CNRI/ Photo Researchers, Inc.; **117** *left* © David Young-Wolff/PhotoEdit, *top right* © Francine Prokoski/Photo Researchers, Inc., *bottom right* © Glenn Oakley/ImageState/PictureQuest; 118 Ken O'Donoghue; 121 © Dennis Kunkel/Phototake; 123 background © Yoav Levy/Phototake, *top left* © Dr. Yorgos Nikas/ Photo Researchers, Inc.; **124** © David Degnan/Corbis; **125** *left* © Christopher Brown/Stock Boston, Inc./PictureQuest, *right* © Nissim Men/Photonica; **130–131** © Brooklyn Productions/Corbis; 131 top Ken O'Donoghue; 133 Frank Siteman; 134 © Tom Galliher/Corbis; 135 © Tom Stewart/Corbis; **136** © Novastock/Index Stock Imagery, Inc.; **138** *left* © Spencer Grant/PhotoEdit, *right* © Michael Newman/PhotoEdit; **139** *all* from STAGE MAKEUP, STEP BY STEP. Courtesy of Quarto Publishing, Inc.; 140 © Ed Young/Corbis; **141** © Ronnie Kaufman/Corbis; **142** © Photodisc/Getty Images; **144** © Don Smetzer/PhotoEdit; **146** © Brett Coomer/HO/AP Wide World Photos; **147** *left* © Eric Kamp/Index Stock Photography, Inc., *top right* © Ariel Skelley/Corbis; 148 Frank Siteman; 149 © Mediscan/Visuals Unlimited; 151 top left © Dr. Kari Lounatmaa/Science Photo Library/Photo Researchers, Inc., *top right* © Dr. Gopal Murti/Photo Researchers, Inc., *bottom left* © Professors P.M. Motta & F.M. Magliocca/PhotoResearchers, Inc., *bottom right* © Microworks/Phototake, *bottom right inset* © Andrew Spielman/Phototake; **152** © Mary Steinbacher/PhotoEdit; **153** © Srulik Haramary/Phototake; **154** *top left* © Kwame Zikomo/Superstock, bottom left Ken O'Donoghue, *bottom right* Frank Siteman; **155** Ken O'Donoghue; **156** © Mediscan/Visuals Unlimited; **r28** © Photodisc/Getty Images.

Illustration

Debbie Maizels **15, 26, 84, 90, 102, 104, 128**
Linda Nye **11, 30, 46, 66, 70, 90, 92, 119, 120, 121, 122, 123, 126**
Steve Oh/KO Studios **78, 105**
Bart Vallecoccia **11,17, 30, 32, 41, 42, 44, 46, 49, 53, 54, 58, 60, 68, 103, 105, 106, 113, 119, 120, 126**
Dan Stuckenschneider/Uhl Studios **r11-r19, r22, r32**

Content Standards: 5–8

A. Science as Inquiry

As a result of activities in grades 5–8, all students should develop

Abilities Necessary to do Scientific Inquiry

A.1 Identify questions that can be answered through scientific investigations. Students should develop the ability to refine and refocus broad and ill-defined questions. An important aspect of this ability consists of students' ability to clarify questions and inquiries and direct them toward objects and phenomena that can be described, explained, or predicted by scientific investigations. Students should develop the ability to identify their questions with scientific ideas, concepts, and quantitative relationships that guide investigation.

A.2 Design and conduct a scientific investigation. Students should develop general abilities, such as systematic observation, making accurate measurements, and identifying and controlling variables. They should also develop the ability to clarify their ideas that are influencing and guiding the inquiry, and to understand how those ideas compare with current scientific knowledge. Students can learn to formulate questions, design investigations, execute investigations, interpret data, use evidence to generate explanations, propose alternative explanations, and critique explanations and procedures.

A.3 Use appropriate tools and techniques to gather, analyze, and interpret data. The use of tools and techniques, including mathematics, will be guided by the question asked and the investigations students design. The use of computers for the collection, summary, and display of evidence is part of this standard. Students should be able to access, gather, store, retrieve, and organize data, using hardware and software designed for these purposes.

A.4 Develop descriptions, explanations, predictions, and models using evidence. Students should base their explanation on what they observed, and as they develop cognitive skills, they should be able to differentiate explanation from description—providing causes for effects and establishing relationships based on evidence and logical argument. This standard requires a subject matter knowledge base so the students can effectively conduct investigations, because developing explanations establishes connections between the content of science and the contexts within which students develop new knowledge.

A.5 Think critically and logically to make the relationships between evidence and explanations. Thinking critically about evidence includes deciding what evidence should be used and accounting for anomalous data. Specifically, students should be able to review data from a simple experiment, summarize the data, and form a logical argument about the cause-and-effect relationships in the experiment. Students should begin to state some explanations in terms of the relationship between two or more variables.

A.6 Recognize and analyze alternative explanations and predictions. Students should develop the ability to listen to and respect the explanations proposed by other students. They should remain open to and acknowledge different ideas and explanations, be able to accept the skepticism of others, and consider alternative explanations.

A.7 Communicate scientific procedures and explanations. With practice, students should become competent at communicating experimental methods, following instructions, describing observations, summarizing the results of other groups, and telling other students about investigations and explanations.

A.8 Use mathematics in all aspects of scientific inquiry. Mathematics is essential to asking and answering questions about the natural world. Mathematics can be used to ask questions; to gather, organize, and present data; and to structure convincing explanations.

Understandings about Scientific Inquiry

A.9.a Different kinds of questions suggest different kinds of scientific investigations. Some investigations involve observing and describing objects, organisms, or events; some involve collecting specimens; some involve experiments; some involve seeking more information; some involve discovery of new objects and phenomena; and some involve making models.

A.9.b Current scientific knowledge and understanding guide scientific investigations. Different scientific domains employ different methods, core theories, and standards to advance scientific knowledge and understanding.

A.9.c Mathematics is important in all aspects of scientific inquiry.

A.9.d Technology used to gather data enhances accuracy and allows scientists to analyze and quantify results of investigations.

A.9.e Scientific explanations emphasize evidence, have logically consistent arguments, and use scientific principles, models, and theories. The scientific community accepts and uses such explanations until displaced by better scientific ones. When such displacement occurs, science advances.

A.9.f Science advances through legitimate skepticism. Asking questions and querying other scientists' explanations is part of scientific inquiry. Scientists evaluate the explanations proposed by other scientists by examining evidence, comparing evidence, identifying faulty reasoning, pointing out statements that go beyond the evidence, and suggesting alternative explanations for the same observations.

A.9.g Scientific investigations sometimes result in new ideas and phenomena for study, generate new methods or procedures for an investigation, or develop new technologies to improve the collection of data. All of these results can lead to new investigations.

B. Physical Science

As a result of their activities in grades 5–8, all students should develop an understanding of

Properties and Changes of Properties in Matter

B.1.a A substance has characteristic properties, such as density, a boiling point, and solubility, all of which are independent of the amount of the sample. A mixture of substances often can be separated into the original substances using one or more of the characteristic properties.

B.1.b Substances react chemically in characteristic ways with other substances to form new substances (compounds) with different characteristic properties. In chemical reactions, the total mass is conserved. Substances often are placed in categories or groups if they react in similar ways; metals is an example of such a group.

B.1.c Chemical elements do not break down during normal laboratory reactions involving such treatments as heating, exposure to electric current, or reaction with acids. There are more than 100 known elements that combine in a multitude of ways to produce compounds, which account for the living and nonliving substances that we encounter.

Motions and Forces

B.2.a The motion of an object can be described by its position, direction of motion, and speed. That motion can be measured and represented on a graph.

B.2.b An object that is not being subjected to a force will continue to move at a constant speed and in a straight line.

B.2.c If more than one force acts on an object along a straight line, then the forces will reinforce or cancel one another, depending on their direction and magnitude. Unbalanced forces will cause changes in the speed or direction of an object's motion.

Transfer of Energy

B.3.a Energy is a property of many substances and is associated with heat, light, electricity, mechanical motion, sound, nuclei, and the nature of a chemical. Energy is transferred in many ways.

B.3.b Heat moves in predictable ways, flowing from warmer objects to cooler ones, until both reach the same temperature.

B.3.c Light interacts with matter by transmission (including refraction), absorption, or scattering (including reflection). To see an object, light from that object—emitted by or scattered from it—must enter the eye.

B.3.d Electrical circuits provide a means of transferring electrical energy when heat, light, sound, and chemical changes are produced.

B.3.e In most chemical and nuclear reactions, energy is transferred into or out of a system. Heat, light, mechanical motion, or electricity might all be involved in such transfers.

B.3.f The sun is a major source of energy for changes on the earth's surface. The sun loses energy by emitting light. A tiny fraction of that light reaches the earth, transferring energy from the sun to the earth. The sun's energy arrives as light with a range of wavelengths, consisting of visible light, infrared, and ultraviolet radiation.

C. Life Science

As a result of their activities in grades 5–8, all students should develop understanding of

Structure and Function in Living Systems

C.1.a Living systems at all levels of organization demonstrate the complementary nature of structure and function. Important levels of organization for structure and function include cells, organs, tissues, organ systems, whole organisms, and ecosystems.

C.1.b All organisms are composed of cells—the fundamental unit of life. Most organisms are single cells; other organisms, including humans, are multicellular.

C.1.c Cells carry on the many functions needed to sustain life. They grow and divide, thereby producing more cells. This requires that they take in nutrients, which they use to provide energy for the work that cells do and to make the materials that a cell or an organism needs.

C.1.d Specialized cells perform specialized functions in multicellular organisms. Groups of specialized cells cooperate to form a tissue, such as a muscle. Different tissues are in turn grouped together to form larger functional units, called organs. Each type of cell, tissue, and organ has a distinct structure and set of functions that serve the organism as a whole.

C.1.e The human organism has systems for digestion, respiration, reproduction, circulation, excretion, movement, control, and coordination, and for protection from disease. These systems interact with one another.

C.1.f Disease is a breakdown in structures or functions of an organism. Some diseases are the result of intrinsic failures of the system. Others are the result of damage by infection by other organisms.

Reproduction and Heredity

C.2.a Reproduction is a characteristic of all living systems; because no individual organism lives forever, reproduction is essential to the continuation of every species. Some organisms reproduce asexually. Other organisms reproduce sexually.

C.2.b In many species, including humans, females produce eggs and males produce sperm. Plants also reproduce sexually—the egg and sperm are produced in the flowers of flowering plants. An egg and sperm unite to begin development of a new individual. That new individual receives genetic information from its mother (via the egg) and its father (via the sperm). Sexually produced offspring never are identical to either of their parents.

C.2.c Every organism requires a set of instructions for specifying its traits. Heredity is the passage of these instructions from one generation to another.

C.2.d Hereditary information is contained in genes, located in the chromosomes of each cell. Each gene carries a single unit of information. An inherited trait of an individual can be determined by one or by many genes, and a single gene can influence more than one trait. A human cell contains many thousands of different genes.

C.2.e The characteristics of an organism can be described in terms of a combination of traits. Some traits are inherited and others result from interactions with the environment.

National Science Education Standards

Regulation and Behavior

C.3.a All organisms must be able to obtain and use resources, grow, reproduce, and maintain stable internal conditions while living in a constantly changing external environment.

C.3.b Regulation of an organism's internal environment involves sensing the internal environment and changing physiological activities to keep conditions within the range required to survive.

C.3.c Behavior is one kind of response an organism can make to an internal or environmental stimulus. A behavioral response requires coordination and communication at many levels, including cells, organ systems, and whole organisms. Behavioral response is a set of actions determined in part by heredity and in part from experience.

C.3.d An organism's behavior evolves through adaptation to its environment. How a species moves, obtains food, reproduces, and responds to danger are based in the species' evolutionary history.

Populations and Ecosystems

C.4.a A population consists of all individuals of a species that occur together at a given place and time. All populations living together and the physical factors with which they interact compose an ecosystem.

C.4.b Populations of organisms can be categorized by the function they serve in an ecosystem. Plants and some microorganisms are producers—they make their own food. All animals, including humans, are consumers, which obtain food by eating other organisms. Decomposers, primarily bacteria and fungi, are consumers that use waste materials and dead organisms for food. Food webs identify the relationships among producers, consumers, and decomposers in an ecosystem.

C.4.c For ecosystems, the major source of energy is sunlight. Energy entering ecosystems as sunlight is transferred by producers into chemical energy through photosynthesis. That energy then passes from organism to organism in food webs.

C.4.d The number of organisms an ecosystem can support depends on the resources available and abiotic factors, such as quantity of light and water, range of temperatures, and soil composition. Given adequate biotic and abiotic resources and no disease or predators, populations (including humans) increase at rapid rates. Lack of resources and other factors, such as predation and climate, limit the growth of populations in specific niches in the ecosystem.

Diversity and Adaptations of Organisms

C.5.a Millions of species of animals, plants, and microorganisms are alive today. Although different species might look dissimilar, the unity among organisms becomes apparent from an analysis of internal structures, the similarity of their chemical processes, and the evidence of common ancestry.

C.5.b Biological evolution accounts for the diversity of species developed through gradual processes over many generations. Species acquire many of their unique characteristics through biological adaptation, which involves the selection of naturally occurring variations in populations. Biological adaptations include changes in structures, behaviors, or physiology that enhance survival and reproductive success in a particular environment.

C.5.c Extinction of a species occurs when the environment changes and the adaptive characteristics of a species are insufficient to allow its survival. Fossils indicate that many organisms that lived long ago are extinct. Extinction of species is common; most of the species that have lived on the earth no longer exist.

D. Earth and Space Science

As a result of their activities in grades 5–8, all students should develop an understanding of

Structure of the Earth System

D.1.a The solid earth is layered with a lithosphere; hot, convecting mantle; and dense, metallic core.

D.1.b Lithospheric plates on the scales of continents and oceans constantly move at rates of centimeters per year in response to movements in the mantle. Major geological events, such as earthquakes, volcanic eruptions, and mountain building, result from these plate motions.

D.1.c Land forms are the result of a combination of constructive and destructive forces. Constructive forces include crustal deformation, volcanic eruption, and deposition of sediment, while destructive forces include weathering and erosion.

D.1.d Some changes in the solid earth can be described as the "rock cycle." Old rocks at the earth's surface weather, forming sediments that are buried, then compacted, heated, and often recrystallized into new rock. Eventually, those new rocks may be brought to the surface by the forces that drive plate motions, and the rock cycle continues.

D.1.e Soil consists of weathered rocks and decomposed organic material from dead plants, animals, and bacteria. Soils are often found in layers, with each having a different chemical composition and texture.

D.1.f Water, which covers the majority of the earth's surface, circulates through the crust, oceans, and atmosphere in what is known as the "water cycle." Water evaporates from the earth's surface, rises and cools as it moves to higher elevations, condenses as rain or snow, and falls to the surface where it collects in lakes, oceans, soil, and in rocks underground.

D.1.g Water is a solvent. As it passes through the water cycle it dissolves minerals and gases and carries them to the oceans.

D.1.h The atmosphere is a mixture of nitrogen, oxygen, and trace gases that include water vapor. The atmosphere has different properties at different elevations.

D.1.i Clouds, formed by the condensation of water vapor, affect weather and climate.

D.1.j Global patterns of atmospheric movement influence local weather. Oceans have a major effect on climate, because water in the oceans holds a large amount of heat.

D.1.k Living organisms have played many roles in the earth system, including affecting the composition of the atmosphere, producing some types of rocks, and contributing to the weathering of rocks.

Earth's History

D.2.a The earth processes we see today, including erosion, movement of lithospheric plates, and changes in atmospheric composition, are similar to those that occurred in the past. Earth history is also influenced by occasional catastrophes, such as the impact of an asteroid or comet.

D.2.b Fossils provide important evidence of how life and environmental conditions have changed.

Earth in the Solar System

D.3.a The earth is the third planet from the sun in a system that includes the moon, the sun, eight other planets and their moons, and smaller objects, such as asteroids and comets. The sun, an average star, is the central and largest body in the solar system.

D.3.b Most objects in the solar system are in regular and predictable motion. Those motions explain such phenomena as the day, the year, phases of the moon, and eclipses.

D.3.c Gravity is the force that keeps planets in orbit around the sun and governs the rest of the motion in the solar system. Gravity alone holds us to the earth's surface and explains the phenomena of the tides.

D.3.d The sun is the major source of energy for phenomena on the earth's surface, such as growth of plants, winds, ocean currents, and the water cycle. Seasons result from variations in the amount of the sun's energy hitting the surface, due to the tilt of the earth's rotation on its axis and the length of the day.

E. Science and Technology

As a result of activities in grades 5–8, all students should develop

Abilities of Technological Design

E.1 Identify appropriate problems for technological design. Students should develop their abilities by identifying a specified need, considering its various aspects, and talking to different potential users or beneficiaries. They should appreciate that for some needs, the cultural backgrounds and beliefs of different groups can affect the criteria for a suitable product.

E.2 Design a solution or product. Students should make and compare different proposals in the light of the criteria they have selected. They must consider constraints—such as cost, time, trade-offs, and materials needed—and communicate ideas with drawings and simple models.

E.3 Implement a proposed design. Students should organize materials and other resources, plan their work, make good use of group collaboration where appropriate, choose suitable tools and techniques, and work with appropriate measurement methods to ensure adequate accuracy.

E.4 Evaluate completed technological designs or products. Students should use criteria relevant to the original purpose or need, consider a variety of factors that might affect acceptability and suitability for intended users or beneficiaries, and develop measures of quality with respect to such criteria and factors; they should also suggest improvements and, for their own products, try proposed modifications.

E.5 Communicate the process of technological design. Students should review and describe any completed piece of work and identify the stages of problem identification, solution design, implementation, and evaluation.

Understandings about Science and Technology

E.6.a Scientific inquiry and technological design have similarities and differences. Scientists propose explanations for questions about the natural world, and engineers propose solutions relating to human problems, needs, and aspirations. Technological solutions are temporary; technologies exist within nature and so they cannot contravene physical or biological principles; technological solutions have side effects; and technologies cost, carry risks, and provide benefits.

E.6.b Many different people in different cultures have made and continue to make contributions to science and technology.

E.6.c Science and technology are reciprocal. Science helps drive technology, as it addresses questions that demand more sophisticated instruments and provides principles for better instrumentation and technique. Technology is essential to science, because it provides instruments and techniques that enable observations of objects and phenomena that are otherwise unobservable due to factors such as quantity, distance, location, size, and speed. Technology also provides tools for investigations, inquiry, and analysis.

E.6.d Perfectly designed solutions do not exist. All technological solutions have trade-offs, such as safety, cost, efficiency, and appearance. Engineers often build in back-up systems to provide safety. Risk is part of living in a highly technological world. Reducing risk often results in new technology.

E.6.e Technological designs have constraints. Some constraints are unavoidable, for example, properties of materials, or effects of weather and friction; other constraints limit choices in the design, for example, environmental protection, human safety, and aesthetics.

E.6.f Technological solutions have intended benefits and unintended consequences. Some consequences can be predicted, others cannot.

F. Science in Personal and Social Perspectives

As a result of activities in grades 5–8, all students should develop understanding of

Personal Health

F.1.a Regular exercise is important to the maintenance and improvement of health. The benefits of physical fitness include maintaining healthy weight, having energy and strength for routine activities, good muscle tone, bone strength, strong heart/lung systems, and improved mental health. Personal exercise, especially developing cardiovascular endurance, is the foundation of physical fitness.

F.1.b The potential for accidents and the existence of hazards imposes the need for injury prevention. Safe living involves the development and use of safety precautions and the recognition of risk in personal decisions. Injury prevention has personal and social dimensions.

F.1.c The use of tobacco increases the risk of illness. Students should understand the influence of short-term social and psychological factors that lead to tobacco use, and the possible long-term detrimental effects of smoking and chewing tobacco.

F.1.d Alcohol and other drugs are often abused substances. Such drugs change how the body functions and can lead to addiction.

F.1.e Food provides energy and nutrients for growth and development. Nutrition requirements vary with body weight, age, sex, activity, and body functioning.

F.1.f Sex drive is a natural human function that requires understanding. Sex is also a prominent means of transmitting diseases. The diseases can be prevented through a variety of precautions.

F.1.g Natural environments may contain substances (for example, radon and lead) that are harmful to human beings. Maintaining environmental health involves establishing or monitoring quality standards related to use of soil, water, and air.

Populations, Resources, and Environments

F.2.a When an area becomes overpopulated, the environment will become degraded due to the increased use of resources.

F.2.b Causes of environmental degradation and resource depletion vary from region to region and from country to country.

Natural Hazards

F.3.a Internal and external processes of the earth system cause natural hazards, events that change or destroy human and wildlife habitats, damage property, and harm or kill humans. Natural hazards include earthquakes, landslides, wildfires, volcanic eruptions, floods, storms, and even possible impacts of asteroids.

F.3.b Human activities also can induce hazards through resource acquisition, urban growth, land-use decisions, and waste disposal. Such activities can accelerate many natural changes.

F.3.c Natural hazards can present personal and societal challenges because misidentifying the change or incorrectly estimating the rate and scale of change may result in either too little attention and significant human costs or too much cost for unneeded preventive measures.

Risks and Benefits

F.4.a Risk analysis considers the type of hazard and estimates the number of people that might be exposed and the number likely to suffer consequences. The results are used to determine the options for reducing or eliminating risks.

F.4.b Students should understand the risks associated with natural hazards (fires, floods, tornadoes, hurricanes, earthquakes, and volcanic eruptions), with chemical hazards (pollutants in air, water, soil, and food), with biological hazards (pollen, viruses, bacterial, and parasites), social hazards (occupational safety and transportation), and with personal hazards (smoking, dieting, and drinking).

F.4.c Individuals can use a systematic approach to thinking critically about risks and benefits. Examples include applying probability estimates to risks and comparing them to estimated personal and social benefits.

F.4.d Important personal and social decisions are made based on perceptions of benefits and risks.

Science and Technology in Society

F.5.a Science influences society through its knowledge and world view. Scientific knowledge and the procedures used by scientists influence the way many individuals in society think about themselves, others, and the environment. The effect of science on society is neither entirely beneficial nor entirely detrimental.

F.5.b Societal challenges often inspire questions for scientific research, and social priorities often influence research priorities through the availability of funding for research.

F.5.c Technology influences society through its products and processes. Technology influences the quality of life and the ways people act and interact. Technological changes are often accompanied by social, political, and economic changes that can be beneficial or detrimental to individuals and to society. Social needs, attitudes, and values influence the direction of technological development.

F.5.d Science and technology have advanced through contributions of many different people, in different cultures, at different times in history. Science and technology have contributed enormously to economic growth and productivity among societies and groups within societies.

F.5.e Scientists and engineers work in many different settings, including colleges and universities, businesses and industries, specific research institutes, and government agencies.

F.5.f Scientists and engineers have ethical codes requiring that human subjects involved with research be fully informed about risks and benefits associated with the research before the individuals choose to participate. This ethic extends to potential risks to communities and property. In short, prior knowledge and consent are required for research involving human subjects or potential damage to property.

F.5.g Science cannot answer all questions and technology cannot solve all human problems or meet all human needs. Students should understand the difference between scientific and other questions. They should appreciate what science and technology can reasonably contribute to society and what they cannot do. For example, new technologies often will decrease some risks and increase others.

G. History and Nature of Science

As a result of activities in grades 5–8, all students should develop understanding of

Science as a Human Endeavor

G.1.a Women and men of various social and ethnic backgrounds—and with diverse interests, talents, qualities, and motivations—engage in the activities of science, engineering, and related fields such as the health professions. Some scientists work in teams, and some work alone, but all communicate extensively with others.

G.1.b Science requires different abilities, depending on such factors as the field of study and type of inquiry. Science is very much a human endeavor, and the work of science relies on basic human qualities, such as reasoning, insight, energy, skill, and creativity—as well as on scientific habits of mind, such as intellectual honesty, tolerance of ambiguity, skepticism, and openness to new ideas.

National Science Education Standards

Nature of Science

G.2.a Scientists formulate and test their explanations of nature using observation, experiments, and theoretical and mathematical models. Although all scientific ideas are tentative and subject to change and improvement in principle, for most major ideas in science, there is much experimental and observational confirmation. Those ideas are not likely to change greatly in the future. Scientists do and have changed their ideas about nature when they encounter new experimental evidence that does not match their existing explanations.

G.2.b In areas where active research is being pursued and in which there is not a great deal of experimental or observational evidence and understanding, it is normal for scientists to differ with one another about the interpretation of the evidence or theory being considered. Different scientists might publish conflicting experimental results or might draw different conclusions from the same data. Ideally, scientists acknowledge such conflict and work towards finding evidence that will resolve their disagreement.

G.2.c It is part of scientific inquiry to evaluate the results of scientific investigations, experiments, observations, theoretical models, and the explanations proposed by other scientists. Evaluation includes reviewing the experimental procedures, examining the evidence, identifying faulty reasoning, pointing out statements that go beyond the evidence, and suggesting alternative explanations for the same observations. Although scientists may disagree about explanations of phenomena, about interpretations of data, or about the value of rival theories, they do agree that questioning, response to criticism, and open communication are integral to the process of science. As scientific knowledge evolves, major disagreements are eventually resolved through such interactions between scientists.

History of Science

G.3.a Many individuals have contributed to the traditions of science. Studying some of these individuals provides further understanding of scientific inquiry, science as a human endeavor, the nature of science, and the relationships between science and society.

G.3.b In historical perspective, science has been practiced by different individuals in different cultures. In looking at the history of many peoples, one finds that scientists and engineers of high achievement are considered to be among the most valued contributors to their culture.

G.3.c Tracing the history of science can show how difficult it was for scientific innovators to break through the accepted ideas of their time to reach the conclusions that we currently take for granted.

Project 2061
Benchmarks for Science Literacy

1. The Nature of Science

By the end of the 8th grade, students should know that

1.A The Scientific World View

1.A.1 When similar investigations give different results, the scientific challenge is to judge whether the differences are trivial or significant, and it often takes further studies to decide. Even with similar results, scientists may wait until an investigation has been repeated many times before accepting the results as correct.

1.A.2 Scientific knowledge is subject to modification as new information challenges prevailing theories and as a new theory leads to looking at old observations in a new way.

1.A.3 Some scientific knowledge is very old and yet is still applicable today.

1.A.4 Some matters cannot be examined usefully in a scientific way. Among them are matters that by their nature cannot be tested objectively and those that are essentially matters of morality. Science can sometimes be used to inform ethical decisions by identifying the likely consequences of particular actions but cannot be used to establish that some action is either moral or immoral.

1.B Scientific Inquiry

1.B.1 Scientists differ greatly in what phenomena they study and how they go about their work. Although there is no fixed set of steps that all scientists follow, scientific investigations usually involve the collection of relevant evidence, the use of logical reasoning, and the application of imagination in devising hypotheses and explanations to make sense of the collected evidence.

1.B.2 If more than one variable changes at the same time in an experiment, the outcome of the experiment may not be clearly attributable to any one of the variables. It may not always be possible to prevent outside variables from influencing the outcome of an investigation (or even to identify all of the variables), but collaboration among investigators can often lead to research designs that are able to deal with such situations.

1.B.3 What people expect to observe often affects what they actually do observe. Strong beliefs about what should happen in particular circumstances can prevent them from detecting other results. Scientists know about this danger to objectivity and take steps to try and avoid it when designing investigations and examining data. One safeguard is to have different investigators conduct independent studies of the same questions.

1.C The Scientific Enterprise

1.C.1 Important contributions to the advancement of science, mathematics, and technology have been made by different kinds of people, in different cultures, at different times.

1.C.2 Until recently, women and racial minorities, because of restrictions on their education and employment opportunities, were essentially left out of much of the formal work of the science establishment; the remarkable few who overcame those obstacles were even then likely to have their work disregarded by the science establishment.

1.C.3 No matter who does science and mathematics or invents things, or when or where they do it, the knowledge and technology that result can eventually become available to everyone in the world.

1.C.4 Scientists are employed by colleges and universities, business and industry, hospitals, and many government agencies. Their places of work include offices, classrooms, laboratories, farms, factories, and natural field settings ranging from space to the ocean floor.

1.C.5 In research involving human subjects, the ethics of science require that potential subjects be fully informed about the risks and benefits associated with the research and of their right to refuse to participate. Science ethics also demand that scientists must not knowingly subject coworkers, students, the neighborhood, or the community to health or property risks without their prior knowledge and consent. Because animals cannot make informed choices, special care must be taken in using them in scientific research.

1.C.6 Computers have become invaluable in science because they speed up and extend people's ability to collect, store, compile, and analyze data, prepare research reports, and share data and ideas with investigators all over the world.

1.C.7 Accurate record-keeping, openness, and replication are essential for maintaining an investigator's credibility with other scientists and society.

3. The Nature of Technology

By the end of the 8th grade, students should know that

3.A Technology and Science

3.A.1 In earlier times, the accumulated information and techniques of each generation of workers were taught on the job directly to the next generation of workers. Today, the knowledge base for technology can be found as well in libraries of print and electronic resources and is often taught in the classroom.

3.A.2 Technology is essential to science for such purposes as access to outer space and other remote locations, sample collection and treatment, measurement, data collection and storage, computation, and communication of information.

3.A.3 Engineers, architects, and others who engage in design and technology use scientific knowledge to solve practical problems. But they usually have to take human values and limitations into account as well.

3.B Design and Systems

3.B.1 Design usually requires taking constraints into account. Some constraints, such as gravity or the properties of the materials to be used, are unavoidable. Other constraints, including economic, political, social, ethical, and aesthetic ones, limit choices.

3.B.2 All technologies have effects other than those intended by the design, some of which may have been predictable and some not. In either case, these side effects may turn out to be unacceptable to some of the population and therefore lead to conflict between groups.

3.B.3 Almost all control systems have inputs, outputs, and feedback. The essence of control is comparing information about what is happening to what people want to happen and then making appropriate adjustments. This procedure requires sensing information, processing it, and making changes. In almost all modern machines, microprocessors serve as centers of performance control.

3.B.4 Systems fail because they have faulty or poorly matched parts, are used in ways that exceed what was intended by the design, or were poorly designed to begin with. The most common ways to prevent failure are pretesting parts and procedures, overdesign, and redundancy.

3.C Issues in Technology

3.C.1 The human ability to shape the future comes from a capacity for generating knowledge and developing new technologies—and for communicating ideas to others.

3.C.2 Technology cannot always provide successful solutions for problems or fulfill every human need.

3.C.3 Throughout history, people have carried out impressive technological feats, some of which would be hard to duplicate today even with modern tools. The purposes served by these achievements have sometimes been practical, sometimes ceremonial.

3.C.4 Technology has strongly influenced the course of history and continues to do so. It is largely responsible for the great revolutions in agriculture, manufacturing, sanitation and medicine, warfare, transportation, information processing, and communications that have radically changed how people live.

3.C.5 New technologies increase some risks and decrease others. Some of the same technologies that have improved the length and quality of life for many people have also brought new risks.

3.C.6 Rarely are technology issues simple and one-sided. Relevant facts alone, even when known and available, usually do not settle matters entirely in favor of one side or another. That is because the contending groups may have different values and priorities. They may stand to gain or lose in different degrees, or may make very different predictions about what the future consequences of the proposed action will be.

3.C.7 Societies influence what aspects of technology are developed and how these are used. People control technology (as well as science) and are responsible for its effects.

4. The Physical Setting

By the end of the 8th grade, students should know that

4.A The Universe

4.A.1 The sun is a medium-sized star located near the edge of a disk-shaped galaxy of stars, part of which can be seen as a glowing band of light that spans the sky on a very clear night. The universe contains many billions of galaxies, and each galaxy contains many billions of stars. To the naked eye, even the closest of these galaxies is no more than a dim, fuzzy spot.

4.A.2 The sun is many thousands of times closer to the earth than any other star. Light from the sun takes a few minutes to reach the earth, but light from the next nearest star takes a few years to arrive. The trip to that star would take the fastest rocket thousands of years. Some distant galaxies are so far away that their light takes several billion years to reach the earth. People on earth, therefore, see them as they were that long ago in the past.

4.A.3 Nine planets of very different size, composition, and surface features move around the sun in nearly circular orbits. Some planets have a great variety of moons and even flat rings of rock and ice particles orbiting around them. Some of these planets and moons show evidence of geologic activity. The earth is orbited by one moon, many artificial satellites, and debris.

4.A.4 Large numbers of chunks of rock orbit the sun. Some of those that the earth meets in its yearly orbit around the sun glow and disintegrate from friction as they plunge through the atmosphere—and sometimes impact the ground. Other chunks of rocks mixed with ice have long, off-center orbits that carry them close to the sun, where the sun's radiation (of light and particles) boils off frozen material from their surfaces and pushes it into a long, illuminated tail.

4.B The Earth

4.B.1 We live on a relatively small planet, the third from the sun in the only system of planets definitely known to exist (although other, similar systems may be discovered in the universe).

4.B.2 The earth is mostly rock. Three-fourths of its surface is covered by a relatively thin layer of water (some of it frozen), and the entire planet is surrounded by a relatively thin blanket of air. It is the only body in the solar system that appears able to support life. The other planets have compositions and conditions very different from the earth's.

4.B.3 Everything on or anywhere near the earth is pulled toward the earth's center by gravitational force.

4.B.4 Because the earth turns daily on an axis that is tilted relative to the plane of the earth's yearly orbit around the sun, sunlight falls more intensely on different parts of the earth during the year. The difference in heating of the earth's surface produces the planet's seasons and weather patterns.

4.B.5 The moon's orbit around the earth once in about 28 days changes what part of the moon is lighted by the sun and how much of that part can be seen from the earth—the phases of the moon.

4.B.6 Climates have sometimes changed abruptly in the past as a result of changes in the earth's crust, such as volcanic eruptions or impacts of huge rocks from space. Even relatively small changes in atmospheric or ocean content can have widespread effects on climate if the change lasts long enough.

4.B.7 The cycling of water in and out of the atmosphere plays an important role in determining climatic patterns. Water evaporates from the surface of the earth, rises and cools, condenses into rain or snow, and falls again to the surface. The water falling on land collects in rivers and lakes, soil, and porous layers of rock, and much of it flows back into the ocean.

4.B.8 Fresh water, limited in supply, is essential for life and also for most industrial processes. Rivers, lakes, and groundwater can be depleted or polluted, becoming unavailable or unsuitable for life.

4.B.9 Heat energy carried by ocean currents has a strong influence on climate around the world.

4.B.10 Some minerals are very rare and some exist in great quantities, but—for practical purposes—the ability to recover them is just as important as their abundance. As minerals are depleted, obtaining them becomes more difficult. Recycling and the development of substitutes can reduce the rate of depletion but may also be costly.

4.B.11 The benefits of the earth's resources—such as fresh water, air, soil, and trees—can be reduced by using them wastefully or by deliberately or inadvertently destroying them. The atmosphere and the oceans have a limited capacity to absorb wastes and recycle materials naturally. Cleaning up polluted air, water, or soil or restoring depleted soil, forests, or fishing grounds can be very difficult and costly.

4.C Processes that Shape the Earth

4.C.1 The interior of the earth is hot. Heat flow and movement of material within the earth cause earthquakes and volcanic eruptions and create mountains and ocean basins. Gas and dust from large volcanoes can change the atmosphere.

4.C.2 Some changes in the earth's surface are abrupt (such as earthquakes and volcanic eruptions) while other changes happen very slowly (such as uplift and wearing down of mountains). The earth's surface is shaped in part by the motion of water and wind over very long times, which act to level mountain ranges.

4.C.3 Sediments of sand and smaller particles (sometimes containing the remains of organisms) are gradually buried and are cemented together by dissolved minerals to form solid rock again.

4.C.4 Sedimentary rock buried deep enough may be reformed by pressure and heat, perhaps melting and recrystallizing into different kinds of rock. These re-formed rock layers may be forced up again to become land surface and even mountains. Subsequently, this new rock too will erode. Rock bears evidence of the minerals, temperatures, and forces that created it.

4.C.5 Thousands of layers of sedimentary rock confirm the long history of the changing surface of the earth and the changing life forms whose remains are found in successive layers. The youngest layers are not always found on top, because of folding, breaking, and uplift of layers.

4.C.6 Although weathered rock is the basic component of soil, the composition and texture of soil and its fertility and resistance to erosion are greatly influenced by plant roots and debris, bacteria, fungi, worms, insects, rodents, and other organisms.

4.C.7 Human activities, such as reducing the amount of forest cover, increasing the amount and variety of chemicals released into the atmosphere, and intensive farming, have changed the earth's land, oceans, and atmosphere. Some of these changes have decreased the capacity of the environment to support some life forms.

4.D Structure of Matter

4.D.1 All matter is made up of atoms, which are far too small to see directly through a microscope. The atoms of any element are alike but are different from atoms of other elements. Atoms may stick together in well-defined molecules or may be packed together in large arrays. Different arrangements of atoms into groups compose all substances.

4.D.2 Equal volumes of different substances usually have different weights.

4.D.3 Atoms and molecules are perpetually in motion. Increased temperature means greater average energy, so most substances expand when heated. In solids, the atoms are closely locked in position and can only vibrate. In liquids, the atoms or molecules have higher energy, are more loosely connected, and can slide past one another; some molecules may get enough energy to escape into a gas. In gases, the atoms or molecules have still more energy and are free of one another except during occasional collisions.

4.D.4 The temperature and acidity of a solution influence reaction rates. Many substances dissolve in water, which may greatly facilitate reactions between them.

4.D.5 Scientific ideas about elements were borrowed from some Greek philosophers of 2,000 years earlier, who believed that everything was made from four basic substances: air, earth, fire, and water. It was the combinations of these "elements" in different proportions that gave other substances their observable properties. The Greeks were wrong about those four, but now over 100 different elements have been identified, some rare and some plentiful, out of which everything is made. Because most elements tend to combine with others, few elements are found in their pure form.

4.D.6 There are groups of elements that have similar properties, including highly reactive metals, less-reactive metals, highly reactive nonmetals (such as chlorine, fluorine, and oxygen), and some almost completely nonreactive gases (such as helium and neon). An especially important kind of reaction between substances involves combination of oxygen with something else—as in burning or rusting. Some elements don't fit into any of the categories; among them are carbon and hydrogen, essential elements of living matter.

4.D.7 No matter how substances within a closed system interact with one another, or how they combine or break apart, the total weight of the system remains the same. The idea of atoms explains the conservation of matter: If the number of atoms stays the same no matter how they are rearranged, then their total mass stays the same.

4.E Energy Transformations

4.E.1 Energy cannot be created or destroyed, but only changed from one form into another.

4.E.2 Most of what goes on in the universe—from exploding stars and biological growth to the operation of machines and the motion of people—involves some form of energy being transformed into another. Energy in the form of heat is almost always one of the products of an energy transformation.

4.E.3 Heat can be transferred through materials by the collisions of atoms or across space by radiation. If the material is fluid, currents will be set up in it that aid the transfer of heat.

4.E.4 Energy appears in different forms. Heat energy is in the disorderly motion of molecules; chemical energy is in the arrangement of atoms; mechanical energy is in moving bodies or in elastically distorted shapes; gravitational energy is in the separation of mutually attracting masses.

4.F Motion

4.F.1 Light from the sun is made up of a mixture of many different colors of light, even though to the eye the light looks almost white. Other things that give off or reflect light have a different mix of colors.

4.F.2 Something can be "seen" when light waves emitted or reflected by it enter the eye—just as something can be "heard" when sound waves from it enter the ear.

4.F.3 An unbalanced force acting on an object changes its speed or direction of motion, or both. If the force acts toward a single center, the object's path may curve into an orbit around the center.

4.F.4 Vibrations in materials set up wavelike disturbances that spread away from the source. Sound and earthquake waves are examples. These and other waves move at different speeds in different materials.

4.F.5 Human eyes respond to only a narrow range of wavelengths of electromagnetic radiation—visible light. Differences of wavelength within that range are perceived as differences in color.

4.G Forces of Nature

4.G.1 Every object exerts gravitational force on every other object. The force depends on how much mass the objects have and on how far apart they are. The force is hard to detect unless at least one of the objects has a lot of mass.

4.G.2 The sun's gravitational pull holds the earth and other planets in their orbits, just as the planets' gravitational pull keeps their moons in orbit around them.

4.G.3 Electric currents and magnets can exert a force on each other.

5. The Living Environment

By the end of the 8th grade, students should know that

5.A Diversity of Life

5.A.1 One of the most general distinctions among organisms is between plants, which use sunlight to make their own food, and animals, which consume energy-rich foods. Some kinds of organisms, many of them microscopic, cannot be neatly classified as either plants or animals.

5.A.2 Animals and plants have a great variety of body plans and internal structures that contribute to their being able to make or find food and reproduce.

5.A.3 Similarities among organisms are found in internal anatomical features, which can be used to infer the degree of relatedness among organisms. In classifying organisms, biologists consider details of internal and external structures to be more important than behavior or general appearance.

5.A.4 For sexually reproducing organisms, a species comprises all organisms that can mate with one another to produce fertile offspring.

5.A.5 All organisms, including the human species, are part of and depend on two main interconnected global food webs. One includes microscopic ocean plants, the animals that feed on them, and finally the animals that feed on those animals. The other web includes land plants, the animals that feed on them, and so forth. The cycles continue indefinitely because organisms decompose after death to return food material to the environment.

5.B Heredity

5.B.1 In some kinds of organisms, all the genes come from a single parent, whereas in organisms that have sexes, typically half of the genes come from each parent.

5.B.2 In sexual reproduction, a single specialized cell from a female merges with a specialized cell from a male. As the fertilized egg, carrying genetic information from each parent, multiplies to form the complete organism with about a trillion cells, the same genetic information is copied in each cell.

5.B.3 New varieties of cultivated plants and domestic animals have resulted from selective breeding for particular traits.

5.C Cells

5.C.1 All living things are composed of cells, from just one to many millions, whose details usually are visible only through a microscope. Different body tissues and organs are made up of different kinds of cells. The cells in similar tissues and organs in other animals are similar to those in human beings but differ somewhat from cells found in plants.

5.C.2 Cells repeatedly divide to make more cells for growth and repair. Various organs and tissues function to serve the needs of cells for food, air, and waste removal.

5.C.3 Within cells, many of the basic functions of organisms—such as extracting energy from food and getting rid of waste—are carried out. The way in which cells function is similar in all living organisms.

5.C.4 About two-thirds of the weight of cells is accounted for by water, which gives cells many of their properties.

5.D Interdependence of Life

5.D.1 In all environments—freshwater, marine, forest, desert, grassland, mountain, and others—organisms with similar needs may compete with one another for resources, including food, space, water, air, and shelter. In any particular environment, the growth and survival of organisms depend on the physical conditions.

5.D.2 Two types of organisms may interact with one another in several ways: They may be in a producer/consumer, predator/prey, or parasite/host relationship. Or one organism may scavenge or decompose another. Relationships may be competitive or mutually beneficial. Some species have become so adapted to each other that neither could survive without the other.

5.E Flow of Matter and Energy

5.E.1 Food provides molecules that serve as fuel and building material for all organisms. Plants use the energy in light to make sugars out of carbon dioxide and water. This food can be used immediately for fuel or materials or it may be stored for later use. Organisms that eat plants break down the plant structures to produce the materials and energy they need to survive. Then they are consumed by other organisms.

5.E.2 Over a long time, matter is transferred from one organism to another repeatedly and between organisms and their physical environment. As in all material systems, the total amount of matter remains constant, even though its form and location change.

5.E.3 Energy can change from one form to another in living things. Animals get energy from oxidizing their food, releasing some of its energy as heat. Almost all food energy comes originally from sunlight.

5.F Evolution of Life

5.F.1 Small differences between parents and offspring can accumulate (through selective breeding) in successive generations so that descendants are very different from their ancestors.

5.F.2 Individual organisms with certain traits are more likely than others to survive and have offspring. Changes in environmental conditions can affect the survival of individual organisms and entire species.

5.F.3 Many thousands of layers of sedimentary rock provide evidence for the long history of the earth and for the long history of changing life forms whose remains are found in the rocks. More recently deposited rock layers are more likely to contain fossils resembling existing species.

6. The Human Organism

By the end of the 8th grade, students should know that

6.A Human Identity

6.A.1 Like other animals, human beings have body systems for obtaining and providing energy, defense, reproduction, and the coordination of body functions.

6.A.2 Human beings have many similarities and differences. The similarities make it possible for human beings to reproduce and to donate blood and organs to one another throughout the world. Their differences enable them to create diverse social and cultural arrangements and to solve problems in a variety of ways.

6.A.3 Fossil evidence is consistent with the idea that human beings evolved from earlier species.

6.A.4 Specialized roles of individuals within other species are genetically programmed, whereas human beings are able to invent and modify a wider range of social behavior.

6.A.5 Human beings use technology to match or excel many of the abilities of other species. Technology has helped people with disabilities survive and live more conventional lives.

6.A.6 Technologies having to do with food production, sanitation, and disease prevention have dramatically changed how people live and work and have resulted in rapid increases in the human population.

6.B Human Development

6.B.1 Fertilization occurs when sperm cells from a male's testes are deposited near an egg cell from the female ovary, and one of the sperm cells enters the egg cell. Most of the time, by chance or design, a sperm never arrives or an egg isn't available.

6.B.2 Contraception measures may incapacitate sperm, block their way to the egg, prevent the release of eggs, or prevent the fertilized egg from implanting successfully.

6.B.3 Following fertilization, cell division produces a small cluster of cells that then differentiate by appearance and function to form the basic tissues of an embryo. During the first three months of pregnancy, organs begin to form. During the second three months, all organs and body features develop. During the last three months, the organs and features mature enough to function well after birth. Patterns of human development are similar to those of other vertebrates.

6.B.4 The developing embryo—and later the newborn infant—encounters many risks from faults in its genes, its mother's inadequate diet, her cigarette smoking or use of alcohol or other drugs, or from infection. Inadequate child care may lead to lower physical and mental ability.

6.B.5 Various body changes occur as adults age. Muscles and joints become less flexible, bones and muscles lose mass, energy levels diminish, and the senses become less acute. Women stop releasing eggs and hence can no longer reproduce. The length and quality of human life are influenced by many factors, including sanitation, diet, medical care, sex, genes, environmental conditions, and personal health behaviors.

6.C Basic Functions

6.C.1 Organs and organ systems are composed of cells and help to provide all cells with basic needs.

6.C.2 For the body to use food for energy and building materials, the food must first be digested into molecules that are absorbed and transported to cells.

6.C.3 To burn food for the release of energy stored in it, oxygen must be supplied to cells, and carbon dioxide removed. Lungs take in oxygen for the combustion of food and they eliminate the carbon dioxide produced. The urinary system disposes of dissolved waste molecules, the intestinal tract removes solid wastes, and the skin and lungs rid the body of heat energy. The circulatory system moves all these substances to or from cells where they are needed or produced, responding to changing demands.

6.C.4 Specialized cells and the molecules they produce identify and destroy microbes that get inside the body.

6.C.5 Hormones are chemicals from glands that affect other body parts. They are involved in helping the body respond to danger and in regulating human growth, development, and reproduction.

6.C.6 Interactions among the senses, nerves, and brain make possible the learning that enables human beings to cope with changes in their environment.

6.D Learning

6.D.1 Some animal species are limited to a repertoire of genetically determined behaviors; others have more complex brains and can learn a wide variety of behaviors. All behavior is affected by both inheritance and experience.

6.D.2 The level of skill a person can reach in any particular activity depends on innate abilities, the amount of practice, and the use of appropriate learning technologies.

6.D.3 Human beings can detect a tremendous range of visual and olfactory stimuli. The strongest stimulus they can tolerate may be more than a trillion times as intense as the weakest they can detect. Still, there are many kinds of signals in the world that people cannot detect directly.

6.D.4 Attending closely to any one input of information usually reduces the ability to attend to others at the same time.

6.D.5 Learning often results from two perceptions or actions occurring at about the same time. The more often the same combination occurs, the stronger the mental connection between them is likely to be. Occasionally a single vivid experience will connect two things permanently in people's minds.

6.D.6 Language and tools enable human beings to learn complicated and varied things from others.

6.E Physical Health

6.E.1 The amount of food energy (calories) a person requires varies with body weight, age, sex, activity level, and natural body efficiency. Regular exercise is important to maintain a healthy heart/lung system, good muscle tone, and bone strength.

6.E.2 Toxic substances, some dietary habits, and personal behavior may be bad for one's health. Some effects show up right away, others may not show up for many years. Avoiding toxic substances, such as tobacco, and changing dietary habits to reduce the intake of such things as animal fat increases the chances of living longer.

6.E.3 Viruses, bacteria, fungi, and parasites may infect the human body and interfere with normal body functions. A person can catch a cold many times because there are many varieties of cold viruses that cause similar symptoms.

6.E.4 White blood cells engulf invaders or produce antibodies that attack them or mark them for killing by other white cells. The antibodies produced will remain and can fight off subsequent invaders of the same kind.

6.E.5 The environment may contain dangerous levels of substances that are harmful to human beings. Therefore, the good health of individuals requires monitoring the soil, air, and water and taking steps to keep them safe.

6.F Mental Health

6.F.1 Individuals differ greatly in their ability to cope with stressful situations. Both external and internal conditions (chemistry, personal history, values) influence how people behave.

6.F.2 Often people react to mental distress by denying that they have any problem. Sometimes they don't know why they feel the way they do, but with help they can sometimes uncover the reasons.

8. The Designed World

By the end of the 8th grade, students should know that

8.A Agriculture

8.A.1 Early in human history, there was an agricultural revolution in which people changed from hunting and gathering to farming. This allowed changes in the division of labor between men and women and between children and adults, and the development of new patterns of government.

8.A.2 People control the characteristics of plants and animals they raise by selective breeding and by preserving varieties of seeds (old and new) to use if growing conditions change.

8.A.3 In agriculture, as in all technologies, there are always trade-offs to be made. Getting food from many different places makes people less dependent on weather in any one place, yet more dependent on transportation and communication among far-flung markets. Specializing in one crop may risk disaster if changes in weather or increases in pest populations wipe out that crop. Also, the soil may be exhausted of some nutrients, which can be replenished by rotating the right crops.

8.A.4 Many people work to bring food, fiber, and fuel to U.S. markets. With improved technology, only a small fraction of workers in the United States actually plant and harvest the products that people use. Most workers are engaged in processing, packaging, transporting, and selling what is produced.

8.B Materials and Manufacturing

8.B.1 The choice of materials for a job depends on their properties and on how they interact with other materials. Similarly, the usefulness of some manufactured parts of an object depends on how well they fit together with the other parts.

8.B.2 Manufacturing usually involves a series of steps, such as designing a product, obtaining and preparing raw materials, processing the materials mechanically or chemically, and assembling, testing, inspecting, and packaging. The sequence of these steps is also often important.

8.B.3 Modern technology reduces manufacturing costs, produces more uniform products, and creates new synthetic materials that can help reduce the depletion of some natural resources.

8.B.4 Automation, including the use of robots, has changed the nature of work in most fields, including manufacturing. As a result, high-skill, high-knowledge jobs in engineering, computer programming, quality control, supervision, and maintenance are replacing many routine, manual-labor jobs. Workers therefore need better learning skills and flexibility to take on new and rapidly changing jobs.

8.C Energy Sources and Use

8.C.1 Energy can change from one form to another, although in the process some energy is always converted to heat. Some systems transform energy with less loss of heat than others.

8.C.2 Different ways of obtaining, transforming, and distributing energy have different environmental consequences.

8.C.3 In many instances, manufacturing and other technological activities are performed at a site close to an energy source. Some forms of energy are transported easily, others are not.

8.C.4 Electrical energy can be produced from a variety of energy sources and can be transformed into almost any other form of energy. Moreover, electricity is used to distribute energy quickly and conveniently to distant locations.

8.C.5 Energy from the sun (and the wind and water energy derived from it) is available indefinitely. Because the flow of energy is weak and variable, very large collection systems are needed. Other sources don't renew or renew only slowly.

8.C.6 Different parts of the world have different amounts and kinds of energy resources to use and use them for different purposes.

8.D Communication

8.D.1 Errors can occur in coding, transmitting, or decoding information, and some means of checking for accuracy is needed. Repeating the message is a frequently used method.

8.D.2 Information can be carried by many media, including sound, light, and objects. In this century, the ability to code information as electric currents in wires, electromagnetic waves in space, and light in glass fibers has made communication millions of times faster than is possible by mail or sound.

8.E Information Processing

8.E.1 Most computers use digital codes containing only two symbols, 0 and 1, to perform all operations. Continuous signals (analog) must be transformed into digital codes before they can be processed by a computer.

8.E.2 What use can be made of a large collection of information depends upon how it is organized. One of the values of computers is that they are able, on command, to reorganize information in a variety of ways, thereby enabling people to make more and better uses of the collection.

8.E.3 Computer control of mechanical systems can be much quicker than human control. In situations where events happen faster than people can react, there is little choice but to rely on computers. Most complex systems still require human oversight, however, to make certain kinds of judgments about the readiness of the parts of the system (including the computers) and the system as a whole to operate properly, to react to unexpected failures, and to evaluate how well the system is serving its intended purposes.

8.E.4 An increasing number of people work at jobs that involve processing or distributing information. Because computers can do these tasks faster and more reliably, they have become standard tools both in the workplace and at home.

8.F Health Technology

8.F.1 Sanitation measures such as the use of sewers, landfills, quarantines, and safe food handling are important in controlling the spread of organisms that cause disease. Improving sanitation to prevent disease has contributed more to saving human life than any advance in medical treatment.

8.F.2 The ability to measure the level of substances in body fluids has made it possible for physicians to make comparisons with normal levels, make very sophisticated diagnoses, and monitor the effects of the treatments they prescribe.

8.F.3 It is becoming increasingly possible to manufacture chemical substances such as insulin and hormones that are normally found in the body. They can be used by individuals whose own bodies cannot produce the amounts required for good health.

9. The Mathematical World

By the end of the 8th grade, students should know that

9.A Numbers

9.A.1 There have been systems for writing numbers other than the Arabic system of place values based on tens. The very old Roman numerals are now used only for dates, clock faces, or ordering chapters in a book. Numbers based on 60 are still used for describing time and angles.

9.A.2 A number line can be extended on the other side of zero to represent negative numbers. Negative numbers allow subtraction of a bigger number from a smaller number to make sense, and are often used when something can be measured on either side of some reference point (time, ground level, temperature, budget).

9.A.3 Numbers can be written in different forms, depending on how they are being used. How fractions or decimals based on measured quantities should be written depends on how precise the measurements are and how precise an answer is needed.

9.A.4 The operations + and – are inverses of each other—one undoes what the other does; likewise x and ÷ .

9.A.5 The expression *a/b* can mean different things: *a* parts of size *1/b* each, *a* divided by *b*, or *a* compared to *b*.

9.A.6 Numbers can be represented by using sequences of only two symbols (such as 1 and 0, on and off); computers work this way.

9.A.7 Computations (as on calculators) can give more digits than make sense or are useful.

9.B Symbolic Relationships

9.B.1 An equation containing a variable may be true for just one value of the variable.

9.B.2 Mathematical statements can be used to describe how one quantity changes when another changes. Rates of change can be computed from differences in magnitudes and vice versa.

9.B.3 Graphs can show a variety of possible relationships between two variables. As one variable increases uniformly, the other may do one of the following: increase or decrease steadily, increase or decrease faster and faster, get closer and closer to some limiting value, reach some intermediate maximum or minimum, alternately increase and decrease indefinitely, increase or decrease in steps, or do something different from any of these.

9.C Shapes

9.C.1 Some shapes have special properties: triangular shapes tend to make structures rigid, and round shapes give the least possible boundary for a given amount of interior area. Shapes can match exactly or have the same shape in different sizes.

9.C.2 Lines can be parallel, perpendicular, or oblique.

9.C.3 Shapes on a sphere like the earth cannot be depicted on a flat surface without some distortion.

9.C.4 The graphic display of numbers may help to show patterns such as trends, varying rates of change, gaps, or clusters. Such patterns sometimes can be used to make predictions about the phenomena being graphed.

9.C.5 It takes two numbers to locate a point on a map or any other flat surface. The numbers may be two perpendicular distances from a point, or an angle and a distance from a point.

9.C.6 The scale chosen for a graph or drawing makes a big difference in how useful it is.

9.D Uncertainty

9.D.1 How probability is estimated depends on what is known about the situation. Estimates can be based on data from similar conditions in the past or on the assumption that all the possibilities are known.

9.D.2 Probabilities are ratios and can be expressed as fractions, percentages, or odds.

9.D.3 The mean, median, and mode tell different things about the middle of a data set.

9.D.4 Comparison of data from two groups should involve comparing both their middles and the spreads around them.

9.D.5 The larger a well-chosen sample is, the more accurately it is likely to represent the whole. But there are many ways of choosing a sample that can make it unrepresentative of the whole.

9.D.6 Events can be described in terms of being more or less likely, impossible, or certain.

9.E Reasoning

9.E.1 Some aspects of reasoning have fairly rigid rules for what makes sense; other aspects don't. If people have rules that always hold, and good information about a particular situation, then logic can help them to figure out what is true about it. This kind of reasoning requires care in the use of key words such as if, and, not, or, all, and some. Reasoning by similarities can suggest ideas but can't prove them one way or the other.

9.E.2 Practical reasoning, such as diagnosing or troubleshooting almost anything, may require many-step, branching logic. Because computers can keep track of complicated logic, as well as a lot of information, they are useful in a lot of problem-solving situations.

9.E.3 Sometimes people invent a general rule to explain how something works by summarizing observations. But people tend to overgeneralize, imagining general rules on the basis of only a few observations.

9.E.4 People are using incorrect logic when they make a statement such as "If *A* is true, then *B* is true; but *A* isn't true, therefore *B* isn't true either."

9.E.5 A single example can never prove that something is always true, but sometimes a single example can prove that something is not always true.

9.E.6 An analogy has some likenesses to but also some differences from the real thing.

10. Historical Perspectives

By the end of the 8th grade, students should know that

10.A Displacing the Earth from the Center of the Universe

10.A.1 The motion of an object is always judged with respect to some other object or point and so the idea of absolute motion or rest is misleading.

10.A.2 Telescopes reveal that there are many more stars in the night sky than are evident to the unaided eye, the surface of the moon has many craters and mountains, the sun has dark spots, and Jupiter and some other planets have their own moons.

10.F Understanding Fire

10.F.1 From the earliest times until now, people have believed that even though millions of different kinds of material seem to exist in the world, most things must be made up of combinations of just a few basic kinds of things. There has not always been agreement, however, on what those basic kinds of things are. One theory long ago was that the basic substances were earth, water, air, and fire. Scientists now know that these are not the basic substances. But the old theory seemed to explain many observations about the world.

10.F.2 Today, scientists are still working out the details of what the basic kinds of matter are and of how they combine, or can be made to combine, to make other substances.

10.F.3 Experimental and theoretical work done by French scientist Antoine Lavoisier in the decade between the American and French revolutions led to the modern science of chemistry.

10.F.4 Lavoisier's work was based on the idea that when materials react with each other many changes can take place but that in every case the total amount of matter afterward is the same as before. He successfully tested the concept of conservation of matter by conducting a series of experiments in which he carefully measured all the substances involved in burning, including the gases used and those given off.

10.F.5 Alchemy was chiefly an effort to change base metals like lead into gold and to produce an elixir that would enable people to live forever. It failed to do that or to create much knowledge of how substances react with each other. The more scientific study of chemistry that began in Lavoisier's time has gone far beyond alchemy in understanding reactions and producing new materials.

10.G Splitting the Atom

10.G.1 The accidental discovery that minerals containing uranium darken photographic film, as light does, led to the idea of radioactivity.

10.G.2 In their laboratory in France, Marie Curie and her husband, Pierre Curie, isolated two new elements that caused most of the radioactivity of the uranium mineral. They named one radium because it gave off powerful, invisible rays, and the other polonium in honor of Madame Curie's country of birth. Marie Curie was the first scientist ever to win the Nobel prize in two different fields—in physics, shared with her husband, and later in chemistry.

10.I Discovering Germs

10.I.1 Throughout history, people have created explanations for disease. Some have held that disease has spiritual causes, but the most persistent biological theory over the centuries was that illness resulted from an imbalance in the body fluids. The introduction of germ theory by Louis Pasteur and others in the 19th century led to the modern belief that many diseases are caused by microorganisms—bacteria, viruses, yeasts, and parasites.

10.I.2 Pasteur wanted to find out what causes milk and wine to spoil. He demonstrated that spoilage and fermentation occur when microorganisms enter from the air, multiply rapidly, and produce waste products. After showing that spoilage could be avoided by keeping germs out or by destroying them with heat, he investigated animal diseases and showed that microorganisms were involved. Other investigators later showed that specific kinds of germs caused specific diseases.

10.I.3 Pasteur found that infection by disease organisms—germs—caused the body to build up an immunity against subsequent infection by the same organisms. He then demonstrated that it was possible to produce vaccines that would induce the body to build immunity to a disease without actually causing the disease itself.

10.I.4 Changes in health practices have resulted from the acceptance of the germ theory of disease. Before germ theory, illness was treated by appeals to supernatural powers or by trying to adjust body fluids through induced vomiting, bleeding, or purging. The modern approach emphasizes sanitation, the safe handling of food and water, the pasteurization of milk, quarantine, and aseptic surgical techniques to keep germs out of the body; vaccinations to strengthen the body's immune system against subsequent infection by the same kind of microorganisms; and antibiotics and other chemicals and processes to destroy microorganisms.

10.I.5 In medicine, as in other fields of science, discoveries are sometimes made unexpectedly, even by accident. But knowledge and creative insight are usually required to recognize the meaning of the unexpected.

10.J Harnessing Power

10.J.1 Until the 1800s, most manufacturing was done in homes, using small, handmade machines that were powered by muscle, wind, or running water. New machinery and steam engines to drive them made it possible to replace craftsmanship with factories, using fuels as a source of energy. In the factory system, workers, materials, and energy could be brought together efficiently.

10.J.2 The invention of the steam engine was at the center of the Industrial Revolution. It converted the chemical energy stored in wood and coal, which were plentiful, into mechanical work. The steam engine was invented to solve the urgent problem of pumping water out of coal mines. As improved by James Watt, it was soon used to move coal, drive manufacturing machinery, and power locomotives, ships, and even the first automobiles.

11. Common Themes

By the end of the 8th grade, students should know that

11.A Systems

11.A.1 A system can include processes as well as things.

11.A.2 Thinking about things as systems means looking for how every part relates to others. The output from one part of a system (which can include material, energy, or information) can become the input to other parts. Such feedback can serve to control what goes on in the system as a whole.

11.A.3 Any system is usually connected to other systems, both internally and externally. Thus a system may be thought of as containing subsystems and as being a subsystem of a larger system.

11.B Models

11.B.1 Models are often used to think about processes that happen too slowly, too quickly, or on too small a scale to observe directly, or that are too vast to be changed deliberately, or that are potentially dangerous.

11.B.2 Mathematical models can be displayed on a computer and then modified to see what happens.

11.B.3 Different models can be used to represent the same thing. What kind of a model to use and how complex it should be depends on its purpose. The usefulness of a model may be limited if it is too simple or if it is needlessly complicated. Choosing a useful model is one of the instances in which intuition and creativity come into play in science, mathematics, and engineering.

11.C Constancy and Change

11.C.1 Physical and biological systems tend to change until they become stable and then remain that way unless their surroundings change.

11.C.2 A system may stay the same because nothing is happening or because things are happening but exactly counterbalance one another.

11.C.3 Many systems contain feedback mechanisms that serve to keep changes within specified limits.

11.C.4 Symbolic equations can be used to summarize how the quantity of something changes over time or in response to other changes.

11.C.5 Symmetry (or the lack of it) may determine properties of many objects, from molecules and crystals to organisms and designed structures.

11.C.6 Cycles, such as the seasons or body temperature, can be described by their cycle length or frequency, what their highest and lowest values are, and when these values occur. Different cycles range from many thousands of years down to less than a billionth of a second.

11.D Scale

11.D.1 Properties of systems that depend on volume, such as capacity and weight, change out of proportion to properties that depend on area, such as strength or surface processes.

11.D.2 As the complexity of any system increases, gaining an understanding of it depends increasingly on summaries, such as averages and ranges, and on descriptions of typical examples of that system.

12. Habits of Mind

By the end of the 8th grade, students should know that

12.A Values and Attitudes

12.A.1 Know why it is important in science to keep honest, clear, and accurate records.

12.A.2 Know that hypotheses are valuable, even if they turn out not to be true, if they lead to fruitful investigations.

12.A.3 Know that often different explanations can be given for the same evidence, and it is not always possible to tell which one is correct.

12.B Computation and Estimation

12.B.1 Find what percentage one number is of another and figure any percentage of any number.

12.B.2 Use, interpret, and compare numbers in several equivalent forms such as integers, fractions, decimals, and percents.

12.B.3 Calculate the circumferences and areas of rectangles, triangles, and circles, and the volumes of rectangular solids.

12.B.4 Find the mean and median of a set of data.

12.B.5 Estimate distances and travel times from maps and the actual size of objects from scale drawings.

12.B.6 Insert instructions into computer spreadsheet cells to program arithmetic calculations.

12.B.7 Determine what unit (such as seconds, square inches, or dollars per tankful) an answer should be expressed in from the units of the inputs to the calculation, and be able to convert compound units (such as yen per dollar into dollar per yen, or miles per hour into feet per second).

12.B.8 Decide what degree of precision is adequate and round off the result of calculator operations to enough significant figures to reasonably reflect those of the inputs.

12.B.9 Express numbers like 100, 1,000, and 1,000,000 as powers of 10.

12.B.10 Estimate probabilities of outcomes in familiar situations, on the basis of history or the number of possible outcomes.

12.C Manipulation and Observation

12.C.1 Use calculators to compare amounts proportionally.

12.C.2 Use computers to store and retrieve information in topical, alphabetical, numerical, and key-word files, and create simple files of their own devising.

12.C.3 Read analog and digital meters on instruments used to make direct measurements of length, volume, weight, elapsed time, rates, and temperature, and choose appropriate units for reporting various magnitudes.

12.C.4 Use cameras and tape recorders for capturing information.

12.C.5 Inspect, disassemble, and reassemble simple mechanical devices and describe what the various parts are for; estimate what the effect that making a change in one part of a system is likely to have on the system as a whole.

12.D Communication Skills

12.D.1 Organize information in simple tables and graphs and identify relationships they reveal.

12.D.2 Read simple tables and graphs produced by others and describe in words what they show.

12.D.3 Locate information in reference books, back issues of newspapers and magazines, compact disks, and computer databases.

12.D.4 Understand writing that incorporates circle charts, bar and line graphs, two-way data tables, diagrams, and symbols.

12.D.5 Find and describe locations on maps with rectangular and polar coordinates.

12.E Critical-Response Skills

12.E.1 Question claims based on vague attributions (such as "Leading doctors say...") or on statements made by celebrities or others outside the area of their particular expertise.

12.E.2 Compare consumer products and consider reasonable personal trade-offs among them on the basis of features, performance, durability, and cost.

12.E.3 Be skeptical of arguments based on very small samples of data, biased samples, or samples for which there was no control sample.

12.E.4 Be aware that there may be more than one good way to interpret a given set of findings.

12.E.5 Notice and criticize the reasoning in arguments in which (1) fact and opinion are intermingled or the conclusions do not follow logically from the evidence given, (2) an analogy is not apt, (3) no mention is made of whether the control groups are very much like the experimental group, or (4) all members of a group (such as teenagers or chemists) are implied to have nearly identical characteristics that differ from those of other groups.